The Properties of Engineering Materials

Raymond A. Higgins BSc (Birm), FIM

Senior Lecturer in Materials Science, West Bromwich College of Commerce and Technology; former Chief Metallurgist, Messrs Aston Chain and Hook Co. Ltd., Birmingham; and Examiner in Metallurgy to the Institution of Production Engineers, The City and Guilds of London Institute, The Union of Lancashire and Cheshire Institutes and The Union of Educational Institutes.

HODDER AND STOUGHTON

LONDON SYDNEY AUCKLAND TORONTO

by the same author

Materials for the Engineering Technician
Engineering Metallurgy

Part 1: Applied Physical Metallurgy
Part 2: Metallurgical Process Technology

ISBN 0 340 179082 (boards edition)
ISBN 0 340 179090 (paperback edition)

First published 1977

Printed in Great Britain for
Hodder and Stoughton Educational
A division of Hodder and Stoughton Ltd.,
Mill Road, Dunton Green, Sevenoaks, Kent
by Richard Clay (The Chaucer Press) Ltd,
Bungay, Suffolk

Preface

This book was written primarily as an introduction to Materials Science for those engineering students who become involved with the subject at advanced technical college level or, as first-year undergraduates, feel the need of an introduction to its basic principles. It may also prove useful to those students of metallurgy who find it necessary to extend their knowledge to include the properties of non-metallic materials.

The treatment of materials science in this book is of a descriptive and qualitative nature. The author has little regard for those writers who, shirking the task of setting down a lucid explanation of some principle or dogma, proclaim that '. . . This is governed by the following mathematical expression . . .'. Similarly the temptation to write a textbook which claims to be an 'introduction' to the subject yet lambastes the luckless reader with differential equations on page 2, has also been resisted.

Some years ago the author was lecturing on Fick's Laws of Diffusion to a class of engineering students and, having given a description and explanation of the principles involved, remarked that for the benefit of devotees of the differential calculus, these Laws could be expressed mathematically as follows. . . . At this point one particular student visage became illuminated with the light of understanding and, in response to a doubtlesly acrid remark from the lecturer, its owner admitted: 'I've just realised what calculus is all about.' The student in question had been exposed to the calculus in a presumably sterile pure-mathematical form for at least three years previously.

When the author began to teach metallurgy to engineers he quickly discovered that their conception of the nature of solids was one in which minute billiard balls—called 'atoms'—were stuck together with some mysterious 'celestial glue'. We now appreciate that even a mildly academic study of the mechanical properties of engineering materials is meaningless if it does not seek to relate the strengths of these materials to the forces which operate between the fundamental particles contained in them. Most modern syllabuses have therefore been written along these lines. Nevertheless many students still enter engineering courses without having offered chemistry even as an 'O-level' subject. With this problem in mind the relationship between atomic structure and chemical bonding is discussed in simple terms in the first three chapters. Similarly an elementary knowledge of van der Waal's forces has been deemed necessary in order that the reader can compare and contrast the mechanical properties of metals with those of plastics materials. It is hoped that the treatment of these topics under the headings of 'The Atom', 'The Molecule' and 'The Crystal' will help the reader to clarify his ideas sufficiently so that he can embark on the study of materials science which follows.

No doubt in an effort to 'get into the market' quickly many authors of otherwise excellent textbooks of metallurgy have, in recent years, added to their books an extra chapter on 'Plastics' and have then considered that they were justified in describing their works as dealing with 'Materials Science'. Such a course of action is rarely fruitful in that the non-metallics chapter remains totally divorced from the rest and this author resisted the temptation to proceed in this way with his 'Engineering Metallurgy' which will continue as a separate publication dealing solely with metals.

Finally the author wishes to thank his friend and colleague of many years standing, E. A. Boyce Esq., F.R.I.C., of the West Bromwich College, for reading the first three chapters of this book and for making helpful suggestions relating to the basic chemistry they contain.

R. A. Higgins
Division of Materials Technology,
West Bromwich College of Commerce and Technology,
Wednesbury, West Midlands.

Contents

To
THE WOMEN OF MY LIFE—
My Wife, Helen and my
Daughter, Alice.

Chapter One
The Atom

1.1 Until a few hundred years ago—a mere blink of the eye in the time scale of the Universe—educated Man believed that our Earth was the fixed hub about which the whole of Creation revolved. The modern astronomer, however, visualises this planet as but a speck of dust in a galaxy, one of billions of other galaxies, that go to make up an ever-expanding Universe, the visible edge of which is some 10^{23} km distant. Meanwhile the atomic scientist concerns himself with the architecture of the atom itself and in so doing is dealing with minute particles of the order of 10^{-31} kg in mass. Somewhere between these two extremes in size ordinary Man gazes out into the apparently limitless void and tries to make sense of it all.

The culture of the Ancient Greeks was literary rather than scientific, philosophical rather than technological. Certainly they built impressive temples but the building blocks of these they fixed together with iron clamps in order to save themselves the trouble of inventing cement. Although we still remember the intellectual observations of Archimedes, experimental research was not really the fashion in those days and so we find Aristotle proclaiming that the Universe consisted of four 'elements'

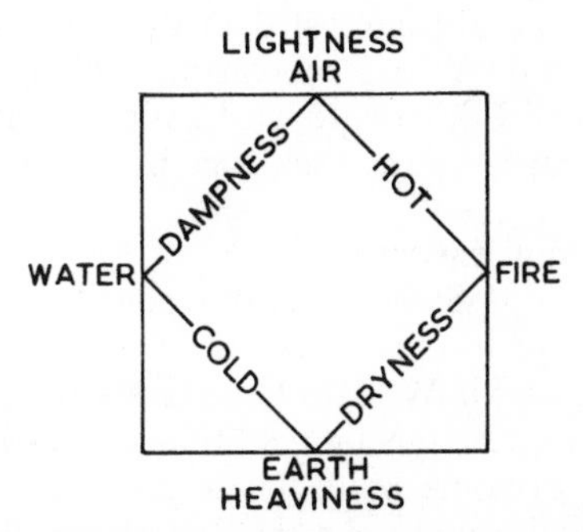

Fig. 1.1—Aristotle's idea of the nature of the Universe. Even much later men were still obsessed with ideas of simple geometrical patterns as governing natural laws. Students of the occult still play around with 'pentacles' and the like.

and four 'essences' (Fig. 1.1). Unfortunately this line of thought persisted even as late as the Renaissance, often with Man trying to 'bend' what facts he had accumulated in order to fit them into some preconceived romantic idea. The mathematician, Johannes Kepler, for example, tried to calculate the planetary orbits so that they were based on some harmonic relationship thus providing the totally imaginary 'music of the spheres'. Though, on the credit side, one must admit that this was an indication that thinking men were using their knowledge of geometry to produce order out of chaos.

1.1.1 Possibly the most important hypothesis left to us by the Greeks was Democritus' idea of the atom. He suggested that all matter was composed of small but indivisible particles which he called 'atoms'. Our modern word is in fact derived from the Greek *atomos*, meaning 'indivisible', and though we now know that these atoms can be subdivided into a series of even smaller particles, by this fundamental concept Man had laid the cornerstone of modern chemistry.

The ideas of Democritus and his friends appear to have been neglected both during the Dark Ages and the Renaissance which followed and it was not until 1803 that the English chemist John Dalton enunciated his Atomic Theory. From then onwards progress in the field of pure chemistry was rapid and by 1868 the Russian chemist Dimitri Mendeléef produced his first great classification of the known elements, a system by which he was able to predict the properties of other elements at that time still unknown.

Science deals largely with systems of classification. Thus, the biologist classifies living things first as either animal or vegetable. The former he subdivides into vertebrate and invertebrate—and so on. Chemistry is, above all, a science of classification. Both elements and their compounds are placed into groups depending upon their chemical and physical properties. In this book we shall be classifying substances mainly because, as engineers, we are interested in their mechanical properties, but first it will be necessary to seek reasons which explain why some materials are strong whilst others are weak, and, in particular, what sort of sub-atomic 'glue' makes atoms stick together at all.

The nature of fundamental particles

1.2 As long ago as 1833 Michael Faraday reached the conclusion that a flow of electricity was due to the movement of separate particles through a conductor whilst, later, Sir William Crooke showed that cathode rays could be bent by an electromagnetic field thus suggesting that these rays consisted of particles of some kind. These particles ultimately became known as 'electrons' and were in fact the first of the sub-atomic particles to be discovered. Today we make considerable use of the electromagnetic deflection of electrons, as for example, in the cathode-ray tube of a television set.

1.2.1 *The electron*, which is a very small particle carrying a charge of what we arbitrarily refer to as 'negative electricity', was examined quantitatively in 1897 by Sir J. J. Thomson. He assigned the first value for the ratio 'charge/mass' (e/m) to the electron by observing the deflections of cathode rays in magnetic and electrical fields of known strengths. His early results were inaccurate—as are those of any explorer with few 'stars to guide'—and the currently accepted values are:

Charge: $1{\cdot}6 \times 10^{-19}$ coulombs (negative electricity)
Mass: $9{\cdot}106 \times 10^{-31}$ kg (0·000 548 8 a.m.u.)

(Since the masses of these particles of atomic and sub-atomic size are inconveniently small when related to the kilogram, the *atomic mass scale* is generally used to express the masses of fundamental particles and atoms. The atom of the carbon isotope $^{12}_{6}C$ is taken as the standard at 12·000 00 atomic mass units (a.m.u.).)

During the course of its history the electron has acquired other names such as the 'negatron' and the 'β-particle' and is represented by various symbols such as β, e^- and $_{-1}^{0}e$.

The electron has another special property which influences its effect on atomic properties. Besides having mass and electrical charge it behaves as though it were spinning on an axis. This 'spin' causes the electron to act as though it were a minute magnet. It also means that the electron possesses angular momentum and this must be taken into account in spectrographic measurements.

1.2.2 *The proton.* In 1886 Eugene Goldstein noticed luminous rays emerging from holes in a perforated cathode which he had used in a cathode-ray tube. These rays were travelling in a direction *opposite* to that of the cathode rays. Later it was shown that they were deflected by both magnetic and electrical fields but in the opposite direction to that of a stream of electrons (cathode rays). Consequently it was realised that the particles of which these new rays consisted were positively charged. Initially they were known as 'canal rays' since they passed through channels in the cathode but in 1907 J. J. Thomson suggested the name 'positive rays' and when a determination of the value charge/mass (e/m) was made it showed them to consist of particles which were much heavier than electrons. Ultimately it was found that the mass of the lightest of these new particles was roughly that of the ordinary hydrogen atom (1.3.2) when stripped of its lone electron. The name *proton*, derived from the Greek word meaning 'first', was suggested by Rutherford in 1920.

The proton carries an equal but opposite (hence positive) charge to that of the electron. Its mass is $1{\cdot}6725 \times 10^{-27}$ kg (1·007 263 a.m.u.) making it some 1836 times heavier than the electron. It is represented by a number of symbols such as p, ^{1}H and $^{1}_{1}H$; the latter being the most commonly used and indicating that it is the nucleus of the hydrogen atom with a mass of one and a positive charge of one.

1.2.3 *The neutron.* During the writer's school days chemists got along quite well without knowledge of the existence of this intriguing particle, the neutron. It was realised, however, that all atoms with the exception of that of ordinary hydrogen were more massive than they should be when taking into account the numbers of positively charged protons in their nuclei. This difficulty was overcome by assuming that also present in the nucleus were a number of protons which were 'paired' with electrons on a 'one for one' basis thus rendering the composite unit electrically neutral. It was later demonstrated that in fact another particle was present in the nucleus. This was called the *neutron* because it has no resultant electrical charge. It has a mass of $1{\cdot}675 \times 10^{-27}$ kg (1·008 665 a.m.u.) making it very slightly more massive than the proton so that earlier chemists can be forgiven for assuming that these units of 'dead mass' in the atomic nucleus were protons 'neutralised' by the presence of comparatively light electrons.

Our knowledge of the rôle of the neutron in nuclear science is still far from complete. It appears to be a relatively stable particle which influences both radioactivity and other forms of nuclear reaction (19.4.3) but it has little effect on ordinary chemical and physical properties (other than atomic mass).

A number of other fundamental particles exist including stable ones like the positron, antiproton and neutrino, as well as unstable ones such as mesons, but since they have little direct effect on physical and mechanical properties of materials we shall not discuss them in this book.

Table 1.1—Summary of properties of the principal fundamental particles.

Particle	*Approximate Mass (relative to the proton)*	*Actual Mass (a.m.u.)*	*Resultant Electrical Charge*
Electron	1/1836	0·000 548 8	Negative } Equal but opposite
Proton	1	1·007 263	Positive } Equal but opposite
Neutron	1	1·008 665	Zero

The structure of the atom

1.3 Everything is built up of atoms. They are very tiny particles and the 1p piece in your pocket contains some 3 489 000 000 000 000 000 000 atoms of copper and tin—give or take a few millions. Each atom has at its centre a nucleus consisting of a group of protons and neutrons whilst 'in orbit' around the nucleus are electrons. An individual atom consists almost entirely of empty space. Although most of the mass is concentrated in the nucleus both the nucleus and the individual electrons in 'orbitals' around it are of approximately the same diameter. If we imagine the nucleus of an atom as consisting of a sphere the size of a golf ball then the outermost electron will be represented by a golf ball in an orbit of some 100 metres radius from the nucleus.

We must be careful, however, not to accept too literally the ideas which diagrams representing atoms present. These diagrams (Fig. 1.2) are merely a convenient way of summarising the general make-up of an atom rather than giving an actual picture of what the atom would look like were it possible for it to be made visible. Thus, such diagrams bear

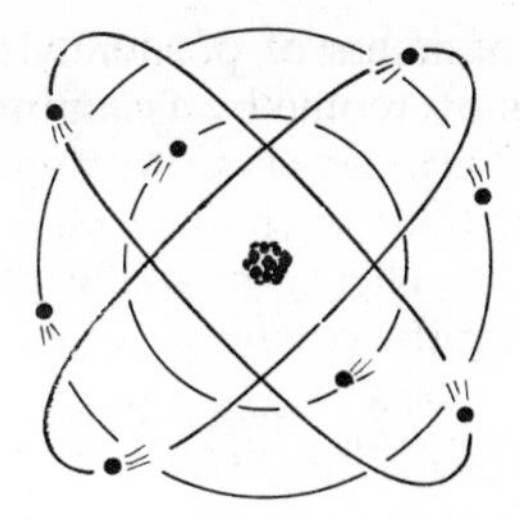

Fig. 1.2—A 'particle' diagram of the oxygen atom.

the same relationship to the actual nature of an atom as does a paper street guide to the pictorial view of a town—it shows the locality of the 'Cock and Magpie' but gives no hint of the actual appearance of that excellent hostelry.

1.3.1 Although the original picture of the atom as visualised by the Danish physicist Niels Bohr in 1913 is now known to give an incomplete picture of the structure of the atom it is still of considerable use in showing, subject to the sort of limitations just mentioned, the general particle make-up of the atom.

Briefly, Bohr suggested that electrons orbited the nucleus in fixed 'shells' but that when an electron received energy it moved from a lower to a higher level within the shell. Conversely when it gave up that energy it fell back again to its former level (Fig. 1.3). Thus Bohr applied a quantum theory to electron orbits in roughly the way in which Max Planck and Albert Einstein had quantised light.

Knowledge of the grouping of electrons was derived originally from spectrometric studies hereby the wavelength (λ) of light emitted was

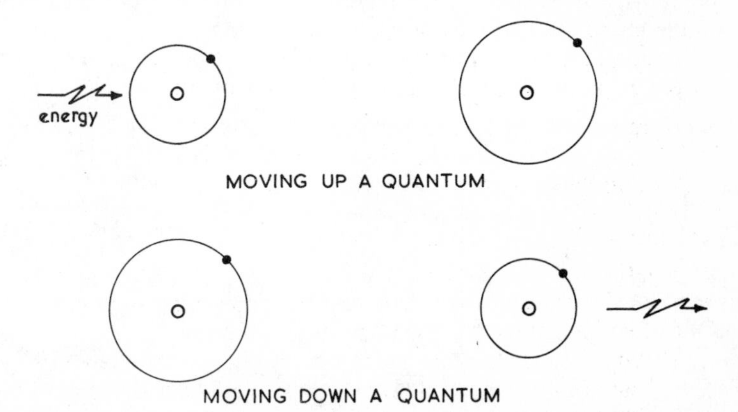

Fig. 1.3—The hydrogen atom contains one proton in the nucleus and one electron in orbit around it. The orbit in which the electron exists depends upon the energy state of the electron.

measured. From these studies it was concluded that a small 'packet' or *quantum* of energy is required to make an electron move from one energy level to the next-higher energy level. Conversely a quantum of energy, known as a *photon*, was emitted when an electron fell back from a higher to the next-lower level. The energy E of the photon can be calculated from its wavelength λ using the equation:

$$E = \frac{h \, . \, c}{\lambda}$$

where h is Planck's constant, c is the velocity of light and c/λ becomes the wave frequency. In practical terms Planck's constant is $6{\cdot}62 \times 10^{-34}$ joule . seconds.

1.3.2 The somewhat tenuous nature of the electron has already been touched upon. In fact physicists no longer think of an electron orbit as consisting of a specific path. The definite 'orbit' has been replaced by the term *orbital* which refers to a mathematical function representing the distribution of the electron within the space occupied by the atom. The electron can be visualised as a sort of 'mist' of electricity rather than as a single particle; the orbital prescribing the density of that 'mist' at any point within the atom. Alternatively the electron may be regarded almost in the nature of a 'resultant particle' (here the term 'resultant' is used in the mechanical sense) and the orbital interpreted as denoting the statistical probability of finding that particle at any given point around the nucleus. Consequently it must be stressed again that pictures showing an atom as consisting of a central nucleus about which a number of electrons are moving in definite orbits must not be interpreted too literally. They are purely symbolic and offer a simple diagrammatic means of showing the electron complement of the various quantum shells. Such diagrams will be used in the pages which follow in order to explain in simple terms the methods of chemical bonding.

The modern idea of the structure of the atom, as applied to that of hydrogen, is indicated in Fig. 1.4. Instead of an electron occupying a

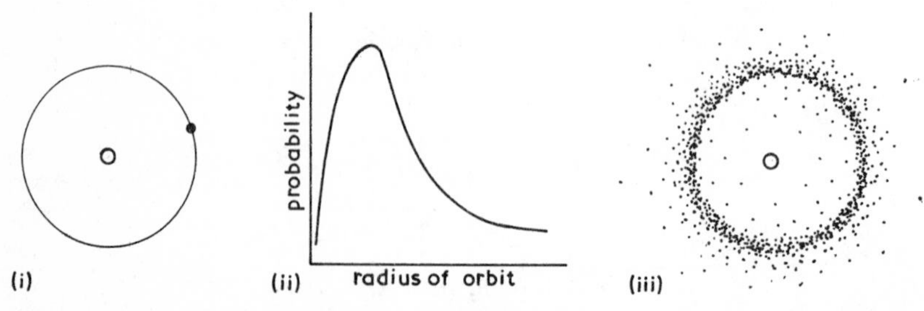

Fig. 1.4—A pictorial interpretation of the possible position of the lone electron in the hydrogen atom. (i) The 'Bohr atom'. (ii) A graph indicating the probability of the orbit being of a particular radius. (iii) The resultant idea of the position occupied by the electron.

single fixed orbit the probability of it existing at any location is indicated by the graph (ii) and the resultant diagram (iii).

Mendeléef had observed the periodicity of chemical properties of the then-known elements and this knowledge ultimately gave rise to the Periodic Table as we know it today (Table 1.3). Not only is there a repeating pattern with regard to chemical properties of the elements in the same group but also periodicity in connection with their electrical, magnetic and mechanical properties. Modern atomic theory successfully explains why the arrangement of elements within the Periodic Table is governed by the manner in which additional electrons are situated in elements of ascending 'Z number'.

1.3.3 Electrons surrounding the atomic nucleus are not all at the same energy level. Consequently it is convenient to group electrons into shells with different energy characteristics. The first *quantum shell* contains only two electrons, whilst the second shell contains a maximum of eight, the third eighteen, and the fourth thirty-two. The greatest number of electrons in a given shell is $2n^2$ where n is known as the *quantum number* of the shell. As already suggested this electron shell concept is somewhat over-simplified. Amongst other things it suggests that all electrons within a shell are equal whereas in practice they are not. The application of what is known as *Pauli's exclusion principle* is of help here. This principle states that there are definite rules governing the energy levels and probable positions of electrons which are in orbit around an atom. Thus the single electron of a hydrogen atom is normally at the lowest possible energy level so that the most probable position of this electron is that indicated in Fig. 1.4 (ii). No more than two interacting electrons can have the same orbital quantum numbers. Even these two are not entirely identical since they exhibit inverted magnetic behaviours or 'spins'.

Fig. 1.5 shows the electron complement—depicted in the manner of the Bohr atom—for the first eleven elements of ascending atomic number

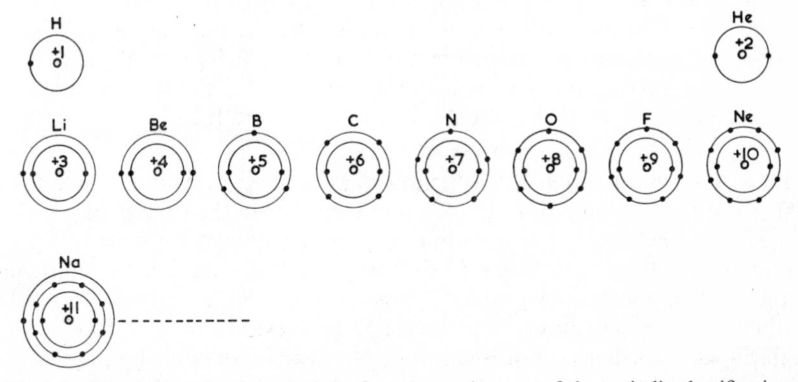

Fig. 1.5—The electron 'make-up' of the first eleven elements of the periodic classification. The atoms are depicted in the Bohr manner.

(Z). An atom of the first element, hydrogen, contains a single proton in the nucleus and a single electron in orbit around it. The atom is of course electrically neutral. Helium contains two protons in the nucleus and two electrons in the first shell. These two electrons fill the first shell making helium a very stable element which does not react with other elements. Lithium has three protons ($Z = 3$), and since the first quantum shell is already filled (with the helium atom) the third electron in the lithium atom must go into the second quantum shell. In atoms of each of the elements which follow—beryllium, boron, carbon, nitrogen, oxygen, fluorine and neon—one more electron is added to the second quantum shell. However, as indicated by *Pauli's exclusion principle* no more than two electrons can have the same energy levels and, hence, the same probability of position. Additional location probabilities are therefore introduced and *subshells* established in the manner indicated in Fig. 1.2 showing a simple 'particle diagram' of the oxygen atom.

The fact that spectrometric methods were used in early work on atomic structures led to the adoption of spectrographic nomenclature. Thus, since the most definite spectral lines are obtained from those electrons which fall to the lowest energy level within a given shell, the notation s has been used for those electrons in each quantum shell which have the lowest energy levels. Hence $1s^2$ indicates that two electrons (of opposite magnetic spins) are present in the low-energy first quantum shell (formerly referred to as the 'K shell'). Similarly $2s^2$ indicates that two electrons are present in the lowest energy level of the second quantum shell (formerly the 'L shell'). Since two is the maximum number of electrons which can exist in this s-subshell, further subshells are created to hold the other six electrons which are members of this second principal shell.

Each of the succeeding shells contains two or more subshells which are designated p, d and f. The greatest number of electrons in these subshells is six, ten and fourteen respectively. We will consider again the example of oxygen (Fig. 1.2) which has six electrons in its second quantum (or L) shell. It has the electron notation $1s^2$; $2s^2$, $2p^4$ which indicates that two electrons are in the first (or K) shell and six electrons in the second (or L) shell (two in the lower subshell and four in the next-higher subshell). The complete electron notation system for the chemical elements in the first four periods of the periodic table is given in Table 1.2.

1.3.4 A study of Table 1.2 will show that in several instances quantum shells of higher number begin to accept electrons in their levels before the preceding shells of lower quantum number have been filled. This is due fundamentally to overlapping of energy levels of succeeding subshells and also of succeeding principal shells. Since electrons are most stable when they are in a position of least energy they naturally seek out suitable spaces in shells or subshells where vacancies of lowest energy level occur. For these reasons there are groups of 'transition' elements in the periodic classification in which an outer—or valency—subshell begins to

Table 1.2—Electron notation of the elements in the first four periods of the periodic table.

Element	Atomic Number (Z)	Shell 1	Shell 2		Shell 3			Shell 4			
		1s	2s	2p	3s	3p	3d	4s	4p	4d	4f
H	1	1									
He	2	2									
Li	3	2	1								
Be	4	2	2								
B	5	2	2	1							
C	6	2	2	2							
N	7	2	2	3							
O	8	2	2	4							
F	9	2	2	5							
Ne	10	2	2	6							
Na	11	2	2	6	1						
Mg	12	2	2	6	2						
Al	13	2	2	6	2	1					
Si	14	2	2	6	2	2					
P	15	2	2	6	2	3					
S	16	2	2	6	2	4					
Cl	17	2	2	6	2	5					
Ar	18	2	2	6	2	6					
K	19	2	2	6	2	6		1			
Ca	20	2	2	6	2	6		2			
Sc	21	2	2	6	2	6	1	2			
Ti	22	2	2	6	2	6	2	2			
V	23	2	2	6	2	6	3	2			
Cr	24	2	2	6	2	6	5	1			
Mn	25	2	2	6	2	6	5	2			
Fe	26	2	2	6	2	6	6	2			
Co	27	2	2	6	2	6	7	2			
Ni	28	2	2	6	2	6	8	2			
Cu	29	2	2	6	2	6	10	1			
Zn	30	2	2	6	2	6	10	2			
Ga	31	2	2	6	2	6	10	2	1		
Ge	32	2	2	6	2	6	10	2	2		
As	33	2	2	6	2	6	10	2	3		
Se	34	2	2	6	2	6	10	2	4		
Br	35	2	2	6	2	6	10	2	5		
Kr	36	2	2	6	2	6	10	2	6		
etc.											

(2s, 2p ← sub-shells; 3s, 3p, 3d ← sub-shells; 4s, 4p, 4d, 4f sub-shells)

4s begins to fill before 3d

4f is not filled until element no.71 (Lu) by which time 6s has already been filled.

fill before a full quota of electrons is present in the previous subshell. For example in the scandium–nickel series the 4*s* subshell begins to fill before the 3*d* subshell has received its full complement of ten electrons. Similar situations occur with other transition series such as yttrium-palladium and in the lanthanide ('rare earth') and actinide (trans-uranic) elements.

The atomic nucleus

1.4 The nucleus of an atom consists of an association of protons and neutrons. Since these are particles some 1836 times heavier than the electron it follows that the mass of an atom tends to be concentrated in its nucleus. The number of protons in the nucleus of a stable atom is

Table 1.3—The Periodic Classification of the Elements.

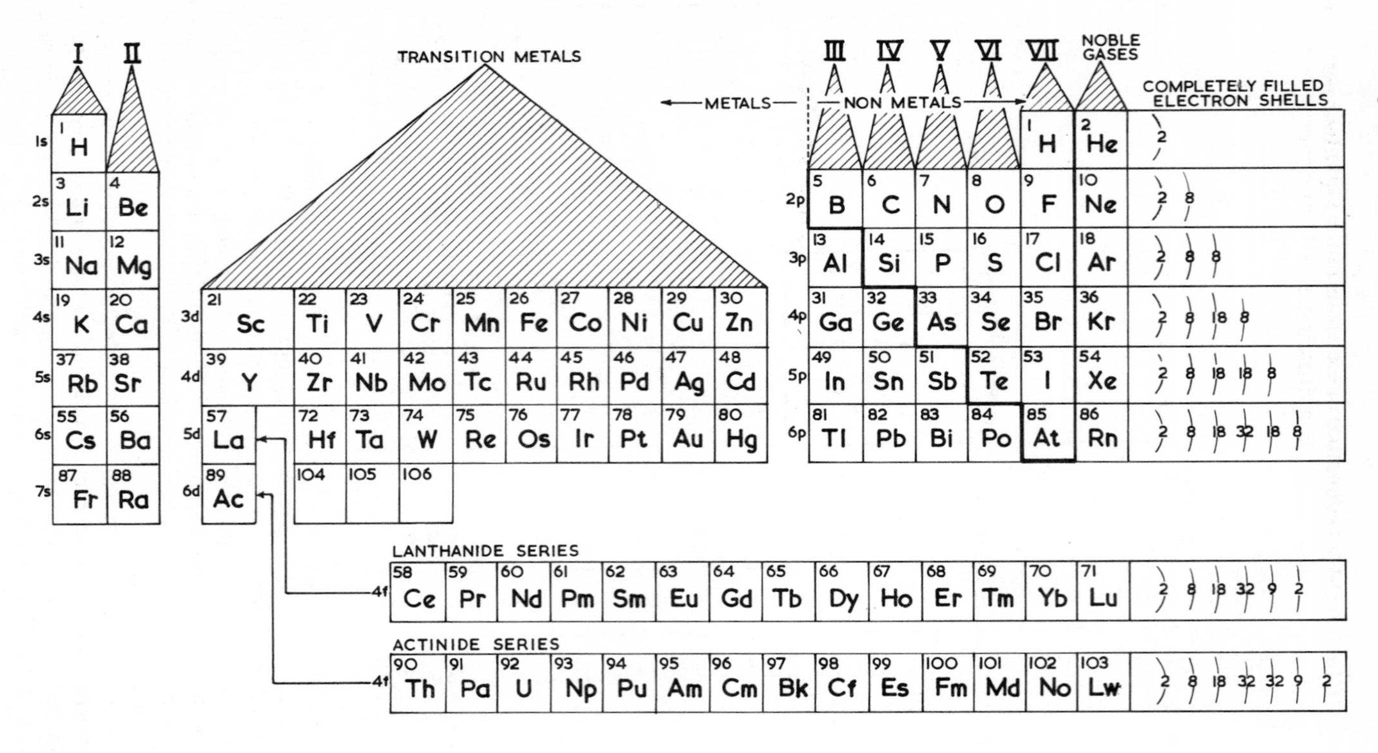

always equal to the total number of electrons in the various shells around the nucleus. Since the charge on a proton is equal but opposite to that on an electron it follows that a stable atom is electrically neutral. The number of protons in the nucleus of an atom is called the Atomic Number (Z) of the element.

The neutrons, also present in the nucleus, have little effect on the *chemical* properties of the element since they carry no resultant electrical charge. They behave as a form of 'nuclear ballast' increasing the total mass of the atom. The number of neutrons present in the nucleus may vary even from one atom to another *of the same element*. Consequently when we measure the *relative atomic mass* of an element we find that it is rarely a whole number since we are in fact measuring the average mass of a large number of atoms of several different masses. Originally the relative atomic mass* of an element was the average specific mass of an atom of that element as compared with the mass of an atom of hydrogen. For various reasons a carbon atom of atomic mass 12·000 0 is now used as the standard. However, in an understanding of the properties of the elements the atomic number (Z) is of much greater significance than the relative atomic mass.

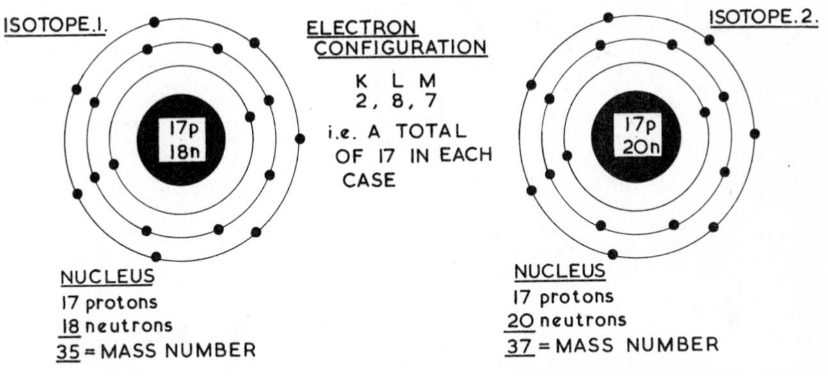

Fig. 1.6—Diagrammatic representation of two isotopes of chlorine.

1.4.1 *Isotopes.* Whilst the number of protons in the nucleus of any atom of a single element is fixed, the number of neutrons can vary from one atom to another. Thus the nucleus of any chlorine atom will always contain 17 protons, but about 75% of all chlorine atoms contain 18 neutrons whilst roughly 25% contain 20 neutrons. These two different chlorine atoms are known as *isotopes* and chlorine is said to be *isotopic*. The relative atomic masses of these two chlorine isotopes will be 35 and 37 respectively (Fig. 1.6) but since the lighter isotope is the more plentiful this explains why the average relative atomic mass of chlorine is

* Previously known as 'atomic weight'.

35·46. It is important to remember that these two isotopes will have exactly similar chemical properties since both have the same atomic number and consequently the same electron configuration.

Hydrogen too is isotopic and occurs as three isotopes (Fig. 1.7). The 'ordinary' hydrogen atom (sometimes called *protium*) contains a single proton in its nucleus, whilst the nucleus of isotope 2 (generally called *deuterium*) contains a neutron in addition to the proton. The nucleus of isotope 3 (known as *tritium*) consists of two neutrons in addition to the proton. Although the relative atomic masses of the three isotopes are 1, 2 and 3 respectively the average relative atomic mass of hydrogen is no

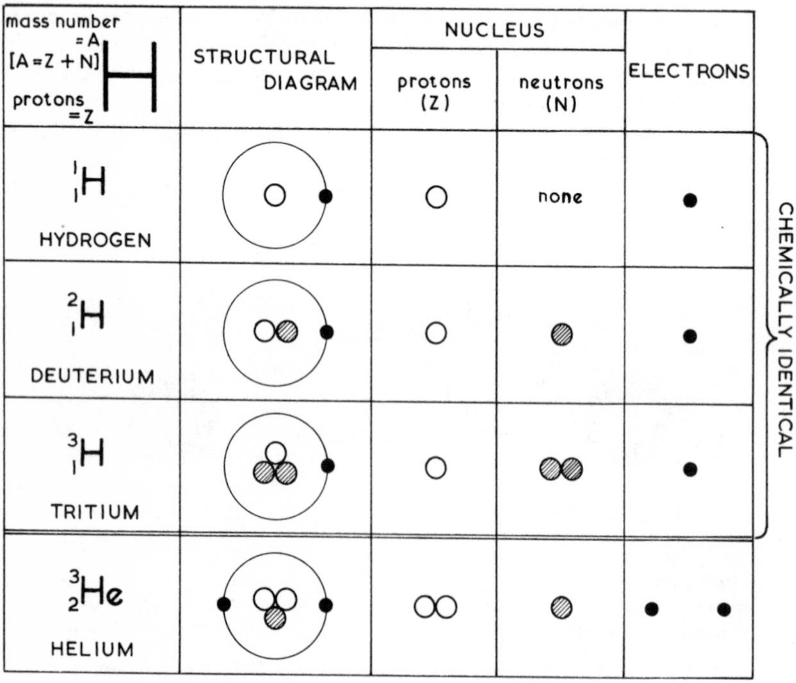

mass number = A [A = Z + N] H protons = Z	STRUCTURAL DIAGRAM	NUCLEUS		ELECTRONS
		protons (Z)	neutrons (N)	
^1_1H HYDROGEN			none	
^2_1H DEUTERIUM				
^3_1H TRITIUM				
^3_2He HELIUM				

CHEMICALLY IDENTICAL

Fig. 1.7—The structural make-up of the three isotopes of hydrogen. The isotope of helium, ^3_2He, has a similar atomic mass to tritium but is of a very different nature because of its 'filled' electron shell.

more than 1·007 97 because isotopes 2 and 3 are extremely rare. The relationship between the isotopic nature of both hydrogen and uranium and their respective rôles in the release of nuclear energy will be discussed elsewhere in this book (19.5 and 19.6).

1.4.2 *Nuclear binding forces.* The nucleus of an atom consists of a collection of protons and neutrons. Now like-charged bodies repel each other with a force which is inversely proportional to the square of their distance apart (the Coulomb Law). Since protons are positively charged particles why does not the nucleus disintegrate due to repulsion between these protons? The answer to this question is that forces which bind together the particles of the nucleus are not in fact of an electrostatic nature. Neither are they magnetic or gravitational in character but appear to be separate forces entirely and about which little is yet known. These forces act not only between the protons but also between the neutrons and are known as nuclear binding forces.

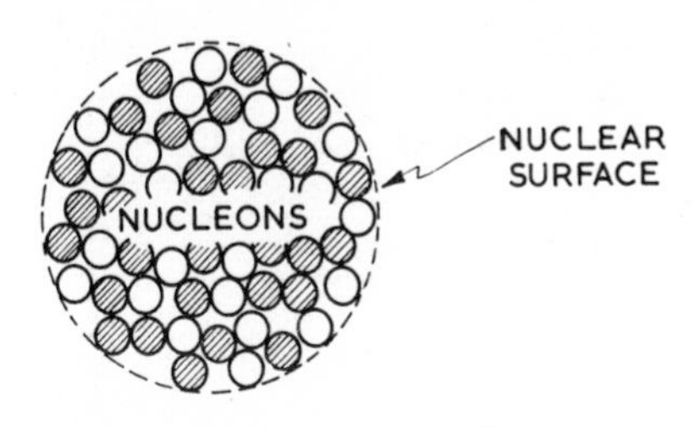

Fig. 1.8—Nucleons (i.e. protons ○ and neutrons ◍) at the surface are attracted inwards by those at the core.

An atomic nucleus may contain a large number of *nucleons* (i.e. protons and neutrons) bound together as a coherent unit. Nucleons in the core are subjected to their special force of attraction by all neighbouring nucleons but those on the surface are attracted inwards towards the core by the bulk of the nucleons situated there. This produces something in the nature of a 'surface tension' effect at the outer limits of the nucleus so that nuclei can be regarded as being constituted after the fashion of droplets of moisture.

Whilst much remains to be discovered about these nuclear binding forces it is clear that they are effective over an extremly limited range, often less than the radius of the nucleus. Fig. 1.9 is a graph in which the potential energy possessed by a proton, due to forces acting between it and a nearby nucleus, is plotted as a function of the distance separating them. When the proton is at *A* simple electrostatic repulsion tends to push it away from the nucleus which has its centre at *O*. Thus the proton can be considered to possess potential energy in the same way that a compressed spring possesses potential energy and if released it would recoil from the nucleus. As the proton gets closer to *O* the electrostatic force—and hence the potential energy of the proton—increase predictably according to the Coulomb Law, and potential energy would be expected to go on increasing in the direction of *X* until contact was achieved between proton and nucleus. At *B*, however, the curve flattens off at some potential energy value, *E*, because the proton now comes close enough to be under the influence of the nuclear binding force. Beyond *C* the nuclear binding force is much greater than the electrostatic force of repulsion and so the proton is now attracted to the nucleus

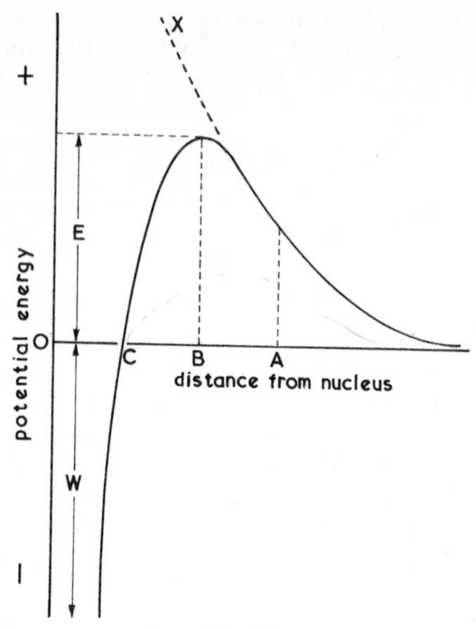

Fig. 1.9.

instead of being repelled by it. Relative to its previous condition it can now be considered to possess 'negative potential energy'. The captured proton is said to be in a 'potential well' of depth $E + W$ and this energy would be required to remove it from the nucleus.

1.4.3 An important point to appreciate is that whilst *electrostatic repulsion acts over fairly large distances, the nuclear binding force operates over very small distances*. Therefore a small number of nucleons can hold together because it is geometrically possible for the nuclear binding force to overcome the electrostatic repulsion between protons, but there is a limit to this number. A very large nucleus is certain to possess many nucleons which are outside the influence of the short-range nuclear binding force but any protons will still be influenced by the mutual repulsion between each other's long-range electrostatic fields. Therefore as a nucleus becomes large the total electrostatic repulsion between protons increases whilst beyond a certain size the total nuclear binding force between nucleons does not.

Thus in a heavy nucleus the total binding force is not strong enough to hold the nucleus together against the total repulsion force between all of the protons. This is one reason why 'extra' neutrons are present in these heavy nuclei of uranium and radium—to provide extra binding force to the total whilst adding nothing to the electrostatic forces of repulsion, since neutrons have no charge. However, these extra neutrons make the nucleus more massive still and the total binding force still weaker so that in spite of its extra neutrons a massive nucleus tends to break up.

For a nucleus to be stable it needs to be less than a certain maximum size, determined by the relative magnitude of the opposing forces of attraction and repulsion and containing an optimum ratio of charged particles (protons) to uncharged particles (neutrons). All naturally occurring nuclei (with the exception of hydrogen ^1_1H) are a mixture of protons and neutrons and the ratio of neutrons to protons increases from 1:1 for the lightest nuclei to roughly 1·5:1 for the most massive nuclei. Those nuclei with atomic mass numbers greater than 209 are unstable to some degree and no atom with a mass number greater than 238 occurs naturally. Those massive nuclei therefore undergo radioactive decay (19.4).

1.4.4 *Nuclear binding energy.* By using an apparatus known as the mass spectrograph it is possible to determine atomic masses with a precision of about 1 in 100 000. In this way the mass of the proton has been determined as 1·007 263 a.m.u. and the mass of the neutron as 1·008 665 a.m.u. Since the numbers of protons and neutrons in any atomic nucleus are known and since the masses of the elementary constituents have been accurately determined it would seem possible to calculate the mass of a nucleus with precision from these values. For example the helium nucleus ^4_2He contains two protons and two neutrons. Hence its mass should be:

$$\begin{aligned} \text{2 protons} &= 2 \times 1{\cdot}007\,263 = 2{\cdot}014\,526 \text{ a.m.u.} \\ \text{2 neutrons} &= 2 \times 1{\cdot}008\,665 = \underline{2{\cdot}017\,330} \text{ a.m.u.} \\ \text{Total} &= \underline{4{\cdot}031\,856} \text{ a.m.u.} \end{aligned}$$

But in fact the mass of the helium nucleus is 4·002 764 a.m.u. The difference between the expected mass and the actual mass is:

$$\begin{aligned} & 4{\cdot}031\,856 - 4{\cdot}002\,764 \text{ a.m.u.} \\ = {} & 0{\cdot}029\,092 \text{ a.m.u.} \end{aligned}$$

This difference is known as the *mass defect*. It follows that in the formation of the helium nucleus from its constituent protons and neutrons a portion of matter amounting to 0·029 092 a.m.u. is converted into energy. This energy is present in the form of the binding force which holds the particles together as suggested in the previous section. The amount of energy released by the conversion of this amount of matter may be derived from Einstein's mass-energy equation:

$$E = mc^2$$

where c is the velocity of light (3×10^8 ms^{-1}); m is the mass involved and E the energy produced. Hence the substitution of above values in this equation indicates the energy released:

$$\begin{aligned} E &= 0{\cdot}029\,092 \times 1{\cdot}661 \times 10^{-27} \times (3 \times 10^8)^2 \text{ J} \\ &= 4{\cdot}35 \times 10^{-12} \text{ J} \end{aligned}$$

(where 1 atomic mass unit $= 1{\cdot}661 \times 10^{-27}$ kg)

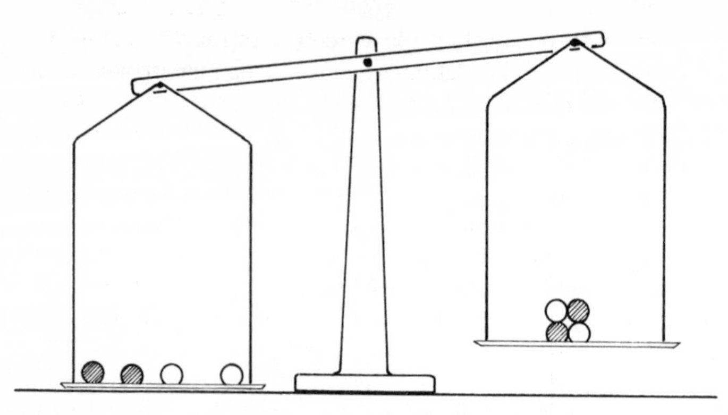

Fig. 1.10—The nucleus of the helium isotope ^4_2He weighs less than the four nucleons (2 protons and 2 neutrons) taken separately.

More will be said regarding the implications of nuclear binding energy and its connection with nuclear fission and fusion processes in Chapter Nineteen.

States of matter

1.5 All elements can exist as either gases, liquids or solids. The *state* in which an element exists depends upon the combination of temperature and pressure which prevails at that time.

Electrostatic forces operate between atoms and the total force involved may be the resultant of a number of forces. First, forces of repulsion will operate between the nuclei of two atoms since both nuclei are positively charged. Similarly the electrons of one atom will repel those of another atom since electrons are negatively charged. Opposing these forces of

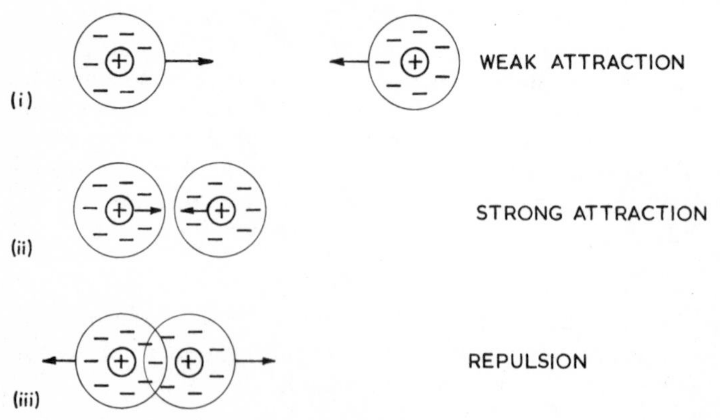

Fig. 1.11—Forces operating between single atoms. (i) Here the atoms are so far apart that only weak forces of attraction operate between them. (ii) Now the atoms are close together and strong forces tend to pull them closer together. (iii) When the electron orbitals are close to overlapping a strong force of repulsion builds up.

repulsion are forces of attraction in which the positively charged nucleus of one atom is attracted by the negatively charged electrons of the other atom. Whether the resultant force between the two atoms is of attraction or repulsion depends upon which of the forces in these two groups is greater.

When separated by very great distances the force between two atoms is very small but as the atoms approach each other a force of attraction develops. This force increases until the orbitals of electrons of each atom begin to overlap. At this point a strong force of repulsion begins to build up (Fig. 1.11). Thus the resultant of the forces which operate between atoms is a function of their distance apart (Fig. 1.12). The maximum value of such a resistant force is small and rarely more than 0·5 mN.

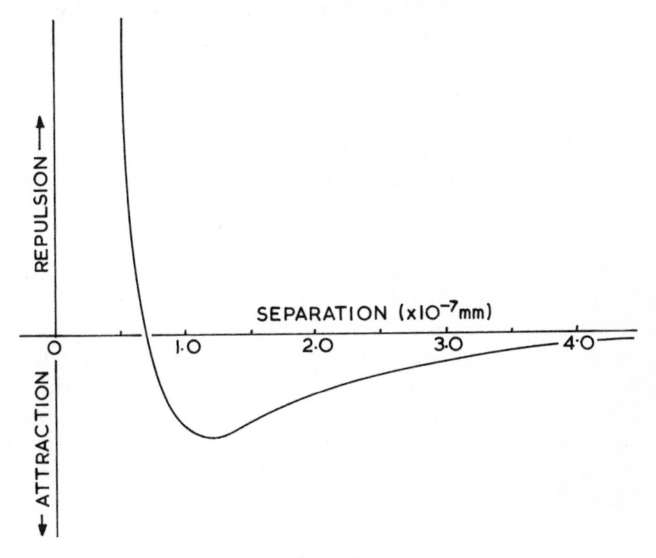

Fig. 1.12.

In a gas at relatively high temperatures the average energy of the atoms is sufficient for the forces of attraction between them to be rendered negligible. They approach each other on a collision path and only at short distances will strong forces of repulsion lead to separation. As the temperature of the gas is decreased so is the kinetic energy of each atom. Consequently forces of attraction between atoms become more significant. At some temperature a stage is reached where large groups of atoms are held together since their thermal activation is insufficient to pull them apart, i.e. the gas condenses to form a liquid. At a lower temperature still the force of attraction predominates to the extent that atoms fall into a fixed pattern and so form a solid.

Chapter Two
The Molecule

2.1 In the previous chapter we considered some of the properties of the atom as a single free particle. Nevertheless, apart from those of the 'noble gases', atoms rarely occur as single particles but are generally attached to other atoms forming small or large groups. Which is just as well otherwise our Earth—if it existed at all—would presumably consist of a mass of rarefied gas.

What is the nature of the force of cohesion which apparently binds atoms together? In this chapter and the next we shall attempt a simplified answer to this question.

We have seen (1.3.3) that in the second and third periods of the Periodic Table the outer electron shell of an atom is generally complete when the number of electrons in it reaches eight.* Consequently as the result of a chemical reaction between two atoms in these periods there is a general tendency for the outer shells of electrons in each atom to finish with eight electrons.

Most metallic atoms contain one, two or three electrons in their outermost shells (or subshells in the case of metals in period four onwards) whilst in the outer shells of atoms of non-metallic elements the number of electrons is nearer to eight. Thus, when one atom is strongly metallic (the element occupies a position on the left-hand side of the Periodic Table) and the other strongly non-metallic (the element is situated on the extreme right of the Table), chemical bonding is achieved by the metallic atom *donating* its outer-shell electrons to fill the outer shell of the non-metallic atom. This produces what is variously termed an *ionic*, *polar* or *electrovalent* bond. Its formation gives rise to the production of *ions*, the building blocks of some crystal structures which will be dealt with in the next chapter.

If the atoms concerned are of elements near to each other in the centre of the Periodic Table they can combine most easily by *sharing* electrons from their outer shells forming common orbitals which are the basis of the *covalent* bond.

The covalent bond

2.2 Two atoms of the gas fluorine will combine by sharing a pair of electrons (Fig. 2.1). In this way the outer shell of each atom now has a complete 'neon octet' and the two atoms are held together by the electrical forces involved by sharing electrons belonging to the outer orbitals

* In the first period, i.e. H → He, the number of electrons involved is of course only two, whilst in the case of periods four and onwards the outer shells increase in size, i.e. 18, 18, 32, etc., each terminating with the appropriate 'noble gas' structure.

Fig. 2.1—The covalent bond in a molecule of fluorine formed by the sharing of a pair of electrons.

of both atoms. For this reason the covalent bond is a strong and stable chemical bond.

Two hydrogen atoms combine in a similar way except that the 'helium duplet' of each is completed by sharing the pair of electrons thus brought together (Fig. 2.2).

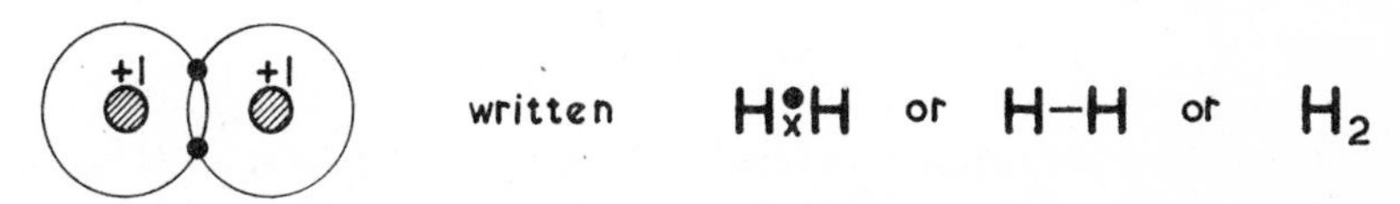

Fig. 2.2—The covalent bond in a molecule of hydrogen—in this case sharing of the electrons completes the 'helium duplet' of each atom.

Often more than two electrons are involved in the formation of the covalent bond. Thus, the outer shell of the nitrogen atom contains five electrons and two nitrogen atoms will combine by sharing six electrons—three donated to the bond by each atom (Fig. 2.3).

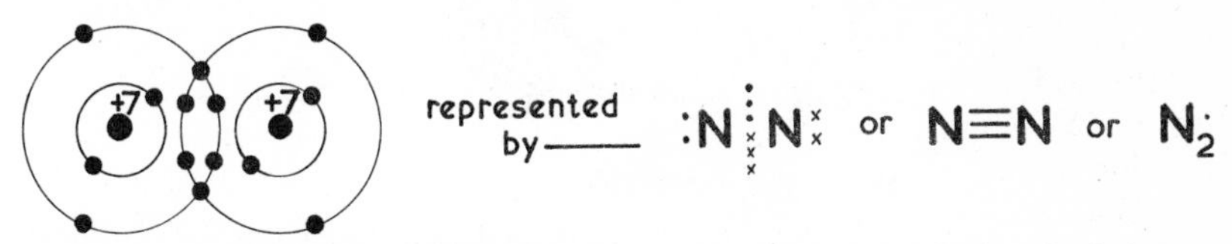

Fig. 2.3—The covalent bond in a molecule of nitrogen formed by sharing six electrons.

In each case the particle formed by the combination of two atoms in this manner is called a *molecule*. Molecules can contain more than two atoms—in fact some molecules of the more complex carbon compounds contain many thousands of atoms. Most of the common non-metallic gases, however, exist as molecules which are diatomic—that is, the molecule contains two atoms. The 'noble gases' (helium, neon, argon, etc.) are exceptions to this general rule. Since the outer electron shells of their atoms are already complete they will not normally combine with other atoms, either of their own type or of other elements. Hence, they are monatomic, that is they exist as single atoms and do not form molecules.

2.2.1 Atoms of unlike elements combine to form molecules in a similar manner. Thus two hydrogen atoms are covalently bonded to an oxygen atom to produce a molecule of the compound we call water. A pair of electrons is shared between the oxygen atom and each of the two hydrogen atoms (Fig. 2.4) so that the neon octet of the oxygen atom is completed, as are the helium duplets of each of the hydrogen atoms.

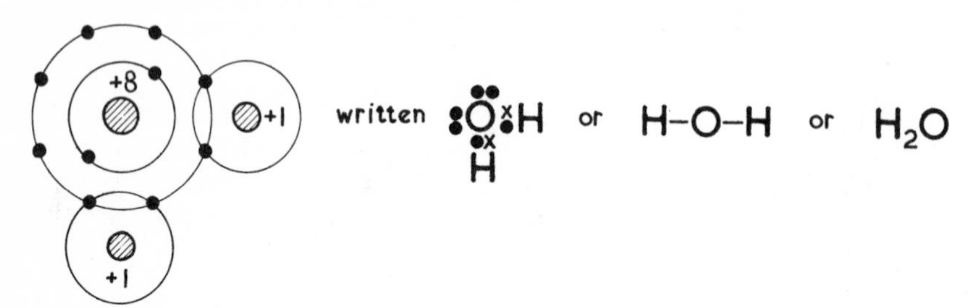

Fig. 2.4—The molecular structure of water, H_2O.

The structure of the water molecule is such that the *bond angle* between the two hydrogen atoms is 105° (Fig. 2.5). As we shall see later (2.4.4), this particular shape of the water molecule offers an explanation of many of the properties of water.

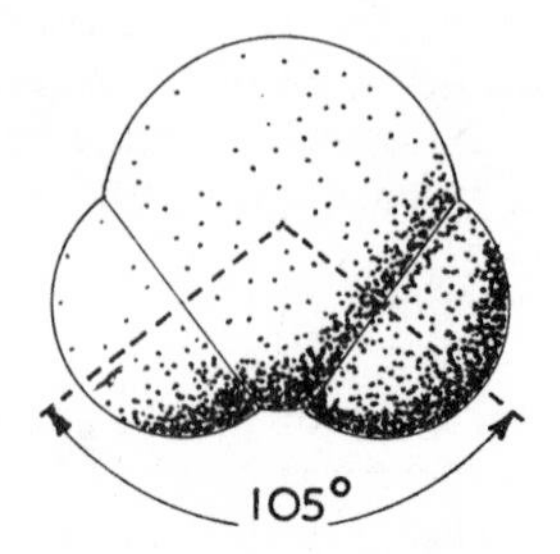

Fig. 2.5—A 'model' representation of the water molecule.

Carbon and its compounds

2.3 The element carbon occupies a position in the centre of the Periodic Table. Thus it may be expected to form covalently bonded compounds more readily than most elements. This it does, particularly with hydrogen, oxygen, nitrogen and the halogens and so produces a vast range of thousands of substances we call 'organic compounds'. These include many complex substances found in living organisms, both animal and vegetable, as well as many materials extracted from coal and petroleum (which are derived from living organisms). The groups of organic compounds of particular interest to the materials scientist are those known as 'super-polymers' or, more commonly, as 'plastics materials'.

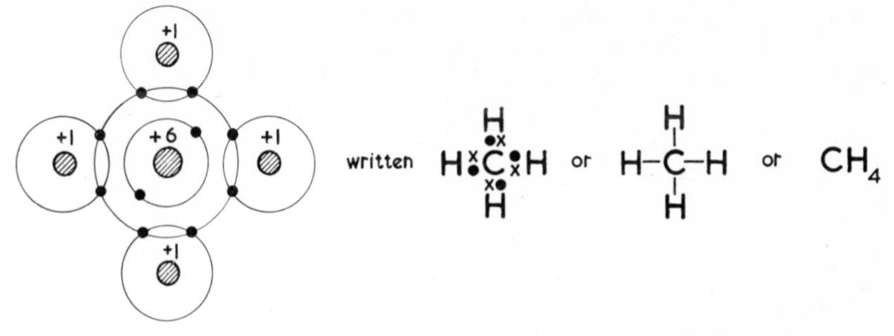

Fig. 2.6—The methane (CH_4) molecule represented diagrammatically.

2.3.1 Possibly the simplest of the organic compounds is the gas methane, the main constituent of 'natural gas'. A molecule of methane consists of an atom of carbon covalently bonded to four hydrogen atoms. Fig. 2.6 indicates diagrammatically the electron distribution within the molecule.

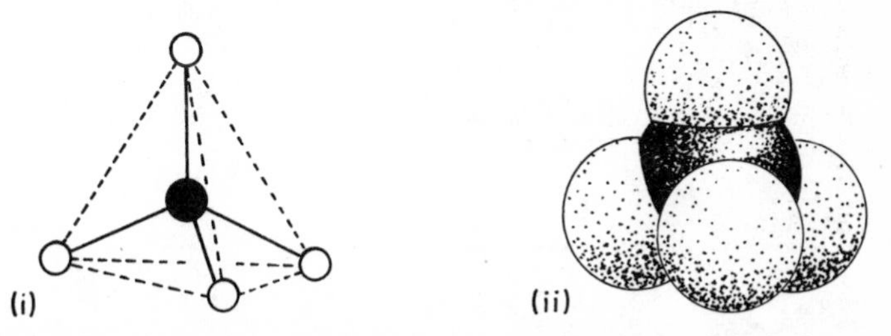

Fig. 2.7—The methane molecule. (i) Indicates the spatial positions of the hydrogen atoms at the points of a tetrahedron at the centre of which is the carbon atom. (ii) A 'model' of the molecule.

The true three-dimensional shape of the molecule will be something like that shown in Fig. 2.7. Methane is in fact the simplest compound in a series of compounds known collectively as alkanes (formerly the 'paraffins'). This is an example of what is called a *homologous series*. As will be seen from Fig. 2.8 the length of the carbon 'chain' increases by the unit:

```
  H
  |
—C—
  |
  H
```

from one compound to the next. As molecular size increases so do forces of intermolecular attraction (2.4). Hence, whilst the compounds methane, ethane and propane are gases at normal temperatures and pressures, butane is a gas which is readily compressed to form a liquid ('Camping Gaz'). Successive members of the series are volatile liquids followed by

Table 2.1—The alkane series.

Number of carbon atoms	*Name*	*Formula* (C_nH_{2n+2})	*Molecular mass*	*Relative density as liquid*	*State at* 0°C	*M. pt.* (°C)	*B. pt.* (°C)	*Product obtained from petroleum*	*Uses*
1	Methane	CH_4	16	0·4240		−183	−162	'Natural gas'	
2	Ethane	C_2H_6	30	0·5462		−172	−87		
3	Propane	C_3H_8	44	0·5826	gas	−187	−42	Calor gas	Heating and lighting
4	n-Butane	C_4H_{10}	58	0·5788		−135	−1	'Camping Gaz'	
5	n-Pentane	C_5H_{12}	72	0·6264		−130	36	'Benzine'	Dry cleaning
6	n-Hexane	C_6H_{14}	86	0·6594		−94	69		
7	n-Heptane	C_7H_{16}	100	0·6837		−90	98		
8	n-Octane	C_8H_{18}	114	0·7028		−57	126	Petrol (Gasoline)	Automobile fuel
9	n-Nonane	C_9H_{20}	128	0·7179	liquid	−54	151		
10	n-Decane	$C_{10}H_{22}$	142	0·7298		−30	174		
16	n-Hexadecane	$C_{16}H_{34}$	226	0·7794		18	280	Paraffin oil Kerosene	Heating and lighting
24	n-Tetracosane	$C_{24}H_{50}$	338	0·7786	solid	51	De-compose before b. pt.	'Mineral oils' Greases	Lubricating oils 'Vaseline'
50	n-Pentacontane	$C_{50}H_{102}$	702	0·7940		92		Paraffin wax (hard)	Candles, waxed paper
								Residue	Asphaltic bitumen

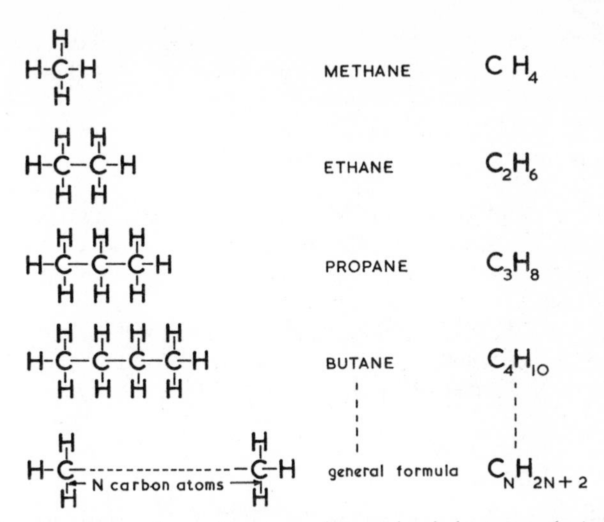

Fig. 2.8—The *alkane* series of organic compounds—previously known as the 'paraffins'.

less volatile liquids such as paraffin oil ('kerosene' in the USA); whilst at the upper end of the series are solids such as paraffin wax with a formula of the order $C_{50}H_{102}$ (Table 2.1). Incidentally all of these compounds in the series are derived from crude petroleum.

2.3.2 There is only one basic pattern in which the carbon atoms can be arranged in the first three compounds in the series, namely, methane, ethane and propane. With the compound butane, however, carbon

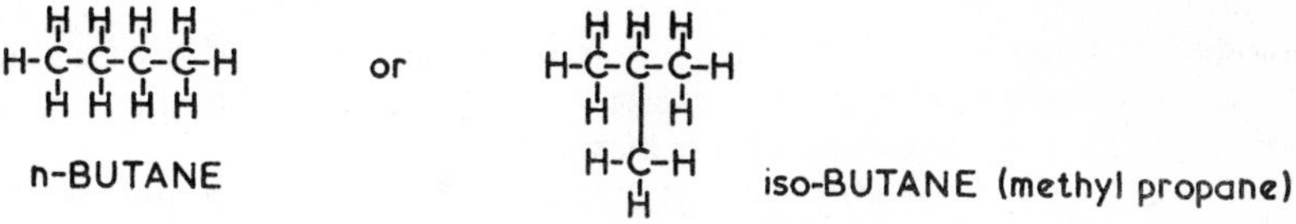

n-BUTANE or iso-BUTANE (methyl propane)

Fig. 2.9—The two isomers of butane.

atoms can be arranged in two different patterns as indicated in Fig. 2.9. In fact two different substances do exist with different physical properties (Table 2.2) even though a molecule of each compound contains the same numbers of atoms of carbon and hydrogen. These two substances, each

Table 2.2—The isomers of butane.

Isomer	*Melting point* (° C)	*Boiling point* (° C)
n-Butane	−138	−0·5
iso-Butane	−159	−10·2

having the same molecular formulae, yet different molecular *structures* are known as *isomers*. In the case of pentane C_5H_{12} there will be *three* isomers; the number of isomers increasing as the molecular formula increases. In fact the molecule Tetracontane, $C_{40}H_{82}$, will be represented by no less than 62 491 178 805 831 possible isomers.

2.3.3 The alkane series described above consists of hydrocarbons which are said to be 'saturated'. Briefly, this means that carbon atoms are joined to their neighbours by means of single bonds (usually called 'valency' bonds). Another series exists in which the compounds are 'unsaturated', that is, double valency bonds connect certain of the carbon atoms. These compounds are of the general formula C_nH_{2n} and the first member of the series is ethylene (Fig. 2.10). Collectively known as the

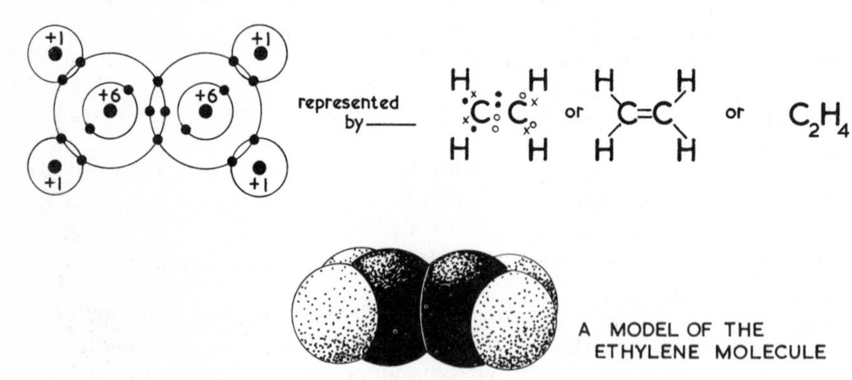

Fig. 2.10—The structural formula of ethylene, C_2H_4.

alk*e*nes (formerly the 'olefins'), these hydrocarbons tend to be less chemically stable than do the alk*a*nes and the double bond can be 'broken' relatively easily by chemical means.

Ethylene is a by-product of the petroleum refinery since, during the 'cracking' of some of the heavier oils to produce light and more volatile hydrocarbons (such as octane), an unsaturated hydrocarbon group is left over—in this case ethylene, C_2H_4. If ethylene is suitably treated the

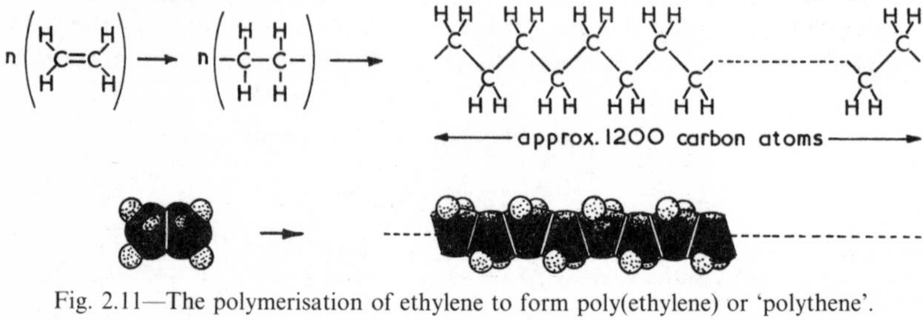

approx. 1200 carbon atoms

Fig. 2.11—The polymerisation of ethylene to form poly(ethylene) or 'polythene'.

double bond between the carbon atoms is broken and simultaneously large numbers of these units are formed. These immediately link up—or *polymerise*—to form long chain-like molecules of the substance 'polyethylene', or polythene as it is commonly known. The sequence in this process is indicated in Fig. 2.11.

Polythene is one of a large number of these super-polymers (commonly called 'plastics') which are finding increasing use as engineering materials. They will be discussed in more detail in Chapter Twelve.

Intermolecular forces

2.4 The *atoms* in a molecule are, as we have seen, held together by covalent bonds. These are strong primary bonds which are dependent upon relatively powerful electrostatic forces. At the same time weaker electrostatic forces of a secondary nature give rise to attractions between any one *molecule* and its neighbours. Only in this way can we explain the fact that a gas condenses to form a liquid and that the liquid subsequently crystallises to form a solid. The secondary electrostatic forces operating between molecules influences such important properties as melting point, solubility and, in plastics, tensile strength. Powerful modern adhesives of the 'impact' variety rely on these secondary intermolecular forces and illustrate the collective effect of large numbers of such secondary bonds operating simultaneously.

As early as 1873 Johannes Diderik van der Waal, a Dutch physicist, sought to explain deviations from the Gas Equation ($PV = RT$). As readers may know, this relationship is true only for a *perfect* gas. No gas is 'perfect' in this sense and van der Waal introduced corrections into the Gas Equation such that:

$$\left(P + \frac{a}{V^2}\right)(V - b) = RT$$

where b is a constant related to the actual volume of the molecules of the gas; a is also a constant and a/V^2 represents the loss of pressure due to *forces of attraction between the molecules of the gas.*

A stable molecule or a stable atom (e.g. an atom of a noble gas) is, taken as a whole, electrically neutral since the number of protons present is balanced by an equal number of electrons. However, the resultant 'centre' of the electropositive charges does not necessarily coincide with the centre of the negative charges. Then the particle—whether molecule or single uncombined atom—can be assumed to possess a resultant

Fig. 2.12—In (i) the centres of positive and negative charges coincide, but in (ii) they are separated so that the molecule (or single atom in the case of the noble gases) will have a resultant 'dipole moment'.

'dipole moment' (Fig. 2.12). These intermolecular forces, often referred to as van der Waal's forces, are of four main types:

2.4.1 *Attraction between permanent dipoles.* A molecule of the gas hydrogen chloride contains one atom of hydrogen and one of chlorine, covalently bonded (H—Cl). However, the electrons tend to be concentrated rather more densely in the region of the chlorine nucleus (which contains a larger positive charge) than around the hydrogen nucleus

Fig. 2.13—The molecular orbital of hydrogen chloride (i); showing, (ii) the formation of a dipole moment.

(which contains only a single positive charge due to the presence of one proton). The uneven distribution of electrons in the molecular orbital is equivalent to a separation of charges. For this reason the hydrogen nucleus tends to be rather 'exposed' so that that part of the molecule has a resultant small positive charge (δ^+). Conversely the part of the molecule containing the chlorine nucleus will carry an equal negative charge (δ^-) (Fig. 2.13). Thus the hydrogen chloride molecule will possess a resultant dipole moment.

Such a molecule is said to be *polar* whereas a hydrogen molecule, in which the charges were equally distributed, would be *non-polar* since it possessed no resultant di-pole moment, because of its relative symmetry. Polar molecules will repel each other or attract each other according to the way in which they are orientated (Fig. 2.14).

Fig. 2.14—Attraction and repulsion between permanent dipoles follows the well-known laws of electrostatic attraction and repulsion.

The degree of mutual dipole alignment will govern the extent of intermolecular attraction. Since this alignment is opposed by thermal motion, dipole forces are dependent largely upon temperature. Thus, as temperature rises these forces decrease. For example, the plasticity of thermoplastic polymers increases with the temperature.

2.4.2 *Attraction between permanent dipoles and induced dipoles.* In some cases the orbital of a non-polar atom or molecule may be affected by the close proximity of a strongly polar molecule. In Fig. 2.15 it is assumed that the 'positive end' of a permanent dipole has approached a non-polar atom or molecule in which the electron cloud is large, diffused and capable of easy distortion. The electron cloud of the non-polar particle will be attracted towards the positive end of the permanent dipole. This distorts the electron cloud of the non-polar particle to the extent that a

Fig. 2.15—The formation of an induced dipole.

temporary dipole is formed. Such a distortion of an electron cloud is referred to as *polarisation*. The positive pole of the permanent dipole attracts the electrons of the non-polar atom or molecule and so builds up a negative charge facing it. Hence the permanent dipole and the induced dipole are in a position in which they will attract each other.

2.4.3 *Dispersion forces.* Although there are many molecules which appear to be non-polar, that is the centres of positive and negative charges would seem to be coincident, all molecules—and the single atoms of noble gases—have time-varying dipole moments the nature of which will depend upon the position of the electrons at any particular instant.

Consider the case of two atoms of the noble gas, argon, in close proximity to each other (Fig. 2.16). If the electron clouds are evenly

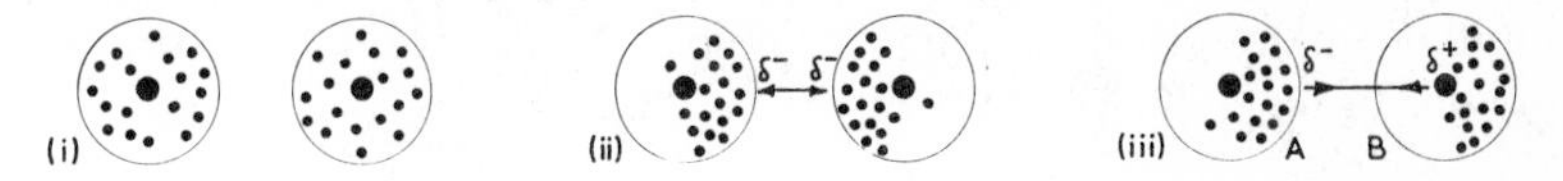

Fig. 2.16—The formation of dispersion forces between two atoms of argon.

distributed as in (i) then there will be no resultant dipole moment in either atom and they will neither attract nor repel each other. If, however, an uneven distribution of the electron clouds is present as in (ii) then the atoms will repel each other; but if the electron clouds are distributed as in (iii) then the electron cloud of atom *A* will be attracted by the 'unprotected' nucleus of atom *B*.

The displacement of electrical charges necessary to produce a dipole is temporary and random so that a molecule, as a whole, over a period of time has no resultant dipole. However, if it has a dipole moment at any one instant it will induce another in a near neighbour and a force of attraction will be set up between the two. It is thought that such interactions must be largely responsible for van der Waal's bonding between non-polar atoms and molecules. The functioning of these dispersion forces was postulated by London as long ago as 1930 and, except in a few cases where very strong dipole moments exist, these forces exceed in magnitude all other intermolecular forces, when considered collectively.

The dispersion effect is evident when considering the relationship between molecular mass and boiling point. Massive molecules generally contain more electrons than do light ones and so the attraction due to dispersion effects is greater in the massive molecules. Heat of evaporation is a measure of the energy required to overcome forces of attraction between molecules so that they can separate. Since the boiling point is proportional to the heat of evaporation it is a convenient criterion in assessing these intermolecular forces. Thus normal covalently bonded compounds with massive molecules have higher boiling points than those with small molecules. An excellent example of this principle is given by the homologous series, the alkanes, mentioned earlier in this chapter. The lower members are gases at ordinary temperatures and pressures, then, as the series is ascended, liquids of increasing boiling point are followed by higher members which are waxy solids (Table 2.1).

Dispersion forces, along with permanent dipoles figure largely in the intermolecular forces which prevail between the chain-like molecules in super-polymers.

2.4.4 *The hydrogen bond.* When a hydrogen atom is covalently bonded to a relatively large atom such as nitrogen, oxygen or fluorine, a powerful permanent dipole is set up. This is because the electron cloud tends to become concentrated around that part of the molecule containing the nitrogen, oxygen or fluorine nucleus, thus leaving the positively charged hydrogen nucleus relatively 'unprotected'. In this way particularly strong dipole–dipole forces of attraction can be set up. These are fundamentally similar in nature to the permanent dipole–dipole forces described earlier but are of much greater magnitude.

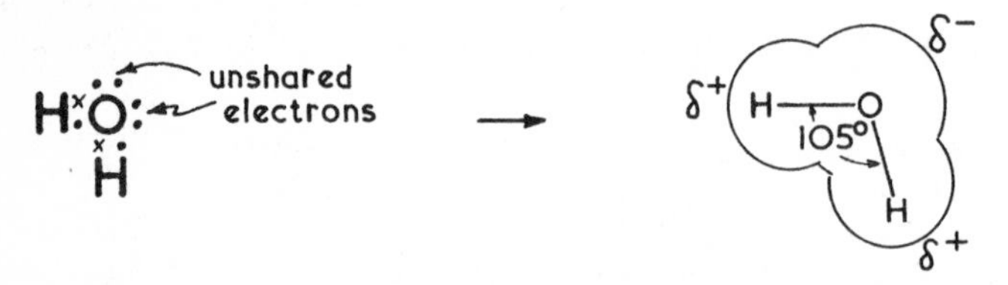

Fig. 2.17—The basis of the 'hydrogen bond' (or 'bridge') in the water molecule.

This effect is apparent in the case of the water molecule. Here the oxygen part of the molecule, carrying the un-paired electrons, will be negatively charged whilst the hydrogen portions will be correspondingly positively charged. The hydrogen atoms form 'bridges' (shown thus ----- in Fig. 2.18) between themselves and oxygen atoms of other molecules thus linking the molecules together. It must be remembered that Fig. 2.18 shows a two-dimensional diagram whereas in practice a three-dimensional spatial pattern exists.

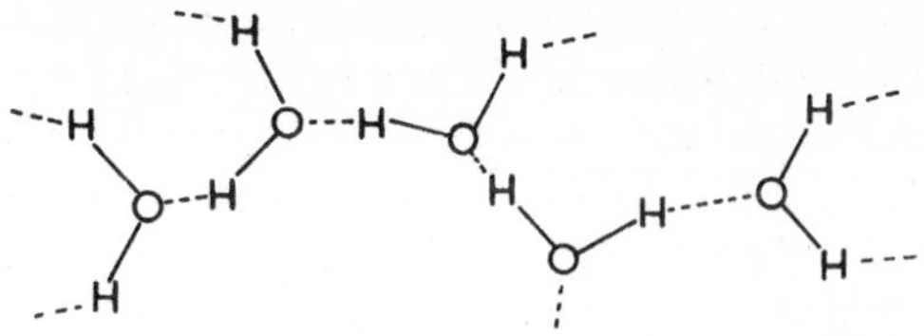

Fig. 2.18—The formation of hydrogen bonds between molecules in water.

Approximately two-thirds of the heat of vaporisation of water at its boiling point is used up in disrupting these hydrogen bonds or 'bridges'. Sufficient hydrogen bonds are formed to account for the abnormally high boiling point of water. For example, since the sulphur atom is more massive than that of oxygen the compound hydrogen sulphide, H_2S, might be expected to have a higher boiling point than that of water, H_2O. In fact the boiling point of hydrogen sulphide is $-60°$ C, some 160° C *lower* than that of water. This is apparently due to the relatively *large* size of the sulphur atom causing a lesser concentration of the electrons so that a weaker dipole moment is produced in the molecule. In the same way the molecule of hydrogen chloride, HCl, does not exhibit a strong hydrogen bond because of the large size of the chlorine atom, whereas the hydrogen fluoride, HF, molecule does possess a powerful hydrogen bond because of the smaller size of the fluorine atom and also its very strongly electronegative character.

The hydrogen 'bridge' associated with the hydroxyl (—OH) and other groups in many organic polymers contributes considerably to their overall cohesion. Whilst it is much stronger than other van der Waal's forces it is nevertheless weaker than the ordinary covalent bond. In terms of bond energy a covalent bond may represent as much as 800 kJ/mole* whilst the hydrogen bond would be equivalent to some 40 kJ/mole. Bond energies of the weaker van der Waal's forces vary between 1 kJ and 15 kJ/mole.

* The SI unit 'mole' refers to the relative molecular mass of a substance expressed in grams.

Chapter Three
The Crystal

3.1 When used in a non-scientific sense the term 'crystal' conveys an impression of a material which is geometrically regular in shape and which is both lustrous and transparent—as the expression 'as clear as crystal' suggests. In fact crystals of many materials have an irregular external shape though the particles within these crystals are arranged in some regular pattern. Many crystalline materials are opaque and do not have a lustrous appearance. In general terms a substance can be described as crystalline when the particles from which it is built are arranged in a definite three-dimensional pattern which repeats over a long range. Metals and indeed all truly solid substances are crystalline in structure. Non-crystalline 'solids' such as glass, pitch and many polymers can often more conveniently be considered as extremely viscous liquids in so far as their physical and mechanical properties are concerned.

Again the question arises—what is the nature of the forces which bind together the units from which a crystal is built? In some types of crystal the binding forces consist of covalent bonds, in others hydrogen 'bridges' (2.4.4); whilst in mineral crystals and those of inorganic salts generally, the force of attraction is due to what is described variously as an ionic, polar or electrovalent bond.

The electrovalent bond

3.2 In the previous chapter we saw that a covalent bond is formed when electrons from the outer shell of two (or more) atoms are *shared* in such a way that the appropriate noble-gas structures are attained in the outer electron shells concerned. However, chemical combination can also take place by the *transfer* of electrons from one atom to another such that a noble-gas structure is attained in the valency shells of each resultant particle. Thus, a metallic atom which contains only one or two valency electrons tends to give up these electrons to a non-metallic atom which already has a valency shell of six or seven electrons. In this way the outer electron shell of each resultant particle attains a noble-gas structure.

3.2.1 The metal sodium reacts vigorously with the poisonous gas chlorine to produce a very stable substance, sodium chloride (common salt). In this instance the sodium atom which carries a single electron in its outer shell donates this electron to a chlorine atom which previously had seven electrons in its outer shell. By means of this transfer the sodium particle is left with a complete outer shell (the 'neon shell') and the chlorine particle also has a completed shell (the 'argon shell'). Since an electron transfer has taken place from one atom to another we can no longer refer to the resultant particles as 'atoms'. The sodium particle is obviously something less than a sodium atom whilst the chlorine particle is something more than a chlorine atom. Particles produced by electron transfer in this manner are known as *ions*.

Prior to the chemical reaction during which the electron transfer took place, the atoms were electrically neutral because in each case the number of positively charged protons in the nucleus was balanced by an equal number of negatively charged electrons in orbitals around the nucleus.

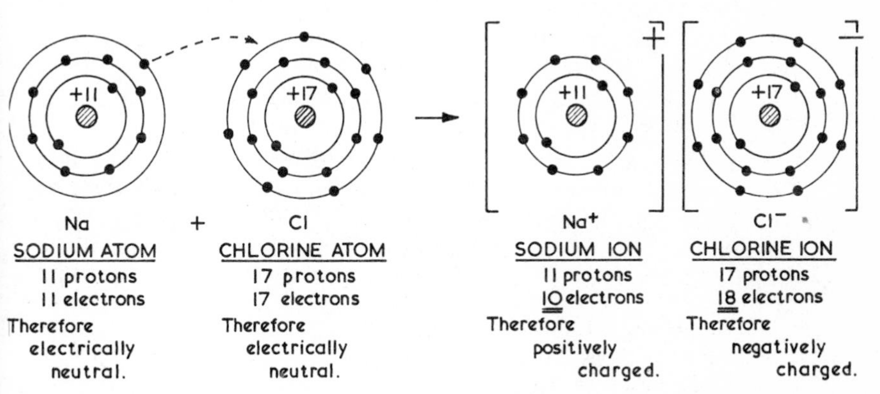

Fig. 3.1—The formation of an electrovalent—or ionic—bond between a sodium atom and a chlorine atom.

During the reaction the sodium atom *lost* an electron and the ion so formed must therefore carry a resultant positive charge. Similarly, since the chlorine atom *received* this electron the chlorine ion produced must carry a resultant negative charge (Fig. 3.1).

Since the two ions formed as a result of this chemical reaction carry equal but opposite charges they will attract each other and it is this force of coulombic attraction which constitutes the electrovalent bond. Although generally weaker than a covalent bond, this electrovalent bond is much stronger than any of the relatively feeble van der Waal's forces.

3.2.2 When sodium and chlorine react with each other even on the laboratory scale many millions of atoms are involved and an equally large number of ions is produced. Whilst there is mutual attraction of sodium ions for chlorine ions, forces of repulsion operate between all like ions. That is, each sodium ion is repelled by all other sodium ions whilst each chlorine ion is repelled by all other chlorine ions. As a result the system attains equilibrium when each chlorine ion is surrounded by six sodium ions and, conversely, when each ion of sodium is at the same time surrounded by six chlorine ions (Fig. 3.2). In this instance the relative sizes of the chlorine ions and sodium ions are such that a cubic type of pattern

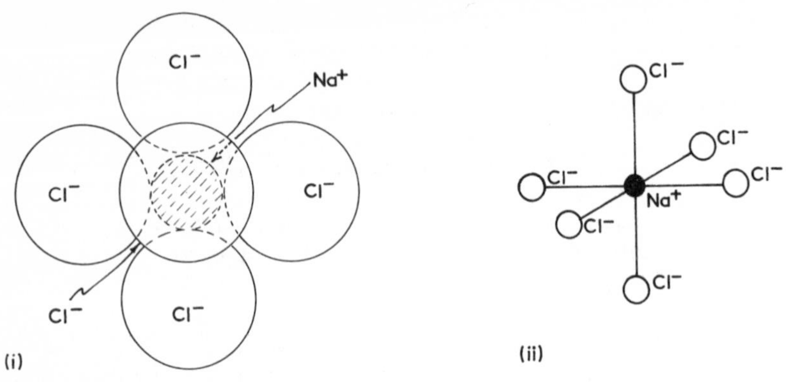

Fig. 3.2—Each Na^+ ion is surrounded by six Cl^- ions.

is formed. Since each sodium ion is surrounded by six chlorine ions it is said to have a co-ordination number of six. For a similar reason chlorine also has a co-ordination number of six.

Diagrams representing crystal structures are somewhat clearer if the centre of each ion is represented by a small circle (Fig. 3.2 (ii) and Fig.

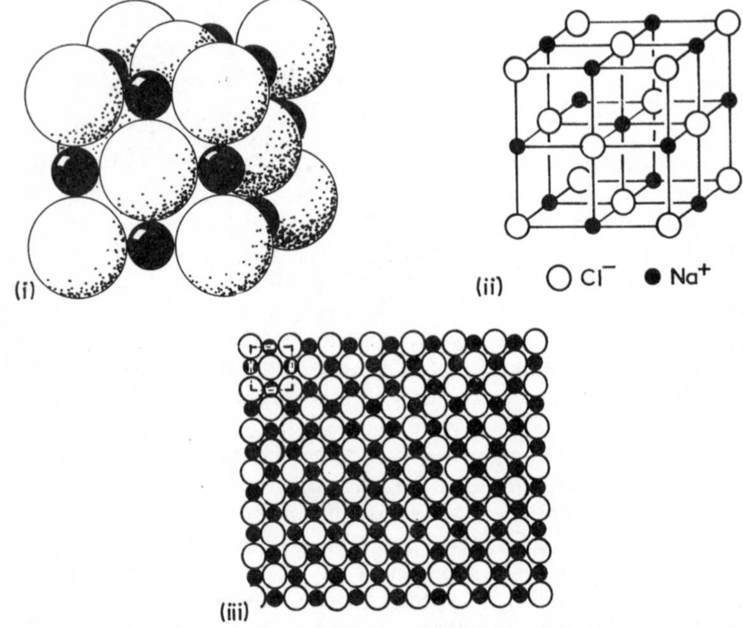

Fig. 3.3—The crystal structure of sodium chloride. (i) and (ii) show the unit cell of sodium chloride, whilst (iii) indicates the continuous structure of the crystal such that each face of a unit cell is shared by the next cell. In fact we have two interpenetrating *face-centred cubic* structures, one of Na^+ and one of Cl^- (3.6.2.).

3.3 (ii)). In this cubic type of structure the three axes of symmetry are at right angles to each other. The chlorine ions take up positions as far as possible from other chlorine ions and since they are equally charged they are also equally spaced. Each chlorine ion has six sodium ions equally spaced around it and in contact with it. Since there is an electrostatic field extending into the space outside this unit other sodium ions and chlorine ions will be attracted so that the structure depicted in Fig. 3.3 builds up and extends as long as ions are available. The framework on which the units of a crystal arrange themselves according to some fixed pattern is known as a *space lattice*.

3.2.3 The ions arrange themselves in positions where the force of attraction between positive and negative ions is a maximum. The coulombic force F acting between charged particles is governed by the formula:

$$F = \frac{C_1 C_2}{k \,.\, d^2}$$

where C_1 and C_2 are the charges on each body respectively; k is the dielectric constant of the medium separating the bodies ($k = 1$ for air) and d is the distance between the centres of the particles. Since the charges on all ions we are considering are of the same magnitude (e),

$$F = \frac{e^2}{k \,.\, d^2}$$

$$\text{and} \quad F \propto \frac{1}{d^2}$$

The distances between the centre of any chlorine ion and the centres of the six nearest sodium ions is less than the distances between the centre of that same chlorine ion and the centres of the six nearest chlorine ions. Consequently the forces of attraction between any ion and its neighbours are greater than the forces of repulsion, and the crystal is therefore a stable structure.

The unit structure in the case of the substance sodium chloride is the crystal. Sodium chloride does not exist in the form of molecules and it is therefore not appropriate to write the formula 'NaCl' since this would imply the existence of one molecule of sodium chloride. Nevertheless busy chemists are in the habit of representing sodium chloride by the empirical formula, NaCl, as a type of 'chemical shorthand'.

3.2.4 Sodium chloride will dissolve readily in water. This is due to two important properties of water:

1. It forms a polar molecule (2.4.4);
2. It has a high dielectric constant.

The hydrogen 'end' of the water molecule is attracted to the negatively charged chlorine ions and the oxygen 'end' of the molecule to the

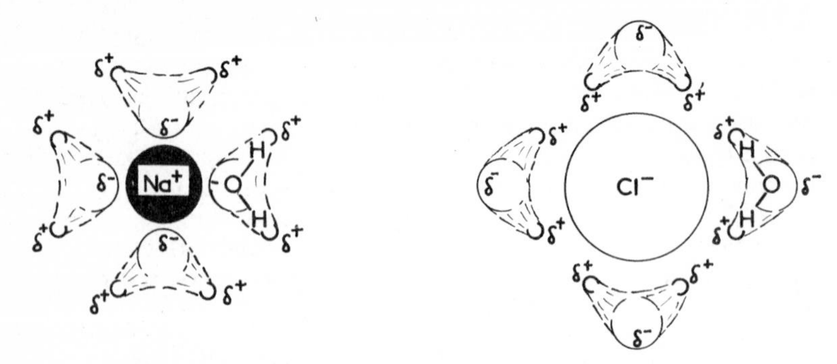

Fig. 3.4—The attraction between ions and water molecules.

positively charged sodium ions. This causes water molecules to cluster round ions at the surface of a sodium chloride crystal until their forces of attraction are sufficient to overcome the electrovalent bond between sodium ions and chlorine ions. The ions then mix freely with water molecules in solution. This process of separation of sodium ions from chlorine ions on solution is termed *dissociation*. It is sometimes incorrectly called ionisation. The latter process in fact took place during the original reaction between sodium and chlorine when atoms of each element became ions.

Space lattices

3.3 A space lattice can be defined as a distribution of particles in three dimensions such that every particle has identical surroundings. It is generally convenient to imagine a space lattice as continuing to infinity in all directions, for although crystals themselves are of finite size the units from which they are built are exceedingly small. Thus a small crystal of copper some 0·1 mm in diameter will nevertheless contain roughly 10^{16} atoms of copper.

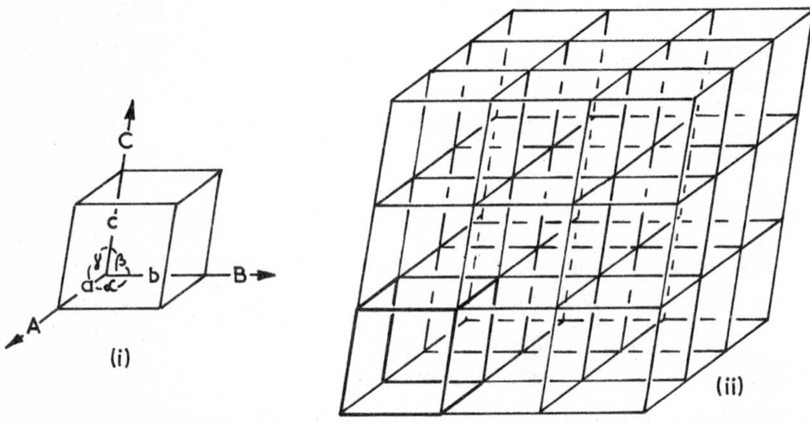

Fig. 3.5—The relationship between (i) a unit cell and (ii) a continuous space lattice.

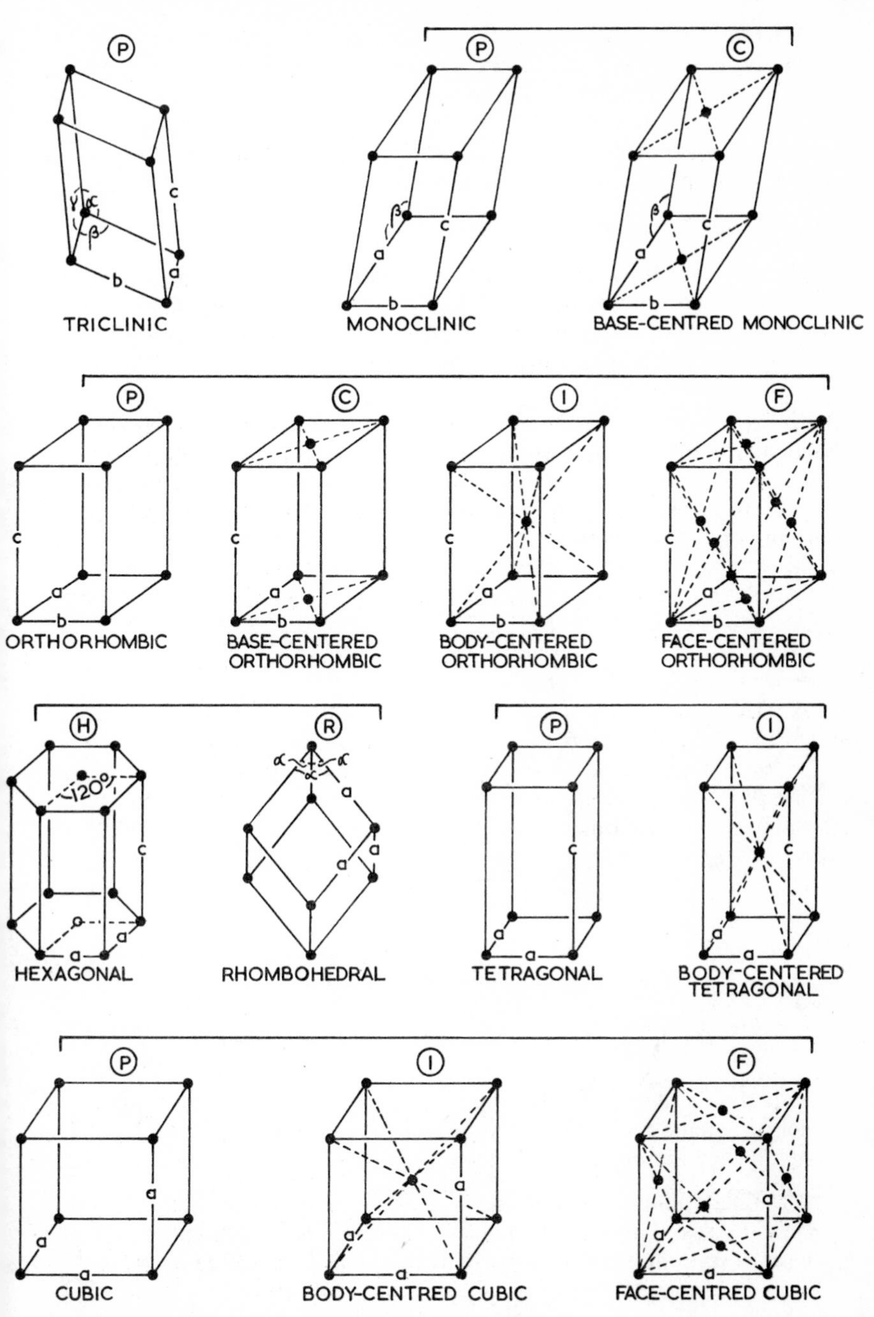

Fig. 3.6—The fourteen Bravais space lattices. (P = primitive; C = centred on 'ab' face; I = body centred and F = face centred.)

A section of a space lattice of general form is illustrated in Fig. 3.5. Since the points at which particles are considered to exist are regularly distributed, the fundamental geometry of the space lattice can be described by three lattice vectors A, B and C. The vectors define the unit cell and the geometry of the space lattice is specified satisfactorily by the lattice constants (vector lengths) a, b and c and the inter-axial angles α, β and γ.

3.3.1 There are only fourteen different patterns in which points can be arranged in order to satisfy the definition of a space lattice. These are the *Bravais space lattices* (Fig. 3.6). Of these not more than six are met with in metallic structures and in fact only three are really common.

3.3.2 Frequently in a study of crystals it becomes necessary to identify a particular plane of atoms running through the structure. Such planes and directions are best described by a set of numbers known as *Miller indices*. These are widely used in X-ray analyses and other methods used in deriving crystal structures.

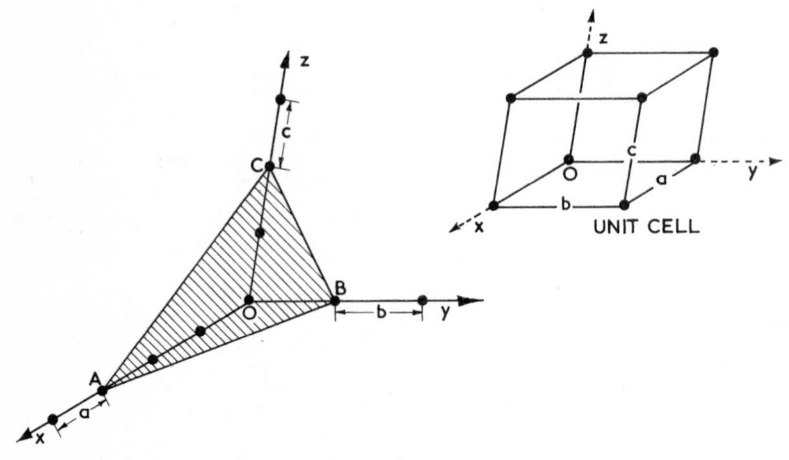

Fig. 3.7—The derivation of Miller indices for plane ABC.

Miller indices are proportional to the reciprocals of the intercepts which the plane makes with the three principal axes (x, y and z) of the system. They are derived as follows:

(*i*) Determine the intercepts which the plane makes with the three crystal axes. These intercepts are expressed as the number of axial lengths (a, b or c) from the origin. In Fig. 3.7 these intercepts are:

$$x = 3;\ y = 1;\ z = 2$$

(*ii*) Take reciprocals of these intercept values, i.e. $\frac{1}{3}$; 1; $\frac{1}{2}$.

(*iii*) Convert these reciprocals to the smallest integers which are in the same ratio, i.e. 2; 6; 3. This is written in the form (263). Thus plane ABC in Fig. 3.7 is represented by the Miller indices (263).

These indices represent not only the plane ABC but all planes parallel to it. For example a plane with the intercepts 6, 2, 4 will be parallel to plane ABC and will be represented by the same Miller indices.

A plane on the opposite side of the origin would have negative intercepts on the axes and would be denoted $(\bar{2}\ \bar{6}\ \bar{3})$.

The above derivation of Miller indices represents the general case (a triclinic structure in which $a \neq b \neq c$ and $\alpha \neq \beta \neq \gamma$). We will now

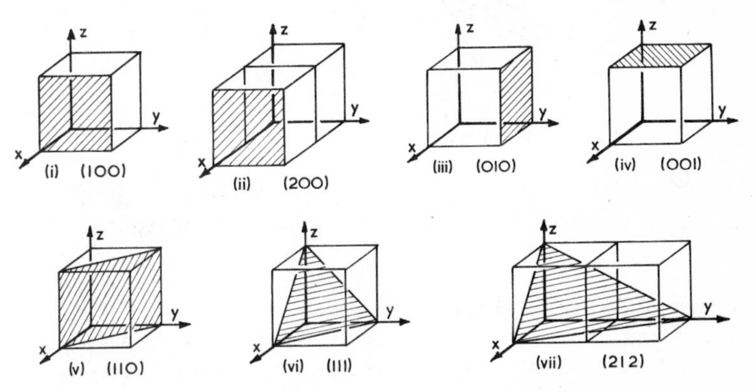

Fig. 3.8—Miller indices for some typical planes in a simple cubic system.

consider some typical planes for a simple cubic system. In Fig. 3.8 (i) the plane indicated has the following intercepts:

$$x = 1;\ y = \infty;\ z = \infty.$$

Consequently the Miller indices will be (100)—the reciprocals of 1, ∞, ∞. The plane indicated in Fig. 3.8 (v) has intercepts:

$x = 1;\ y = 1;\ z = \infty$, and hence Miller indices (110).

Fig. 3.8 (vi) having equal intercepts on all axes is represented by Miller indices (111).

In the case of the simple cubic structure, planes which form parallel groups are enclosed in brackets, e.g. {hkl}. Thus the following are equivalent parallel planes:

$$111 = (111);\ (\bar{1}11);\ (1\bar{1}1);\ (11\bar{1});\ (\bar{1}\bar{1}1);\ (\bar{1}1\bar{1});\ (1\bar{1}\bar{1});\ (\bar{1}\bar{1}\bar{1})$$

3.3.3 Directions through a crystal can be specified in terms of the indices which give the integral co-ordinates of a point on a line drawn between the point and the origin. Direction indices are written in square brackets [ ] to distinguish them from Miller indices written in parentheses ( ) which represent a plane. In the case of the simple cubic cell the *direction*

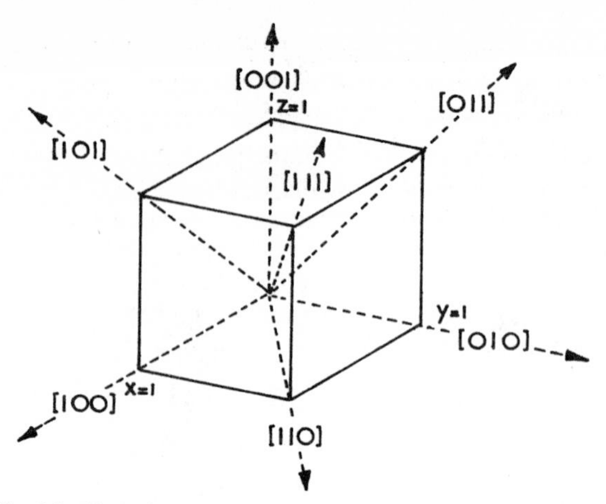

Fig. 3.9—Some important directions in a simple cubic cell.

normal to the *plane* (hkl) is [hkl]. Fig. 3.9 indicates some of the important directions for a simple cubic cell. A general *family* of directions is written ⟨hkl⟩.

Experimental derivation of crystallographic planes

3.4 Interplanar distances for crystalline materials are generally determined by X-ray methods. The X-rays used are of short wavelength and

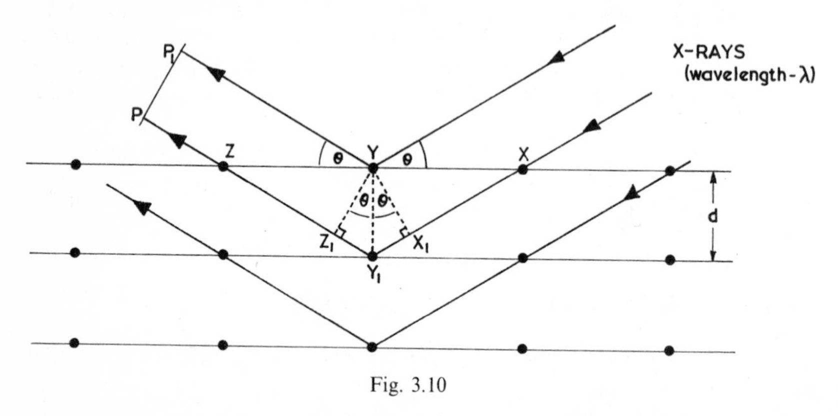

Fig. 3.10

hence great penetrating power. They are also as nearly as possible 'monochromatic', that is, of a single wavelength. The method used depends upon the fact that in a crystal, planes occur which are occupied by atoms according to a regular pattern and that X-rays will be either reflected or transmitted at these planes.

Fig. 3.10 represents diagrammatically two layers of atoms in a crystal structure. The layers are distance d apart. It is assumed that two incident waves strike the crystal at X and Y. The wave which strikes the first layer at X will either be reflected or transmitted to a point Y_1 in the second layer. Here the process is repeated. The wave reflected from Y interferes constructively at PP_1 with the wave which is transmitted from X to Y_1 and subsequently reflected providing that $X_1Y_1Z_1$ is an integral multiple of the wavelength λ of the incident X-rays. Using simple geometry it can be seen that:

$$X_1Y_1 = Y_1Z_1 = d \,.\, \sin\theta$$

3.4.1 *Bragg's Law*. For constructive interference of X-rays, Bragg's Law states that:

$$n\lambda = 2d \,.\, \sin\theta$$

where θ is the incident angle of X-radiation. For the principal diffraction line, $n = 1$ and so d can be calculated:

$$d = \frac{\lambda}{2\sin\theta}$$

Interplanar spacings are very small and since the maximum value of $\sin\theta$ is one, short wavelength monochromatic X-rays must be used so that λ is of the same order as the interplanar spacings, i.e. approximately 10^{-10} m.

In practice if a beam of X-rays falls on a crystal reflection from a given plane can only take place if that plane is inclined to the 'pencil' of X-rays at the necessary angle, θ. This condition is unlikely to be satisfied and a powdered sample of the crystalline material is used so that, inevitably, a few particles will be suitably orientated to make reflection possible. The powdered sample is placed at the geometric centre of a circular strip of photographic film and subjected to a pencil of X-rays (Fig. 3.11). Diffraction occurs from planes that are more or less randomly orientated

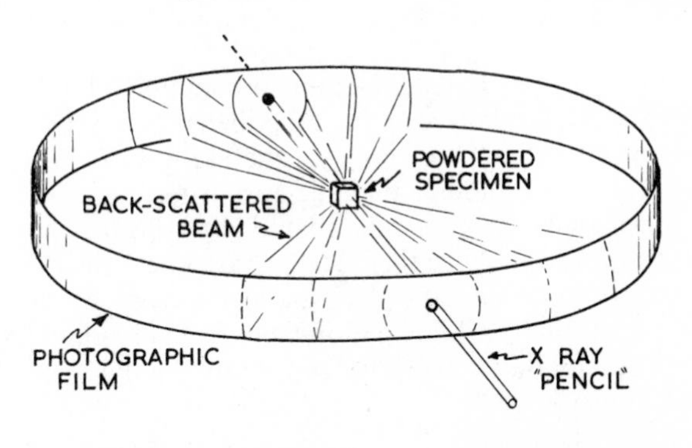

Fig. 3.11—The principles of a method for X-ray crystallography.

with respect to the source. Hence the diffraction pattern will be spherically distributed such that the circular arrangement of the film intercepts also the back-diffraction patterns.

This is a very sensitive method for the determination of interplanar spacings, giving errors of no more than 0·05% in measurements of the order of 2×10^{-10} m.

The metallic bond

3.5 About three-quarters of the elements are metals, only about one-eighth are non-metals, whilst the remainder, the metalloids, have properties which are intermediate between metals and non-metals. The non-metals are characterised by relatively high electronegativities, by being poor conductors of heat and electricity and by having melting points and boiling points which are generally low. Metals, however, have relatively high melting points and boiling points and are good conductors of heat and electricity. Moreover they possess plasticity, the property which permits permanent deformation without accompanying fracture. All of these properties are related to the nature of the metallic bond, and are connected with the fact that metallic atoms have only one or two electrons in their valency subshells. These electrons are bound to the atom much less strongly than are those in the atoms of non-metals.

The atoms of a metal are arranged such that their ions conform to some regular crystal pattern (Fig. 3.12) whilst their valency electrons

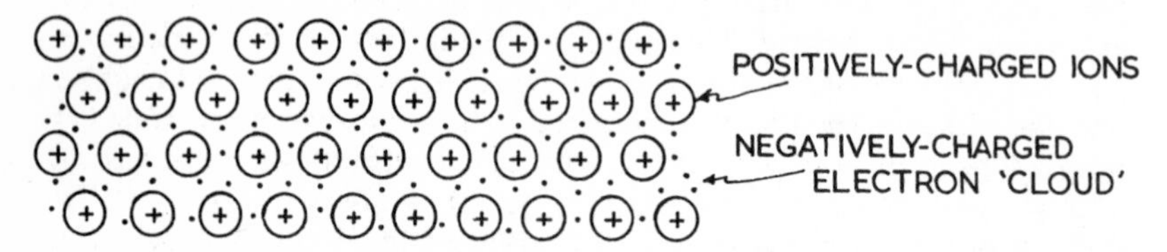

Fig. 3.12—A very much simplified picture of the nature of the metallic bond.

behave as though they are mobile, forming a common 'cloud' surrounding the ions. These electrons can be described as *delocalised* and no particular electron 'belongs' to any particular atom. Although the individual electrons are bound to metallic atoms much less strongly than are those in non-metallic elements, the shared electrons bind metallic ions very tightly into the lattice since there is attraction between the positively charged nuclear protons and the delocalised electrons of the common 'cloud'. This results in the high melting points and boiling points of metals, whilst the mobility of the electrons confers high conductivity of both heat and electricity.

3.5.1 It must be admitted that the above interpretation of the metallic bond is oversimplified and fails under more detailed quantitative examination. Nevertheless this simplified picture is useful at this stage in considering some of the mechanical properties of metals, and a slightly more detailed examination of the nature of the metallic bond will be presented later in this book (18.3.1).

3.5.2 *Electrical properties.* Electrons in the electron 'cloud' are not bound to fixed orbitals but are able to move freely throughout the body of metal, that is their movement is random. If a relatively small electrical potential difference is applied to a piece of metal the random electron movements are converted into an orderly electron stream towards the higher potential. Thus metals are excellent conductors of electricity and this flow of electrons produced by the application of a potential difference is what we call a 'current' of electricity.

In non-metallic elements the outer-shell electrons of atoms are confined in covalent bonds. They are therefore unaffected by the application of an electrical potential difference unless it is very great. Consequently most non-metallic elements and compounds are bad conductors, particularly those of the organic polymer type where covalent bonding exists throughout large molecules (2.3.3). Many such substances are in fact used as insulators.

Some elements, often referred to as 'metalloids', possess properties which are intermediate between those of metals and non-metals. This is reflected in their electrical properties. In silicon and germanium for example the energy necessary to detach electrons from their covalent bonds and move them through the structure of the solid is much less than with most non-metals. These elements are termed semi-conductors (18.6).

Crystal structures in metals

3.6 As we have seen, the forces binding the atoms in a metal are non-directional, and since in a pure metal all atoms are of the same kind and size, they will arrange themselves in the closest possible packing patterns which are associated with positions of minimum potential energy. Most of the metals of engineering importance crystallise in one of three principal types of structure in all of which the atoms are held together solely by the metallic bond. The spheres used in the illustrations which follow represent atomic cores between which the electron cloud will be distributed.

3.6.1 *The close-packed hexagonal structure* (CPH). When we place the snooker balls in the triangular frame at the beginning of a game we are in fact packing them in the closest possible arrangement in a single plane. The arrangment is that indicated in Fig. 3.13 (i) and it will be seen that any sphere, *X*, is touched by six immediate neighbours. Lines joining the centres of these spheres form a regular hexagon. If we consider any three spheres, lines joining their centres form an equilateral triangle and at the centre of this triangle is a small space between the spheres. Now consider a sphere put on top of these first three so that it rests in the hollow formed by them. The space so enclosed by the four spheres is termed a *tetrahedral site* since it is enclosed by four spheres whose centres lie on the points of a regular tetrahedron. This type of site is of great importance in a study of crystallography.

If a second layer of spheres is now placed on top of the first layer and allowed to find its own level, each sphere in this second layer will come to rest in one of the hollows formed in the first layer as indicated in Fig. 3.13 (ii).

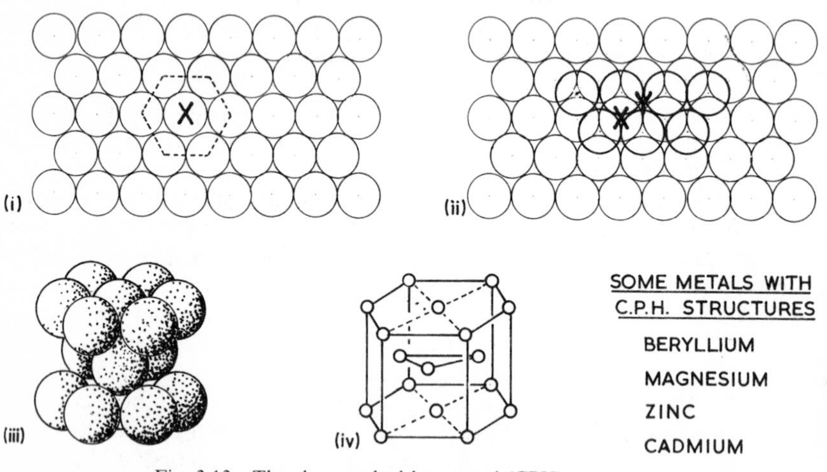

Fig. 3.13—The close-packed hexagonal (CPH) structure.

There are *two* possible positions in which spheres of a third layer can be placed—either in those positions marked *X* or in those marked *Y*. If we place them in positions marked *X*, the centre of each sphere in the third layer will be vertically above the centre of a sphere in the first layer. This is the close-packed hexagonal (CPH) structure and if the arrangement in the first layer is represented by *A* and that in the second layer by *B*, the building order involved here is *ABABABAB* . . . etc. Any atom in the system is touched by six atoms in its own plane (Fig. 3.13 (i)) and by three atoms in the plane beneath. Hence the co-ordination number is 12 and is characteristic of all metallic structures with the closest possible packing of atoms. The CPH structure is so called because it has an axis of sixfold symmetry perpendicular to the layers of closest packing, that is, perpendicular to the plane of the paper in Fig. 3.13 (i). A unit cell of this structure is shown in Figs. 3.13 (iii) and (iv); Fig. 3.13 (iii) indicating the atomic cores in contact and Fig. 3.13 (iv) the space lattice.

Plastic deformation in metals takes place by a process known as 'slip', that is by layers of atoms gliding over each other. This occurs most frequently by the close-packed planes slipping in close-packed directions. In the CPH structure such slip is limited since it will occur mainly along basal planes of the hexagon. Thus, most metals with a CPH structure are much less ductile than those with a face-centred cubic structure.

3.6.2 *The face-centred cubic structure* (FCC). In this structure the basic positions of spheres in the *first two* layers are the same as those indicated in Fig. 3.13 (ii). For the *third* layer, however, the spheres are placed in the hollows marked *Y* (Fig. 3.13 (ii)). Thus in this third layer the spheres are in different positions from those in the first layer (thus differing from the CPH structure). If we represent this third layer by *C* then the building order will be *ABCABCABCABC* . . . etc. This is the closest possible packing in a cubic form (Fig. 3.14 (i)).

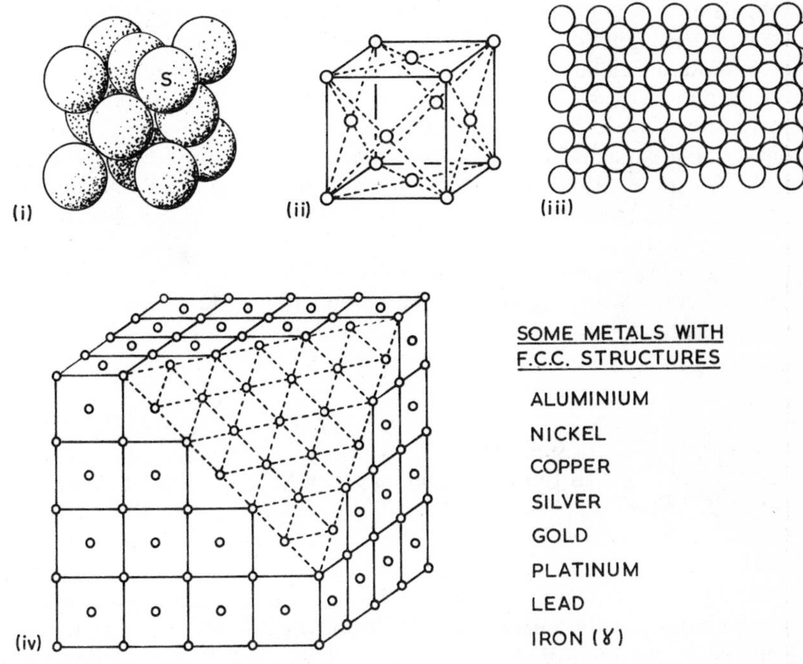

Fig. 3.14—The face-centred cubic (FCC) structure. (i) and (ii) depict the unit cell whilst (iii) shows the continuous structure of an 'end face'. (iv) Illustrates the arrangement of atoms on the (111) close-packed plane.

If the unit cell of this structure is examined it will be found that there is a sphere in the centre of each of six faces of the cube. The structure is therefore known as *face-centred cubic*. It will also be noted that the spheres forming the faces of the unit cell are not closely packed, and if the building up of the crystal is continued (Fig. 3.14 (iii)) it will be observed that the spaces in this crystal face are bounded in each case by four spheres. If a sphere is now placed at the front of one of these spaces and another at the rear, the gap so enclosed is called an *octahedral* site since it is bounded by six touching spheres with their centres at the points of a regular octahedron. Octahedral sites, like tetrahedral sites, are important in a study of crystallography.

If the sphere marked *S* in Fig. 3.14 (i) is now removed the exposed layer, which is along the (111) planes, has six spheres (Fig. 3.14 (iv)) which reveal the close-packed arrangement formed in the building up of the structure. These close-packed layers are in (111) planes which are at right angles to the body-diagonal of the cube and the sequence of layers along that diagonal is *ABCABCABCABC* . . . etc. The co-ordination number is the same as with the CPH system, i.e. 12.

Fig. 3.14 (iv) reveals that the (111) plane is one of close-packing. The direction of closest packed rows of atoms is [110]. Altogether there are four sets of closely packed (111) planes in the face-centred cubic structure and twelve sets of close-packed directions. As already mentioned, plastic deformation of metallic crystals generally occurs by the slipping of close-packed planes over each other and in the close-packed directions. For this reason the face-centred cubic structure would be expected to show a considerable capacity for plastic deformation. This is indeed supported by the fact that all of the more ductile and malleable metals have face-centred cubic structures. They include gold, silver, copper, aluminium, platinum, lead and nickel.

3.6.3 *The body-centred cubic structure* (BCC). The unit cell for this structure consists of a sphere at the centre of a cube with eight other spheres at the corners of the cube and touching the central sphere (Fig. 3.15). It thus has a co-ordination number of 8 and packing is not so close

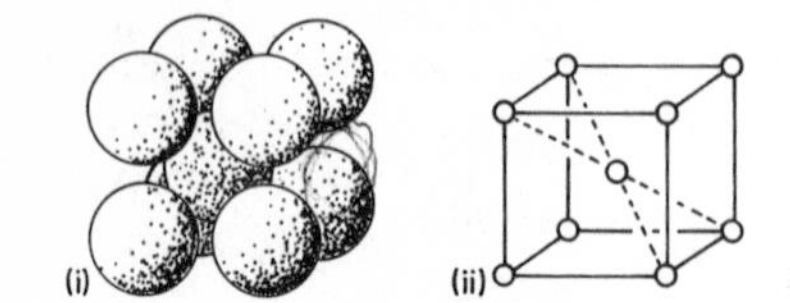

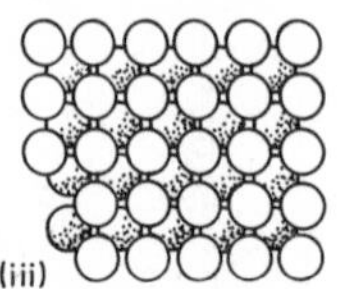

Fig. 3.15—The body-centred cubic (BCC) structure.

as in the other two systems. Thus when iron, which is an allotropic (3.6.4) metal, changes from a BCC structure to one which is FCC on heating to 910° C, a measurable contraction occurs spontaneously.

The most closely packed planes are (110) of which there are six. The closest packed direction is that of (111) of which there are four. Since the closest packed planes are still more open than those of FCC and there are less close-packed directions, it follows that the BCC structure is less ductile than the FCC structure. For example, iron in its FCC form is more malleable than in its BCC form.

3.6.4 *Allotropy* is defined as the ability of a single substance to exist in more than one physical form. Carbon is allotropic since it can exist both as diamond and graphite, yet each form consists solely of carbon atoms.

It is in the crystal structure and the method of bonding of the atoms where the difference lies.

Allotropes are stable over different temperature ranges. Thus iron (α-iron) is body-centred cubic below a temperature of 910° C, whilst between 910° C and 1400° C it exists as γ-iron which is face-centred cubic. Above 1400° C it becomes body-centred cubic again and is termed δ-iron to distinguish it from the body-centred cubic (or α-) iron which exists below 910° C. These allotropic changes are reversible:

$$\underset{\text{(BCC)}}{\alpha\text{-iron}} \underset{}{\overset{910^\circ\text{ C}}{\rightleftharpoons}} \underset{\text{(FCC)}}{\gamma\text{-iron}} \overset{1400^\circ\text{ C}}{\rightleftharpoons} \underset{\text{(BCC)}}{\delta\text{-iron}}$$

The allotropy of solids which relies solely on differences in crystal structure is known as *polymorphism*.

Many of the metals are allotropic and sometimes two allotropes will be face-centred cubic and close-packed hexagonal respectively. This is feasible since the structure can change from FCC to CPH, and vice versa, with relative ease.

The crystallisation of metals

3.7 All pure elements exist either as gases, liquids or solids and the *state* in which an element* exists depends upon the prevailing combination of temperature and pressure. Of the metallic elements, only mercury is a liquid at ordinary temperatures, though a number of metals melt below red heat. A few, like zinc, cadmium and mercury, boil at relatively low temperatures and can be purifed by distillation.

In 1827 botanist Robert Brown was using a microscope to examine some minute pollen particles floating on the surface of water. He was surprised to observe that these particles were in a state of continuous motion which was agitated and irregular. At the time he was completely baffled by the phenomenon but later it was realised that the motion produced in the particles was the result of impact by moving invisible water molecules at the surface of the water. The same kind of motion, known as 'Brownian motion', can be seen in the solid particles comprising smoke. When these particles, suspended in air, are viewed through a microscope they are seen to pursue irregular jerky paths due to bombardment by fast-moving oxygen and nitrogen molecules. Observations of this kind showed that molecules in gases and liquids were in a state of continuous motion and this led ultimately to the formulation of the *kinetic theory of matter*.

3.7.1 Metals form monatomic gases, that is, the vapour consists of single atoms of the metal. These atoms are in constant motion and are colliding continually with the walls of their container. The collisions give rise to

* The same is not necessarily true of compounds since many of these decompose at a temperature below their boiling point, and some below their melting point.

the pressure exerted by the gas. If temperature is increased atoms move more quickly as heat energy is transformed into kinetic energy. Hence collisions with the sides of the container will increase and so pressure will increase, assuming that the volume of the gas is kept constant by restraining expansion.

If the temperature is allowed to fall the average kinetic energy of the atoms decreases and is lost as heat energy (specific heat). Eventually an average energy content is reached where van der Waal's forces acting between atoms are able to overcome the kinetic movement of many of the atoms. Hence those with less than the average kinetic energy are attracted to each other so that they stick to each other, that is they condense to form droplets of liquid which fall under gravity. At this point the remainder of the kinetic energy is given up as heat (latent heat of vaporisation). No orderly arrangement of atoms exists in the liquid metal and atoms are still free to move with respect to each other, that is, a liquid possesses *mobility*.

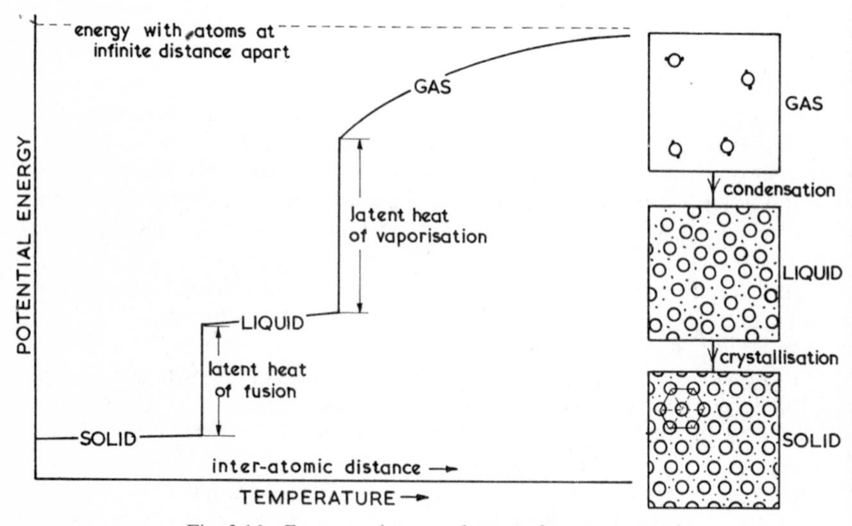

Fig. 3.16—Energy and states of matter for a pure metal.

At a lower temperature still the liquid begins to solidify. This is a process where atoms change from a disordered or *amorphous* state to an ordered *crystalline* state. As this occurs potential energy is lost (Fig. 3.16)—this time as latent heat of fusion—and since the atoms arrange themselves according to some orderly pattern, shrinkage generally accompanies solidification. For this reason the upper portion of a metal ingot is 'piped' (Fig. 3.23).

3.7.2 Pure metals, like other pure substances, solidify at a fixed single temperature. Ideally this will occur in accordance with the type of cooling curve shown in Fig. 3.17 (i). However, in most cases there is some degree of undercooling of the liquid before the onset of crystallisation. This is due to a lack of nucleation of the system (Fig. 3.17 (ii) and (iii)). Once a crystal nucleus forms it provides a solid/liquid interface where

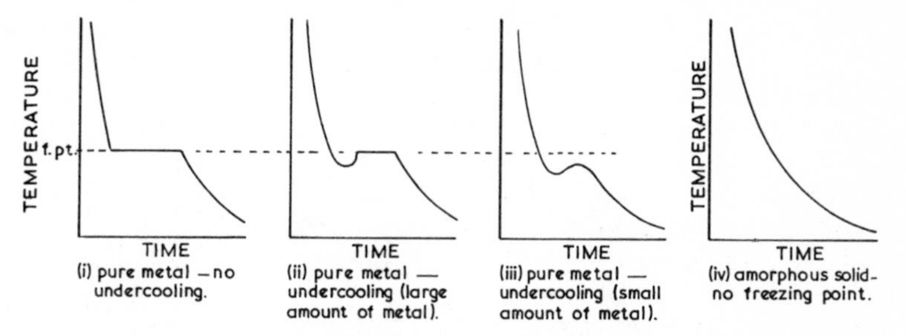

Fig. 3.17—Typical cooling curves for crystalline and amorphous solids. In (iii) there was insufficient molten metal to provide latent heat which would otherwise have caused a return to equilibrium as in (ii).

crystallisation can proceed (Fig. 3.18). Under industrial conditions nucleation may occur on a particle of slag or dross, but with very pure liquid metal some degree of undercooling will occur. When nucleation ultimately takes place the greater the extent of undercooling the greater will be the rate of crystallisation and, consequently, the rate at which latent heat is liberated. As a result the temperature of the melt will rise to reach equilibrium at the freezing point of the metal (Fig. 3.17 (ii)). If only small amounts of molten metal are involved insufficient latent heat may be available to allow the freezing point to be regained (Fig. 3.17 (iii)).

The nuclei which form will be simple units of the crystal pattern in which the metal solidifies, generally either CPH, FCC or BCC. Crystal growth will tend to occur in the opposite direction to that in which heat

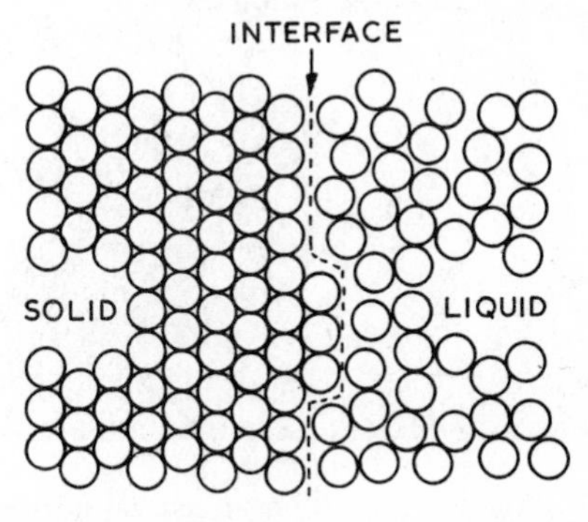

Fig. 3.18—Schematic representation of a solid/liquid interface.

is being conducted from the melt. Thus a 'spike' begins to grow from the nucleus into a region of under-cooled liquid, but as this occurs, latent heat is liberated and this warms up the liquid in front of the growing spike. Hence growth of the spike is retarded and as a result secondary spikes begin to grow from the primary one and these are followed by

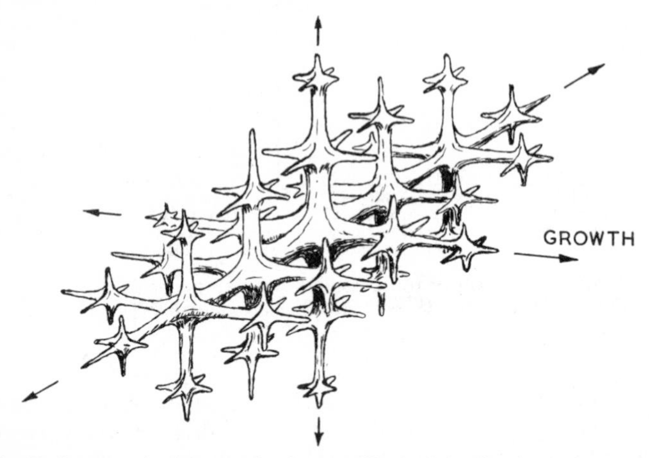

Fig. 3.19—Early stages in the growth of a metallic dendrite showing primary, secondary and tertiary arms.

tertiary spikes (Fig. 3.19). These primary, secondary and tertiary branches follow crystallographic planes and give rise to the general regularity of the structure. The crystal skeleton which develops is called a *dendrite*, a reference to its tree-like growth (Gk. *dendron*, 'a tree'), in which the primary, secondary and tertiary arms resemble the trunk, branches and twigs of a tree.

(*Courtesy of Dr. J. Moore, W.B.C.C.T.*)

Plate 3.1—Dendrite of zinc. This was grown from an aqueous solution of zinc sulphate by electrolysis. × 15.

The main branches of the dendrite continue to grow until the outer fringes make contact with those of neighbouring dendrites (Fig. 3.20). Since they are thus restrained from further outward growth these existing arms thicken as heat flows outwards from the region. Each dendrite develops independently so that the outer branches of neighbouring dendrites make contact at irregular angles. Moreover heat flow is not necessarily uniform and so the final crystals are of irregular shape though the atoms within any one crystal are regularly spaced with respect to each other in a crystal lattice.

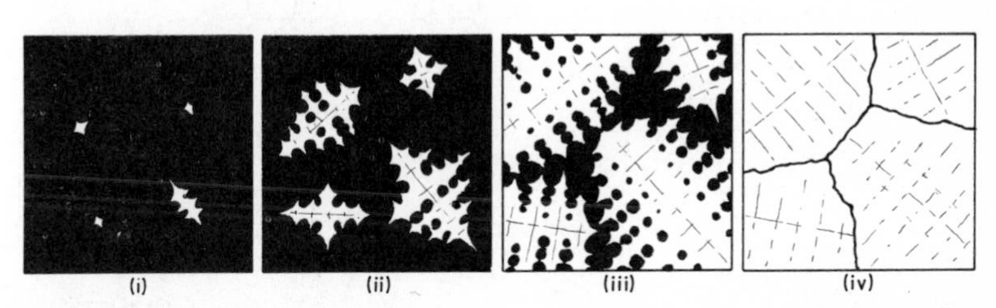

Fig. 3.20—The dendritic solidification of a metal. (i) Dendrites begin to develop from newly formed nuclei by putting out primary and secondary arms. (ii) Tertiary arms form and meet others growing in the opposite direction. (iii) Dendrites continue to grow until their outer arms touch those from neighbouring dendrites. Existing arms then thicken. (iv) When the metal is completely solid there is little evidence of the dendritic method of growth since in a pure metal all atoms are similar. Then only grain boundaries are visible.

If the metal under consideration is pure only crystal boundaries will be visible when a section of the metal is examined under the microscope since all atoms present in the structure are similar. The presence of impurities, however, may reveal the dendritic pattern to some extent as these impurities tend to congregate in that metal which solidifies last, that is, between the dendrite branches and particularly towards the crystal boundaries. Shrinkage cavities, formed due to lack of liquid as solidification contraction occurs, may also outline the dendrite shape to a limited extent (Fig. 3.21).

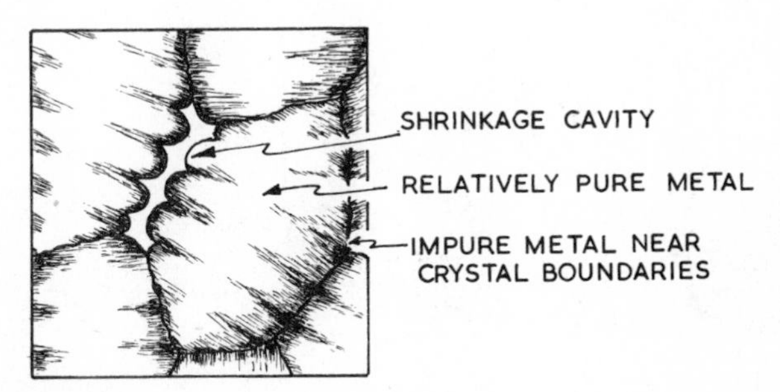

Fig. 3.21—The dendritic structure may be revealed by the presence of impurities and/or shrinkage cavities.

3.7.3 Although the dendrites in Fig. 3.20 are shown as being geometrically perfect in respect of the directionality of their branches, in practice there is often some misorientation between neighbouring arms with respect to each other *within the same crystal*. This may be as much as 4° in some cases. It is due to the pressure of growing arms upon each other as they compete for space, rather like the branches of pine trees in a

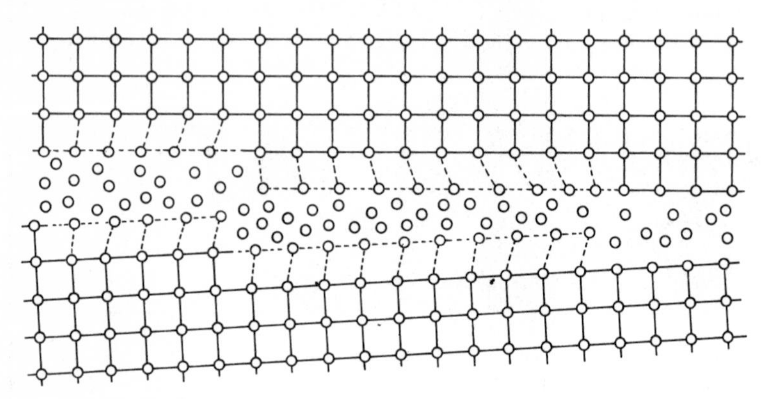

Fig. 3.22—The 'amorphous effect' due to misorientation at grain boundaries.

dense Forestry Commission plantation. Further misorientation of individual atoms occurs at crystal boundaries. Here is a fringe of metal about three atoms thick where the atoms are misfits with respect to both crystals. Years ago this was known as 'grain boundary cement' but fortunately this misleading term has fallen into disuse. The existence of this fringe of misfit atoms explains some of the mechanical properties of metals. Since it is of an amorphous nature (Fig. 3.22) it behaves rather like an extremely viscous liquid. Whilst at low temperatures metals gen-

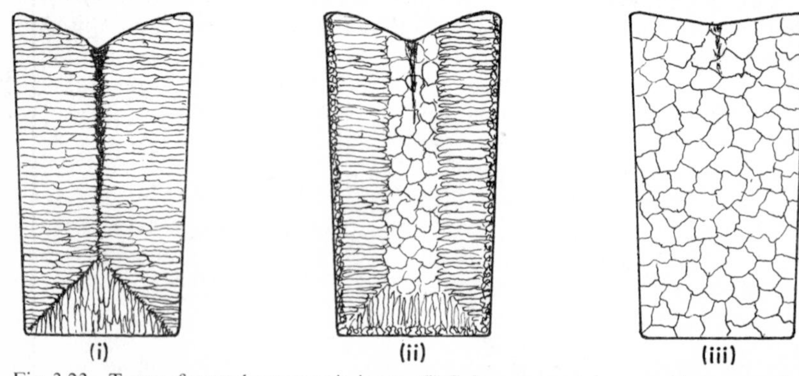

Fig. 3.23—Types of crystal structure in ingots. (i) Columnar crystals—excessive segregation of impurities at the core. (ii) Zones containing chill, columnar and equi-axed crystals. (iii) Large equi-axed crystals—less segregation because the cooling rate was low.

Plate 3.2—Cross-section of a small aluminium ingot. The structure was developed by etching in 2% hydrofluoric acid. Actual size.

erally fail by the propagation of *trans*crystalline cracks, at high temperatures failure invariably follows *inter*crystalline paths, presumably because the amorphous film has become less viscous at the higher temperature.

Fine-grained metals, which will contain more grain boundary and, hence, more misfit atoms per unit volume, creep more readily than coarse-grained metals of similar composition. It is reasonable to suppose that creep (16.4) is a property closely associated with the amorphous state.

3.7.4 The overall size and, to some degree, the shape of crystals in a metal ingot varies with the rate of cooling. Rapid cooling gives rise to considerable undercooling and the presence of a mould surface helps to nucleate the liquid. As a result a dense shower of nuclei will form instantaneously and so a layer of very small or 'chill' crystals will result (Fig. 3.23 (ii)). As the mould warms up progressively and the rate of cooling is reduced a stage is reached when crystal growth inwards is balanced by heat flow outwards. Fresh nuclei are not formed and so the existing crystals grow in columnar shape. With large ingots the rate of cooling at the centre may be so low that few nuclei form since there is little undercooling. Consequently the resultant crystals tend to be large and 'equi-axed', that is, of equal axes.

Crystal structures of non-metallic materials

3.8 So far we have dealt with crystal structures which are dependent upon the formation of the ionic bond or the presence of the metallic bond between cores. The covalent bond and relatively weak van der Waal's forces also play their parts in crystal formation. The covalent bond which is the basis of molecule formation may lead to the production of molecules so large that within these molecules exists a repetitive pattern common to crystalline structures. Similarly, the formation of crystals in which molecules are units relies on van der Waal's forces operating between the molecules. Thus water crystallises to form ice. Here the forces of attraction are supplied by the hydrogen bond (2.4.4).

Giant molecules

3.8.1 *Three-dimensional crystals.* Although in a geometrical sense all crystals will be three-dimensional, the term is used here as a reference to axial directions. The best-known example of this type of structure is that of diamond. The crystal pattern is determined by the fact that in this allotrope carbon exhibits its normal valency of four, such that each carbon atom is joined to four of its immediate neighbours by covalent bonds. In order that this shall occur symmetrically each carbon atom is surrounded by four others whose centres lie at the corners of a regular tetrahedron. The central atom is in contact with the other four (in Fig. 3.24 only the centres of the atomic cores are indicated). If the structure is

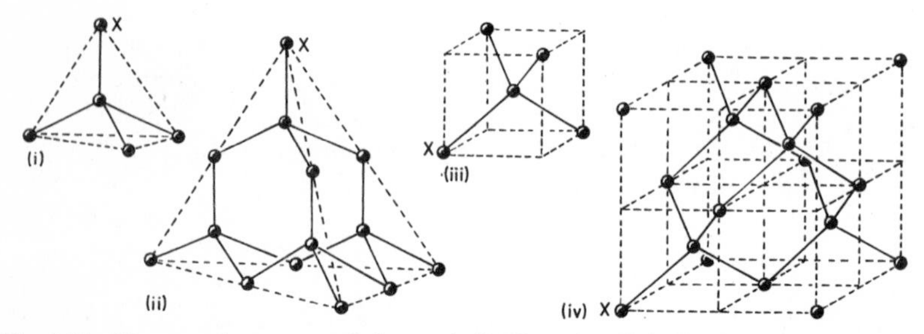

Fig. 3.24—The crystal structure of diamond. (i) The unit cell, indicating that here carbon has a co-ordination number of four; (iii) and (iv) show how diamond can be considered as a *cubic* structure. Note that in (ii) not all valencies are used up in the 'fringe' atoms of the crystal, which will of course be continuous in practice. (*X* represents the same atom in the system in each case.)

extended to include more carbon atoms (Fig. 3.24 (ii)) it will be noticed that the whole figure is tetrahedral and will continue to extend 'downwards' such that each carbon atom is joined to four others by covalent bonds. The structure can also be regarded as being basically cubic (Fig. 3.24 (iii)) and it can be shown that as such the structure is much less closely packed than a body-centred cubic structure. Thus, in Fig. 3.24 (iv), only four out of the eight cubic units has a carbon atom at its centre whilst many of the 'corner sites' are unused.

Diamond can therefore be regarded as a three-dimensional giant molecule (or macromolecule), C_n. In the crystal so produced the carbon atom has a co-ordination number of four. Diamond is mechanically strong and is the hardest substance known. Moreover it has an extremely high melting point—in fact at ordinary pressures it does not melt, but at about 3500° C thermal vibrations of the atoms are strong enough to overcome the powerful covalent bonds and vaporisation begins. Both of these factors illustrate the very great strength of the covalent bond. Since in diamond all carbon atoms are covalently bonded such that all available electrons are used up, diamond is not a conductor of electricity (18.3.2).

Other elements of similar valency (with carbon they constitute Group IV of the periodic classification), e.g. silicon, germanium and the 'grey' allotrope of tin, also crystallise in the diamond-type of structure.

3.8.2 *Layer structures.* Each carbon atom in a diamond crystal achieves a stable octet of electrons by forming four covalent bonds with neighbouring carbon atoms. Another allotrope of carbon, graphite, also exists. In graphite the atoms are bonded in the form of a ring structure so that flat

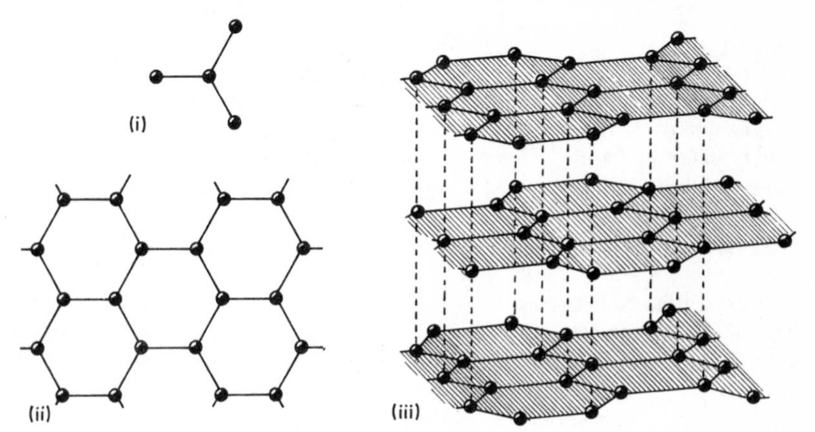

Fig. 3.25—The layer structure of graphite.

layers are produced in a regular hexagonal network (Fig. 3.25). Each carbon atom is surrounded by three other carbon atoms (Fig. 3.25 (i)) so that the co-ordination number of carbon in this instance is three. As Fig. 3.25 (ii) indicates each carbon atom uses only three of its outer shell electrons to form covalent bonds. The fourth electron in each case can be regarded as being 'shared' to some extent in the nature of the electrons in the metallic bond. Thus, whilst diamond does not conduct electricity, graphite does. Whilst the carbon atoms within these layers or sheets are bonded covalently the layers themselves are held together only by weak van der Waal's forces. For this reason whilst the sheets themselves are strong they slide over each other very easily so that graphite is a good lubricant. Materials like clay (un-fired) and talc are also of this type.

3.8.3 *Molecular crystals.* Molecules are separate self-contained units as are atoms so that it is possible for them to arrange themselves in a crystalline form. Whilst the atoms within a molecule are held together by covalent bonding, the molecules themselves are bonded by van der Waal's forces. The element tellurium forms long-chain molecules and these in turn arrange themselves in a pattern with like molecules to produce crystals.

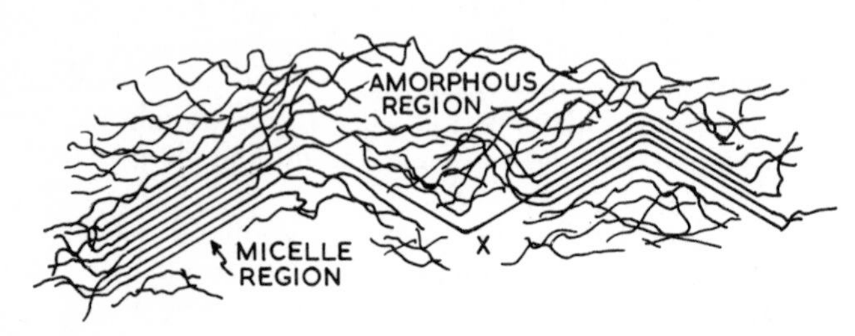

Fig. 3.26—Fringed *micelles* in polythene showing one molecule (X) extending between two micelles.

Many polymers are regarded as being amorphous, that is, non-crystalline, yet many do possess some crystalline regions known as *fringed micelles*. These are often linked together by a chain molecule which extends from one micelle to another (Fig. 3.26). Since inter-molecular forces are only of the weak van der Waal's type, crystallisation is not as perfect over a long range with these cumbersome chain molecules as it is when crystals grow from single atoms.

Non-crystalline substances

3.9 All true solids are crystalline in nature. However, a number of apparently solid materials are amorphous, that is, their atoms or molecules do not conform to any regular geometrical patterns. Solids such as glass and pitch are comparatively rigid and are broken by a sudden impact, but if a steady load is applied over a long period of time the material will flow. A piece of pitch will flow very gradually, over a period of months, to take the shape of a vessel into which it has been placed. A glass rod, supported at each end, will sag under its own weight if left in this position for a long time. The deformation produced will be *permanent* and unlike the temporary elastic deformation produced when an elastic solid is lightly stressed within its elastic limit.

These amorphous 'solids' do not melt at a definite temperature as do pure crystalline materials. Instead they soften gradually and become more mobile thus resembling liquids of very high viscosity. The viscosity of a liquid increases as its temperature falls and if it is cooled below its freezing point without crystallisation taking place its viscosity may reach a value as much as 10^{10} times that of water. Viscosity is related to van der Waal's forces. As temperature falls thermal vibrations of the molecules are reduced and so viscosity increases, since van der Waal's forces will also increase.

Amorphous solids may therefore be regarded as supercooled liquids in which the particles have a much reduced freedom of movement. However, many materials such as powdery metallic oxides which seem to be amorphous are in fact crystalline. The particles in which they exist are so small that the crystalline nature is not immediately apparent until investigated by X-ray analysis.

Chapter Four
Mechanical Properties

4.1 Structural materials used in mechanical and civil engineering practice must generally have *strength.* This value is a measure of the externally applied forces which are necessary to overcome internal forces of attraction between fundamental particles within the material such as we have been discussing in the first three chapters of this book. Briefly, strength is due to the sum of forces of attraction between negatively charged electrons and positively charged protons within the material In small molecules the constituent atoms are held together by the covalent bonds operating between them but van der Waal's forces acting between these small molecules are also small and strength is negligible. Many such substances, e.g H_2S, CO_2 and SO_2 are in fact gases at ordinary temperatures and pressures. When covalent bonds join large numbers of atoms to produce giant molecules as in the case of the carbon atoms in diamond or carbon fibre, the strength is great. The metallic bond can be regarded in some respects as a multiple covalent bond so that metals are very strong.

Many substances rely upon the operation of van der Waal's forces acting between large molecules to give them strength. In plastics materials the sum of the van der Waal's forces acting between the long chain-like molecules is sufficient to give them considerable strength. The atoms within such molecules are of course covalently bonded. The electrovalent bond tends to be weak and once it is broken complete cleavage takes place since, as soon as the position of any ion is disturbed, equilibrium is upset.

4.1.1 Constructional materials generally must be able to withstand the action of considerable forces without undergoing other than very small amounts of distortion. In the field of what is loosely termed production engineering, however, very different properties may be desirable. Here a material must be capable of permanent deformation at the expense of as little energy as possible. That is, it must be *malleable* and *ductile.* In the case of metals a forming process causes the metal to lose its softness and become harder and stronger, that is, it *work hardens.* Thus the production engineer designs his forming process to utilise the malleability or ductility of the material and at the same time generate in it sufficient strength for subsequent service. Other mechanical properties include *elasticity*, *hardness*, *toughness* and also *creep* and *fatigue* properties. In each case the property is associated with the behaviour of the material towards the application of force and the engineer is generally interested in the 'density of force' which is necessary to produce some definite amount of deformation, either temporary or permanent in the material.

4.1.2 *Stress* is a measurement of 'density of force' and is defined as force per unit area. It is expressed in Newtons per square metre (N/m²), though in materials science it is perhaps more conveniently measured in terms of Newtons per square millimetre (N/mm²). This unit, moreover, produces a value which is easier to appreciate, whereas the force necessary to break (for example) a steel bar one square *metre* in cross-section is so large as to be difficult to visualise in ordinary finite terms. Stress, then, is calculated by dividing the force by the area on which it is acting.

Example
A steel rod 6 mm in diameter is under the action of a tensile force of 400 Newtons. Calculate the tensile stress in the bar.

$$\text{Tensile stress} = \frac{\text{Tensile force}}{\text{Area of cross-section of rod}}$$

$$= \frac{\text{Tensile force}}{\pi r^2}$$

$$= \frac{400}{3{\cdot}142 \times 3^2} \text{ N/mm}^2$$

$$= \underline{\underline{14{\cdot}15 \text{ N/mm}^2}}$$

Units

$$\text{Tensile stress} = \frac{\text{Newtons}}{(\text{mm})(\text{mm})}$$

$$= \text{N/mm}^2$$

4.1.3 *Strain* refers to the proportional deformation produced in a material under the influence of stress. It is measured as the number of millimetres of deformation suffered per millimetre of original length and is a numerical ratio.

Example
A 40-mm gauge length is marked on an aluminium test piece. The test piece is strained in tension so that the gauge length becomes 42·3 mm. Calculate the strain.

$$\text{Strain} = \frac{\text{Increase in length}}{\text{Original length}}$$

$$= \frac{42{\cdot}3 - 40}{40}$$

$$= \frac{2{\cdot}3}{40}$$

$$= \underline{\underline{0{\cdot}0575}}$$

Units

$$\text{Strain} = \frac{\text{mm} - \text{mm}}{\text{mm}}$$

$$= \frac{\text{mm}}{\text{mm}}$$

= a numerical ratio

Strain is commonly quoted as a percentage. In this case:

$$\text{Strain} = 0{\cdot}0575 \times 100\%$$

$$= \underline{\underline{5{\cdot}75\%}}$$

Strain may be either *elastic* or *plastic*. Elastic strain is reversible and disappears when the stress is removed. Atoms are displaced from their initial positions by the application of stress but when this stress is removed they return to their initial positions relative to their neighbours provided that the strain has been of an elastic nature. Strain is roughly

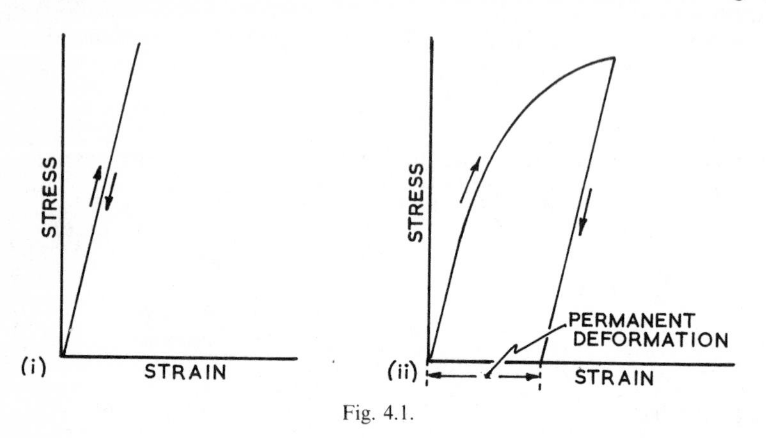

Fig. 4.1.

proportional to the applied stress (Fig. 4.1 (i)), and, for practical purposes, the material obeys Hooke's Law. This states that, for an elastic body, strain produced is directly proportional to stress applied.

4.1.4 *Young's Modulus of Elasticity* (E) is the ratio between the stress applied and the elastic strain it produces. That is, it is the stress required to produce a unit quantity of elastic strain. It is related to the *rigidity* of the material and is a value of supreme importance to the constructional engineer. The modulus of elasticity is expressed in terms of either tensile or compressive stresses and its units are the same as those for stress, since:

$$E = \frac{\text{stress}}{\text{strain}}$$

$$= \frac{\text{N/mm}^2}{\text{mm/mm}}$$

$$= \frac{\text{N}}{\text{mm}^2}$$

$$= \underline{\underline{\text{N/mm}^2}}$$

In view of the numerical magnitude of the value E is commonly expressed as GN/m^2 or MN/mm^2.

(Courtesy of Messrs. Avery-Denison Ltd., Leeds and T. I. Chesterfield Ltd.)

Plate 4.1—A modern Servo Hydraulic Universal Testing Machine of 600 kN capacity (tension or compression). The operator is standing at the control and indicator/display console which is coupled electro-hydraulically to the straining unit on the left. The resultant force/extension characteristics of the specimen are plotted automatically on a display panel on the right-hand side of the desk. Provision is also made for these measurements to be time based.

Example

A steel wire, 0·5 mm² in cross-sectional area, and 10 m long is extended elastically 1·68 mm by a force of 17·24 N. Calculate the modulus of elasticity for the steel.

$$\text{Stress} = \frac{\text{Force}}{\text{Cross-sectional area}}$$

$$= \frac{17{\cdot}24}{0.5}\ \text{N/mm}^2$$

$$= 34{\cdot}48\ \text{N/mm}^2$$

$$\text{Associated strain} = \frac{1{\cdot}68 \times 10^{-3}\ \text{m}}{10\ \text{m}}$$

$$= 0{\cdot}000168$$

$$\text{Modulus of elasticity} = \frac{\text{Stress}}{\text{Strain}}$$

$$= \frac{34{\cdot}48}{0{\cdot}000168}\ \text{N/mm}^2$$

$$= 20{\cdot}52 \times 10^4\ \text{N/mm}^2$$

$$= 0{\cdot}2052\ \text{MN/mm}^2$$

$$\text{or}\quad \underline{\underline{205{\cdot}2\ \text{kN/mm}^2}}$$

The sophisticated technology of these closing decades of the twentieth century often involves consideration of the mass of material required to provide the necessary strength and rigidity in a structure. This is particularly so in the aero-space and other transport industries and in fact in any situation where work done against gravity must be paid for in terms of increasingly expensive fuel. Thus the modulus of elasticity is commonly expressed as a *specific modulus of elasticity* in which E is related to the relative density of the material:

$$\text{Specific modulus of elasticity} = \frac{E}{\text{relative density}}$$

$$\text{Units} = \frac{\text{N/m}^2}{\frac{\cancel{\text{Kg/m}^3}}{\cancel{\text{Kg/m}^3}}} = \text{N/m}^2$$

4.1.5 *Plastic strain* results when a material is stressed to the extent where its elastic limit is exceeded. It coincides with a movement of the atoms within the structure of the material into permanent new positions with respect to neighbouring atoms. When the stress is removed only elastic strain disappears and any plastic strain produced is retained (Fig. 4.1 (ii)). *Malleability* refers to the extent to which a material can undergo

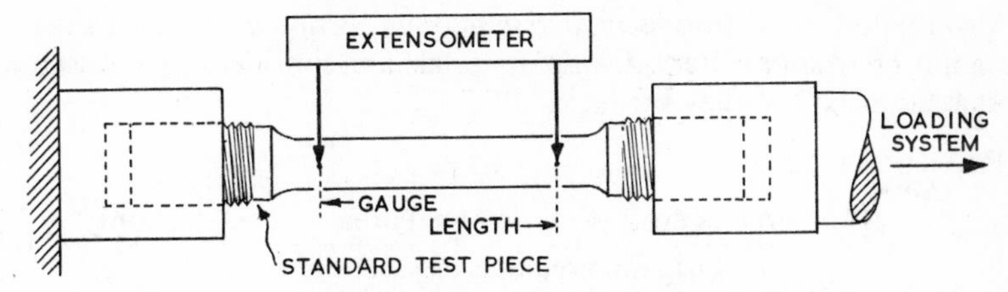

Fig. 4.2—The principle of tensile testing. A threaded test piece is shown here but in most machines the test piece will be plain and held in friction grips.

deformation in compression before failure occurs, whilst *ductility* refers to the degree of extension which takes place before failure of a material in tension. All ductile materials are malleable but malleable materials are not necessarily always ductile since a soft material may lack strength and thus tear apart very easily in tension.

Ductility is commonly expressed in practical terms as the percentage elongation in gauge length a standard test piece suffers during a tensile test to failure. Fig. 4.3 illustrates the necessity for a standard relationship between gauge length and cross-sectional area of the test piece if results

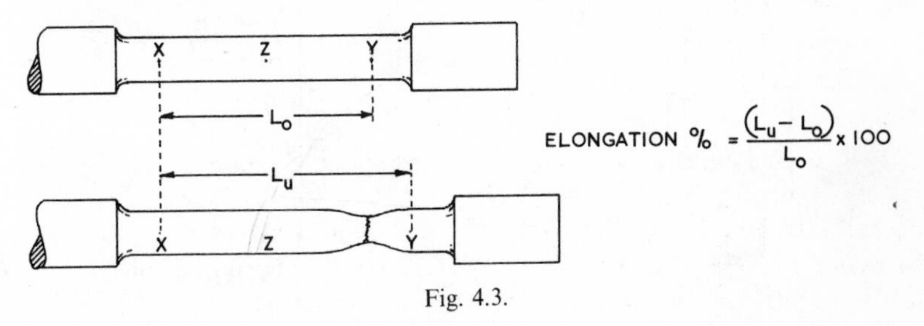

Fig. 4.3.

are to be comparable. Since the bulk of plastic deformation occurs at the 'neck' (between Z and Y) it is clear that a percentage elongation based on ZY as the gauge length would give different results from that based on XY as the gauge length. Consequently tensile test pieces should be geometrically similar and are known as *proportional* test pieces. They are usually circular in cross-section and BSI lays down (B.S. 18: Part 1 and 2) that for proportional test pieces:

$$L_0 = 5{\cdot}65\sqrt{S_0}$$

where L_0 is the gauge length and S_0 the original area of cross-section. This formula was adopted internationally and SI units are used. For test pieces of circular cross-section it gives a relationship:

$$L_0 = 5d$$

where d is the diameter at the gauge length. Thus a test piece of cross-sectional area 200 mm^2 will have a diameter of 15·96 mm (i.e. 16 mm) and a gauge length of 80 mm.

Also related to ductility is *reduction in cross-sectional area* measured at the point of fracture. Very ductile materials are considerably reduced in cross-section before they break.

$$\% \text{ Reduction in area} = \frac{(\text{Original area of cross-section} - \text{final area of cross-section})}{\text{Original area of cross-section}} \times 100$$

4.1.6 *Stress–strain diagrams.* When corresponding values of stress and strain derived during a tensile test are plotted graphically it is found that each type of material is represented by a characteristic curve. Materials of negligible ductility, such as fully hardened steels, cast iron and concrete, undergo little or no plastic deformation before fracture (Fig. 4.4

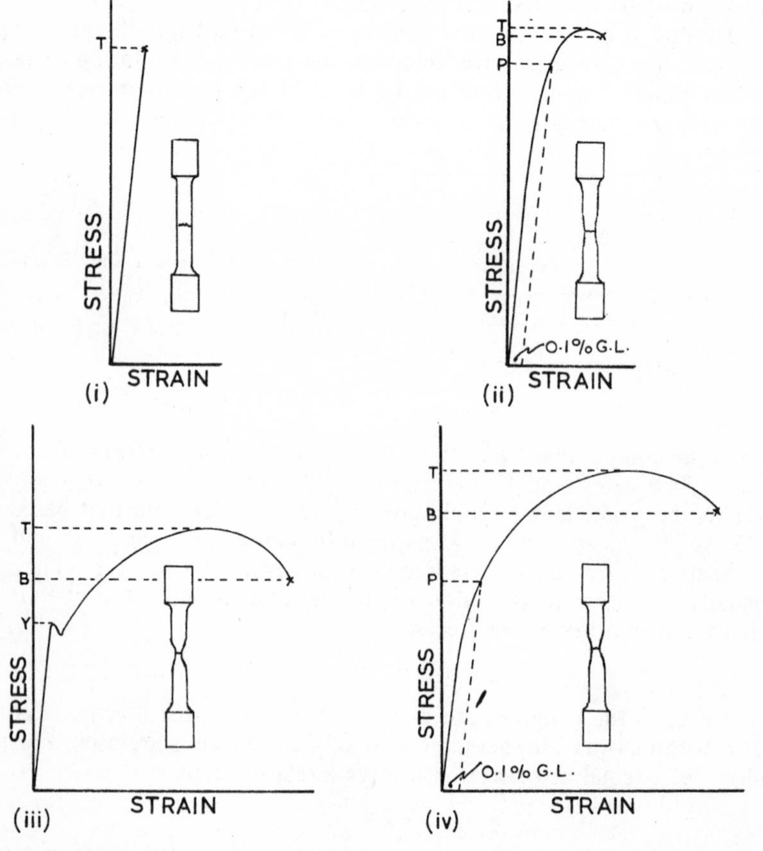

Fig. 4.4—Representative stress/strain diagrams for various types of material. (i) Non-ductile material. (ii) Semi-ductile material. (iii) and (iv) Ductile materials.
T = tensile strength; B = breaking strength; Y = yield stress, and P = proof stress.

(i)). That is, there is no yield point and only elastic extension occurs. A ductile material, on the other hand, exhibits an *elastic limit* (or *limit of proportionality*) beyond which plastic deformation occurs. The maximum stress which a material can withstand before plastic flow sets in is known as its *yield strength*. In softer ferrous materials (wrought-iron and low-carbon steels) and some plastics materials the onset of plastic flow is marked by a very definite yield point (Fig. 4.4 (iii)) and it is therefore a simple matter to calculate the yield stress. In other materials, comprising practically all ductile metals and alloys and most plastics materials, the elastic limit is not well defined (Fig. 4.4 (iv)). In most respects the yield stress of a material is of greater importance to the design-engineer than is the maximum strength attained during plastic flow. Consequently a substitute value for yield strength is derived for those materials which show no obvious yield point. This is known as the *proof stress* and is that stress which will produce a permanent (plastic) extension of 0·1%* in the gauge length of the test piece. It is derived as indicated in Figs. 4.4 (ii) and (iv).

Materials which have received some treatment such as work-hardening or, in the case of some alloys, suitable heat-treatment, are generally stronger but less ductile than those in the fully soft condition. This is indicated in the stress/strain curve Fig. 4.4 (ii).

4.1.7 *The tensile strength* of a material is derived by dividing the maximum force sustained during the test by the *original* cross-sectional area of the test piece. The units involved are those of stress and generally MN/m^2 or N/mm^2 are the most convenient to use; they will of course be numerically the same. It is important to note that at all stages during a tensile test *stress is calculated on the basis of the original area of cross-section*. That is, it does not take into account the diminishing cross-sectional area at the narrowest point in the 'neck' during the final stages of plastic deformation. For this reason these so-called 'stress/strain' diagrams are really modified force/extension diagrams and to plot a *true* stress/strain diagram it would be necessary to take into account the diminishing cross-section by measuring the minimum diameter at the neck with each reading of the force used (Fig. 4.5). Generally measurement of true stress in this manner is impracticable and the value referred to as the *engineering stress* is of more use in practice.

$$\text{Engineering stress} = \frac{\text{Force}}{\text{Original area of cross-section}}$$

However, it should be appreciated that the ordinate usually labelled 'stress' in the majority of published diagrams nearly always refers to this 'engineering stress' rather than to true stress. The reduction in cross-section of ductile materials during plastic flow leads to the apparent anomaly that the breaking strength is less than the tensile strength. In

* Or up to 0·5% for some materials.

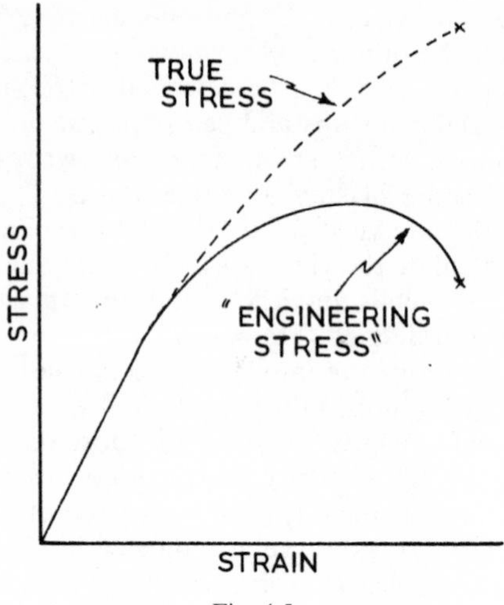

Fig. 4.5.

fact of course the *true* breaking stress is the greater as indicated in Fig. 4.5.

Hardness

4.2 In general terms hardness is defined as the ability of a material to resist surface abrasion. The relative hardnesses of minerals have in fact long been assessed by reference to Moh's Scale (Table 4.1). This consists of a list of materials arranged in order such that any mineral in the list will scratch any one below it. Thus diamond, the hardest substance known, heads the list with a hardness index of 10 whilst talc is at the foot with a hardness index of 1. The surface hardness of any substance can be related to Moh's Scale by determining which of these standard substances will just scratch it.

Table 4.1—Moh's Scale.

Mineral	*Hardness index*
Diamond	10
Corundum	9
Topaz	8
Quartz	7
Orthoclase feldspar	6
Apatite	5
Fluorite	4
Calcite	3
Gypsum	2
Talc	1

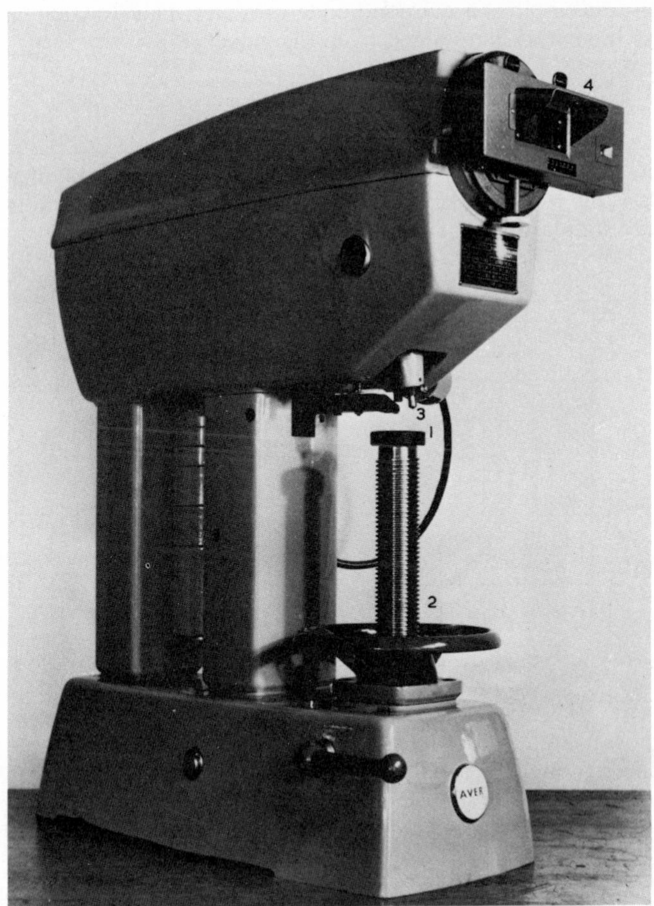

(*Courtesy of Messrs. Avery-Denison Ltd., Leeds*)

Plate 4.2—A Hardness Testing Machine for both Brinell and Vickers Hardness determinations. The specimen is placed on the platform (1) the height of which is adjusted by the screw mechanism (2). An image of the impression made by the indenter (3) is then projected on to the screen assembly (4) and measured visually.

Obviously Moh's Scale would be inadequate in the *accurate* determination of the hardness of such materials as metallic alloys, and rather different types of hardness test have been developed for such substances. Such instruments as the Turner Sclerometer (which attempted to measure surface 'scratchability') were soon abandoned in favour of machines which measure the resistance of the surface layers of a material to penetration by some form of indenter rather than surface hardness defined in terms of abrasion resistance. In the Brinell test the indenter is a steel ball whilst in the Diamond Pyramid test a pyramidal diamond is used. The

Rockwell test employs a diamond cone or a steel ball. In each case the Hardness Index (H) is obtained from the value:

$$\frac{\text{Force used}}{\text{Surface } \textit{area} \text{ of indentation produced}}$$

The units will be those of stress though these are never stated when a hardness value is quoted since on any one hardness scale conditions of testing are standard.

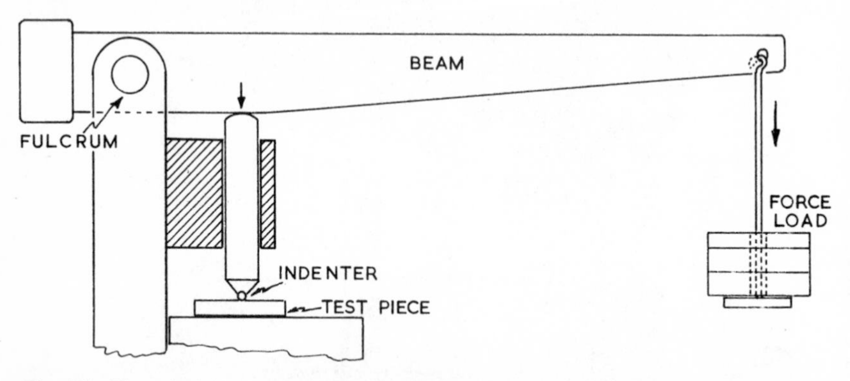

Fig. 4.6—The principle of most hardness-measuring machines. The indenter may be a steel ball as shown; a diamond pyramid or a diamond cone.

For most metallic alloys tensile strength is approximately proportional to hardness though there is no fundamental connection between them other than in the general stiffness of the material.

Toughness

4.3 Toughness is measured in terms of the amount of *energy* required to fracture a standard test piece. For this reason it must not be confused with strength which is measured in terms of *stress* required to break a standard test piece. Since energy is the product of average force and the distance through which it acts, the area under the stress/strain diagram is related directly to the energy necessary to break a material.

In fact some materials, when hardened and strengthened by a suitable process, lack toughness whereas in their softest and most ductile state they are extremely tough. This relationship is indicated by the area under the stress/strain curve (Fig. 4.7) in each case.

4.3.1 The practical methods used for measuring toughness differ from those associated with the stress/strain diagram in that they employ shock loading. A portion of the kinetic energy of a swinging pendulum is absorbed in fracturing a standard, suitably notched, test piece. In both the Izod and Charpy methods of measuring impact toughness the unit of

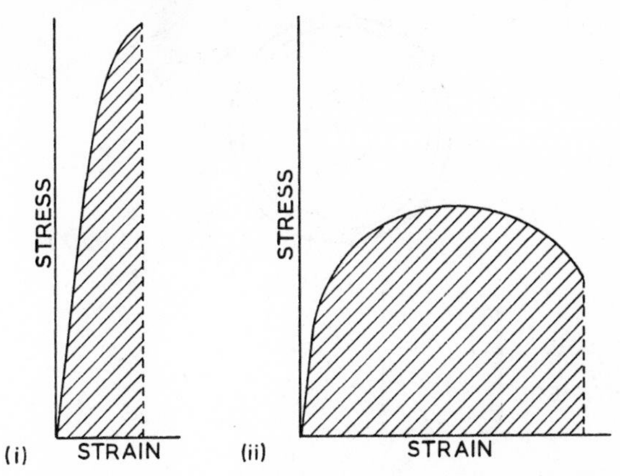

Fig. 4.7—Stress/strain diagrams for (i) an alloy which has been treated to increase its strength, (ii) the same alloy in the soft, ductile condition. The energy, indicated by the area under the curve, required to break the test piece, is greater in the case of the less strong but more ductile material.

(*Courtesy of Messrs. Avery-Denison Ltd., Leeds*)

Plate 4.3—A Universal Impact Testing Machine. This model can be used for both Izod and Charpy impact tests.

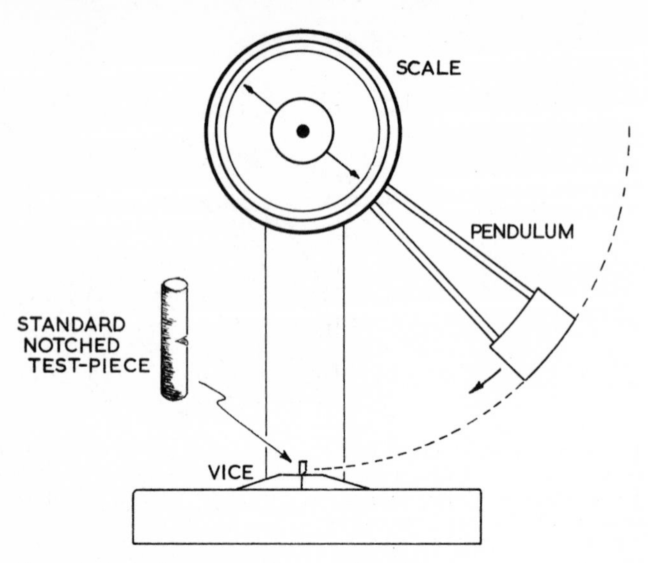

Fig. 4.8—The principal of impact testing machines. The energy required to fracture the specimen is measured on the scale in joules.

energy employed is the Joule. These tests offer a useful practical indication of how a material will react to conditions of shock loading, and in many circumstances toughness will be more important than tensile strength as a criterion of suitability of a material.

Other mechanical tests

4.4 Many other tests have been devised in order to assess specific mechanical properties of interest in practical engineering. For example the Erichsen test gives a rather better estimation of a material's suitability for deep drawing than does a formal % elongation. Various bend tests and compression tests are also useful in some circumstances. Simple descriptions of some of these tests will be found in *Materials for the Engineering Technician* by this author.

It may have been noticed that no mention has been made here of creep- or fatigue-testing. These topics are associated with metal failure and will be dealt with later in this book (16.4 and 16.5).

Chapter Five
The Deformation of Materials

5.1 A well-worn cliché of yesteryear, much too old to have originated from any of today's pseudo-scientific journalists, speculated upon the possible effect of 'an irresistible force meeting an immovable object'. In fact any force may be regarded as irresistible in so far that no object is completely immovable, since all substances when subjected to mechanical stress suffer some change in shape. Many return to their original form when the stress is relaxed, providing the stress has not been too great;

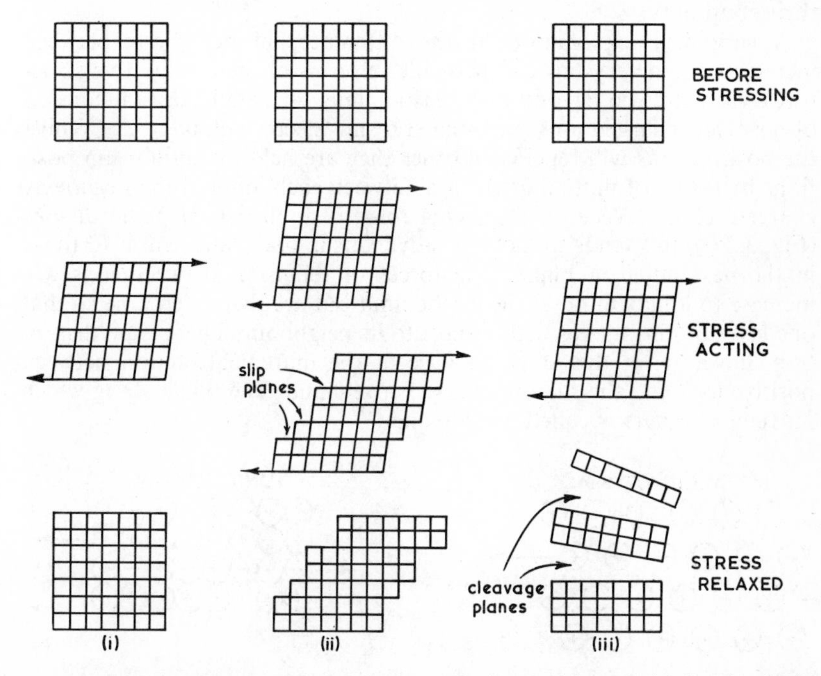

Fig. 5.1—The behaviour of materials in stress. (i) A material stressed below its elastic limit—removal of stress leads to disappearance of strain. (ii) A ductile material, such as a metal, undergoes elastic and plastic strain. The latter does not disappear when stress is removed. (iii) Brittle materials do not strain plastically.

such substances possess *elasticity*. When force is applied the ions or molecules of which the material is composed are moved small distances from their positions of equilibrium and when the displacing force is removed they return to their original positions (Fig. 5.1 (i)).

If the displacing force is increased steadily the substance ultimately reaches its elastic limit. Increase of stress beyond this limit leads to one of two alternative results:

1. The material may shatter into fragments (Fig. 5.1 (iii)).
2. It may remain as a coherent whole but fail to return to its original form when the displacing forces have been removed. It has suffered *plastic deformation* (Fig. 5.1 (ii)).

5.1.1 In general non-metallic elements and most ionically bonded compounds belong to the first group, that is, they possess elasticity but no plasticity. Metals and many organic polymers, however, possess high degrees of both elasticity and plasticity. They can be shaped by compression and tension and after such treatment their structures are substantially unaffected since they have suffered deformation but have retained their continuity.

A simplified explanation of the differences in behaviour between metals and ionic crystals can be made in terms of their respective structures when stressed beyond their elastic limits. A metallic crystal consists of positively charged ions surrounded by an 'electron cloud' (3.5). Whilst the positive ions will repel each other they are held in equilibrium positions by forces of mutual attraction between each ion and the negatively charged 'cloud'. When a displacing force is applied to a plane of ions (Fig. 5.2 (i)) this tends to move positive ions in one plane nearer to those in the next adjacent plane. The forces of repulsion between ions will increase to a maximum at the elastic limit but are then overcome so that one layer of ions moves with respect to its neighbour. There is no disruption, however, at this stage because of the mutual attraction between positive ions and the surrounding electron cloud. The plane along which movement occurs is called a *slip* plane.

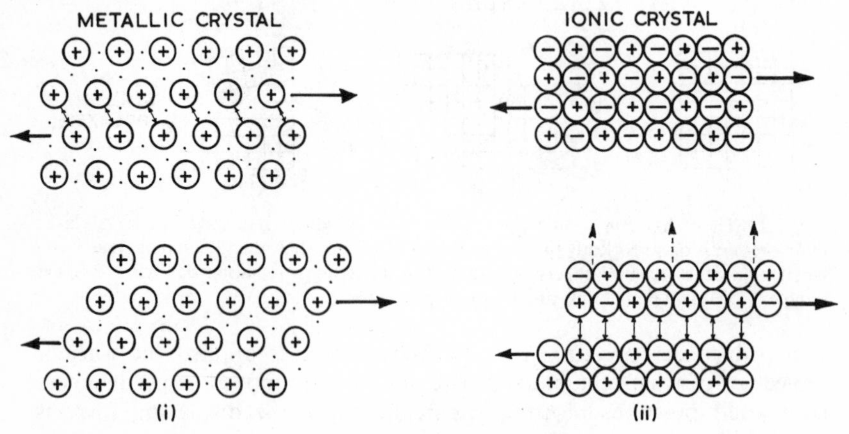

Fig. 5.2—Deformation in crystals. (i) 'Slip' occurs in a metallic crystal. (ii) 'Cleavage' occurs in ionic crystals.

A crystal of a substance such as sodium chloride consists of positively charged sodium ions and negatively charged chlorine ions arranged so that each ion is surrounded by those of opposite charge. Any attempt to produce slip in the [100] direction (Fig. 5.2 (ii)) will fail since this would bring like ions into closer contact. Like ions would immediately repel each other so that the two halves of the crystal separate. Thus the crystal shatters along a *cleavage* plane.

Because of this severe restriction in the availability of slip systems in ionically bonded materials, such materials tend to fracture at fairly low stresses—lower than those at which slip would otherwise occur. In ceramics both ionic and strong covalent bonds are present. Consequently one might expect them to have high strengths and high elastic moduli. Whilst moduli are reasonably high and compressive strengths demonstrate the bond strengths, tensile strengths tend to be low. This is due principally to the propagation of micro-cracks which act as stress raisers.

Plastic deformation in metals

5.2 The fact that a metal can undergo both elastic and plastic deformation is demonstrated during a tensile test in which the test piece is tested to destruction. That plastic deformation takes place by some form of slip is observable using an ordinary metallurgical microscope (Fig. 5.3 and Plate 5.1). If a ductile metal or alloy is polished and etched to reveal its crystal

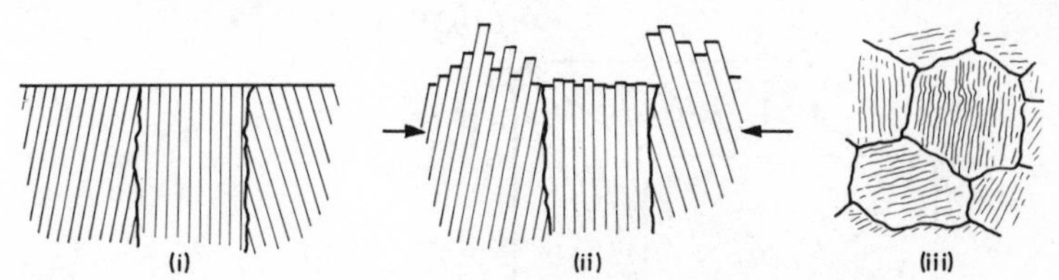

Fig. 5.3—The formation of slip bands in a metal stressed beyond its yield point. (i) Before stressing. (ii) After stressing—blocks of slipped atoms some 40 atoms wide and up to 400 atoms high cast shadows on the surface of the metal giving an appearance as in (iii).

structure and is then squeezed laterally in a vice, taking care not to damage the etched surface, the shadows cast by the ridges formed as slip occurs can be seen as hair-like lines (Fig. 5.3 (iii)) at a magnification of 100 × or so.

Slip takes place on specific crystallographic planes in a given metal. Generally these slip planes are the atom planes of greatest inter-planar spacing since then forces between the sliding planes will be at a minimum. Within the planes the greatest density of atom packing is present. Thus slip occurs in the direction of the greatest density of atom packing.

5.2.1 In FCC metals like aluminium, copper and gold slip usually occurs along the {111} planes (Fig. 3.14 (iv)) since the perpendicular distance between these planes is greater than that for any other set of planes in the crystal and the atomic population in these planes is denser than in any

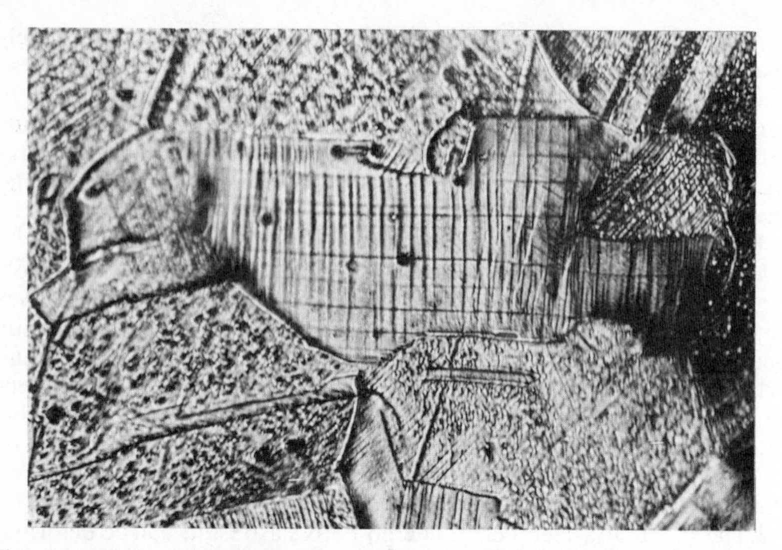

Plate 5.1—'Slip-bands' in copper. The specimen was polished, etched and then squeezed laterally in a vice. Slip-bands are shown clearly in the crystal in the centre of the field of view. × 200.

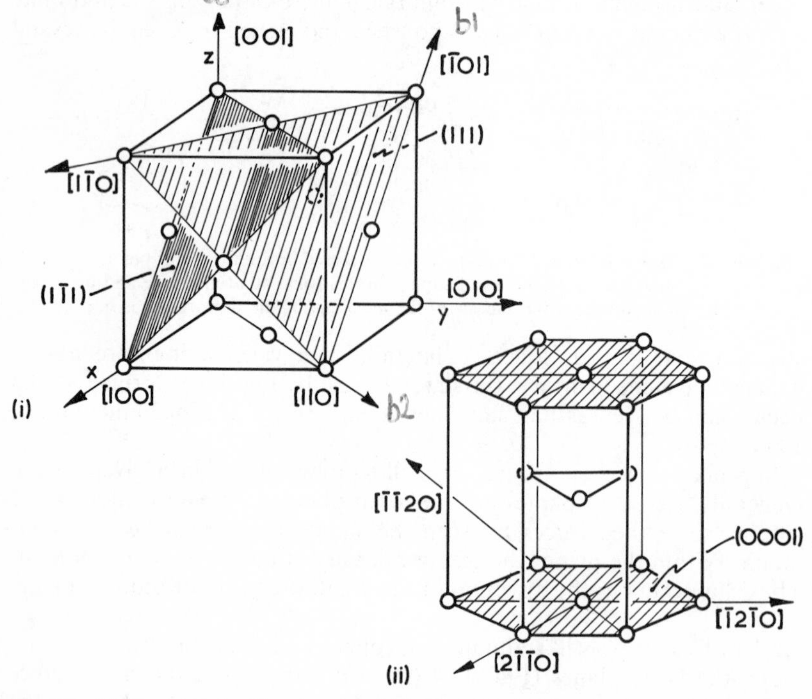

Fig. 5.4—There are twelve directions in which slip can take place in the FCC structure (i), but only three in which slip can take place in CPH structures with equal ease (ii).

other family of planes. Thus the planes of easiest slip are also those of densest packing. The slip *directions* in the FCC structure are the ⟨110⟩ family (face diagonals). There are four effective sets of planes but each plane has three possible slip directions, giving a total of twelve directions in which slip could take place with equal ease in a FCC crystal (Fig. 5.4 (i)).

5.2.2 In the CPH structure the (0001) or basal planes are of similar arrangement to the (111) planes in the FCC structure. However, in CPH there is but one such set of basal planes instead of four (111) planes in FCC. Hence there are only three easy slip systems in CPH (Fig. 5.4 (ii)). This is reflected in the differences in mechanical properties between the malleable and ductile FCC metals like aluminium and copper, and the relatively brittle CPH metals like zinc.

5.2.3 Normally we are dealing with polycrystalline metals but much fundamental information on the nature of slip has been obtained by studying the behaviour of *single* crystals of metals in stress. These single crystals can be grown under carefully controlled laboratory conditions and then machined as test pieces. During a tensile test slip occurs along parallel slip planes (Fig. 5.5) the precise direction of which can be observed. By determining the yield stress of such a crystal experimentally we can then calculate the critical resolved shear stress acting along a slip plane which will just produce slip.

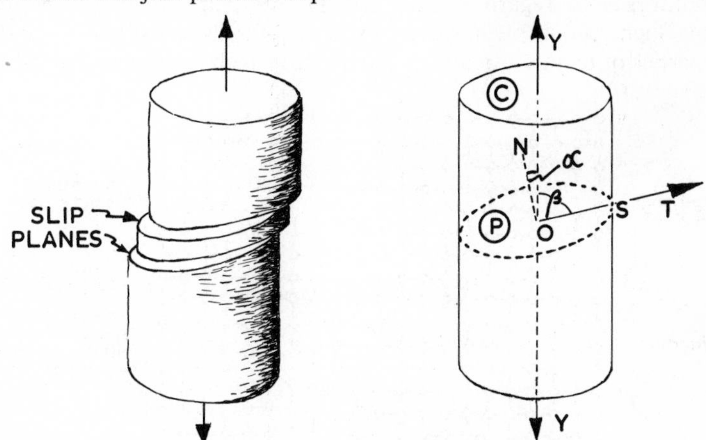

Fig. 5.5—Slip in a single metallic crystal.

If Y is the yield stress and T the critical resolved stress acting along OS on the slip plane Ⓟ then,

Component of Y along $OS = Y \, . \, \cos \beta$

The area of projection on Ⓒ of unit area of Ⓟ $= \cos \alpha$

Hence, $T = Y \, . \, \cos \alpha \, . \, \cos \beta$

In different single crystals of the same material values of α and β will vary, that is crystallographic planes will be differently orientated with respect to the axis of the test piece from one specimen to the next. Thus single crystals of the same material will slip at different angles to the axis and consequently at different values of Y, but when the stress is resolved along the slip plane all crystals of the same material slip at the same critical resolved shear stress.

It is possible to calculate theoretically this critical resolved shear stress from a knowledge of the forces which act between fundamental particles within the metallic bond. If we do this we find that the 'theoretical strength' of a metal is generally *one thousand times or more* the value obtained experimentally as outlined above. Obviously, therefore, slip is not a simple process in which all of the atoms in a complete plane move *simultaneously* with reference to the atoms in an adjoining plane, rather like one penny in a pile sliding over its neighbour. Moreover, the assumption that slip takes place in this simple manner does not explain why metals work harden progressively during a cold-working operation. Only in recent years has the low strength of a metal relative to its 'theoretical strength' been successfully explained.

Imperfections in crystals

5.3 During the early days of this century a great deal of investigation into the crystal structures of metals took place and there was a tendency for metallurgists to regard metallic crystals as being relatively perfect in form. Such a concept, however, made it difficult to explain some of the properties of metals and alloys, particularly such phenomena as diffusion

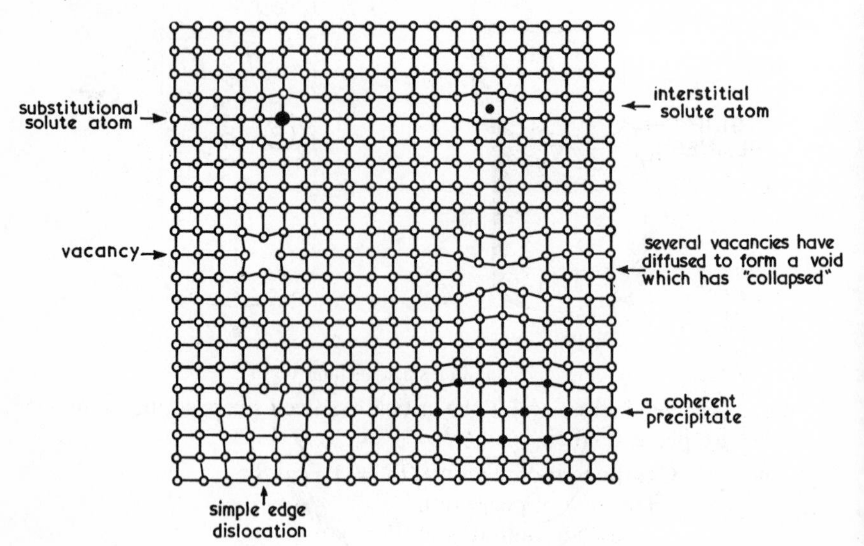

Fig. 5.6—The imperfect nature of a metallic crystal.

in solid solutions (7.4) and the fact mentioned above, namely, that the practical value obtained for the critical shear stress of a metal is several orders of magnitude less than that deduced from theoretical considerations based on instantaneous 'block' slip.

The suggestion that atoms in a metallic crystal are arranged in a rigid pattern from which there is no deviation has long since been rejected as evidence to the contrary has accumulated over the years. Currently the metallurgist thinks of a metallic crystal as having atoms arranged according to some general overall pattern but in which all manner of local faults and deficiencies can exist causing distortions and irregularities in the surrounding crystal lattice (Fig. 5.6). Whilst X-ray investigations (3.4) will reveal the general crystal pattern of a metal they cannot detect single lattice faults and the existence of many of these is still largely a matter of hypothesis.

5.3.1 One of the most important of these lattice faults was postulated more or less independently in 1934 by Taylor, Polanyi and Orowan. In its

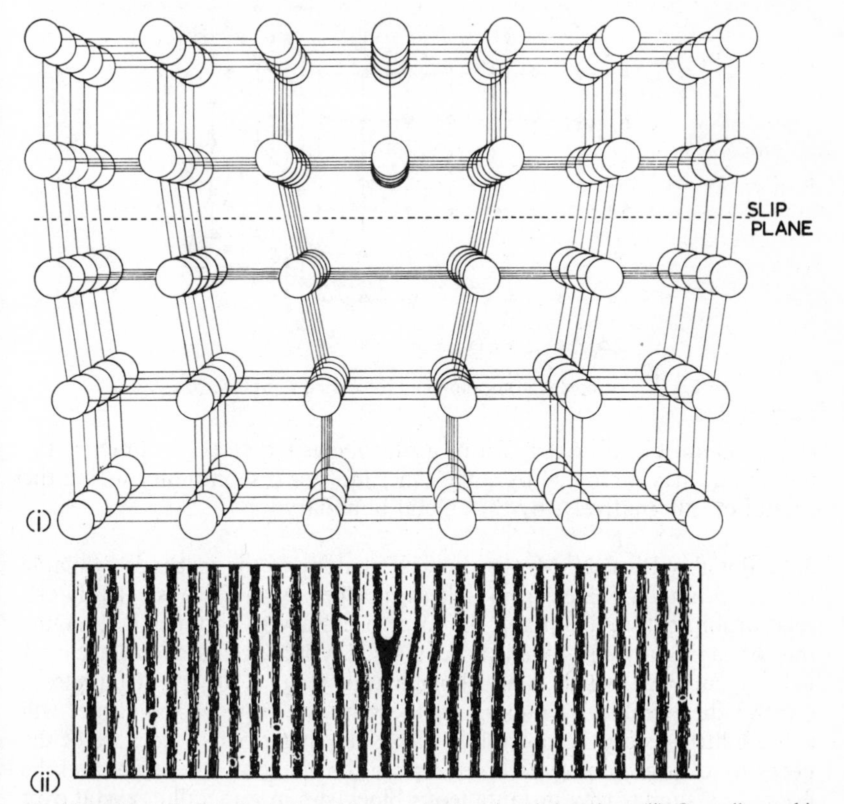

Fig. 5.7—(i) A 'ball-and-wire' model of a simple edge dislocation. (ii) Crystallographic planes containing an edge dislocation as they appear in aluminium at a magnification of several millions.

simplest form it consists of a region containing an extra half-plane of atoms (Fig. 5.7). This is termed an *edge dislocation* and it plays a very important rôle in determining the mechanical properties of metals. In the early days of the Dislocation Theory there was no direct proof that such faults did in fact exist in crystals but over the years more and more supporting evidence accumulated so that during the post-war years the theory blossomed and became widely accepted. More recently the existence of dislocations was verified metallographically using very high-power modifications of the electron microscope.

The arrangement of atoms in the region of an edge dislocation is shown in Figs. 5.7 and 5.8. The presence of dislocations in such diagrams is generally indicated by the sign $\perp$. Under the action of adequate stress the dislocation will move progressively through the crystal to the right

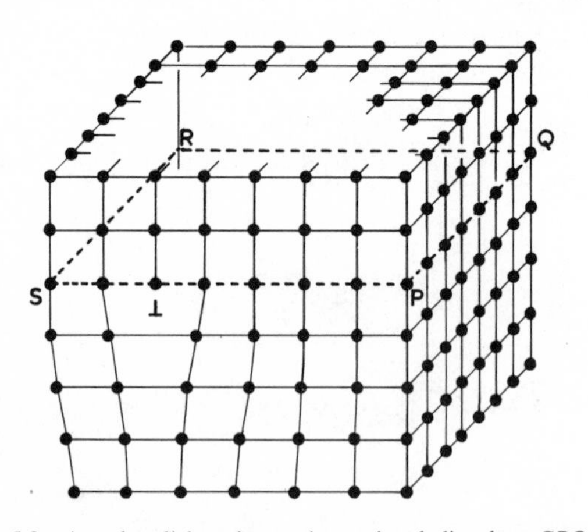

Fig. 5.8—An edge dislocation and associated slip plane *SPQR*.

(Fig. 5.9) until it ultimately forms a *slip step* as indicated. In practice the movement may be halted by some other fault or discontinuity within the crystal or, alternatively, by the crystal boundary.

5.3.2 Possibly Prof. N. Mott's classical analogy of the methods available for smoothing wrinkles from a heavy carpet explains most clearly the relationship between slip and the force necessary to produce it. Imagine that on laying a wall-to-wall carpet a small wrinkle remains in the carpet near to an edge which is touching a wall (Fig. 5.10 (i)). Attempts to remove the wrinkle by pulling on the opposite edge of the carpet will achieve little—save the possibility of broken finger nails—because of the necessity of overcoming friction between the whole of the carpet and the floor. In a similar way instantaneous block slip in a metallic crystal over a complete plane would necessitate overcoming inter-atomic forces over the whole of the plane at the same instant.

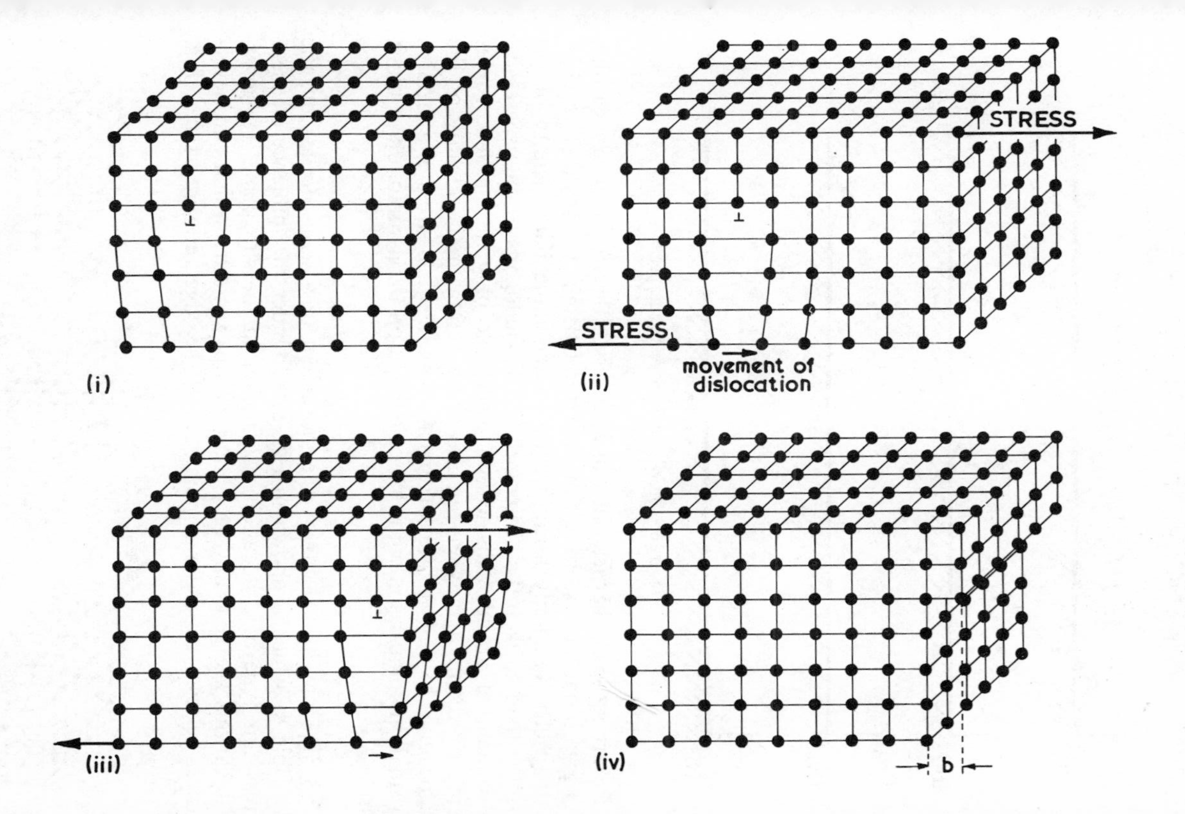

Fig. 5.9—The movement of an edge dislocation under the influence of stress. The unit slip step, *b*, is known as the Burgers vector of slip.

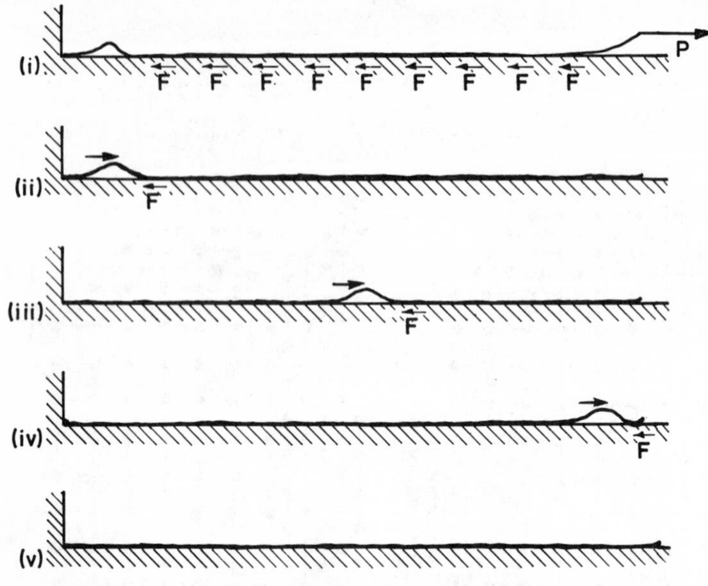

Fig. 5.10—Mott's analogy of the carpet.

Returning to our carpet, one can with very little effort coax the wrinkle along with one's toe so that it moves, step by step, across the floor (Fig. 5.10 (ii) to (v)). Only local friction in the immediate vicinity of the wrinkle then needs to be overcome. This is analogous to the progressive movement of an edge dislocation where the only inter-atomic forces being overcome at any instant are those acting in the locality of the dislocation. In this way we can explain why the yield stress of a metal is so small in comparison with that calculated on the assumption of block slip.

5.3.3 Slip may also occur by the movement of a *screw dislocation*, the principle of which is shown in Fig. 5.11. Under the influence of shear stress this type of dislocation also moves in the slip plane. However, dislocations are generally of a more complex nature which can nevertheless be resolved into a combination of the edge and screw types (Fig. 5.12).

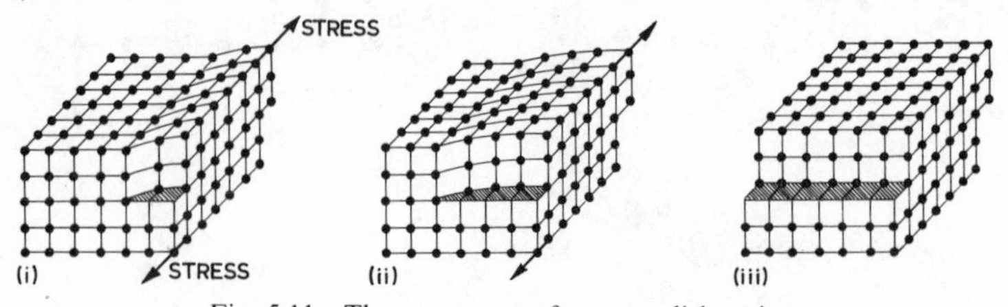

Fig. 5.11—The movement of a screw dislocation.

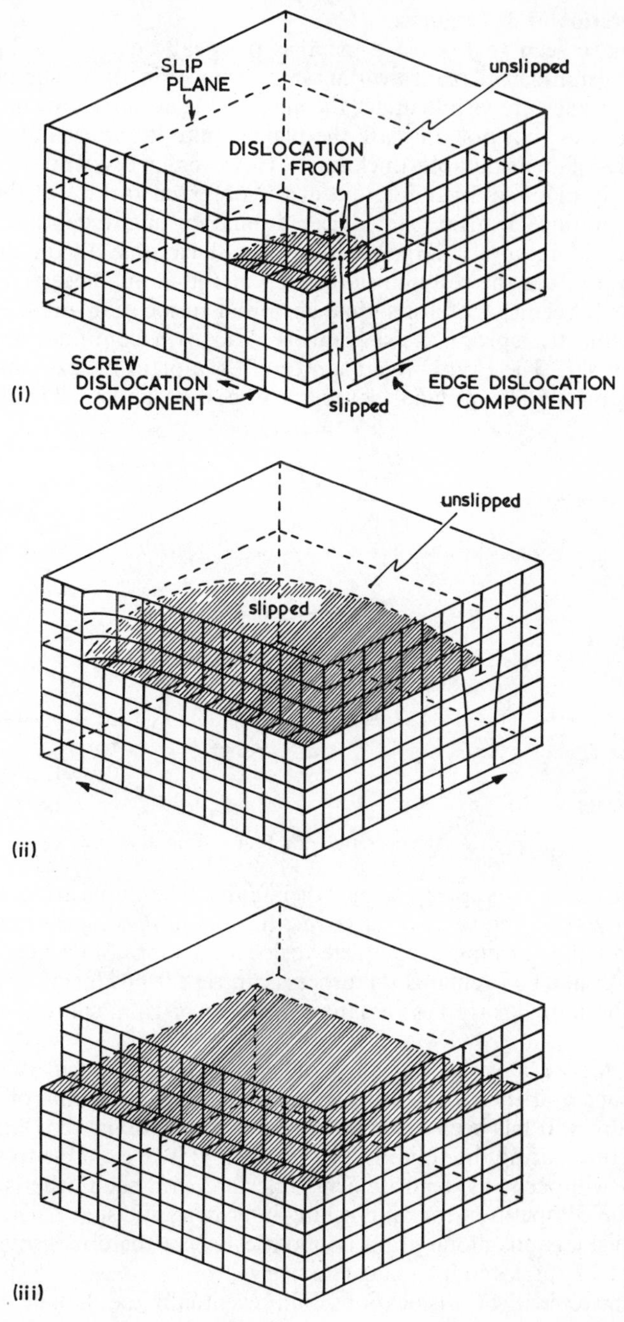

Fig. 5.12—Slip by the movement of a dislocation loop over a slip plane.

The generation of dislocations

5.4 We have seen that when a metal is plastically deformed slip takes place on a number of planes within any crystal (Fig. 5.3). The fact that these slip planes are visible under the microscope at fairly low magnifications indicates that not only are the planes quite far apart but that the 'steps' are of considerable thickness.* However, a dislocation moving along a slip plane and running out at the observed surface of the metal can only produce a surface step of one atomic spacing in depth (Fig. 5.9) and it would require a large number of dislocations, all on the same plane, to produce the resultant large step of 400 atoms observed experimentally. It seems, therefore, that an initial dislocation must in some way be able to reproduce itself and so lead to a continuation of the process of slip. The Frank–Read source offers a plausible explanation of this possibility. The original source is visualised as a dislocation line

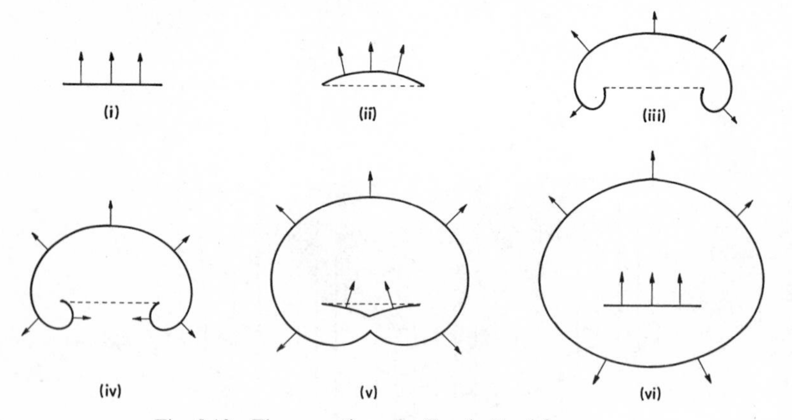

Fig. 5.13—The operation of a Frank–Read Source.

anchored at its ends, possible by other faults. The application of shear stress causes the line to bow outwards, turning in upon itself (Fig. 5.13) and eventually forming a complete dislocation loop. Since the original dislocation line still remains the process can repeat producing a series of dislocation loops or ripples radiating from the original source.

5.4.1 So far we have been dealing only with the movement of existing dislocations and their permitting of extensive slip by means of Frank–Read sources. It has been assumed that such dislocations were 'built into' the structure during the solidification process. It is reasonable to suppose that this will occur as dendrite arms interfere with each other's growth during the competition for diminishing inter-dendritic space. The formation of dislocations along grain boundaries where there is a small angle of misfit (5.5) arises in a similar manner.

The movement of dislocations will eventually be halted by such

* Approximately 400 atoms high as mentioned earlier.

obstacles as grain boundaries, other groups of dislocations and imperfections along the slip plane. If the acting stress is increased new dislocations will be initiated in the structure and slip will continue until all available slip planes have been used up so that the movement of dislocations is no longer possible. Further increase of stress will ultimately lead to fracture.

The stress necessary to produce new dislocations is much greater than that required to move those already present. This relationship is illustrated by the tensile properties of metallic 'whiskers'. These are very small hair-like single crystals grown under carefully controlled conditions and generally containing only a single dislocation which runs along the central axis. A tensile stress along this axis will cause no movement of the dislocation. As no other dislocations are available, slip cannot take place until a new dislocation is initiated at *E* (Fig. 5.14). Immediately, the

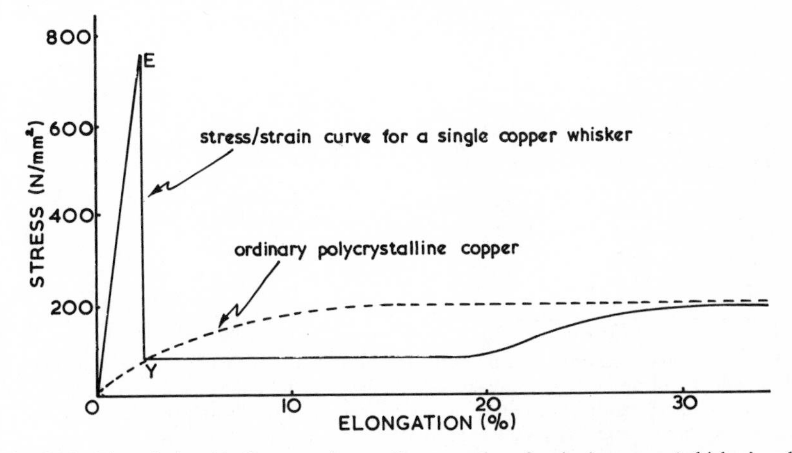

Fig. 5.14—The relationship between the tensile properties of a single copper 'whisker' and ordinary polycrystalline copper. The whisker was stressed parallel to the {111} direction. The whisker yielded suddenly at *E* and the load was relaxed and then re-applied at *Y*.

stress falls to that necessary to move this new dislocation (*Y*) and further dislocations are propagated by the Frank–Read mechanism so permitting plastic flow to proceed.

5.4.2 In 1956 Marsh developed an extremely sensitive micro testing machine capable of determining the tensile properties of specimens of such small dimensions as 10^{-7} mm² cross-sectional area and 0·25 mm long. This machine could detect extensions of the order of 0·5 nm. In addition to being useful for determinations on single whiskers other small diameter specimens containing a limited number of dislocation centres could also be tested. Fig. 5.15 (i) shows the stress/strain diagram for a small thin specimen tested in a Marsh machine. Here the stress/strain curve is in the form of a series of steps since the machine is sufficiently sensitive to detect the generation of dislocations from an individual source. Whilst *single* dislocations could not be detected even by the sensitive Marsh machine the production of a dislocation source releases a number of dislocations simultaneously so that the total amount of slip associated with one source is measurable.

With further increases of stress there are potential dislocation sources ready to generate dislocations and these are presumably triggered by the inflow of thermal energy. Sources are triggered quite erratically and the slip produced is virtually instantaneous giving a series of irregular steps illustrated. In the case of a large test piece (Fig. 5.15 (ii)) so many centres of dislocation are being triggered continuously throughout the specimen that the typical smooth curve of a stress/strain diagram is produced.

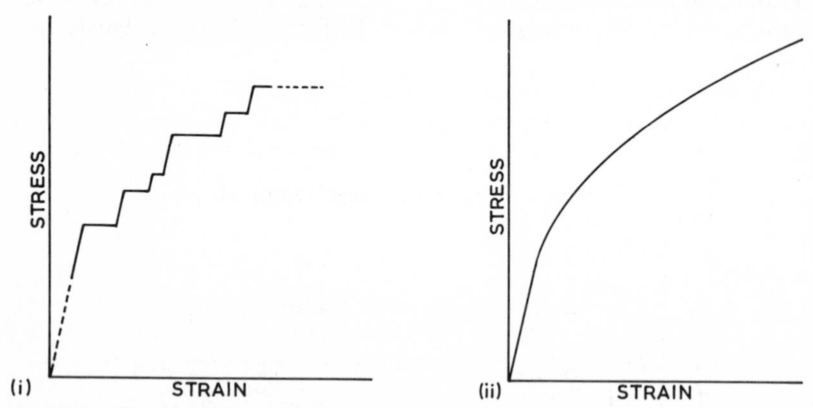

Fig. 5.15—Using a sensitive 'micro' testing machine on a very small specimen containing a finite number of dislocation centres reveals the incremental nature of slip (i). In a normal tensile test on a large specimen (ii) slip is taking place continually at a very large number of centres so that a regular curve is obtained.

5.4.3 Interaction between dislocations moving on the same or on nearby planes must be considered particularly in so far as they affect the distribution of strain energy. The presence of the extra half plane of atoms in a simple edge dislocation gives rise to an increase in strain energy in that region. Imagine an axe blade (representing the extra half plane of atoms) driven into a log of wood (Fig. 5.16). This will produce locked-up strain energy in the surrounding wood in a manner similar to the extra half-plane of atoms in an edge dislocation. Consequently two edge dislocations of the *same* sign (Fig. 5.17), moving on the same or nearby planes, will tend to repel each other in order to reduce the concentration of elastic strain energy in the region. Increased stress will be required to move them closer together so that, under such circumstances, not only does the total strain energy increase but also the tensile strength which is a measure of the stress required to produce slip. Normally dislocations of the same sign will pile up against obstacles such as grain boundaries when the applied stress is great enough.

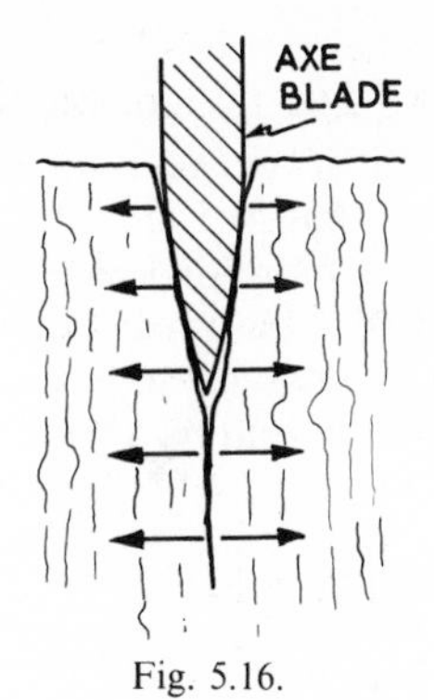

Fig. 5.16.

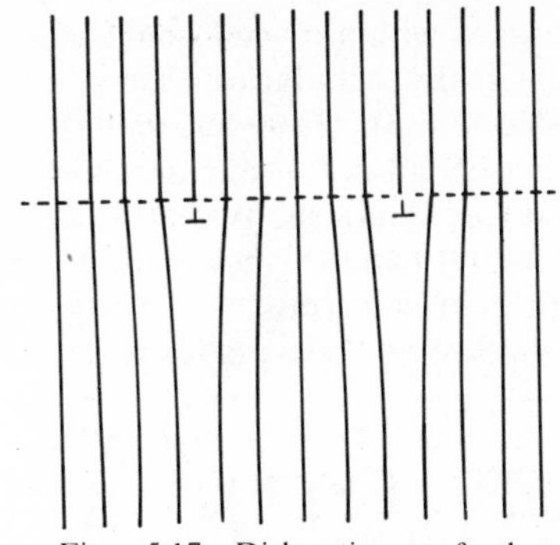

Fig. 5.17—Dislocations of the same sign will repel each other.

Conversely, if edge dislocations of *opposite* sign (Fig. 5.18) move into close proximity they will tend to attract each other. Assuming that they are on the same slip plane they will annihilate each other with a consequent

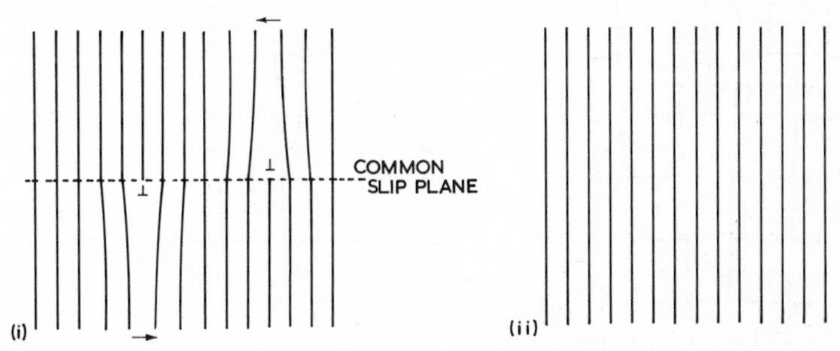

Fig. 5.18—Dislocations of opposite sign will tend to attract (i) and so annihilate each other (ii).

reduction in strain energy. If their slip planes are separated by a few atomic spacings a row of vacancies will remain (Fig. 5.19) but the total strain energy will be reduced due to the annihilation of the dislocations.

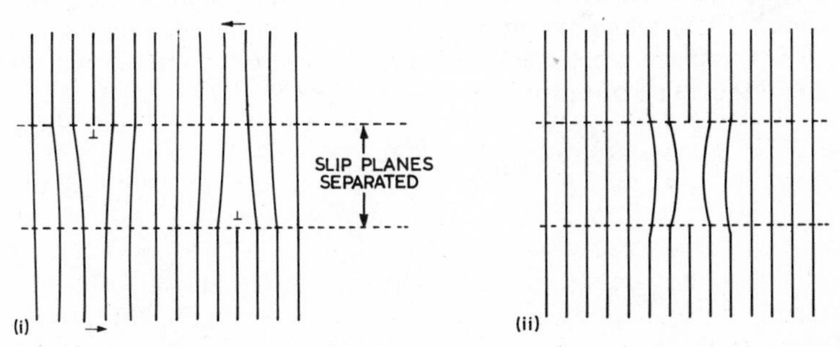

Fig. 5.19—Dislocations of opposite sign not moving on the same plane. Here a row of vacancies (ii) remains.

Dislocations within crystal boundaries

5.5 That grain boundaries are a region of some disorder has already been suggested (3.7.3). However, when the angle of 'misfit' between adjacent crystals is small an atomic arrangement similar to that in Fig. 5.20 may be produced. Here the two crystals are able to join more or less continuously along most of their common boundary with, of course, a certain amount of elastic strain. Since adjacent planes in the two crystals are not parallel some of these planes must terminate at the boundary giving rise

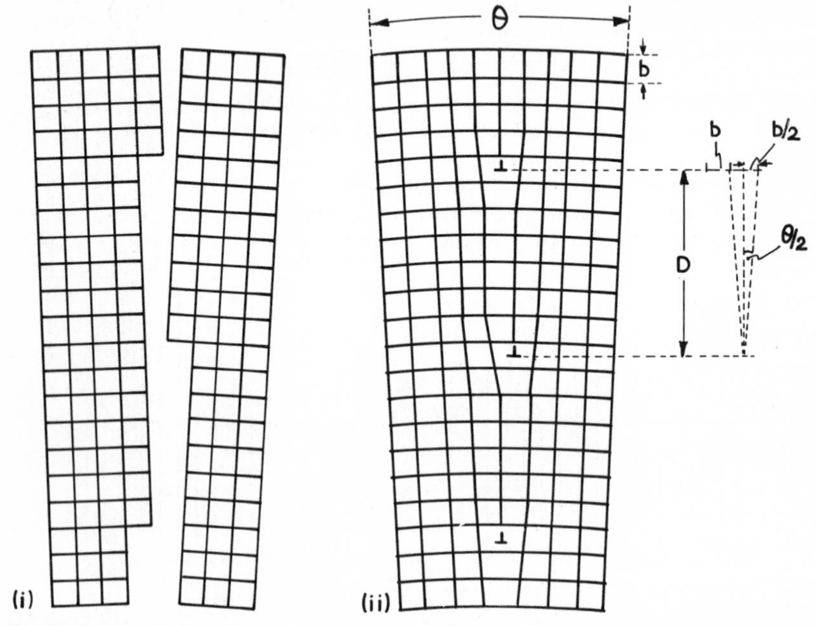

Fig. 5.20—A grain boundary where the angle of misfit is small. (i) Indicating the position of the crystal boundary. (ii) Crystals joined with edge dislocations present. (After Read, 'Dislocations in Crystals'.)

to line imperfections passing through the crystal normal to the plane of the paper. These faults manifest themselves as simple edge dislocations.

If D is the spacing between these dislocations, b the interatomic spacing and θ the angle of misfit, then:

$$\sin\frac{\theta}{2} = \frac{b/2}{D}$$

$$= \frac{b}{2D}$$

$$\text{Thus } D = \frac{b}{2\sin\theta/2}$$

Since θ is small

$$D = \frac{b}{\theta}$$

As θ increases the value of D decreases, i.e. the dislocations move closer together until their individual identity is lost and the boundary becomes a general disordered region (Fig. 3.22).

5.5.1 The passage of a dislocation across a grain boundary must be extremely rare since it would require a high degree of 'coherency' between the adjacent crystal lattices. Normally, grain boundaries of what ever sort will act as barriers to the movement of dislocations and, during plastic deformation, will become regions of very high strain energy as dislocations pile up there to form a complex transcrystalline 'traffic jam'. This high concentration of strain energy at the grain boundaries of cold-worked metal gives rise to the preferred intercrystalline corrosion prevalent in heavily cold-worked metals (15.5).

Deformation by twinning

5.6 Although slip is the more significant process by which plastic deformation occurs in metals brief mention must be made of the phenomenon known as *twinning*. Whilst slip is a process associated with a *line* defect (the dislocation), twinning is related to a *plane* defect (the twin boundary).

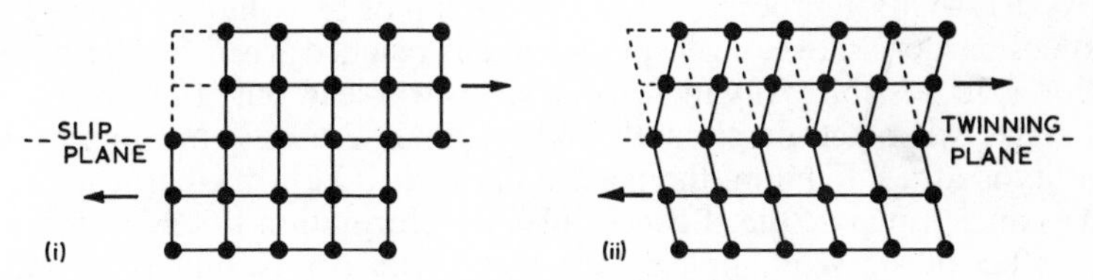

Fig. 5.21—Deformation by (i) slip and (ii) twinning.

In slip all atoms in a block have moved the same distance when slipping is complete (Fig. 5.21 (i)) but in deformation by twinning (Fig. 5.21 (ii)) atoms within each successive plane in a block will have moved different distances. When twinning is complete the lattice direction will have altered such that one half of the twin is a mirror image of the other half, the twinning line corresponding to the position of a mirror. Twinning, like slip, also proceeds by the movement of dislocations. The net result of a twinning process on the lattice of a crystal is depicted in Fig. 5.22.

The stress required to produce deformation by twinning tends to be higher than that necessary to produce deformation by slip. Twinning is more likely to occur in metals which are shock loaded at low temperatures. Thus in BCC iron sudden loading at low temperatures produces

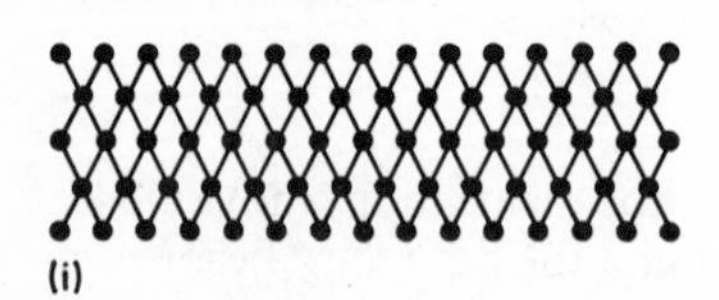

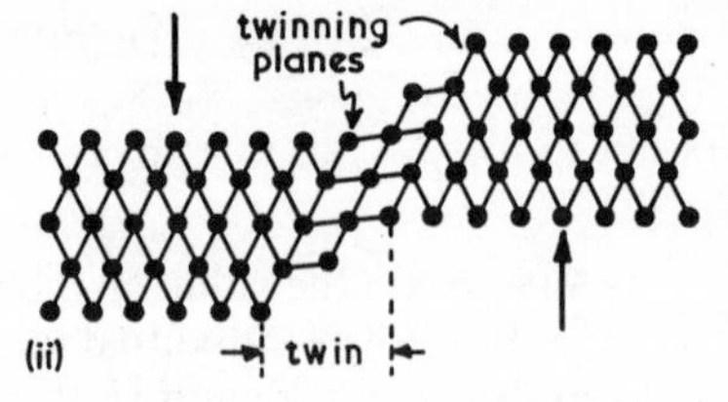

Fig. 5.22—An example of deformation by twinning. (i) Before stressing. (ii) After stressing.

thin lamellar twins which are generally known as Neumann bands. Twinning is commonly encountered in CPH metals like zinc, whilst when a bar of tin is bent suddenly the formation of twins can be heard taking place. The Ancients knew this as 'the cry of tin'.

5.6.1 Twin-formation takes place during the annealing of some cold-worked metals. In this case recrystallisation is initiated at a plane instead of at a point source and twin bands are observed within the new crystals as a result. These 'annealing twins' are common in copper, brasses, bronzes and austenitic steels—alloys with low stacking fault energies—and are different in this respect from the mechanical twins under consideration here. They are formed not by mechanical shear but as part of the process of grain growth.

Work hardening

5.7 As deformation proceeds—whether by slip or by twinning—the metal becomes harder and stronger and a stage is reached when further deformation is impossible. Any increase in stress will lead only to fracture. At this stage, when tensile strength and hardness are at a maximum and ductility is at a minimum, the material is said to be work-hardened.

We can summarise the effects of plastic deformation briefly as follows—if sufficient stress is applied to a metal, slip (or twinning) will take place in individual crystals. As deformation proceeds the capacity for further cold work diminishes as dislocations find themselves in positions from which further movement is impossible. An increase in stress is then necessary to initiate new dislocations and move these until they too become blocked. Ultimately a point is reached when no more mobile dislocations are available. This coincides with the point of maximum resistance to slip (the maximum strength and hardness) where no more deformation is possible and fracture will take place if an increase in stress is attempted. Further plastic deformation can only be carried out if the material is annealed.

5.7.1 During a cold-working process some 90% of the mechanical energy employed is converted to heat whilst the remaining 10% is retained in the material as potential energy in the form of elastically balanced strains in the lattice structure. The introduction of heat energy to the system during an annealing process permits sufficient mobility of the atoms so that dissipation of strain energy can take place. The initial relief of stress is followed by recrystallisation of the metal.

Stress-relief and recrystallisation

5.8 As we have seen a cold-worked metal is in a state of considerable mechanical stress resulting from elastic strains internally balanced. Much of the 'locked up' strain energy is associated with the presence of dislocations. In the region of *P* (Fig. 5.23) the lattice is in tension and since atoms there are displaced further apart the region will possess potential

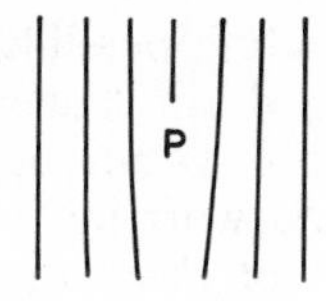

Fig. 5.23.

energy. The high energy associated with the congregation of dislocations at grain boundaries has been mentioned as a reason why corrosion occurs more readily there than elsewhere in the structure. These high-energy regions also initiate recrystallisation during an annealing process so that new 'seed' crystals appear first at the old grain boundaries of the original distorted structure (Fig. 5.27).

In the early stages of an annealing process some degree of stress relief occurs as atoms move over limited distances into positions nearer to equilibrium. At this stage, however, there is no alteration in the distorted appearance of the structure and in fact hardness and tensile strength remain at the high value produced by cold work.

At low temperatures the movement of dislocations is restricted to glide along slip planes, but at higher temperatures an edge dislocation is able to move out of its slip plane by a process known as *climb*. If the terminal row of atoms (normal to the plane of the paper) in the extra half-plane is removed this constitutes *positive climb* and the strain energy of the immediate vicinity will be reduced. If an extra row is added to the half-plane this results in *negative climb* and the strain energy of the region will be increased.

In this instance we are interested mainly in positive climb and the reduction of strain associated with it. It can take place most easily by the diffusion (7.4) of vacancies towards the dislocation as suggested in Fig. 5.24. Here atoms migrate from the end of the half-plane to fill vacancies which approach them by means of the diffusion mechanism.

5.7.2 The diffusion of interstitial atoms towards an edge dislocation (Fig. 5.25 (i)) can give rise to *negative climb* whilst the diffusion of atoms away from the half-plane to become interstitials will cause *positive climb* (Fig. 5.25 (ii)). All of these processes necessitate the mass migration of atoms

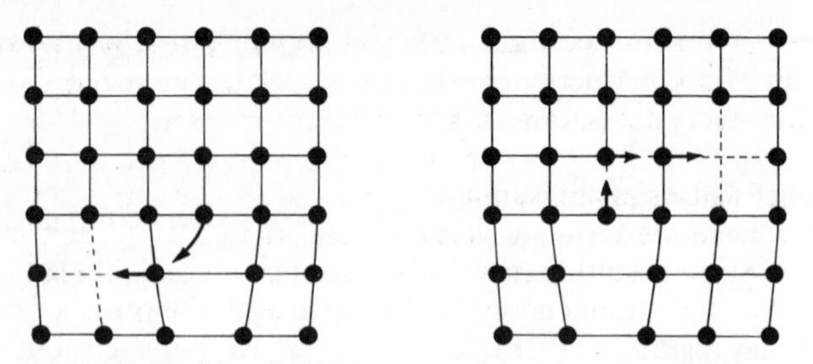

Fig. 5.24—'Positive climb' by two examples of vacancy diffusion.

by diffusion and climb is therefore possible only by thermal activation. The reduction of strain energy during annealing will most likely involve positive climb by the diffusion of vacancies.

The reader may possibly be worried by the fact that in the above description of the mechanism of climb the assumption has been made

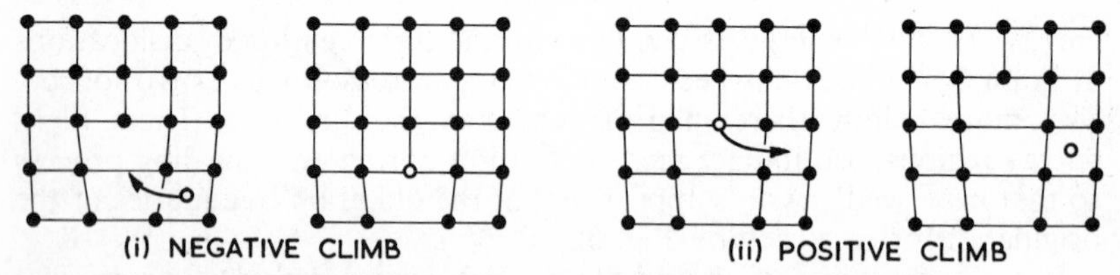

Fig. 5.25—Dislocation climb associated with interstitial atoms.

that a whole row of atoms is removed (or added) simultaneously, whereas in practice, individual vacancies or small groups of vacancies diffuse to (or from) the dislocation. Fig. 5.26 illustrates climb involving a short section of a dislocation line, resulting in the formation of two steps generally described as *jogs*.

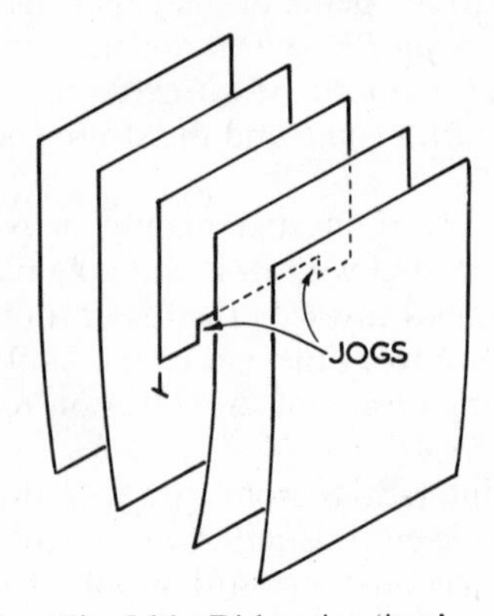

Fig. 5.26—Dislocation 'jogs'.

During an annealing process elastic strains are first dissipated by limited movements of atoms in the manner of those mentioned above. Then, at a higher temperature, wholesale recrystallisation of the distorted structure occurs and is accompanied by a fall in tensile strength and hardness to approximately their original values, whilst the capacity to accept cold-work returns. This type of annealing process, which is used principally when a metal is to undergo further cold deformation, takes place in three stages:

5.8.1 *The relief of stress.* This often takes place at relatively low temperatures at which atoms nevertheless are able to move small distances into positions where they will incur less strain. Local movement of interstitial atoms or vacancies may be involved as outlined above. There is, however, no observable alteration in the distorted structure which was produced by mechanical work and both tensile strength and hardness will remain at a high level.

Low-temperature annealing which does not cause recrystallisation and which consequently does not lead to a fall in strength and stiffness engendered by cold-work, is useful in dissipating local strain which might otherwise give rise to stress corrosion (15.5).

5.8.2 *Recrystallisation.* Although low-temperature annealing is often used to relieve internal stress as outlined above, most annealing processes involve complete recrystallisation of the distorted cold-worked structure. As the annealing temperature is increased a point is reached where new

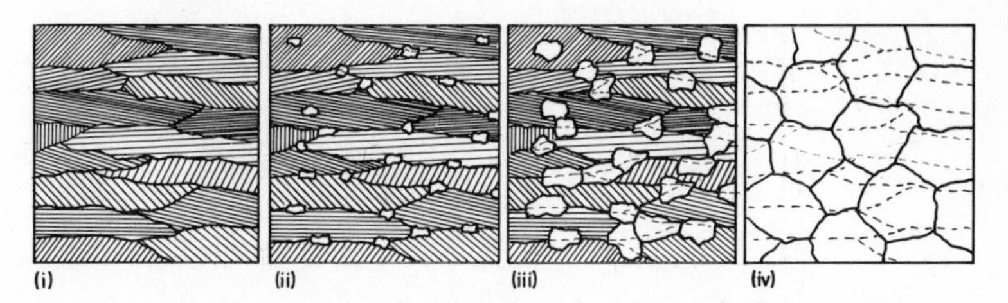

Fig. 5.27—The recrystallisation and grain growth of a cold-worked metal.

crystals begin to grow from nuclei initiated within the most heavily deformed regions. These nuclei form at points of high energy such as crystal boundaries and other localities rich in piled-up and entangled dislocations. The new crystals so formed are very small at first but grow steadily until they have absorbed the whole of the distorted structure produced originally by mechanical work (Fig. 5.27). The new crystals are equi-axed in shape, that is they do not show any directionality as did the distorted crystals which they replace.

The minimum temperature at which recrystallisation will take place is called the *recrystallisation temperature*. This temperature is lowest for pure metals and is raised significantly by even small amounts of impurities. Thus, whilst commercial-grade aluminium recrystallises at about 150° C following mechanical work, that of 'six nines' purity (99·999 9% pure) apparently recrystallises below room temperature and consequently does not cold-work when deformed at ambient temperatures.

Tin and lead normally recrystallise below room temperature so that it is impossible to work-harden them since they recrystallise whilst the working process is taking place. Hot-working is, of course, a process which is carried out above the recrystallisation temperature. Normally the working temperature must be well above the recrystallisation temperature so that recovery is rapid enough to keep pace with the rate of deformation.

It is not possible to assign a precise recrystallisation temperature to any metal since the temperature at which recrystallisation begins is governed largely by the amount of cold-work which the metal has undergone. The greater the amount of cold-work, the greater will be the amount of strain energy locked up in the structure and so the lower the temperature at which recrystallisation may begin when heat energy is supplied. For most metals the recrystallisation temperature is between one third and one half of the melting point (Kelvin scale). That is the mobilities of all metallic atoms are roughly equal at the same fraction of their melting points (K).

5.8.3 *Grain growth.* At temperatures above that of recrystallisation of the metal, the newly formed crystals will tend to grow by absorbing each

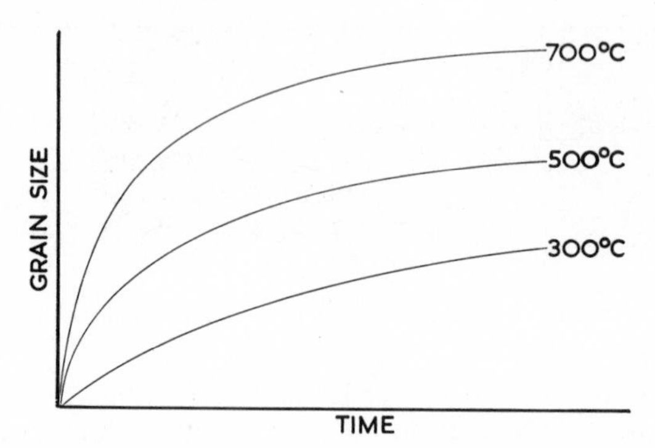

Fig. 5.28—The relationship between grain size, annealing temperature and annealing time.

other in cannibal fashion. This will result in the formation of a relatively coarse-grained structure, the ultimate grain size depending jointly upon the time of treatment and the temperature employed (Fig. 5.28).

Since boundaries have higher energies than any other region of the crystal, a polycrystalline mass will be able to reduce its energy if some of the grain boundaries disappear. Thus at temperatures in excess of that of

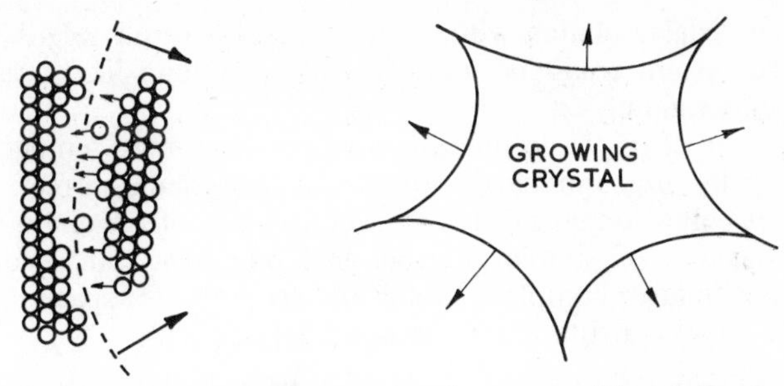

Fig. 5.29—The growth of a crystal at the expense of its neighbours.

recrystallisation large crystals grow by absorbing small ones. A crystal boundary tends to move towards its centre of curvature (Fig. 5.29) in order to shorten its length. To accommodate this atoms move across the boundary to positions of greater stability where they will be surrounded by more neighbours in the concave face of the growing crystal.

The deformation of polymeric materials

5.9 The arrangement of atoms in the crystal structure of a metal follows a regular three-dimensional periodicity but the atoms in a polymer are generally arranged in a far more complex and less-regular manner, particularly since the molecules themselves are often not geometrically aligned. For this reason it has been less easy to explain the deformation of polymers in terms of their structures.

Single bonds in a polymer chain are capable of *rotation*. This is illustrated in Fig. 5.30 which represents the last twelve atoms of a linear polymer molecule. Suppose a rotation of 180° took place about *a* only. The rest of the molecule, containing possibly thousands of carbon atoms, would be displaced relative to the bond *a* and it is unlikely that sufficient space would be available for this even if enough molecular energy were available. If, however, the rotation at *a* were accompanied by a simultaneous 180° rotation at *b*, only a short section of the chain would be displaced. Not only would less space be needed but less molecular energy would be required. Nevertheless the position of a short section of the molecule would be changed with respect to its neighbours. If sufficient rotational energy is supplied thermally this type of rotational transition will be taking place continuously in all molecules throughout the material

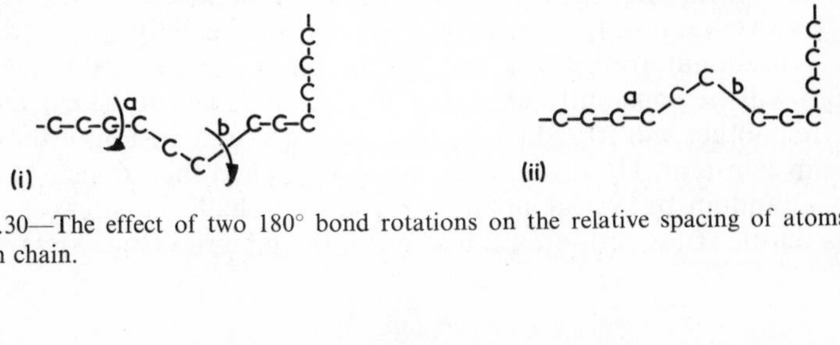

(i) (ii)

Fig. 5.30—The effect of two 180° bond rotations on the relative spacing of atoms in a carbon chain.

and the molecular chains will be in constant motion relative to each other. The system could be visualised as something like a tub full of intertwined wriggling eels.

The activity of the molecular chain will depend upon temperature but also upon the size and nature of the intermolecular forces opposing rotation. It will also depend upon the complexity of the chain units. So, whilst molecules of simple 'section' like polythene can rotate easily, molecules with large complex units would require extra space in which to rotate, thus making rotation correspondingly difficult.

5.9.1 In materials of low molecular weight atoms are held in fixed positions either by chemical bonds or by inter-molecular forces acting within the crystal structure. When stress is applied the atoms are displaced from their equilibrium positions and distances between them alter until the new forces which are set up balance the applied stress. Assuming that distortion is within the elastic range, atoms will return to their original positions when stress is relaxed. Greater stresses cause permanent distortion. In metals this is achieved by the movement of dislocations whilst in other materials, such as ceramics, sudden rupture occurs with little or no plastic deformation.

When stress is applied to a polymer material elastic deformation will occur wherever two atoms are held in fixed positions relative to one another by chemical bonds or inter-molecular forces. Thus in cross-linked polymers (12.3) and in thermoplastics at temperatures below their glass-transition points (12.4.3) all atoms can be regarded as being fixed in this way so that only elastic deformation is possible. Some elastic deformation is also possible in thermoplastics at temperatures above their glass-transition points as the atoms within the molecules are held by covalent bonds, whilst if crystallites are present atoms will be held by inter-molecular forces also. Elastic deformation of this type occurs instantaneously.

In amorphous regions of the polymer material the molecules are held only by the relatively weak van der Waal's forces operating between them. Hence they can move relative to each other without difficulty. This molecular slip—or plastic deformation—does not occur instantaneously but is time dependent since it is related to the viscosity. Consequently the combined elastic and plastic distortion is said to be *viscoelastic*.

5.9.2 In a linear polymer a large number of different chain arrangements are possible varying from the tightly coiled to the fully extended form. Since rotational transitions are taking place continuously the chain shapes will be constantly changing. For each molecule which tends to curl up another will straighten so that the *average* extension of the chains remains constant. However, when stress is applied the extensions are no longer random but occur in the direction in which the stress is acting. Thus tensile stress produces extension in the material (Fig. 5.31).

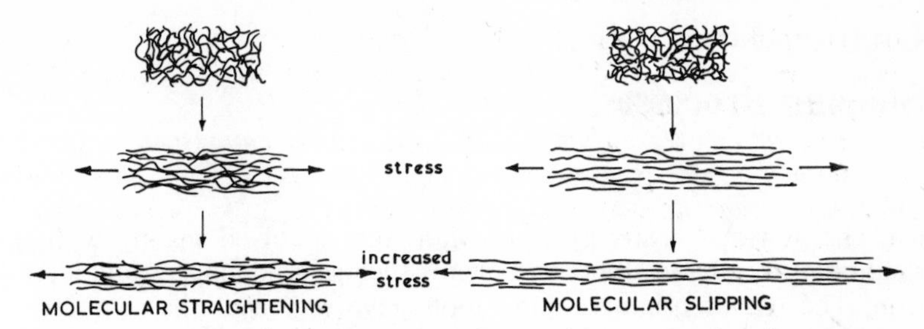

Fig. 5.31—The effect of stress on molecular shape and arrangement in a linear polymer.

When stress is removed the changes in shape of the molecules continue but those changes which involve curling up are now in the preponderance and the original average extension is regained so that the material returns to its original size. Thus chain straightening is an elastic process but it is viscoelastic because the changes in shape are time dependent in so far as the time dependence is determined by the magnitude of those inter-molecular forces opposing the rotational transitions.

The application of stress also causes molecular slip (Fig. 5.31) as the relatively weak van der Waal's forces operating between the linear molecules are overcome. This is a non-reversible process so that the extension produced is non-elastic. It is also a time-dependent process being influenced by the rotational transitions. The extent of rotational transitions to accommodate slip must be greater than that to allow for the type of segment shift which facilitates chain straightening. Straightening is therefore quicker than slipping. The phenomenon is discussed further in Chapter Twelve (12.5).

Chapter Six
Shaping Processes

6.1 The arts and crafts of early Man were involved mainly with the shaping of wood and stone, and metals such as gold, copper and later—iron. The use of primitive hand tools gave way over the centuries to manufacture by relatively sophisticated processes which take into account the accumulated knowledge of physical and chemical properties of the substances being shaped. In the present century forming processes based on the use of moulds and dies—often of complex shape—are used for shaping a wide range of materials. The development of such processes made possible the mass production of large numbers of identical articles with a consequent reduction in the *real* costs of manufactured goods generally.

Many solid substances possess sufficient plasticity to allow them to be shaped by either tensile or compressive forces. Generally plasticity increases with temperature and, so, many such processes are hot-working operations. Some solid materials, however, possess negligible plasticity and are relatively brittle at all temperatures. These materials can most easily be shaped by allowing them to solidify from the liquid state in a suitably shaped mould; or by using some machining process in which particles of the material are chipped away by a lathe or even a mason's chisel. Nevertheless, recent work on processes such as hydrostatic extrusion has shown that the reduction of surface friction between work piece and die, leading to the maximum utilisation of what plasticity a material does possess, allows relatively brittle materials like cast iron to be deformed mechanically without fracture.

Methods of shaping materials are therefore varied and often very ingenious and a single chapter in a book of this type can do little more than catalogue the more important methods used.

Shaping by solidification from the liquid state

6.2 What are commonly called 'casting processes' are used to shape a number of materials but more especially metals. In fact well over 99% of the total mass of metals used are melted and cast at some stage in the manufacturing process. A considerable quantity of both ferrous and non-ferrous alloys are of course cast as ingots which are then forged, rolled or extruded, whilst large amounts of iron, steel and light alloys are cast into some form of shaped mould to produce a component which will be used in substantially this form. Metallic alloys are strongly crystalline and pass from a mobile liquid to a rigid solid over a temperature range which is generally small. The resultant solid is strong because of its crystalline nature and the operation of the metallic bond within crystals.

Organic polymers are generally unsuitable for casting because many of them decompose at temperatures well below those necessary to produce adequate liquid mobility. Van der Waal's forces operating between long-chain molecules oppose mobility. Such plastics materials as are cast therefore depend upon a *chemical* rather than a *physical* change taking place to effect solidification. Cold-setting plastics of the type used with glass-fibre or carbon-fibre reinforcement, consist of *two* mobile liquids which are mixed and which will then proceed to polymerise (12.2.1 and 12.3.2) and so, harden. The same principles apply in many respects to cement, concrete, plaster and other ceramic products which are cast into moulds. Crystallisation, accompanying chemical change, produces rigidity in these cases.

6.2.1 *Casting metals*. Since the bulk of metallic materials is cast as either ingots or finished shapes, a melting process is an integral stage in most metal-forming operations. Furnaces used vary considerably in design but those burning fuel can be classified broadly into three groups:

(*i*) Furnaces, like the foundry cupola, in which the charge is in intimate contact with the burning fuel. Although thermal efficiency is high because of the continuous operation of the furnace and the intimate contact between burning fuel and charge, direct transfer of impurities from fuel to charge can take place.

(*ii*) Furnaces in which the charge is out of contact with the fuel but is in contact with its products of combustion. Transfer of impurities will be limited to those of a gaseous nature and thermal efficiency will still be reasonably high. Reverberatory furnaces (Fig. 6.1 (ii)) of many individual designs—including the open-hearth furnace formerly used so widely in steel making—fall into this category.

(*iii*) Furnaces in which the charge is isolated both from the fuel and its products of combustion. Crucible furnaces constitute this group. Maximum purity of charge is maintained because contact between charge and fuel is at a minimum. Nevertheless thermal efficiency will be very low since most of the heat by-passes the charge and leaves with the flue gases.

Obviously electricity is a chemically clean fuel whether used for resistance heating, induction heating or in an arc furnace, but it is relatively expensive. It is therefore apparent that in any melting process where purity of charge is important melting costs will be high.

6.2.1.1 *Ingot casting*. Steel ingots of several tonnes in mass are commonly cast into cast-iron moulds. These are usually cast 'big-end-down' (Fig. 6.2 (i)) to facilitate the stripping of the mould. This can be lifted off the solidfied ingot with a minimum disturbance of the latter (Fig. 6.2 (ii)), an important factor since the ingot is very easily damaged by cracking when in the hot 'tender' state.

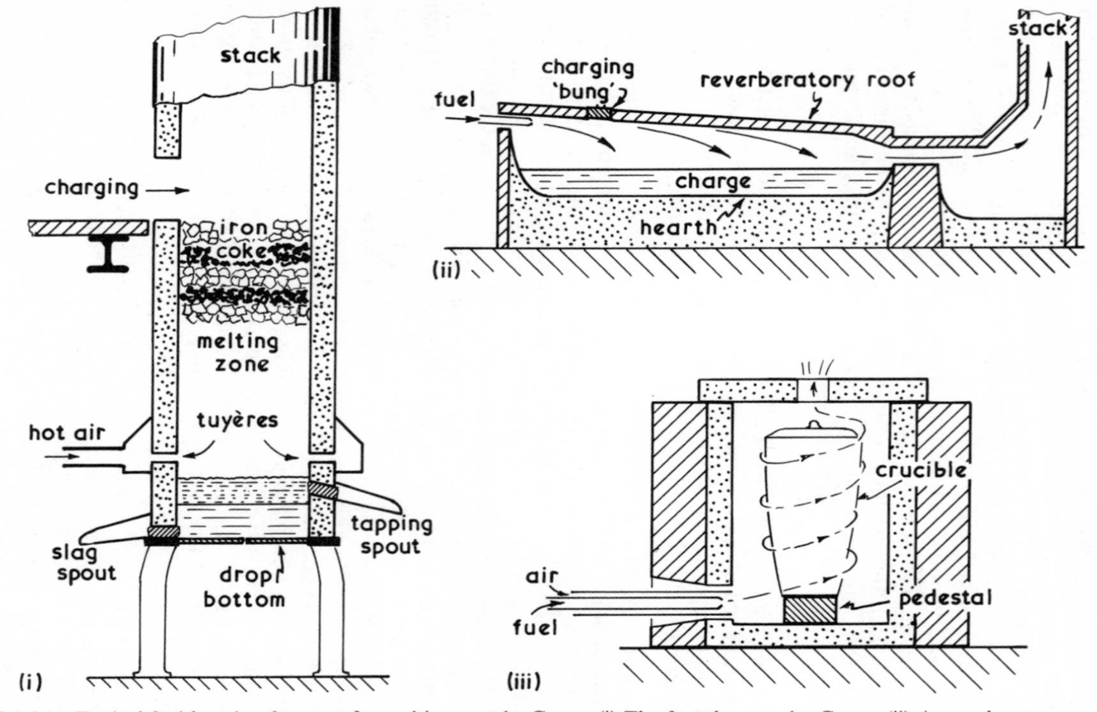

Fig. 6.1—Typical fuel-burning furnaces for melting metals. Group (i) The foundry cupola. Group (ii) A reverberatory furnace using pulverised coal, oil or gas. Group (iii) An oil- or gas-fired crucible furnace.

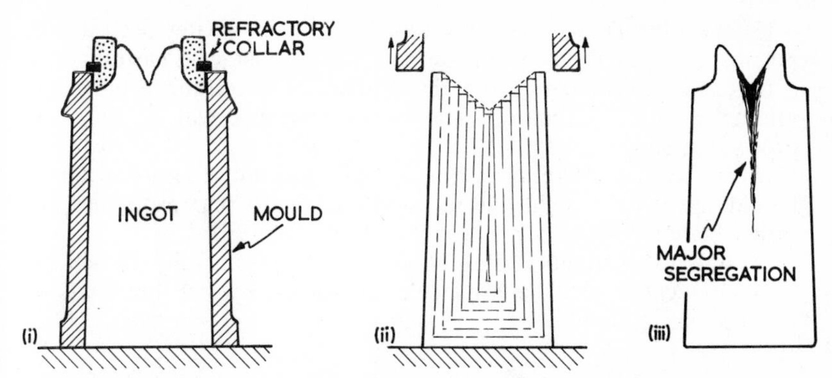

Fig. 6.2—The structure of steel ingots. (i) A typical 'big-end-down' ingot mould showing how the 'pipe' can be restricted. (ii) The development of the 'pipe' by successive solidification of elemental 'shells'. (iii) Major segregation of impurities in the central pipe.

As a metal solidifies it shrinks and so the level of the remaining liquid falls progressively leading to the formation of a *pipe* in the top of the ingot (Fig. 6.2 (ii)). This is minimised by the use of a *hot top*—a refractory collar at the top of the ingot mould which retains a reservoir of molten metal which feeds into the ingot as it solidifies and shrinks. The pipe, as it forms, is filled from the molten reservoir. The absence of any feeding arrangement (Fig. 6.2 (ii)) would lead to excessive piping.

Segregation of impurities is also troublesome in steel ingots. During the initial stages of solidification relatively pure metal crystallises first so that any dissolved impurities tend to accumulate in that metal which solidifies last. Hence the metal in the central pipe will be more impure than that which crystallises initially at the surface of the ingot (Fig. 6.2 (iii)). This is generally termed *major* segregation. *Minor* segregation on the other hand refers to the accumulation—by the coring mechanism (7.3.3)—of impurities at crystal boundaries throughout the ingot. Of the two, minor segregation is the more deleterious in its effects since it causes overall intercrystalline weakness.

Non-ferrous alloys may be cast as flat slabs for rolling to sheet, or as cylinders for extrusion of rod, tube and other sections of varying complexity (6.3.3).

The production of castings to some finished and often intricate shape may be achieved by one of a number of different processes. The particular process used will be governed by such criteria as:

(1) the size of the casting—large castings will be most economically made in sand moulds;

(2) the surface finish desired—a better surface finish is obtained using metal, 'synthetic sand' or plaster moulds, but these processes are more expensive than sand casting;

(3) the dimensional accuracy required—again greater accuracy is possible using methods mentioned in (2) as against sand casting;

(4) complexity of design, e.g. the presence of re-entrant sections and internal cavities. In many instances sand casting will be the only process possible;

(5) the mechanical properties specified—generally the use of metal moulds gives finer grain due to rapid cooling so that properties are rather better;

(6) the total number of castings required—generally it will be uneconomical to produce a metal mould unless several thousands of castings are required.

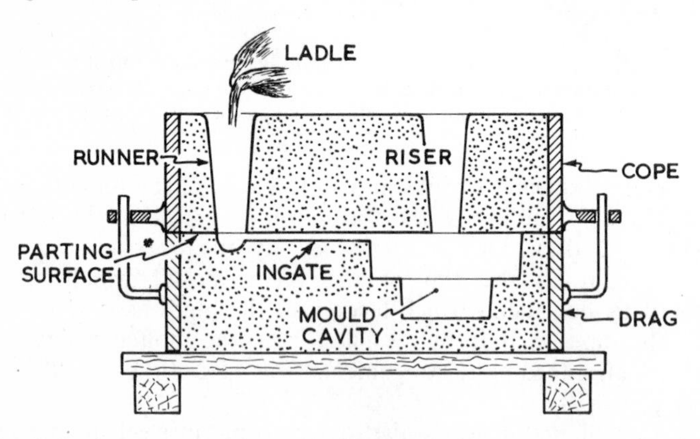

Fig. 6.3—A simple two-part sand mould with the wooden pattern removed and ready to receive the charge of molten metal.

6.2.1.2 *Sand casting*. Most readers will be familiar with the principles of this process. A wooden pattern is made such that when sand is rammed round it in a moulding box (Fig. 6.3) the mould can be split and the pattern withdrawn leaving a cavity into which metal is poured. Castings of great complexity can be produced by this process using multi-part boxes and, where necessary, sand *cores* to form holes and other internal cavities in the component. Moreover the process is cheap to operate particularly when only small numbers of castings are required, since the wooden pattern is relatively inexpensive to produce. As against these advantages dimensional accuracy is inferior to that attainable by other processes whilst surface finish is sometimes poor. The slow rate of solidification which prevails in a sand mould gives rise to coarse grain and this in turn results in a relatively weak brittle structure.

6.2.1.3 *Shell moulding*. In ordinary sand casting the moulding sand contains sufficient natural clay to act as a binding material for the silica particles. In shell-moulding processes this natural clay bond is replaced by a synthetic bonding material of the thermosetting-resin (phenol formaldehyde) type. The moulding mixture therefore consists of clean, clay-free silica sand mixed with 5% of the plastic bonding agent.

Each half of the shell mould is made on a metal pattern plate which is heated to about 250° C before being placed on top of a 'dump box'. The box is then inverted (Fig. 6.4 (i)) so that the plate is covered with the

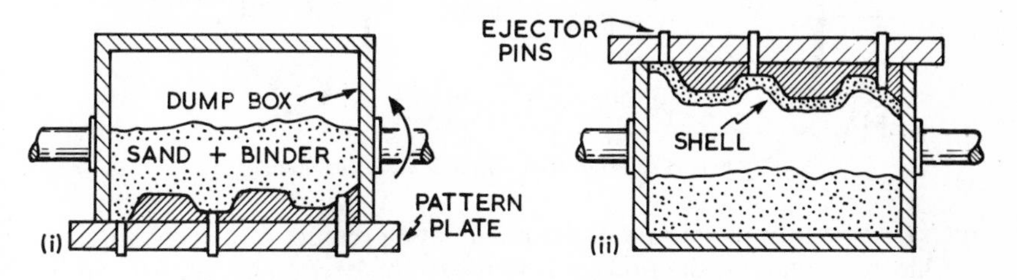

Fig. 6.4—The principles of shell moulding.

sand/resin mixture. The resin melts and in about thirty seconds the hot pattern plate becomes coated with a shell of resin-bonded sand. This shell hardens as the resin sets and the box is turned to its original position (Fig. 6.4 (ii)) so that surplus sand/resin mix falls back into the box. The shell is then stripped from the pattern plate with the help of ejector pins. The two halves of the mould thus produced are then clamped or cemented together to receive their molten charge.

The main advantage of this process lies in the high degree of dimensional accuracy obtained. Surface finish is also far superior to that associated with an ordinary sand casting. Shell moulds store well and are quite rigid so that they can be transported as necessary. The principal disadvantage of the process arises from the high cost of metal patterns though this is somewhat offset by the fact that relatively unskilled labour can be utilised so that labour costs are much less than those associated with green-sand moulding.

6.2.1.4 *Investment casting.* Whilst shell moulding was developed in Germany during the Second World War, the origins of investment casting lie in pre-history when ancient craftsmen produced moulds by pressing clay around a pattern carved from beeswax. Subsequent baking of the clay mould melted out the wax leaving a cavity to receive the metal charge.

For modern investment casting wax patterns are produced in precision metal moulds and then affixed to a 'tree' (Fig. 6.5). The wax assembly is then 'invested' with a mixture of finely powdered sillimanite and ethyl silicate. The action of water and heat on the latter cause it to form a strong silica bond between the sillimanite particles. The heating process also melts out the wax pattern leaving a mould cavity which will receive the charge of metal. To improve the accuracy of the mould impression

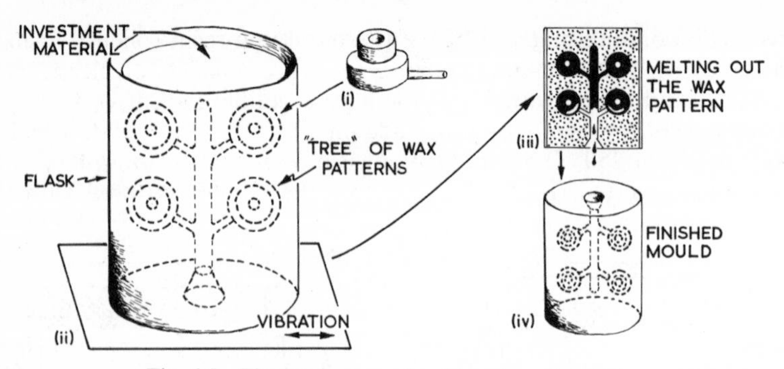

Fig. 6.5—The production of an 'investment' mould.

metal is cast into the mould under pneumatic pressure or by centrifuging.

One advantage of the process is that extremely complex shapes can be cast since the pattern is not withdrawn. Much of the intricate work in gold ornamentation carried out in the sixteenth century by Benvenuto Cellini was by this process. From the aesthetic point of view there is no disfiguring 'parting line' on the finished casting as is apparent with other casting processes. Extreme precision in dimensions is possible and the process is of engineering importance for making small components from very hard, strong materials which cannot be shaped by forging or machining. Blades for gas-turbines and jet engines can be cast by this process. The principal disadvantage of investment casting is its high operational cost.

6.2.1.5 *Die-casting*. This process involves the use of a metal mould. Since the cost of machining the mould cavity is high, die-casting is economical only when large numbers of castings—generally several thousands—are required. In *gravity* die-casting, now termed *permanent-mould* casting, molten metal is poured into the metal mould in a similar manner to ordinary sand casting. In *pressure* die-casting, often called just 'die-casting', the molten metal is injected into the mould cavity under pressure (Fig. 6.6) so that a much more accurate impression of the mould cavity is obtained.

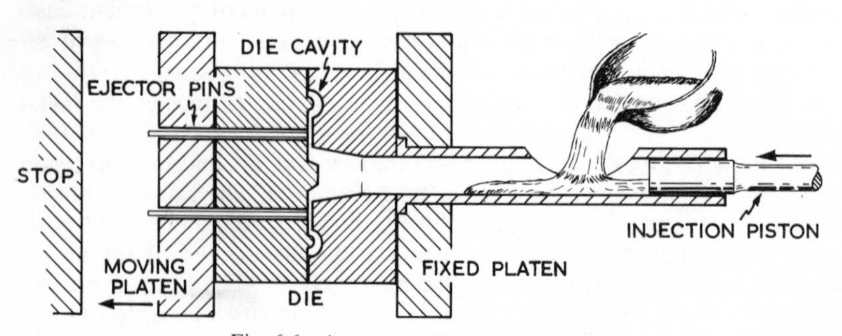

Fig. 6.6—A pressure die-casting machine.

Pressure die-casting is the more common process and 'cycling' is rapid. As soon as the casting is solid the die is parted. Generally it is so designed that the casting moves away with the moving part of the die from which it is detached by a system of ejector pins.

The main advantages of die-casting are accuracy of dimensions, good surface finish and improved mechanical properties arising from the fine grain produced by rapid solidification. Although the cost of the mould is high, labour costs in operation are low relative to sand casting, so that the process is economical for the manufacture of large numbers of zinc-base and aluminium-base components in the automobile and other industries.

6.2.2 *Casting non-metallic materials.* Not many non-metallic materials are suitable for casting because most of them lack a well-defined point where a mobile liquid transforms to a rigid solid as is the case with metals.

6.2.2.1 *Organic polymers.* The disorder→order change which occurs in metals on crystallisation results in a complete and sudden change in properties from a mobile liquid to a very rigid solid. No such well-defined change occurs in plastics materials since only very limited crystallisation may occur. Because of the presence of large chain molecules in a polymer considerable van der Waal's forces operate between them so that mobility in plastics is never very high and most of them in fact decompose on heating to temperatures well below those at which any significant mobility is attained.

Such polymer materials as are cast are in fact those of the cold-setting type (12.3.2). Here two substances of relatively small molecular size are mixed and allowed to polymerise so that long-chain molecules subsequently form and the viscosity of the material increases progressively until a rigid structure is obtained (Fig. 12.2). This change is chemical rather than physical since valency links form between small molecules to produce large ones. Casting processes involving such changes are used for 'potting' small specimens in a transparent plastic case and also for encasing electronics equipment in a protective insulating capsule. The process is also used to produce structures in which cold-setting resins are strengthened with glass- or carbon-fibre.

6.2.2.2 *Ceramics.* Plaster of Paris and sillimanite-ethyl silicate mixtures (6.2.1.4) are often cast for various purposes. Some cement mixtures are also cast, generally mixed with some form of aggregate. In all of these instances hardening is accompanied by a chemical reaction though in a few cases crystallisation may also occur. Again, the change producing rigidity is chemical rather than physical.

Concrete (14.6.3) is cast into simple shapes in building construction. The mould is generally constructed of simple wooden 'shuttering' which is removed when the concrete has set. Such casting is carried out *in situ*. Various concrete mixtures may also be cast in such forms as paving flags, pipes or garden 'furniture' (gnomes, toads or Aphrodites). This is termed *pre-cast* concrete in the sense that it is not cast *in situ*, but is transported after it has set. The more complex shapes and ornamental work are often cast into Plaster of Paris or gelatine moulds. Much of the cast concrete used in building construction is either of the reinforced or the pre-stressed variety (14.6.5/6).

Hot-forming of solids

6.3 As might be expected, these processes are applied to those materials the plasticity of which increases appreciably as the temperature rises. Many metals and all thermoplastic polymers fall into this class. However, whilst increase in temperature gives rise to an increase in malleability, this is usually accompanied—particularly in metals—by a *decrease* in ductility, because tensile strength is reduced and the material tears apart more readily. Consequently most hot-forming processes involve the use of compressive forces so that the material is rolled, forged in a die or forced through a shaped orifice.

A metal is hot-worked above its recrystallisation temperature (5.8.2) so that recrystallisation occurs simultaneously with deformation and the material does not work harden. In metals the as-cast structure is relatively weak and brittle because of the coarse crystal structure present and also the segregation of impurities at the crystal boundaries. Mechanical working redistributes these impurities more uniformly throughout the metal so that the embrittling effect is less marked. At the same time recrystallisation produces much finer grain which further leads to better mechanical properties. Impurities tend to be elongated in a fibrous manner along the main direction of working.

6.3.1 *Forging*. Blacksmiths have been shaping heated pieces of metal during the last 6000 years or so using simple tools such as a hammer and an anvil. Although some hand work is still carried out by smiths particularly in repairs to agricultural machinery and the like, most engineering forgings are produced in some form of 'closed die'.

6.3.1.1 *Drop forging* involves the use of a closed die, one half of which is fixed to a massive anvil, whilst the other half is attached to the 'tup' of a guided hammer (Fig. 6.7). A heated work piece is interposed between the die faces as they come together. In order that the die cavity shall be filled a small excess of metal must be available and this is squeezed outwards to form a *flash* which is easily trimmed away from the resultant forging.

6.3.1.2 *Hot pressing* is a development of drop forging which is generally used for producing simple shapes. Here the hammer is replaced by a hydraulically driven ram so that the work piece is shaped by steady

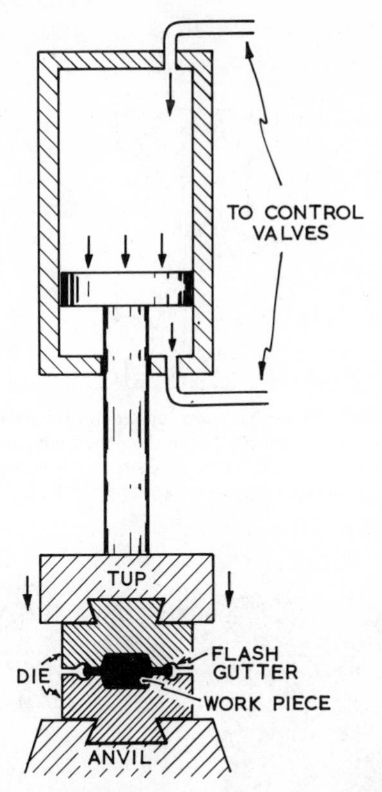

Fig. 6.7—'Closed-die' forging using a double-acting steam hammer.

pressure rather than by a succession of hammer blows. Deformation is then more uniform throughout the work piece. Large forgings can be made direct from ingots of suitable size. During forging refinement of grain and redistribution of impurities lead to a big improvement in mechanical properties. Thus a forging is always stronger and tougher than a casting of similar shape and composition. Naturally a forging is far more expensive to produce than an equivalent casting. Moreover relatively simple shapes only can be produced by forging since solid metal, even under high pressures, flows to the shape of a die much less easily than does a liquid metal.

Smaller forgings are generally produced from stock in the form of hot-rolled bar. In such stock impurities will already have been distributed directionally by the rolling process producing *fibre* as indicated in Fig. 6.8 (i). During forging this fibre will follow the contours of the die and so strength will be maintained. If for some reason of bad die design, fibre 'outcrops' to the surface producing a situation similar to that indicated for the *machined* component in Fig. 6.8 (iii), this may give rise to planes of weakness along which fracture can occur.

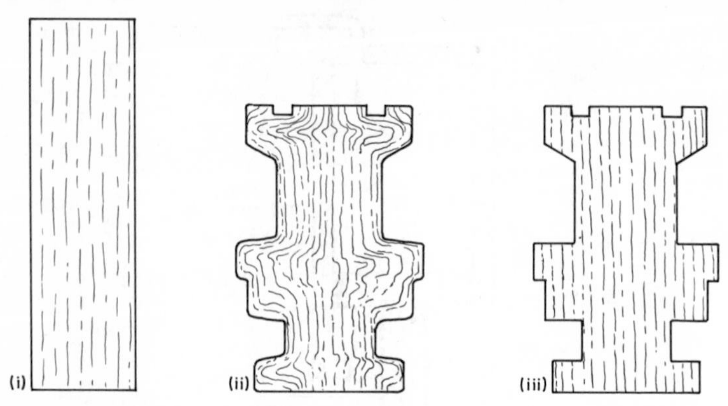

Fig. 6.8—The component in (ii) has been 'up-set' forged from stock bar (i). Flow lines follow the contours of the component. In (iii) the same component has been machined from bar stock and in this case the 'fibre' is cut. High stress concentration is likely to cause failure along these exposed fibres and so weaken the component.

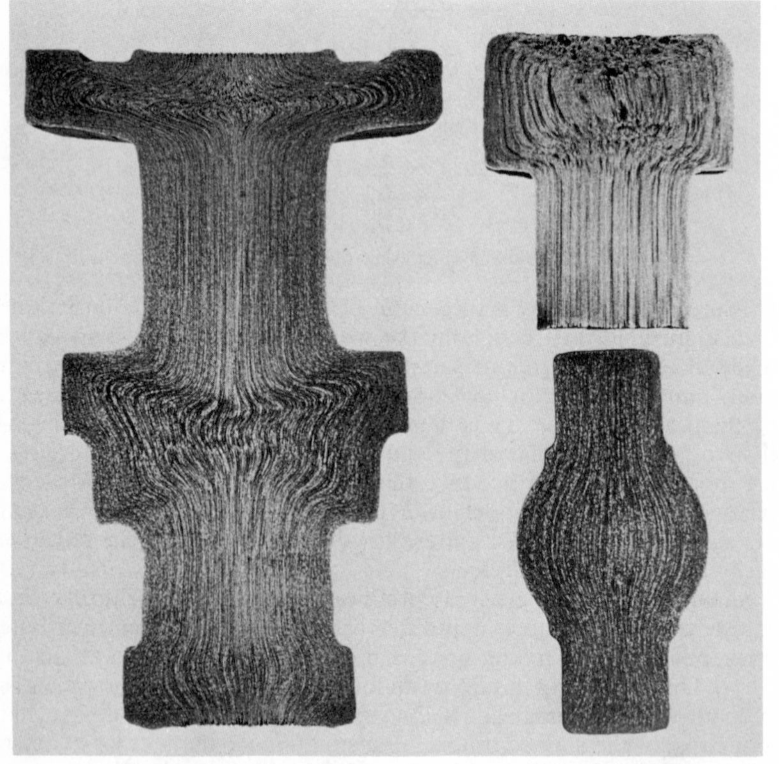

Plate 6.1—'Flow lines' in forged components. The fibrous structures were developed by deep-etching the cross-sections in boiling 50% hydrochloric acid for 30 minutes × 0·6.

Fibre distribution in a forging can be examined quite easily. A representative section is first cut—this is the most tiresome part of the job if no mechanical aid is available—and then ground flat to a reasonably smooth finish (about '280' grade emery paper). The section is then etched for about 15 minutes in hot 50% hydrochloric acid when the fibre will be clearly revealed (Plate 6.1).

6.3.2 *Hot rolling*. Metal plate, sheet, strip and rod are all produced by hot-rolling processes. In most cases a simple 'two-high' reversing mill (Fig. 6.9) is used in the initial stages to 'break down' the ingot. The resultant slab or *bloom* is then successively reduced in cross-section in a suitable train of finishing rolls.

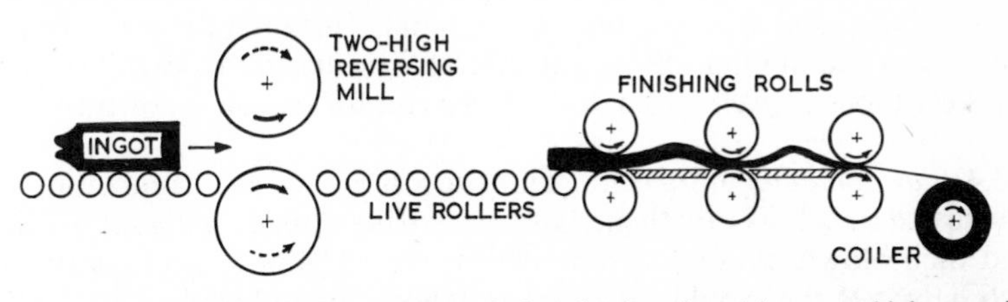

Fig. 6.9—Hot-rolling steel strip. The ingot is first 'broken down' in a two-high reversing mill (the piped top is cropped after several passes through the mill). The resultant work piece then passes, still hot, to the train of finishing rolls which roll it down to strip.

Plate, sheet and other flat material may be produced in plain rolls but rod and other sections such as rails and RSJ must be passed through rolls which are appropriately grooved. Rolled steel products are manufactured by hot-rolling since the FCC structure (present above about 900° C in mild steel) is much more malleable than is the BCC structure. However, most metals and alloys are significantly more malleable at high temperatures and are therefore hot rolled in the initial stages from the ingot. Cold working is only applied as a finishing process in the interests of surface finish, accuracy of dimensions and the control of mechanical properties.

6.3.3 *Extrusion*. The extrusion of both simple and complex sections is achieved by forcing the *solid* material through a suitably shaped die rather in the manner in which toothpaste is forced from a tube. Both ferrous and non-ferrous metals are extruded from heated billets though in many cases cold extrusion is possible. A few metals are extruded from hot or cold powders whilst thermoplastic polymers are extruded as heated powders, sometimes around a steel reinforcement strip as in the case of plastic curtain rail.

The cast billet is heated to the required temperature (350°–500° C for aluminium alloys; 1100°–1250° C for steels) and transferred to the container of the extrusion press. The ram is driven hydraulically with

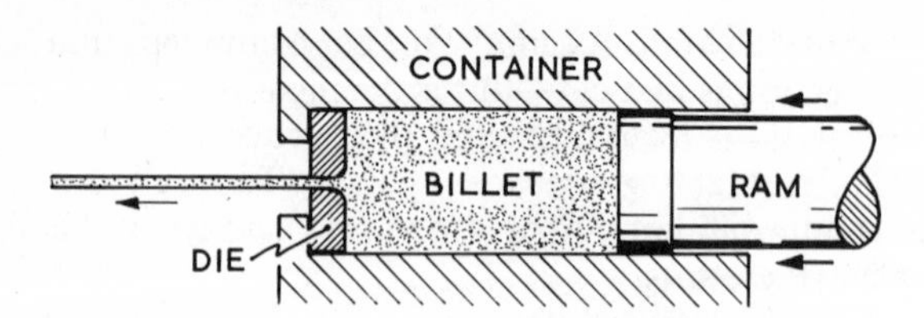

Fig. 6.10—The principles of extrusion of a solid billet.

sufficient force to extrude the metal through an alloy steel die (Fig. 6.10).

The principal advantage of the process is that, in a single operation from a cast ingot, quite complex sections with reasonably accurate dimensions can be obtained. Typical products include round rod; hexagonal brass rod (for parting off as nuts); curtain rail; stress-bearing sections in aluminium alloys (for aircraft construction); carbon and alloy-steel tubes as well as small diameter rod for drawing down to wire.

6.3.4 *Blow-moulding of plastics*. This process is used to manufacture hollow articles such as polythene bottles. Extruded tube (parison) is heated and then blown by air pressure against the walls of a steel mould. The two halves of the mould then part to release the moulding.

6.3.4.1 *Film-blowing of plastics*. Film can be blown continuously from extruded thermoplastics as indicated in Fig. 6.11. Such material has a

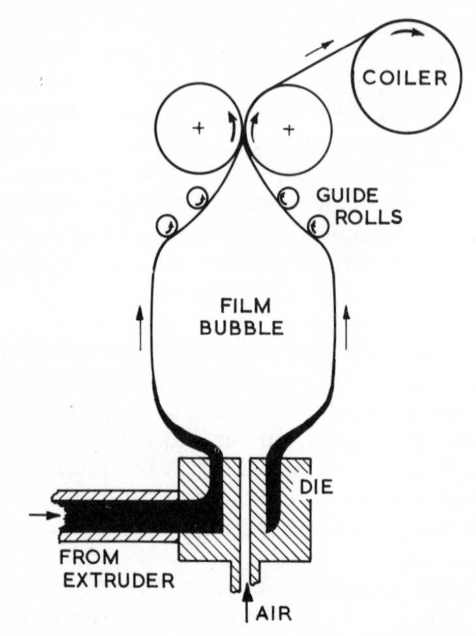

Fig. 6.11—Film-blowing of thermoplastics materials.

nominal thickness of not more than 0·25 mm and the process is widely used in the manufacture of polythene film and tube for parting off as bags and other forms of packaging.

6.3.5 *Calendering*. This process is used to produce thermoplastic sheet, particularly of PVC and polythene. The calendering machine consists of two or more heated rolls into which the plastics material is fed as a

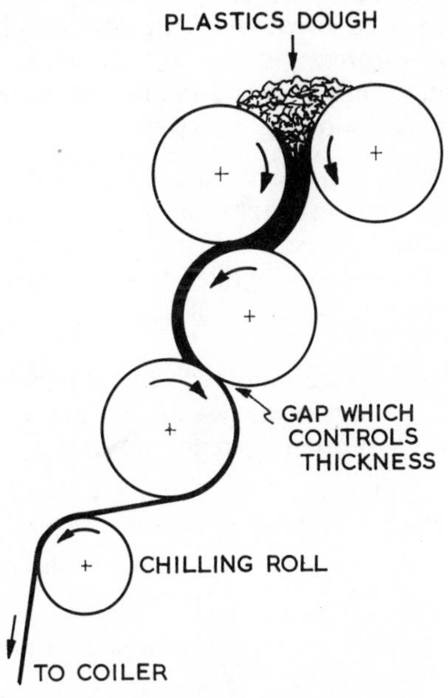

Fig. 6.12—The principles of calendering.

heated dough. The formed sheet is cooled on a chill roll (Fig. 6.12). Calendering machines are also used to coat paper, fabric and foil with a film of plastics material.

6.3.6 *Vacuum-forming of plastics*. This process is used to produce simple shapes from thermoplastic sheet stock. The heated sheet is held at its edges and then stretched by the advancing mould section (Fig. 6.13). The final shape is produced by applying a vacuum so that the work piece is moulded to the shape of the die by external atmospheric pressure.

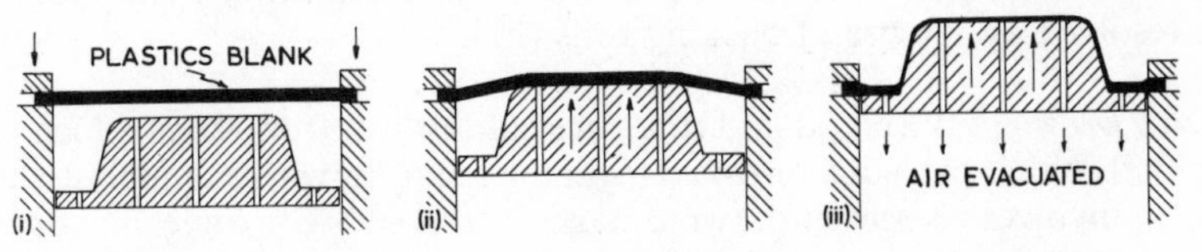

Fig. 6.13—The principles of the vacuum-forming of plastics materials.

Cold-forming of solids

6.4 Most solid materials can be shaped more easily at high temperatures. The cost of the heating process is more than offset by a big saving both in time and mechanical energy required for the subsequent shaping of the work piece. Nevertheless a large number of cold-working processes are used in the forming of ductile metals and alloys. These are generally finishing processes except in the case of drawing and deep-drawing operations where the applied forces are tensile in nature and hot-working cannot therefore be used because of the low tensile strength of most metals at high temperatures.

Cold-working is a terminal operation in many metal-shaping processes for one or more of the following reasons:

(*i*) a clean, smooth or polished finish can be obtained;
(*ii*) closer dimensional tolerances are possible;
(*iii*) the mechanical properties of many alloys can be adjusted only by the amount of cold-work applied to them. Strength, hardness and rigidity increase with the amount of cold-work.

6.4.1 *Cold-rolling*. Most sheet and strip metal is cold-rolled in the final stage for one or all of the reasons just noted. Simple two-high mills with highly polished rolls are often used to produce a smooth dense finish.

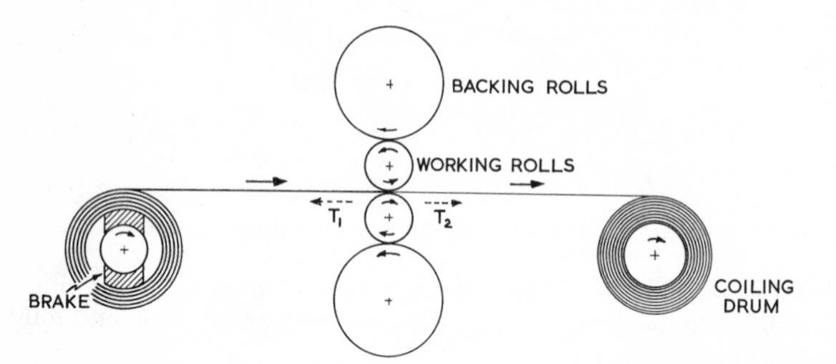

Fig. 6.14—Cold-rolling with a four-high mill (by rolling the strip whilst in tension, T_1 and T_2, roll pressures can be reduced).

When rolling wide sheets of thin foil considerable roll pressures are required. Since small diameter rolls are necessary in order to prevent slip between rolls and work piece, the rolls tend to bend under such pressures thus frustrating the attempt to produce very thin hard foil. The situation is resolved by the use of backing rolls (Fig. 6.14).

6.4.2 *Drawing*. Wire, rod and tubes are cold-drawn from stock which was initially either extruded or rolled from the ingot. In wire manufacture this often involves a series of drawing stages interposed with annealing operations to soften the material as it becomes work-hardened.

The dies used (Fig. 6.15) are of hardened tool steel or tungsten carbide. The drawing force in wire manufacture is provided by winding the wire on to a rotating drum or 'block' (Fig. 6.15 (i)) whilst for the production of rods or tubes some form of draw bench is required (Fig. 6.15 (ii)).

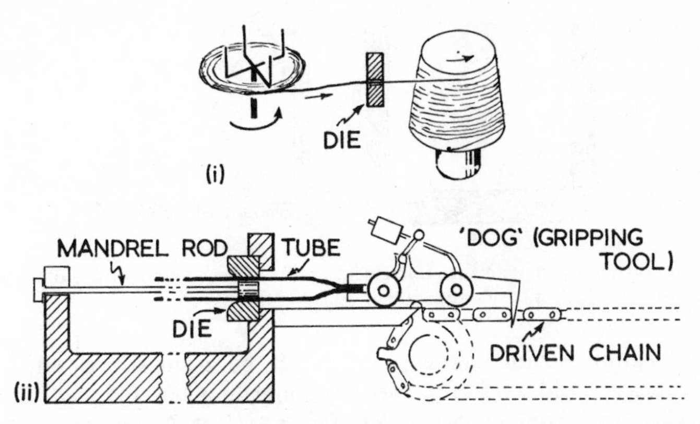

Fig. 6.15—Drawing processes: (i) wire; (ii) tube—a similar draw bench is used for rod.

In tube drawing it is generally necessary to support the internal bore of the tube. This is often done by drawing the tube on to a steel rod—or *mandrel*—which passes through the die along with the tube from which it is subsequently extracted.

6.4.3 *Deep-drawing*. This process differs from simple metal presswork in that some degree of wall-thinning always takes place. Generally some form of cup is first produced from a flat blank and this cup is then drawn so that its walls become progressively thinner. The tools used comprise some form of punch and die system.

Fig. 6.16—Stages in a deep-drawing process.

Deep-drawn components range in size from small brass cartridge cases to aluminium milk churns. Generally the more ductile materials such as 70–30 brass, cupro-nickel, copper, aluminium and mild steel are used.

The forces operating in deep-drawing are largely tensile so that it is inevitably a cold-working process. Inter-stage annealing is therefore used when the work piece becomes work-hardened. As a result of annealing, regions of weakness are often encountered in a deep-drawn component.

These are due to variations in grain size arising during the recrystallisation process which accompanies annealing. Thus, in the drawn cup (Fig. 6.17), the grain size at *A* is satisfactory since the material has received no cold-work and the grain size is more or less that of the original stock, no

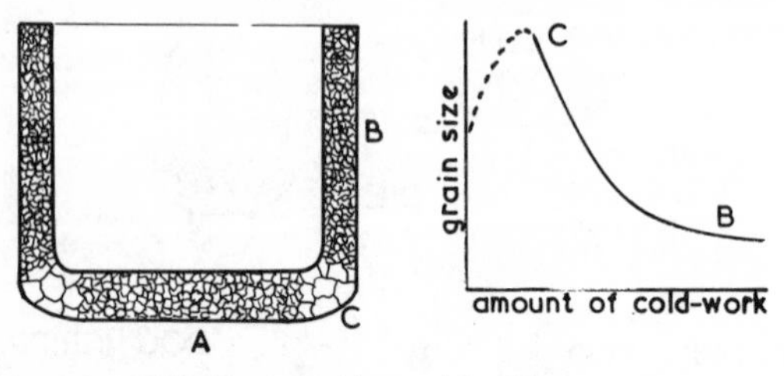

Fig. 6.17—Variations in grain-size coincident with different amounts of cold-work in a deep-drawn cup which was subsequently annealed.

recrystallisation having taken place. At *B* the grain size is satisfactory because heavy cold-work followed by annealing has induced complete recrystallisation from many nuclei. The metal at *C*, however, received only the *critical* amount of cold-work necessary just to initiate recrystallisation. Hence very few nuclei formed in this region of very little deformation so that the grain size was large and ductility correspondingly low.

6.4.4 *Cold-forging*. This includes a number of processes where a cold work piece is forged in some form of closed die.

6.4.4.1 *Coining* (Fig. 6.18) is a process of this type in which variation in thickness occurs in the component during forging and hence some degree

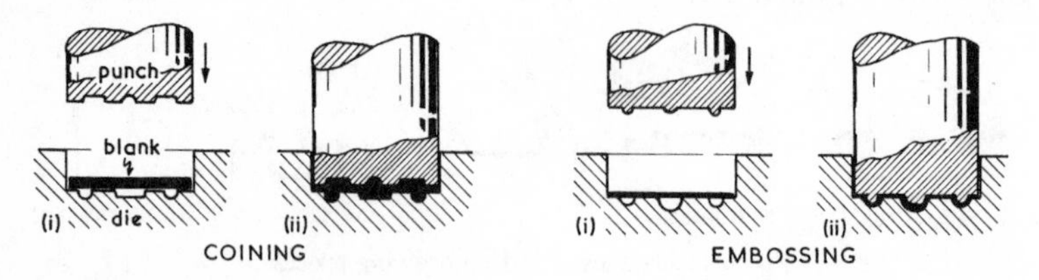

Fig. 6.18—The difference between coining and embossing. In embossing no variation in thickness of the work-piece occurs.

of metal flow relative to the die. This consumes energy as friction and wear on dies is considerable. Nevertheless a hard, wear-resistant component results.

6.4.5 *Impact extrusion*. This is a cold-working process used in a number of variations. Possibly the best known of these is the method used to produce collapsible tubes for toothpaste, shaving cream and the like (Fig. 6.19). These may be in lead, tin or aluminium, though the latter is now the most widely used.

A small unheated blank is fed into the die cavity and as the punch descends very rapidly metal is squeezed upwards into the gap between punch and die to form a cylindrical shell. The threaded nozzle may be formed during impact or it may be produced in a separate process. Other

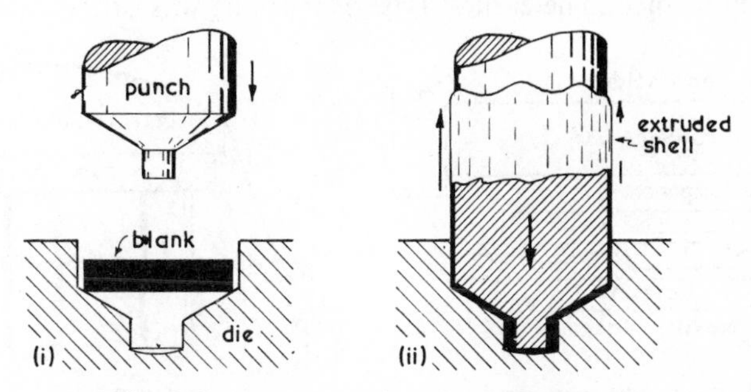

Fig. 6.19—Impact extrusion of a disposable tube.

articles made by impact extrusion include aluminium canisters and capsules for food, medical products and photographic films; and shielding cans for radio components.

6.4.6 *Cold-forming from sheet metal.* In these processes little or no variation in the thickness of the work piece occurs during forming though diverse methods are used to achieve the required shape.

6.4.6.1 *Embossing* (Fig. 6.18) differs from coining in that the thickness of the original stock is more or less maintained. Since little flow of metal occurs much less energy is lost as friction as compared with the coining process. Embossing is therefore used in the manufacture of hollow buttons, badges and the like.

6.4.6.2 *Rubber pressing* (Fig. 6.20) is used for the forming of car-body panels, hub caps and other re-entrant shapes, stainless-steel tableware, aircraft panels and pots and pans. A rubber press usually consists of a stationary upper container which holds the rubber pad and a lower platen which carries the die and work piece. This lower platen is forced upwards into the rubber by means of hydraulic cylinders, the flexible rubber forcing the work piece into the die cavity and thus acting as a universal punch.

For pressing soft aluminium alloys dies can be made relatively cheaply from paper-reinforced synthetic resin, though zinc alloys and improved plywoods are also useful die materials. Some materials such as titanium sheet and magnesium alloy sheet are hot pressed. Then mild-steel or cast-iron dies are used together with heat-resisting rubber pads (generally protected by a 'slave' mat).

6.4.6.3 *Stretch forming* (Fig. 6.21) of panels both in the aircraft industry and in coach building is carried out principally on ductile heat-treatable aluminium alloys, though some stainless steel and titanium alloys are also stretch formed. The earliest type of machine was of the rising-table

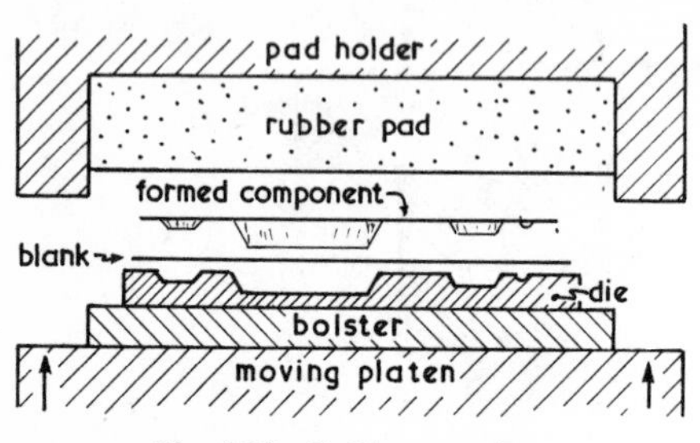

Fig. 6.20—Rubber pressing.

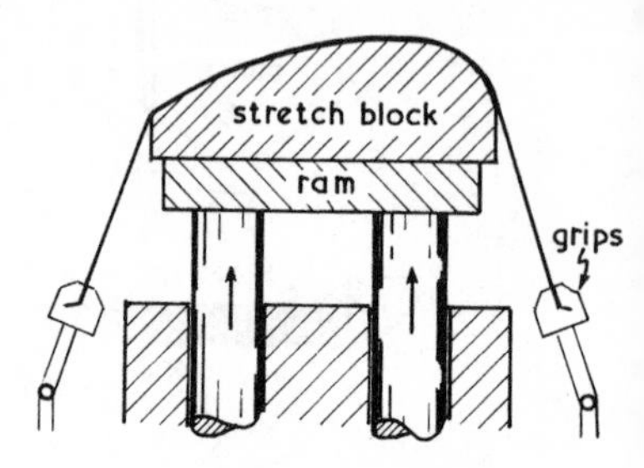

Fig. 6.21—Stretch forming.

model as shown. Here the work piece is gripped between jaws and the forming block is mounted on a rising table actuated by a hydraulic ram.

As with the other cold-forming processes stresses used must exceed the elastic limit so that the work piece is deformed within its plastic range. Stretch blocks are generally of wood or compressed resin-bonded plywoods.

6.4.6.4 *Spinning* is a process used to produce a hollow shape by applying lateral pressure to a revolving blank so that it is forced on to a suitable former which rotates with it (Fig. 6.22). In its simplest form the equipment consists of a lathe in which the blank is held between a chuck and a tail plate, and simple forming tools which are generally bars with rounded ends. For external spinning a former is usually attached to the chuck.

The equipment required is simple and hardwood formers are cheap to produce. Small numbers of components can therefore be produced much more cheaply than by an equivalent deep-drawing operation. Moreover complex components of re-entrant shape can be made easily by spinning, the former (as in Fig. 6.22) then being segmented to facilitate withdrawal. Conical shapes, difficult to make by deep drawing, can also be spun.

Typical products are ornaments in copper and brass, musical instruments (including 'antique' posthorns currently being produced in large numbers in B*rm*ng*am!), cooking utensils and stainless-steel dairy utensils. Mechanised spinning has been developed in recent years under the name of 'flow turning'.

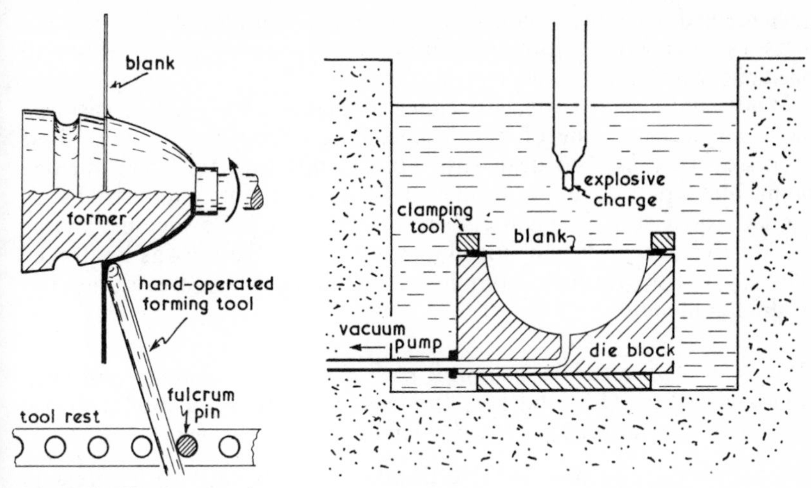

Fig. 6.22—Hand-operated spinning.

Fig. 6.23—Explosion forming.

6.4.6.5 *Explosion forming*. During recent years a number of techniques have been developed for the forming of metals in which the energy used for forming is released over a very short time interval. These are loosely called high-energy-rate forming—or HERF—processes. In these HERF processes energy is released at such a rate that local deformation occurs before energy can be dissipated by 'plastic waves' through the material. Mechanical efficiency is thus improved.

Energy can be released quickly either by a sudden change in pneumatic pressure; by the use of high-voltage electrical discharge; or by the use of high explosives. Thus by detonating no more than 0·1 kg of a suitable high explosive, energy can be dissipated at a rate equivalent to 6×10^6 kW.

A method of explosion forming in which water is used to transmit the shock waves is illustrated in Fig. 6.23. This is generally termed the 'stand-off' method, the charge being suspended above the blank-die assembly. Dies can be of clay or plaster for one-off jobs or of concrete, wood, resin or metal for longer 'runs'. The process is used extensively at present for prototype production or for very short runs. Obviously labour costs of the process are relatively high since skilled and responsible operatives must be employed to handle and prevent the loss of such dangerous explosives as RDX (of 'Dambuster' fame) to unscrupulous organisations.

Forming from powders

6.5 All finished shapes in thermosetting polymers and many of those in thermoplastic polymers are produced by compressing heated powders or granules. In thermosetting polymers cross-linking bonds formed during the setting process render the change irreversible and the component once formed, cannot subsequently be softened. With thermoplastic polymers van der Waal's forces operating between molecules give rise to plasticity which varies with temperature.

Whilst most metals are cast either as ingots or as shaped castings, increasing use is made of powder-metallurgy processes. Initially these processes were used to deal with those metals with very high melting points and which could not be melted satisfactorily on an industrial scale for lack of a suitable high-temperature refractory. Powder metallurgy was further developed to introduce such useful features as controlled porosity and the formation of alloys from metals which are not miscible in the liquid state.

6.5.1 *Powder metallurgy*. The principles of powder metallurgy are mentioned elsewhere (7.8) and here the technical details will be briefly described. A typical process may be divided into four stages:

(1) production of a suitable powder and mixing it with necessary additives;
(2) compacting the powder in a shaped die;
(3) sintering the resulting compact at a temperature high enough to effect continuity between the particles;
(4) sizing and finishing.

Powders may be produced by a number of processes. For example, tungsten powder is obtained by the chemical reduction of powdered tungsten oxide, whilst copper powder can be electro-deposited from solution. Some hard, brittle metals are powdered by grinding or pulverisation.

The powder is then compacted in a die of suitable shape (Fig. 6.24). Pressures used may vary between about 5 N/mm^2 for soft metals like tin and 1500 N/mm^2 for very hard metals like tungsten. Dies of tool steel or tungsten carbide are generally used but even so wear on dies is severe. To reduce this wear some lubricant such as graphite, oil, zinc stearate or paraffin wax (in a solvent) is usually mixed with the powder prior to compacting it.

The compacts are then sintered by heating them to a temperature which is generally above that of recrystallisation. Diffusion and recrystallisation across adjacent particle boundaries knit the particles firmly together so that a strongly continuous structure is produced. In some cases, as in the manufacture of oil-less bearings from mixed copper and tin powders, one of the metals melts and is drawn by capilliary action into the spaces between the particles of the other metal virtually 'soldering' them together. Sintering is carried out in some form of protective atmosphere which will limit surface oxidation to a minimum. Some refractory metals like platinum and tungsten are given treatments in the region of 1250°–2500° C in an atmosphere of hydrogen.

Sintering inevitably produces some distortion and alteration in size. A

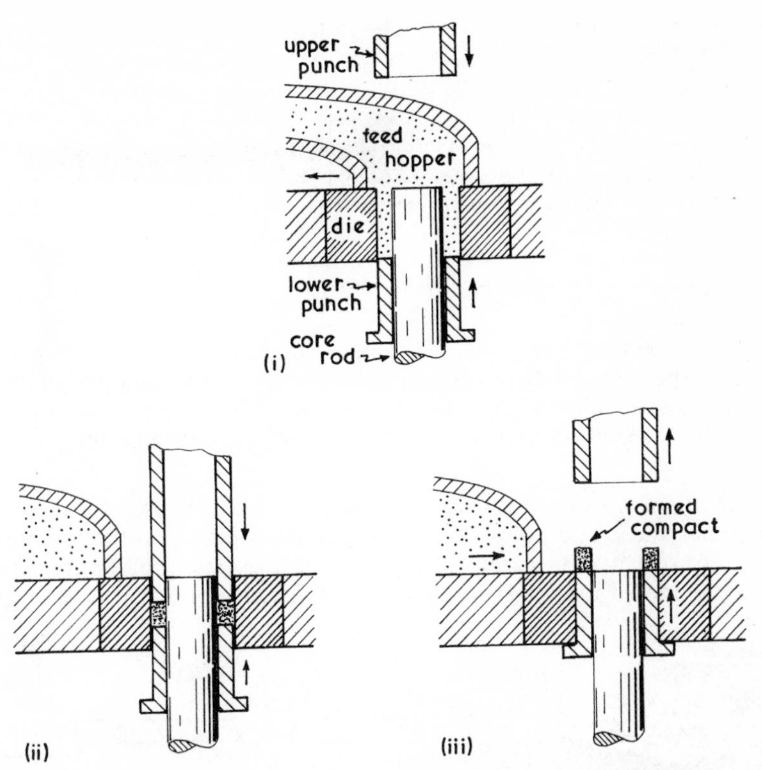

Fig. 6.24—Stages in the compacting of a bronze-bearing bush, from a mixture of copper and tin powders.

sizing operation is therefore used to achieve dimensional precision and improve surface finish of the component. Sizing is carried out on mechanically or hydraulically operated presses, using tools which are basically similar to the original compacting tools.

6.5.2 *Forming polymers from powders.* In most of these processes the powdered or granulated polymer is compressed in a heated mould or die, though in injection moulding the polymer powder is heated in an external compression chamber and transferred to a *cold* mould. This has the advantage that the moulding hardens quickly thus shortening the production cycle.

6.5.2.1 *Extrusion.* Plastics are generally extruded by means of a 'screw pump' (Fig. 6.25). The powder is carried forward by the screw mechanism into the heated zone where it becomes soft enough to be forced through the die, the aperture of which is shaped in accordance with the required cross-section.

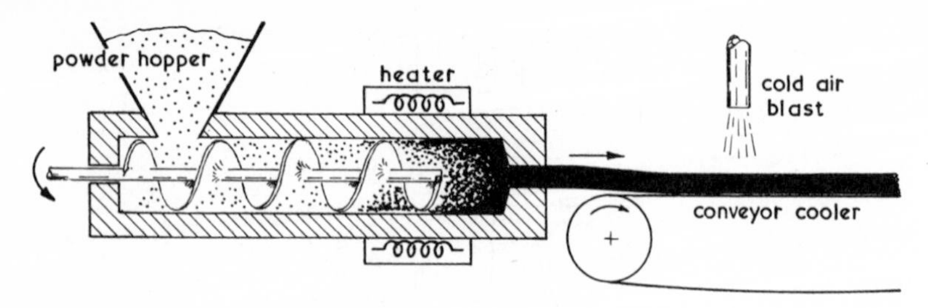

Fig. 6.25—The extrusion of plastics materials—using a 'screw pump'.

Man-made fibres are extruded using a multi-hole die, or *spinneret*. Wire can be coated with plastics by extrusion, the wire being passed into the die aperture in the manner of a mandrel. Plastics curtain rail is similarly extruded on to a steel strip which provides added strength.

6.5.2.2 *Injection moulding*. This is a rapid process widely used for moulding in polythene and polystyrene. The powder is heated in the injection nozzle so that it softens and is transferred to the mould under pressure. Since the mould itself is *cold* the plastics material soon hardens and can

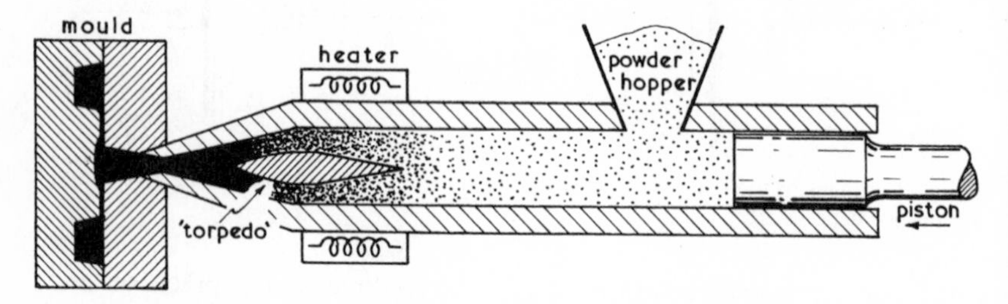

Fig. 6.26—Injection moulding of plastics materials.

be quickly ejected. A 'spray' of components is moulded at one 'shot'. Individual components are then broken away from the system of runners.

6.5.2.3 *Compression moulding* is very widely used for both thermoplastic and thermosetting polymers, but is particularly suitable for the latter. The mould must be heated in either case but must be cooled again for thermoplastic mouldings before they can be ejected. A carefully measured quantity of powder is used and provisions made to squeeze out the slight excess necessary to ensure filling of the mould cavity.

6.5.2.4 *Transfer moulding* (Fig. 6.28) is used for thermosetting polymers. Both powder chamber and die are heated and the resultant moulding allowed to remain in the mould long enough to set. This process is capable of producing rather more intricate shapes than is compression moulding.

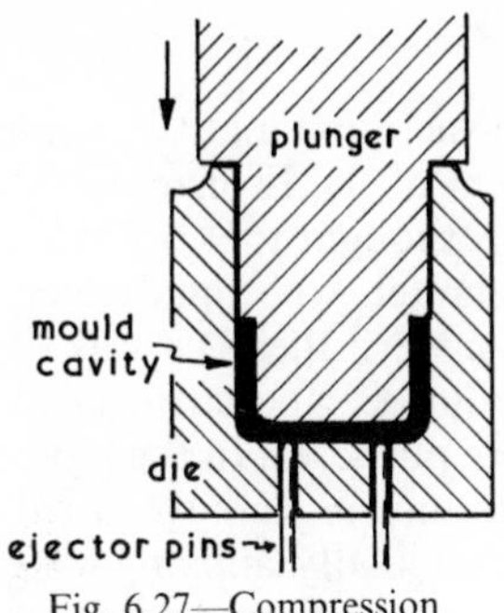

Fig. 6.27—Compression moulding.

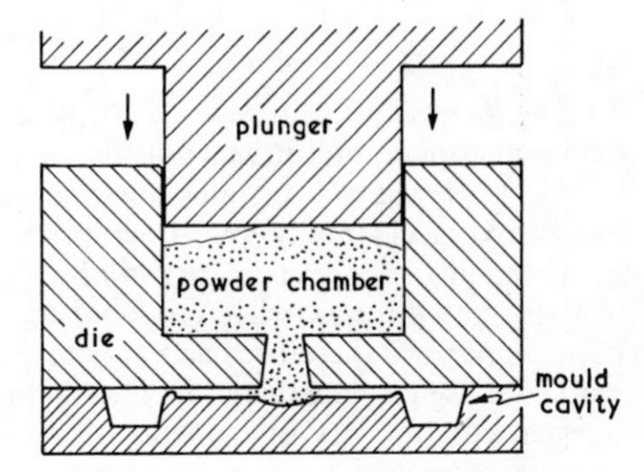

Fig. 6.28—Transfer moulding.

Chapter Seven
Alloys

7.1 The first metals to be used by Man were those found 'native', namely gold and copper, and it was not until the advent of the Bronze Age that the widescale use of a metallic alloy began. It is thought that bronze was discovered accidentally by some primitive hunter who inadvertently smelted a mixed copper–tin ore in his camp fire. Nevertheless it was soon realised that this new material could be much harder and stronger than pure copper and therefore useful for many more purposes. In the modern world we use pure metals only when properties such as high electrical or thermal conductivity or high ductility are required. By alloying we are able to extend the range of useful metallic properties and often to introduce entirely new ones.

7.1.1 An alloy can be defined as a metallic solid or liquid consisting of an intimate association of two or more elements. The elements concerned mingle together on the atomic scale and whilst the principal constituents are generally metals some non-metallic elements are also important constituents of engineering alloys. Thus carbon as the alloying addition to iron forms our most important alloy, steel.

Normally a useful alloy can only be produced if the elements concerned are soluble in each other in the molten state, that is they form a single homogeneous *solution* in the crucible. Some molten metals do not dissolve in each other but instead form two separate layers—as do oil and water when an attempt is made to mix them. Thus molten lead and molten zinc will not dissolve in each other completely (unless the temperature is in excess of 798° C). Instead a layer of molten zinc (containing some dissolved lead) will float on top of a layer of molten lead (which will contain some dissolved zinc). Such a situation cannot be expected to give rise to the formation of a useful alloy. Clearly, when cast the lighter

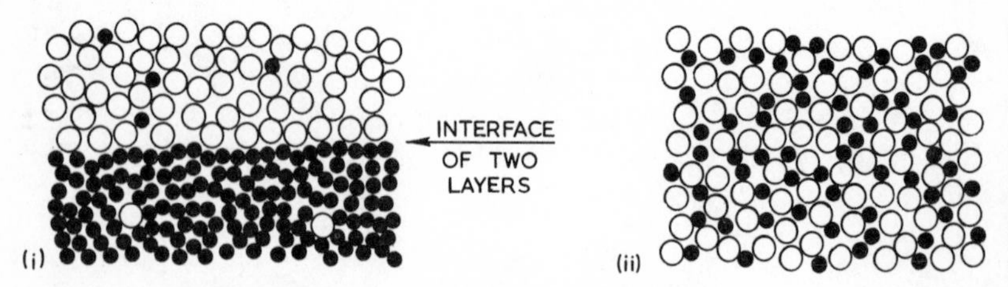

Fig. 7.1—The extent of liquid solubility. (i) Limited solubility in which two separate liquid layers will form but each layer containing a little of the other metal. (ii) Complete mutual solubility in which atoms of each substance are intimately mingled so that a single 'phase' is present in the container.

metal will tend to float to the surface before solidification is complete though there may be limited 'entanglement' of the two component metals. Two such metals may be compounded successfully using the techniques of powder metallurgy (7.8 and 6.5.1).

7.1.2 Generally, then, a prerequisite to the formation of a useful alloy is that all of the components of the alloy shall be mingled intimately together (Fig. 7.1 (ii)) in the liquid state. Whilst an alloy frequently contains more than two elements we shall be dealing in the main with a *binary* mixture, that is one containing *two* components. The term *component* refers in this case to a metallic element though it could equally imply a chemical compound as would be the case if we were dealing with the solution of some substance in water.

The metal which is present in the larger proportion is often referred to as the *parent metal* or *solvent*, whilst the metal (or non-metal) present in the smaller proportion is known as the *solute*. This nomenclature is used in a study of solutions generally, for example, solutions of substances in water or in organic solvents.

When the molten solution begins to solidify the affinities which different types of atoms have for each other govern the type of arrangement which results in the subsequent solid structure. Thus if two different metals are similar in chemical properties so that they have little chemical affinity for each other, they will, provided their atoms are of similar order of size, 'coexist peacefully' forming mixed crystals, or what is termed a *solid solution*. If, on the other hand, they are rather different in respect of chemical characteristics they may attract each other to the extent that they form a type of chemical compound—often referred to as an *intermetallic compound*. In a situation where unlike atoms attract each other less than like atoms, the two types tend to separate into different crystals which meet at a mutual grain boundary. A heterogeneous mixture of this type is termed a *phase mixture*. The *eutectic* structure found in many alloys is such a mixture.

Solid solutions

7.2 A fundamental characteristic of all solid solutions is that the complete intermingling of the atoms of both metals which prevail in the liquid solution is retained in the solid state. The *range* of compositions can vary in both liquid and solid solutions. Some materials are soluble in each other in all proportions, others only partially soluble one in the other.

7.2.1 Alcohol and water furnish an example of two liquids which show complete mutual solubility in all proportions. Thus beer is, alas, a relatively weak solution of alcohol in water whilst whisky and vodka are relatively strong solutions of alcohol and water. In all alcoholic beverages a *single phase* is present in the bottle. The parallel case in metallic

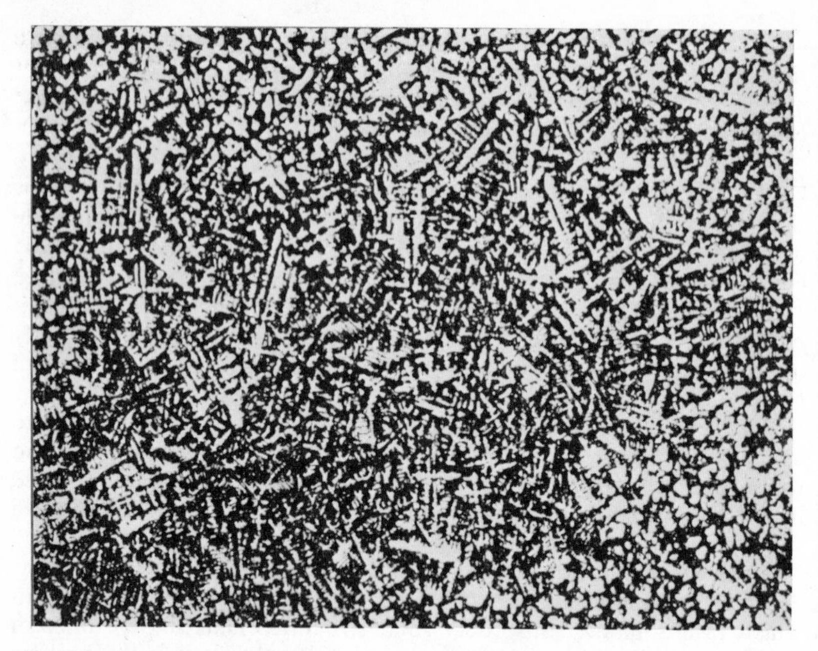

Plate 7.1—Cored dendrites of solid solution in a 60Ni–40Cu alloy which has been chill cast. The dendrite cores (light) are nickel-rich whilst the outer fringes (dark) are copper-rich. (Etched in Acid ferric chloride solution.) × 80.

solid solutions is illustrated by the metals copper and nickel. They are said to be *completely miscible* as solids since they form single-phase alloys in all proportions. That is, whatever the composition of a copper–nickel alloy, its microstructure will consist of crystals of one type only, containing copper and nickel atoms intimately mingled. Only a limited number of pairs of metals are completely mutually soluble in this way. They include, in addition to copper and nickel, gold–silver; gold–platinum and antimony–bismuth.

7.2.2 Some liquids are partially soluble in each other. Thus water will dissolve a small amount of acetone giving a single liquid solution rich in water, whilst acetone will similarly dissolve a small amount of water giving a single liquid solution rich in acetone; but if approximately equal volumes of water and acetone are shaken together, two separate layers will remain. The upper layer will be acetone *saturated* with water and the lower layer will be water *saturated* with acetone. The compositions of these two layers will be different, the upper being acetone-rich and the lower water-rich. A great many pairs of metals show limited solid solubility in parallel manner. They are said to be *partially miscible*.

7.2.3 A very few pairs of metals, mutually soluble as liquids, become completely insoluble as solids. Even then it is possible that some very slight solid solubility exists but is so small as to be difficult to measure.* Thus cadmium and bismuth which form a single homogeneous liquid solution in all proportions, separate out, to all intents and purposes, completely on solidification so that the microstructure contains two phases—crystals of pure cadmium and crystals of pure bismuth.

7.2.4 Solid solutions are either *substitutional* or *interstitial* (Fig. 7.2). Whilst atomic size is not the only criterion governing the type of solid solution to be formed it is clear that interstitial solid solutions are generally produced only when the solute atom is small compared with the solvent atom. The solute ion is then able to fit into the *interstices*, or gaps, between the sites occupied by the solvent ions. Whilst an interstitial solid solution may be formed during the solidification of an alloy it will be apparent that such a solution can also form in a parent metal which is already solid, by 'infiltration' of the spaces between its ions. Non-metallic elements of small atomic size are able to dissolve in this way in a number of metals. Thus steel can be carburised by permitting carbon atoms to diffuse into the surface of FCC iron (9.9.1), whilst it can be nitrided by allowing nitrogen atoms to be absorbed by BCC iron (9.9.2). On the debit side, hydrogen embrittlement is suffered by a number of metals since the very small hydrogen atom can be absorbed interstitially by the lattices of many solid metals.

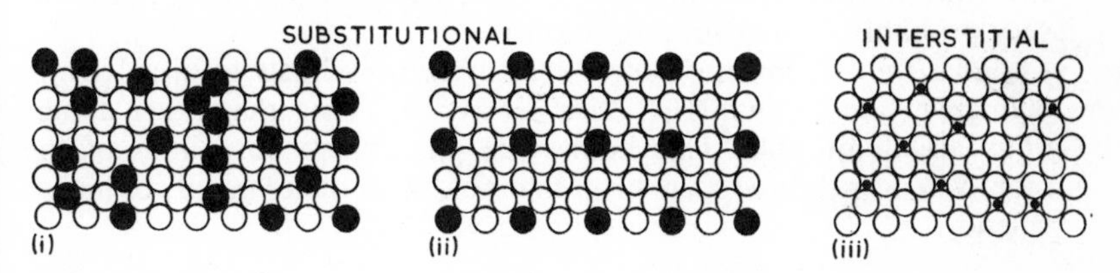

Fig. 7.2—Types of solid solution: (i) Disordered substitutional. (ii) Ordered substitutional. (iii) Interstitial.

In substitutional solid solutions atoms of the solute metal take up sites in the lattice structure of the solvent metal. This substitution can be either *ordered* or *disordered*. In the former, atoms of the solute metal take up certain fixed positions (Fig. 7.2 (ii)). Such solid solutions tend to be hard and brittle, whilst those of the disordered type are tough and ductile and therefore more useful as constituent phases of engineering alloys. Fortunately the greater proportion of substitutional solid solutions are of the disordered type.

* In a similar manner we often loosely refer to substances as being insoluble in water, whereas in fact they are very sparingly soluble. Most substances have some degree of solubility in water.

7.2.5 The formation of solid solutions in metallic alloys is to be expected since atoms in metals are held together principally by the mutual attraction between their positive ions and the common 'electron cloud' (3.5) permeating them. Such a bond is relatively indifferent both to the exact proportions of the different types of component ions and also their precise distribution in the crystalline pattern. Variable solid solutions of wide ranges of composition are therefore common.

7.3 We are indebted to Prof. W. Hume-Rothery of Oxford more than to any other metallurgist for our fundamental knowledge of the structure of alloys. He concluded that the following factors have the greatest influence on the formation of substitutional solid solutions:

(1) *Atomic size*. The greater the difference in size between the atoms of the two metals involved the smaller will be the range over which they are soluble. If atom diameters differ by more than 15% of that of the solvent metal then solid solubility is generally extremely small. Any difference in atomic size will of course produce some strain in the resultant crystal structure.

(2) *Electrochemical properties*. The ions of all metals are electropositive but some are more electropositive than others. Generally, the greater the difference in electropositivity the greater will be the chemical affinity of one ion for another so that they will tend to form a *compound* rather than a solid solution. If on the other hand their electropositivities are similar then, other things being equal, they will very probably form a solid solution.

(3) *Valency*. A metal of lower valency is more likely to dissolve one of higher valency than *vice versa* always assuming that other conditions are favourable. This holds true particularly for alloys of the monovalent metals copper, silver and gold with many metals of higher valency.

7.3.1 From the above it can be concluded that two metals are most likely to form substitutional solid solutions over a wide range of compositions if their atomic sizes are about equal and if their electrochemical properties are similar. These conditions are fulfilled when two metals are very close together in the periodic classification of the elements—usually side by side in the same period (nickel and copper) or one above the other in the same group (silver and gold). Also if singly they crystallise in the same pattern this will assist simple disordered substitution. Thus copper–nickel and silver–gold are all FCC elements.

7.3.2 The valency effect is related to the ability of the metallic bond to accept changes in the electron concentration as compared with the rigid valency laws associated with covalent compounds. If, say, aluminium is dissolved in copper the electron concentration is increased since, whilst each copper atom contributes one electron to the common 'cloud' each aluminium atom will contribute three electrons. Solubility of aluminium in copper reaches a limit with about 20% atoms of aluminium—that is, about 1·4 valency electrons per atom.* This ratio holds good for the solubility of many other metals (of different valencies) in copper.

7.3.3 The atomic size factor plays a significant part in the formation of many alloys of industrial importance. Thus both copper and iron are roughly in the middle of the metals as far as size factor is concerned. Consequently they will dissolve many metals in fairly large quantities, so that copper is the basis of brasses, bronzes and cupro-nickels; whilst iron will dissolve large amounts of nickel, chromium, manganese, tungsten, vanadium and cobalt giving a wide range of alloy steels.

Primary substitutional solid solutions based on FCC metals are very suitable for use as wrought alloys since they can be shaped by plastic deformation because of their high ductilities. The process also increases strength and hardness. When solubility decreases with fall in temperature to a very low value this may form the basis for *precipitation hardening* treatment (7.5.2).

7.3.4 Whilst a pure metal solidifies at a single fixed temperature a solid solution solidifies over a range of temperatures (8.6.3.1). As crystallisation commences there is a tendency for the metal with the higher melting point to solidify more quickly than that of lower melting point. Consequently the core of the resultant crystal will contain rather more atoms of the metal of higher melting point whilst the outer fringes and boundary regions of the crystal will contain a larger proportion of atoms of the metal of lower melting point. This change in composition from core to outer fringes reveals the dendritic structure of such an alloy when a prepared section is examined under the microscope. The difference in composition between the core of the crystal and its boundary regions is referred to as *coring* (Fig. 7.3), and is most pronounced in solid solutions which have solidified rapidly and cooled quickly to room temperature. Slow cooling on the other hand permits diffusion to occur so that coring is totally or partially lost.

7.3.5 The presence of solute atoms in a lattice structure imposes strain. This strain will be at a minimum if the solute atoms are evenly dispersed throughout the crystal. Hence there is a tendency for diffusion to occur until the concentration gradient of solute atoms within the crystal is zero. If rapid cooling of the alloy has resulted in extensive coring then prolonged annealing at a temperature just below the freezing range will generally supply sufficient thermal activation to permit diffusion and the dispersion of coring.

* In every 100 atoms, 80 valency electrons will be provided by 80 atoms of copper and 60 valency electrons by 20 atoms of aluminium.

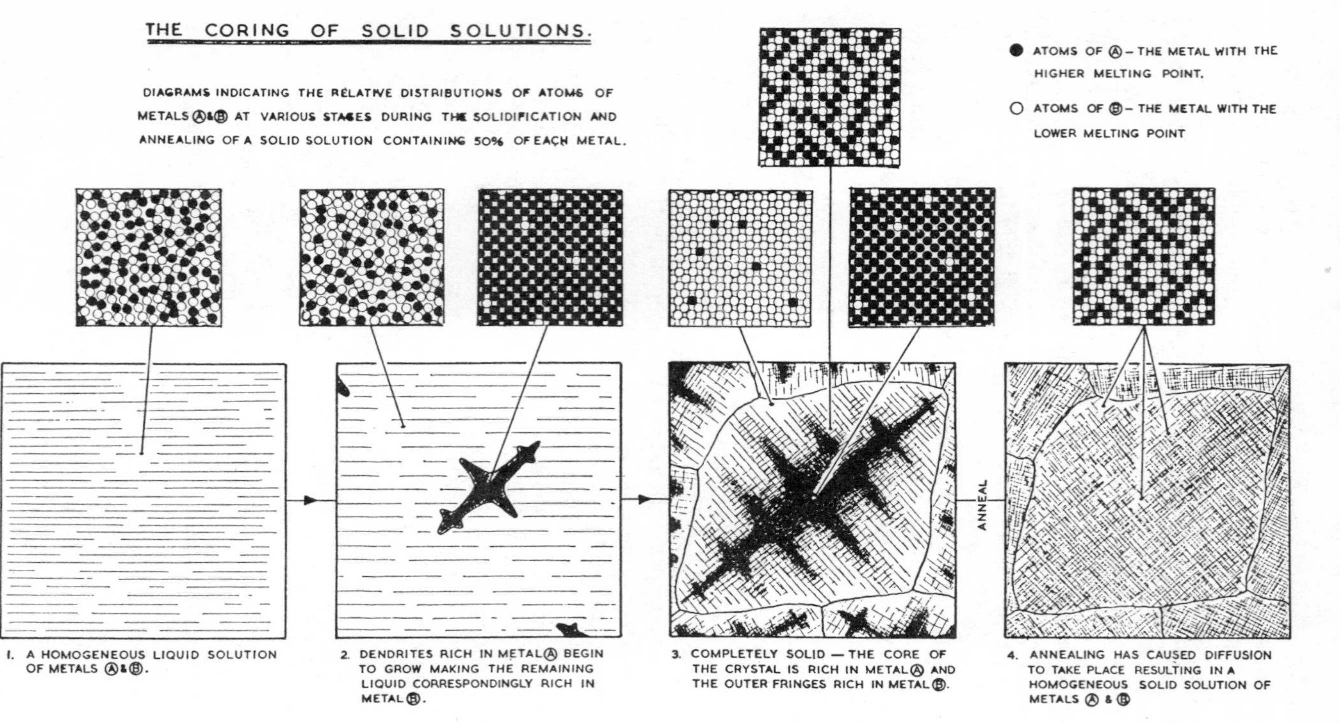

Fig. 7.3—The solidification and coring of a substitutional solid solution.

Diffusion in solid solutions

7.4 It is a feasible assumption that diffusion of interstitial solute atoms will take place with ease since the relatively small atoms involved can move easily through the structure of the parent metal. Carburising and nitriding have already been mentioned in this context. We can only attempt to explain diffusion in substitutional solid solutions, however, if we postulate the presence of *vacant sites* (or *vacancies*) (5.3) in the structure. We have already seen that the existence of dislocations explains some of the mechanical properties of metals and this is a further instance where an otherwise unexplained physical property relies on the speculation that metallic crystals contain many imperfections (5.3).

7.4.1 Since the presence of a solute atom imposes strain it is reasonable to suppose that a solute atom and a vacancy would migrate to each other and so reduce strain to a minimum. The associated solute atom and vacant site could then migrate through the structure, each movement of one site spacing being achieved by a series of 'moves' as indicated in Fig. 7.4. The rate at which diffusion takes place depends presumably upon the

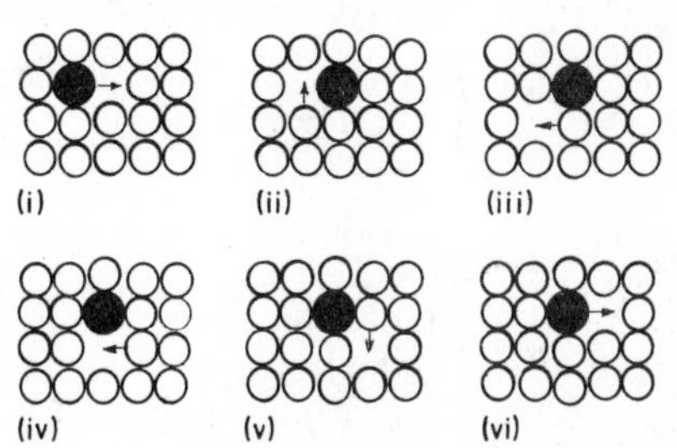

Fig. 7.4—The diffusion of a substitutional solute atom associated with an adjacent vacant site.

availability of vacant sites. It also depends upon the concentration gradient of the solute atoms within the structure of the parent metal and upon the temperature at which diffusion is taking place.

In 1858 Adolph Fick enunciated laws governing the diffusion of substances generally on a quantitative basis. Here we are concerned with the application of these laws to metallic solid solutions.

7.4.2 *Fick's First Law* states that the amount (J) of material moving across a unit area of a plane in unit time is proportional to the concentration gradient ($\partial c/\partial x$) at the same instant but of opposite sign, i.e.

$$J = -D\left(\frac{\partial c}{\partial x}\right) \qquad (1)$$

(This assumes that the x-axis is parallel to the direction in which the concentration gradient is operating.)

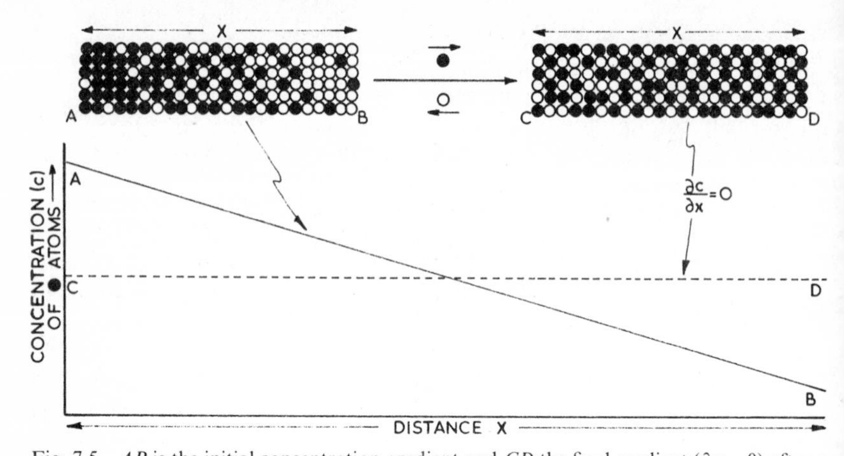

Fig. 7.5—*AB* is the initial concentration gradient and *CD* the final gradient ($\partial x = 0$) after a long time during which diffusion was in progress.

The constant *D* is the *diffusion constant* and varies with different solid solution systems. It is expressed in the units, cm²/s, i.e.

$$\frac{\text{atoms}}{\text{cm}^2 \,.\, \text{s}} = -D\left(\frac{\text{atoms/cm}^3}{\text{cm}}\right) \qquad \left[J = -D \,.\, \frac{\partial c}{\partial x}\right]$$

$$\text{or} \quad -D \equiv \text{cm}^2/\text{s}$$

The negative sign indicates that movement is in a down-gradient direction. When $\partial c/\partial x = 0$, $J = 0$ which satisfies the requirement that diffusion will cease when the concentration gradient is zero (line *CD* in Fig. 7.5).

Fick's First Law is of the same general type which also includes Ohm's Law, the latter dealing with the flow of electrons from a high concentration to a low one whilst the former deals with a flow of ions under similar conditions.

7.4.3 *Fick's Second Law* which is derived from the First Law and from the fact that matter is conserved, relates the change in concentration with time ($\partial c/\partial t$). It can be expressed:

$$\frac{\partial c}{\partial t} = \frac{\partial}{\partial x}\left(D\frac{\partial c}{\partial x}\right) \tag{2}$$

which will delight the hearts of *aficionados* of the differential calculus. The above expression deals with the general case but if *D* refers to a specific case and is constant then:

$$\frac{\partial c}{\partial t} = D\frac{\partial^2 c}{\partial x^2}$$

From this it can be seen why the final stages of diffusion during a homogenisation annealing process are very slow indeed. The rate of diffusion decreases as the concentration gradient diminishes and even after prolonged annealing a trace of coring may be visible in the microstructure of the alloy.

Strengthening mechanisms dependent on solid solubility

7.5 The most common reason for alloying is to increase the strength of a metal and generally the required properties result from the formation of a solid solution. The presence of solute atoms will impede the movement of dislocations and so increase the yield stress of the material.

7.5.1 '*Solute atmospheres.*' In any metal the presence of a dislocation causes distortion of the lattice structure near it. In a solid solution this distortion—and the strain energy associated with it— can be reduced by the presence of solute atoms. Thus, if the solute atoms are larger than

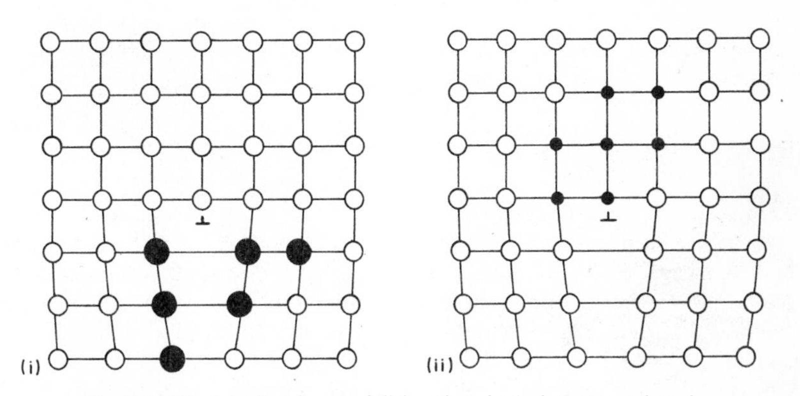

Fig. 7.6—The 'pinning down' of dislocations by 'solute atmospheres'.

those of the parent metal they will reduce strain in the lattice if they take up sites where the lattice is in tension (Fig. 7.6 (i)). If on the other hand the solute atoms are smaller than those of the parent metal they will reduce strain if they occupy sites where the lattice is in compression (Fig. 7.6 (ii)). The necessary migration of solute atoms to suitable sites to produce this situation will occur during solidification or during an annealing process which will permit diffusion. Such an array of solute atoms in the region of a dislocation is often referred to as a *solute atmosphere* or *Cottrell atmosphere* after Sir Alan Cottrell the eminent metallurgist who has done so much to establish dislocation theories.

7.5.1.1 Having thus migrated to these positions the solute atoms, whether large or small, will effectively 'pin down' the dislocation and oppose its movement since any attempt to move the dislocation away from the solute atmosphere will increase considerably the distortion and strain energy of that region of the lattice. This effect is indicated in Fig. 7.7 which shows the distortion remaining when the dislocation has moved away from the solute atmosphere in each case. In this way solute atmospheres increase the yield stress of an alloy. Annealing, following a cold-deformation process, will permit solute atoms to align themselves once more near to a dislocation and so produce a new solute atmosphere.

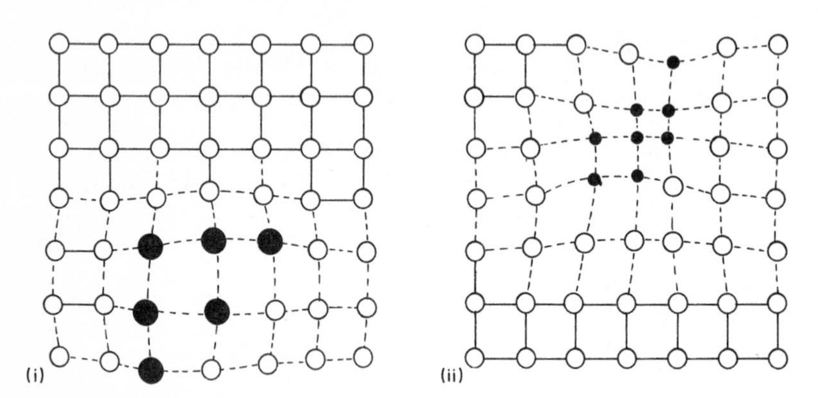

Fig. 7.7—Distortion of the lattice caused by the presence of a residual solute atmosphere.

7.5.1.2 In interstitial solid solutions the small solute atoms will occupy positions where the lattice is in tension (Fig. 7.8) since interstitial gaps will be larger in these regions. This hypothesis affords a convincing explanation of the yield-point effect in mild steel and did much to help establish the dislocation theory at a time when very limited evidence was available.

In Fig. 7.9 the yield point *A* corresponds to the stress required to move dislocations away from their associated solute atmospheres of interstitial

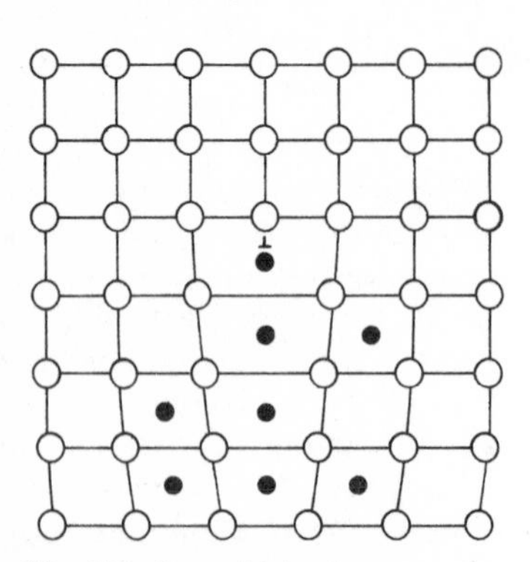

Fig. 7.8—Interstitial solute atmosphere, e.g. carbon in BCC iron.

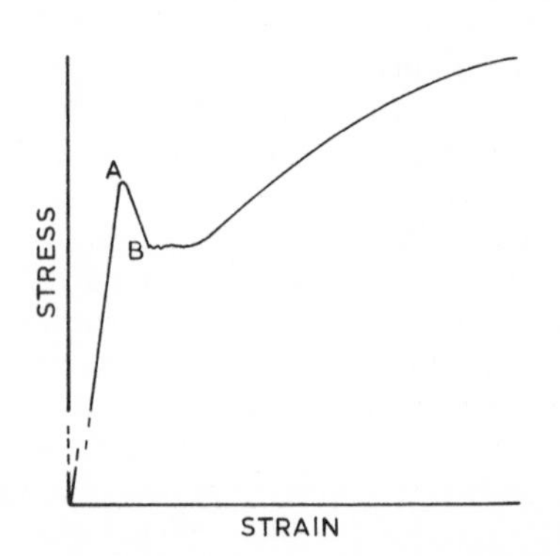

Fig. 7.9—The 'yield point effect' in mild steel.

carbon atoms by which they were pegged down. The stress B is that necessary to keep the dislocations moving once they have been separated from their solute atmospheres. (The stress required for slip to continue subsequently increases as the steel work-hardens, that is, dislocations become jammed against each other or against other obstacles.)

7.5.2 *Coherent precipitates.* If salt is dissolved in water in increasing amounts a *saturated* solution is ultimately formed and no more salt will dissolve at that temperature. As the temperature is raised the solution becomes unsaturated and so more salt will dissolve until saturation is again reached. When the temperature is allowed to fall the solution tends to become super-saturated so that solid salt is precipitated from the solution as small crystals.

7.5.2.1 A similar relationship usually prevails in metallic *solid* solutions in cases where solubility is limited. That is, the solid solubility of one metal in the other increases as the temperature rises (Fig. 7.10). If, however, the

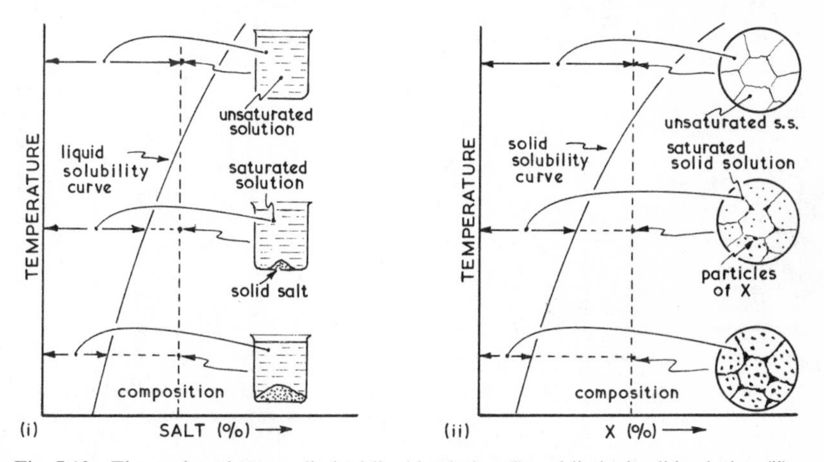

Fig. 7.10—The analogy between limited liquid solution (i) and limited solid solution (ii).

temperature is allowed to fall quickly, precipitation cannot occur as easily as in the case of liquid solutions since diffusion in solids is very slow as compared with liquids. Consequently a supersaturated solid solution may be retained at room temperature. This supersaturated solid solution will not be in a state of equilibrium and the tendency for the solute to precipitate will remain. Whether or not precipitation does occur will depend upon the degree of thermodynamical imbalance in the system. In many cases the supersaturated solid solution will be retained completely and indefinitely whilst in others the solute will separate out as isolated particles visible at high magnifications under the microscope. There is, however, an intermediate stage, where, by means of diffusion,

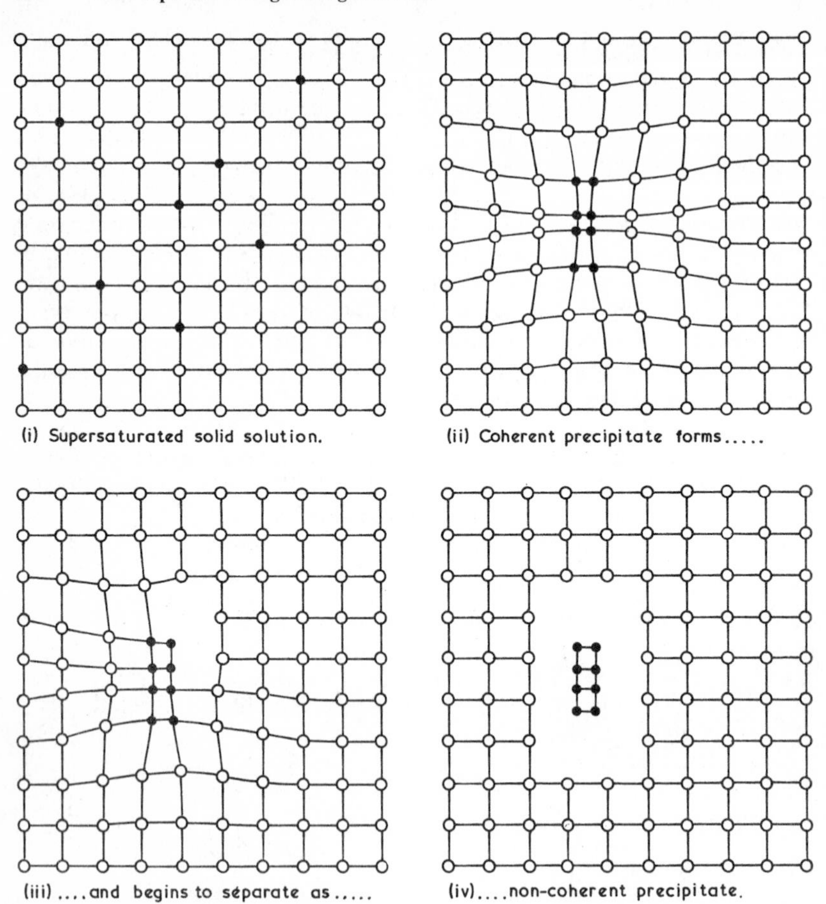

Fig. 7.11—This attempts to illustrate in simple terms precipitation from a *supersaturated* solid solution. First, solute atoms (i) migrate to form a coherent precipitate which is continuous with the lattice of the parent metal (ii). The solute atoms ultimately begin to leave the lattice of the parent metal (iii) and form crystals of non-coherent precipitate (iv) now separated from the parent metal at a grain boundary. (In practice this precipitate will rarely be of pure metal as shown in this simplified diagram. Generally the solute atoms will take with them some atoms of the parent metal to form crystals of an intermetallic compound or another solid solution.)

solute atoms begin to congregate into groups so that their concentration in some regions is greater than it was initially (Fig. 7.11). Nevertheless they still occupy positions within the lattice of the parent metal and the lattice is still *continuous* or *coherent*, albeit distorted by the presence of the high concentration of solute atoms. Such an arrangement of solute atoms is termed a *coherent precipitate* and is known to have a considerable strengthening effect on the structure presumably because it will oppose the movement of dislocations.

7.5.2.2 If precipitation continues the solute atoms ultimately leave the lattice of the parent metal to form their own particular lattice pattern. Since this lattice does not 'match' that of the parent metal the structures are no longer coherent so that a grain boundary forms between the precipitated particle and the 'mother' solid solution (Fig. 7.11 (iv)). Thus a *non-coherent precipitate** is produced and will be observed as separate small crystals, even though they are so small as to require an electron microscope to render them visible. When non-coherent precipitates have been formed the strength of the alloy falls considerably since large numbers of solute atoms have been lost from the initial solid solution. Dislocations will therefore be able to move more easily.

In Fig. 7.11 (iv) the non-coherent precipitate has been depicted, for the sake of simplicity, as being of a pure element. In practice it is much more likely to be either an intermetallic compound or another solid solution. That is, it will take with it some atoms of the parent metal to produce a crystal structure which is not coherent with that of the original solid solution.

Solid solutions in non-metallic materials

7.6 Solid solutions are formed in many ionic compounds. Thus mixed crystals can be grown involving those inorganic salts which have geometrically similar crystal forms. Such salts are said to be *isomorphous* and are typified by the group known as 'alums'. These are double sulphates of sodium, potassium, aluminium and ammonium.

However, not all ionic solid solutions are derived from isomorphous salts. Thus, silver bromide and silver iodide, though of very different crystal structures will nevertheless form solid solutions. Partial solid solubility is common with ionic compounds as it is with metallic solid solutions.

7.6.1 Although organic polymers are not necessarily crystalline in form an analogy can be drawn between co-polymers (12.2.1) and metallic solid solutions. In these co-polymers different 'mers' (12.1.3) are linked together by covalent bonds which correspond to the free electron bonds in metallic solid solutions. By using different mers to build up a polymer chain materials of modified mechanical properties can be produced. The extremely tough plastics material ABS—a co-polymer of acrylonitrile, butadiene and styrene mers—is such a substance. Whilst these co-polymers are *not* solid solutions the use of different mers in the polymer chain modifies the properties as does the introduction of solute atoms into a ductile metal.

Intermediate phases

7.7 These are phases which in general have different lattice structures from those of the parent metals and are formed in the intermediate

* Some writers describe these as '*in*coherent precipitates'.

ranges of composition. Several different forms of intermediate phase occur.

7.7.1 *Secondary solid solutions.* Sometimes solid solutions are formed in intermediate ranges of composition which do not include the pure metals as components. Generally they have crystal patterns which are different from those of the component metals and are termed *secondary solid solutions*. Except that the range of compositions does not include one of the pure metals, the secondary solid solution is not otherwise very different in properties from ordinary primary solid solutions. In some secondary solid solutions the distribution of atoms amongst the available sites follows a disordered arrangement with a fairly wide range of solubility. In others, the atoms are in an ordered pattern and in some respects resemble intermetallic compounds more than disordered secondary solid solutions.

7.7.2 *Size-factor compounds* are those in which composition and crystal structure adjust themselves in order to permit component atoms to pack themselves together in a coherent pattern. This state of affairs is typified by the *Laves phases* in which the general composition is based on the formula AB_2, e.g. $MgCu_2$ (cubic); $MgZn_2$ and $MgNi_2$ (both hexagonal). These compounds form mainly because of a substantial difference in size between the component atoms. This size difference is of the order of 30% and allows the component atoms to pack together in a crystal pattern of higher co-ordination number (3.2.2 and 3.6.1) than the maximum (twelve) which is possible in lattices of close-packed equal spheres. In the formula AB_2 each 'A' atom has sixteen neighbours (4A + 12B) whilst each 'B' atom has twelve neighbours (6A + 6B), giving an average co-ordination number of 13.33. These high co-ordination numbers are made possible by the presence of the free-electron bond in metallic structures.

7.7.3 *Intermetallic compounds* fall into two separate groups:

7.7.3.1 *Valency compounds.* When one metal is very strongly electropositive and the other only weakly electropositive they may combine to form a compound in which the ordinary rules of chemical valency apply. Thus, the strongly electropositive metals, towards the left-hand side of the Periodic Table, form compounds such as these with the very weakly electropositive metals adjacent to the non-metals on the right-hand side of the Table, e.g. Mg_3Sb_2, Mg_2Sn and Mg_2Pb. In these compounds the range of solubility of any excess metal is generally very small and often they have simple lattice structures like ionic crystals (3.2.2). Sometimes such a compound is so stable that it will have a higher melting point than either of its parent metals. Thus Mg_2Sn melts at 780° C whereas the pure parent metals magnesium and tin melt at 650° C and 232° C respectively.

These appear to be true chemical compounds whereas the other intermediate phases mentioned here are virtually transitional between these compounds and primary solid solutions since in all other cases atoms are held in position by the free electrons of the metallic bond.

7.7.3.2 '*Electron compounds*'—or Hume–Rothery compounds—in which the normal valency laws are not observed but in which there is often a fixed ratio between the total number of valency electrons of all the atoms involved and the total number of atoms in the 'molecular formula' of the intermetallic compound in question. There are three such ratios:

Ratio	*Type of structure*	*Representative phases*
(i) 3/2 (21/14)	BCC (β-brass type)	CuZn; Cu_5Sn; Cu_3Al; CuBe; Ag_3Al.
(ii) 21/13	Complex cubic (γ-brass type)	Cu_5Zn_8; $Cu_{31}Sn_8$; Cu_9Al_4; Ag_5Zn_8.
(iii) 7/4 (21/12)	CPH (ϵ-brass type)	$CuZn_3$; Cu_3Sn; $CuBe_3$; Ag_5Al_3.

Thus in the compound CuZn the copper atom provides one valency electron whilst the zinc atom provides two, giving a total of three valency electrons to two atoms, i.e. a ratio of 3/2. Similarly in the compound $Cu_{31}Sn_8$, each copper atom contributes one electron and each tin atom four electrons. Hence 31 electrons are contributed by copper and 32 (4 × 8) by tin making a total of 63 electrons. Altogether 39 atoms are involved in the molecular formula $Cu_{31}Sn_8$. Hence the ratio is 63/39 or 21/13.

These Hume–Rothery ratios have been useful in relating structures which apparently had little else in common.

7.7.4 In general intermetallic compounds are less useful as a basis for engineering alloys than are solid solutions. In particular intermetallic compounds are usually very hard and brittle, having negligible strength or ductility. Often the compound has none of the physical characteristics of its parent metals. Thus $Cu_{31}Sn_8$ is a pale blue phase as seen under the microscope, and a small ingot of it will crumble to powder under quite gentle pressure in the jaws of a vice. The unexpected colour is due to the different lattice characteristics as compared with those of the parent metals.

Since intermetallic compounds are of fixed compositions, and will generally dissolve only small amounts of their parent metals, they do not exhibit coring when viewed under the microscope. This fact—and, often, unusual colour—helps to identify them microscopically.

7.7.5 Very often, small amounts of these intermetallic compounds are present as essential phases in engineering alloys. For example, a bearing metal must be hard and wear-resistant whilst at the same time be tough and ductile. These are diverse properties not to be encountered in a single-phase alloy since, whilst primary solid solutions are tough and ductile they are relatively soft and do not resist wear and, on the other hand, whilst intermetallic compounds are hard and wear-resistant with a low coefficient of friction, they are weak and brittle. The metallurgist, however, formulates an alloy which will contain small isolated particles of intermetallic compound held firmly in a tough ductile matrix of solid solution. The latter wears leaving the particles of intermetallic compound standing proud so that they carry the load at a very low coefficient of friction. Thus bearing bronzes (10–15% Sn) contain particles of $Cu_{31}Sn_8$ embedded in a matrix of primary solid solution α (11.7.2). Similarly 'white' bearing metals (Babbitts) contain crystals of the intermetallic compound SbSn embedded in a matrix of antimony–tin solid solution (11.7.1).

7.7.6 *Interstitial compounds* are formed when the solubility of an interstitially-dissolved element (7.2.4) is exceeded. Then, an intermediate phase is precipitated in which the former solute atoms are again held in interstitial positions, but now in a compound often of more complex crystal structure. Thus, under normal conditions, BCC iron will hold no more than 0.02% carbon in interstitial solid solution and when this amount is exceeded the interstitial compound Fe_3C (*Cementite*) is precipitated.

These interstitial compounds are generally the carbides, nitrides, hydrides and borides of the transition metals, e.g. TaC, TiN, TiH_2, W_2C and WC. Most of them have metallic properties. They include the hardest and most refractory of substances with melting points as high as 3500° C (NbC) and 3800° C (TaC). Some of the carbides are extremely important and form the basis of high-speed cutting tools (WC, W_2C, TiC, MoC and TaC) (10.4.1). Since many of these interstitial compounds are strong at high temperatures they are used in a finely-dispersed form to strengthen many heat-resisting steels (10.5.2) and alloys (11.5.2).

Powder metallurgy

7.8 It is sometimes desirable to make an 'alloy' of two metals which are either completely or partially *in*soluble in each other in both the solid *and liquid* states. This is the oil–water concept mentioned earlier. Obviously any attempt to make an alloy by the usual procedure of melting and casting will fail since the two metals will form separate layers during the fusion process. A wide dispersion of the two metals can be produced however by mixing them thoroughly as extremely fine powders. The powder mixture is then 'compacted' in a closed die using a pressure which varies between 4·0 and 800 N/mm^2 depending upon the strength and ductility of the materials being used. After such treatment the compacted shape is strong enough to permit handling since some adhesion will have taken place between individual particles providing that their surfaces are clean and unoxidised. Metal-to-metal contact under high pressure apparently leads to some degree of 'cold-welding' between sliding adjacent surfaces.

7.8.1 The compact is then 'sintered', that is, heated to some temperature generally below the melting points of both metals (but sometimes above that of one of them as in the case of sintered-bronze bearings where the tin powder melts and infiltrates the copper particles). This treatment produces a mechanically strong material as the particles bond together across the interface where only limited adhesion was effected by compacting. In some cases this bonding is probably due to grain-growth across the cold-worked interfaces but it is more generally agreed that diffusion is the principal mechanism by which surfaces knit together.

7.8.2 Initially, powder metallurgy was used to produce ingots of refractory metals like platinum and tungsten which have such high melting points that they cannot be melted conveniently. Later the process was used as a means of 'alloying' metals, which, as mentioned above, do not dissolve in each other in the liquid state. The controlled degree of porosity possible with compacted powders is also utilised in such components as stainless-steel filters for corrosive liquids, and, of course, oil-less bronze bearings. Useful low-cost oil-less bearings can also be made from mixtures of copper and iron powders since these two metals are only partially soluble in each other in either the liquid or solid states.

As the relative costs of metallic powders continue to fall, powder metallurgy will be used more and more as a standard production method. The negligible amount of process scrap as compared with casting operations will make the method competitive for uses other than the rather specialised ones outlined above.

Chapter Eight
Phase Equilibrium

8.1 In a study of metallic alloys, or indeed other materials which contain more than one component, a most important concept is that concerning the term 'phase'. A *phase* can be defined as a *homogeneous body of matter existing in some prescribed physical form.* Thus the chemical substance water, H_2O, can exist in three different phase forms—solid, liquid and vapour; whilst pure iron can exist as four different phases (Table 8.1). (Note that there are only two different solid phases in iron since α and δ are identical. The terms α and δ are only used to differentiate between the temperature ranges over which BCC iron can exist.)

Table 8.1—Phases in pure iron.

Temperature range (°C) *over which the phase is stable at 760 mm pressure*	*Phase*
Above 3070	gas
1536 to 3070	liquid
1400 to 1536	BCC solid (δ)
910 to 1400	FCC solid (γ)
Below 910	BCC solid (α)

8.1.1 Many solid substances are *multiphase* in character. For example, ordinary concrete contains cement, sand and aggregate as identifiable phases in the structure whilst, on the microscopical scale, the structure of steel consists of two phases—a solid solution we call ferrite, and iron carbide, Fe_3C. The various types of phase likely to be encountered in metallic alloys were discussed in Chapter Seven. These phases can be either pure metals, intermetallic compounds or solid solutions for, despite the effects of coring, a solid solution is a single continuous phase which will become homogeneous if it is allowed to reach thermodynamic equilibrium by a process of diffusion. In any multiphase structure two phases have a common boundary separating them. Thus, a drop of oil placed in water shares a common phase boundary with the water which surrounds it.

The thermodynamic basis of phase equilibrium

8.2 The entire study of phase equilibrium is dependent upon the fundamental laws of thermodynamics. Whilst a comprehensive treatment of the discipline cannot be undertaken here an understanding of some of the terms involved will be useful.

Knowledge of the energy content of a phase is necessary to a study of phase relationships. Although this energy content may be expressed in a number of ways it is perhaps most easily defined in terms of the internal energy, E. This comprises the thermal energy of the body together with vibrational energy, electronic energy and other forms which may be internally stored in the structure of the body.

8.2.1 The kinetic theory of gases proclaims that the constituent particles* are in a state of constant motion and that the pressure exerted by a gas is due to the resultant changes in momentum arising from the sum of impacts which these particles make with the walls of the container. If heat energy is supplied to the system this is stored, since energy is conserved, as an increase in the kinetic energy of the particles. Nevertheless the kinetic energy of the particles is not the only energy possessed by the gas. Consider a quantity of gas enclosed in a container by means of a mass-less, friction-less piston (Fig. 8.1). If external heat energy is supplied

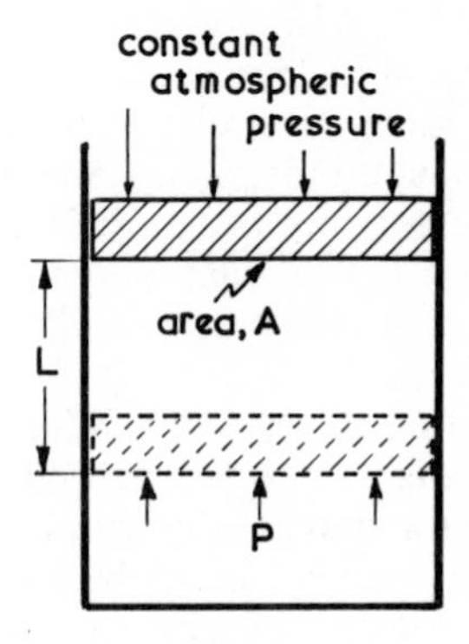

Fig. 8.1.

to the gas then it will expand and the piston will be raised so that the pressure both inside and outside the container remains constant. Since the volume of the gas increases and moves the piston against atmospheric pressure, work will be done. This work will be equal to the product of the total force acting on the end of the piston and the distance through which it moves, i.e.

$$\begin{aligned} \text{Work done by expanding gas} &= P.A.L \\ &= P.V \\ \text{where the volume, } V &= A.L \end{aligned}$$

Thus the heat supplied to the gas not only raises its temperature but also does work equivalent to PV. Hence the total energy supplied is $E + PV$ where E is the increase in internal energy.

* Either atoms or molecules depending upon the nature of the gas.

If we now consider the process operating in reverse so that energy is stored in the gas this energy will be available subsequently. Further if we imagine that this reverse process ends when both volume and internal energy approach zero then the total stored energy is given by:

$$H = E + PV$$

This total stored energy is called the *enthalpy* (H).

8.2.2 The First Law of Thermodynamics deals essentially with the conservation of either internal energy or of enthalpy depending upon how the Law is applied. When we consider an isolated quantity of a substance in a *closed* system only internal energy is involved but when the mass of the substance we are dealing with is allowed to vary, that is, enter or leave the system then enthalpy is involved.

8.2.3 The Second Law of Thermodynamics deals with the conservation of another thermodynamical quantity known as *entropy*, S. Entropy is associated with the degree of 'randomness' or disorder of the particles constituting a mass of matter. The greater the degree of disorder the greater the entropy, the more precise the order the lower the entropy. Thus when a metal melts its atoms pass from a state of order to one of relative disorder and the entropy increases. Whilst most materials expand slightly as they melt a few, such as antimony, bismuth and ice, shrink slightly. Thus entropy is not connected with volume change and the term PV is not involved.

We can conclude therefore that latent heat supplied in melting a solid is stored in the degree of disorder which is produced. This stored energy however is not available for doing work and is referred to as 'transformation energy' or 'entropy energy'. It can be expressed as the product of temperature (T) and the entropy (S). Because the liquid phase which is produced contains non-available energy this must be subtracted from the total energy—or enthalpy (H)—in order to derive the amount of energy available for doing work. Thus the free energy (G) can be derived from:

$$\begin{aligned} G &= H - T.S \\ &= E + PV - TS \end{aligned}$$

The quantity TS represents the energy which is 'bound up' in the solid. The free energy (G) is at a minimum when the system is in equilibrium. It is this thermodynamic quantity which becomes important in studying phase relationships. Since free energy incorporates both the thermal energy content of a material through H and the degree of order (or disorder) of the material through S it defines the nature of the material absolutely.

Phase equilibrium of single substances

8.3 The phenomenon of vapour condensation has already been mentioned in terms of the relationship between the kinetic energy of particles and van der Waal's forces operating between them. As a vapour is cooled the average kinetic energy of its particles decreases. Ultimately a temperature is reached where the kinetic energy of most of the particles is insufficient to resist the van der Waal's forces acting between them. Those particles with less than average kinetic energy stick together, that is, condense to form droplets of the liquid phase. The temperature at which this condensation begins is often referred to as the *dew point* of the vapour.

8.3.1 As further kinetic energy is dissipated particles condense in increasing numbers. This occurs at a constant temperature, the kinetic energy being given up in the form of heat energy (the latent heat of vaporisation). Some of the particles with kinetic energy greater than the average do not condense but are able to remain in the vapour phase, this state of affairs obtaining when these remaining particles exert a pressure equal to that of the prevailing ambient pressure. Once the majority of the particles have condensed the temperature begins to fall again as further heat is withdrawn from the system and further condensation of some of the remaining vapour particles occurs. The process is illustrated by the cooling curve for water (Fig. 8.2) in which the condensing particles will be molecules of H_2O.

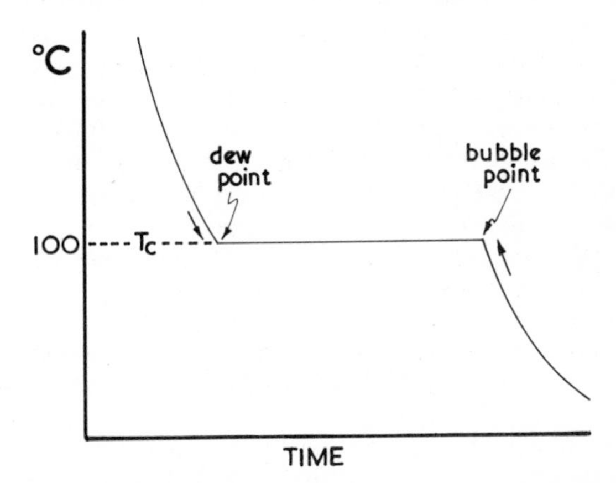

Fig. 8.2—The cooling curve for steam/water at 760 mmHg pressure.

This process is of course reversible. If the liquid is heated the first bubbles of vapour would appear at the temperature T_c. Further addition of heat—the latent heat of vaporisation—would cause the liquid to continue to vaporise isothermally* until all of the liquid had been vaporised. If heat continues to be passed into the system the vapour temperature will rise and the kinetic energy of its molecules will increase. The temperature at which bubbles begin to appear is known as the *bubble point*. For

* At the same temperature.

a *pure* substance both dew point and bubble point are at the same temperature which is referred to generally as the *boiling point*.

8.3.2 At the temperature T_c both water and water vapour can exist together and if the flow of heat to or from the system is stopped then the relative amounts of the two phases will remain constant. This is obvious since heat must flow into the system in order to provide energy to overcome the van der Waal's forces and so separate one molecule from another. Conversely, to condense more vapour the kinetic energy of molecules must be reduced, by extracting heat energy, in order that van der Waal's forces can overcome the reduced kinetic energy and so operate between molecules to bind them together.

Although the relative amounts of the two phases remain constant as long as energy neither enters nor leaves the system, a state of dynamic equilibrium exists. Those vapour molecules of higher energy give up their kinetic energy and condense if they strike the surface of the liquid. Since this energy cannot escape from the system it increases the energy level of the neighbouring molecules in the liquid phase so that they are likely to overcome the van der Waal's forces operating between them. Hence as they acquire sufficient kinetic energy they join the vapour phase. Thus a small-scale dynamic equilibrium exists between the condensing molecules from the vapour and the vaporising molecules from the liquid. Since the proportions of the phases remain constant the rates of condensation and vaporisation must remain equal. This state of affairs constitutes the basis of *phase equilibrium*. The vapour phase is in equilibrium with the liquid phase so that thermodynamically the free energies of the two phases are equal. The dynamic balance corresponds to a balance between changes in enthalpy and entropy.

In the above it has been assumed that pressure remained constant (at 760 mmHg) and that only temperature was altered. If the pressure also were altered then the equilibrium of the system would be upset. Since enthalpy would change the relative quantities of the two phases present would also change until equilibrium was once more established.

8.3.3 A state of equilibrium will exist between the liquid and vapour phases at temperatures other than the normal* boiling point. At a temperature below the boiling point the liquid exerts a vapour pressure in the space above it because the vapour molecules are in dynamic equilibrium with it. If the *overall* pressure to which the liquid phase is subjected is reduced to the vapour pressure at any given temperature, the liquid will begin to boil. For this reason it is impossible to make a decent cup of tea on Mount Everest. At high altitudes the atmospheric pressure is so low that water boils at temperatures well below 100° C and cannot of course be raised to a higher temperature.

* The boiling point at 'atmospheric pressure' (760 mmHg).

Phase diagrams

8.4 The relationship between temperature, pressure and the phases in which pure water exists is shown in the simple phase diagram, Fig. 8.3. Using this diagram we can, for any given combination of temperature and pressure, find in what phase state 'H_2O' exists, and also determine what effect any alterations of temperature and pressure will have on the stability of a given phase. For example the 'fusion curve' is an almost

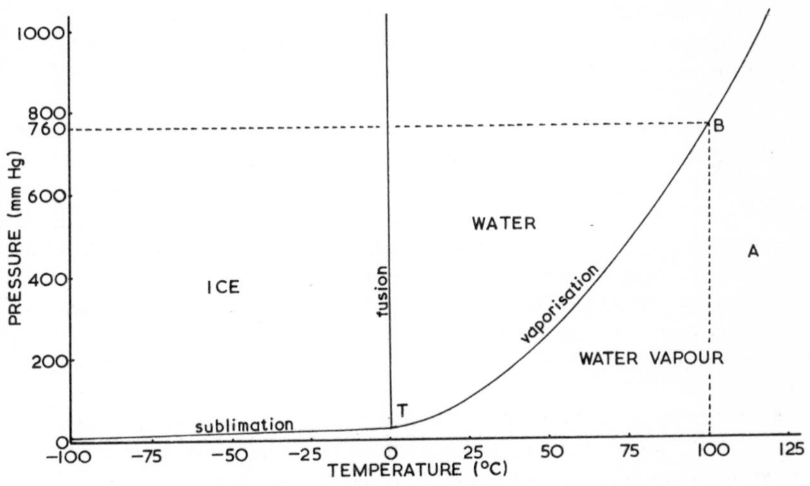

Fig. 8.3—The P/T phase diagram for pure water.

vertical line indicating that pressure variations have little effect on the freezing temperature of water. The same is true of the freezing points of metals, for which reason we can generally neglect the effects of pressure in our studies of alloy phase equilibrium.

8.4.1 The 'vaporisation curve', however, indicates that the boiling point of water is very dependent upon pressure. Water boils at 100° C at a pressure of 760 mmHg (point B). Increase in pressure above 760 mmHg will raise the boiling point whilst reduction below 760 mm will considerably lower it. The 'sublimation curve' indicates that at low pressures solid ice can vaporise—or *sublime*—without passing through the liquid phase. A number of substances, notably carbon dioxide ('dry ice') and solid iodine, sublime at ordinary atmospheric pressures.

8.4.2 Possibly the most interesting feature of this simple phase diagram is the *triple point*, *T*, at which water can exist in the solid, liquid and vapour forms simultaneously. This point *T* represents a pressure of 4·579 mmHg and a temperature of 0·009 8° C and only under these precise conditions of temperature and pressure can the three phases co-exist. Any variation in either temperature or pressure will cause at least one of the phases to disappear.

The triple point in phase equilibrium is of course not confined to water only. For example, solid, liquid and gaseous hydrogen can co-exist at −259·2° C and 54·1 mmHg.

The phase rule

8.5 First it will be necessary to define more precisely some of the terms used in connection with phase equilibrium. Thus the whole aggregate of different bodies with which we are dealing is described as a *system*. If this system is completely uniform throughout, having the same composition and properties at any point, it is said to be *homogeneous*. Either ice, water or water vapour existing separately would represent such a system. If, however, the system consists of two or more different parts, each of which is homogeneous in itself but whose properties differ from those of the other parts, then the system is said to be *heterogeneous*. Hence, at the triple point, *T* (Fig. 8.3), ice, water and water vapour existing together constitute a *heterogeneous three-phase system*.

At any point on the curve *TB* two phases can co-exist, viz. water and water vapour, so that the complete system is heterogeneous and two-phase. Point *A* represents a single homogeneous phase, water vapour.

8.5.1 A general law dealing with phase equilibrium was developed in 1876–78 by Josiah Willard Gibbs, professor of mathematical physics at Yale. Known as the *Phase Rule* it can be of considerable use in solving some of the more difficult problems of phase equilibrium.

Suppose we again consider point *A* in Fig. 8.3. This indicates that at a temperature of approximately 110° C and a pressure of 450 mmHg the chemical compound, H_2O, exists as a homogeneous single phase, water vapour. We can alter *both temperature and pressure quite independently* (within limits) without altering the nature of the system which will remain a homogeneous single phase. Thus a system of this type can be said to possess *two degrees of freedom* or to be *bivariant*. The same conditions of course apply to ice and water when they exist as single phases by themselves. The term 'degree of freedom' (or *variance*) describes any variable property such as temperature, pressure, volume or other values which are interrelated in the system.

Point B on Fig. 8.3 represents the temperature (100° C) at which water boils under normal atmospheric pressure (760 mmHg). Thus at that point two phases can co-exist, viz. water and water vapour. However, the number of degrees of freedom have now been reduced to one since we can only vary either temperature *or* pressure. That is, the new temperature (or pressure) must be represented by a point on the vaporisation curve otherwise one of the phases will disappear (either the vapour phase will condense if we move to a point on the left of the curve; or the liquid phase will vaporise if we move to a point on the right of the curve).

Consequently if, for the sake of argument, we raise or lower the temperature then the pressure must be raised or lowered *in sympathy* with it. Hence we have only one degree of freedom since in this case pressure is tied to temperature and cannot be varied independently without the disappearance of one or other of the two phases. Such a system is *univariant*.

At point T on Fig. 8.3 all three phases, ice, water and water vapour, can exist together but clearly if we alter either temperature or pressure then at least one of the phases will disappear. Consequently in this three-phase system there are no degrees of freedom and it is said to be *nonvariant*.

8.5.2 If the number of degrees of freedom are represented by f and the number of phases by p, then the phenomena described above can be related by a single rule:

$$f + p = 3 \tag{1}$$

Thus in a single-phase system such as a solid pure metal, $f = 2$, i.e. there are two degrees of freedom whilst at the melting point (where $p = 2$) then $f = 1$, and the system is univariant. If three phases—solid, liquid and vapour—are present at the same time then no degrees of freedom remain and at least one of the phases will disappear if either temperature or pressure are altered.

In an alloy containing two metals, melting occurs over a *range* of temperatures rather than at a single, fixed temperature. Thus, two phases (liquid and solid) can be present over a range of temperatures instead of at a single temperature. Although two phases are present we are able to vary temperature and pressure *independently* without a loss of equilibrium so that there are two degrees of freedom.

We can now make appropriate alterations to equation (1):

$$f + p = 4 \tag{2}$$

This equation can be written in the general form:

$$f + p = n + 2 \tag{3}$$

where n is the number of *components* in the system. As far as metallic alloy systems are concerned the term 'component' signifies a pure metal. Thus an alloy containing two metals is a two-component system. In practice we often deal with components which are not elements. For example, in the system, ice-salt-solution, the components are, effectively, salt and water both of which are chemical compounds.

8.5.3 Although Gibb's Phase Rule as represented above by (3) deals with the general case in which pressure constitutes one of the degrees of freedom, a study of metallic alloys does not normally involve a consideration of the effects of pressure. We are normally dealing only with liquid and solid phases and pressure variation then has a negligible effect on their equilibrium temperature, generally even less in fact than it has on the slope of the 'fusion curve' in the ice–water–vapour system (Fig. 8.3).

If pressure is assumed to be kept constant the number of degrees of freedom falls to one since *one* possibility of variance has been removed from the system. At *constant pressure*, therefore, the Phase Rule can be expressed:

$$f + p = n + 1 \qquad (4)$$

8.5.4 This modification of the Phase Rule will be generally useful to us when we are studying alloy systems at a constant atmospheric pressure. A single-phase molten alloy containing two metals—that is, two components—in a uniform liquid solution in each other, can now be examined using the modified Rule:

$$f + 1 = 2 + 1$$
$$\therefore \quad f = 2$$

Thus there are two degrees of freedom—one being temperature and the other composition—so that, within certain limits, both temperature and composition can be varied independently without introducing a second phase into the system.

Phase equilibrium in two-component systems

8.6 This constitutes by far the most important group of systems to be studied since it includes all binary metallic alloys. We shall begin, however, by considering a system involving two liquids which are only partially *miscible** in each other since a number of important features can be discussed in terms of this system.

8.6.1 Two liquids which are incompletely miscible in each other. Fig. 8.4 represents the phase-equilibrium conditions for the system lead–zinc in the liquid state above 500° C. Diagrams of this type are commonly known as *thermal equilibrium diagrams*, and indicate the relationships which exist between temperature, composition and phase constitution in a two-component system. The base line—or *abscissa*—of the diagram always shows the composition varying between 100% of one component at one side of the diagram and 100% of the other component at the opposite end. In most diagrams the percentages are by mass but in some cases it is more relevant to show the percentages in terms of the numbers of atoms. The *ordinate* represents temperature. Thus, point P (Fig. 8.4) represents a mixture consisting of 45% by mass of lead and 55% by mass of zinc at a temperature of 600° C.

At 500° C molten zinc will dissolve about 2% lead (point A) giving a single solution which becomes saturated at this amount of 2% lead. At

* Literally 'mixable'—or in this case 'soluble'.

the same temperature molten lead will dissolve up to about 5% zinc (point *B*). Between these two limits given by *A* and *B* two separate molten layers will co-exist in the container. The upper layer will consist of molten zinc *saturated* with lead (composition A) and the lower layer will consist of molten lead saturated with zinc (composition B).

The slope of the line *ADJC* indicates that, as the temperature rises, the solubility of lead in molten zinc increases. Similarly the slope of BEC

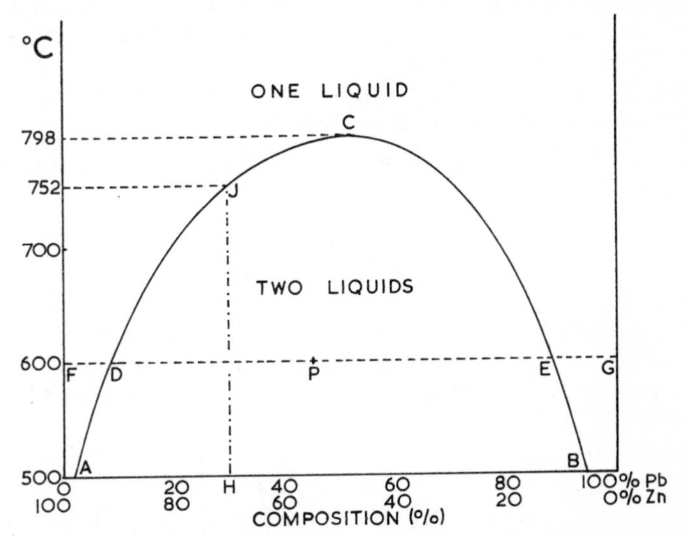

Fig. 8.4—The equilibrium diagram for lead and zinc above 500° C.

indicates that the solubility of zinc in molten lead also increases under similar conditions. The two solubility curves meet in *C* at a temperature of 798° C indicating that above that temperature the two components, lead and zinc, become completely soluble in each other so that only one layer (phase) will be present whatever the composition of the mixture as long as the temperature remains above 798° C.

Suppose we have a molten mixture consisting of 30% lead and 70% zinc (composition *H*). At, say, 500° C this will consist of a saturated solution of lead in zinc (composition *A*—2% lead/98% zinc) together with a saturated solution of zinc in lead (composition *B*—95% lead/5% zinc). Thus the mixture will consist of two phases, i.e. two separate saturated solutions. If the temperature is now raised the amount of lead which dissolves in molten zinc will increase (as will the amount of zinc which dissolves in molten lead), until at *J* (752° C) the molten zinc will have absorbed the whole 30% lead present and so one phase disappears and a single homogeneous solution remains.

Now let us consider a mixture containing 45% lead/55% zinc at 600° C, represented by point *P* (Fig. 8.4). Assuming that the molten mixture has had time to reach equilibrium at this temperature—stirring will have assisted this process—then by drawing a 'temperature horizontal' through P we are able to obtain information from the diagram. This temperature horizontal, $FDPEG$, cuts the boundaries between the single- and double-phase fields at D and E respectively. Since the temperatures of the phases in equilibrium must be the same, then D and E must represent the compositions of the two phases in equilibrium at that temperature (600° C). Thus at 600° C the liquid phase, D, contains 8·5% lead/91·5% zinc (from Fig. 8.4) whilst the other liquid phase, E, contains 88·5% lead 11·5% zinc.

From Fig. 8.4 we can also determine the precise proportions in which each phase (compositions D and E respectively) is present:

Let d, e and p be the proportions of lead in the mixtures of compositions D, E and P respectively.

Then:

$$d = \frac{FD}{FG}; \quad e = \frac{FE}{FG}; \quad \text{and} \quad p = \frac{FP}{FG}$$

If we let the mixture contain a fraction x of the phase of composition D, then it will contain a fraction $(1 - x)$ of the other phase (composition E). The proportion of lead in the total molten mixture is p and:

$$\begin{aligned}
p &= dx + e(1 - x) \\
&= dx + e - ex \\
&= e + x(d - e) \\
\therefore x &= \frac{p - e}{d - e} \\
&= \frac{-1 \,.\, (p - e)}{-1 \,.\, (d - e)} \\
&= \frac{e - p}{e - d} \\
&= \frac{\dfrac{FE}{FG} - \dfrac{FP}{FG}}{\dfrac{FE}{FG} - \dfrac{FD}{FG}} \\
&= \frac{FE - FP}{FE - FD} \\
&= \frac{PE}{DE} \text{ (from Fig. 8.4)}
\end{aligned}$$

Similarly:

$$1 - x = \frac{DP}{DE}$$

Therefore the ratio:

$$\frac{\text{Weight of phase of composition } D}{\text{Weight of phase of composition } E} = \frac{x}{1-x}$$

$$= \frac{\frac{PE}{DE}}{\frac{DP}{DE}}$$

$$= \frac{PE}{DE} \cdot \frac{DE}{DP}$$

$$\frac{\text{Weight of phase of composition } D}{\text{Weight of phase of composition } E} = \frac{PE}{DP} \qquad (1)$$

This is commonly known as the *Lever Rule* for fairly obvious reasons (Fig. 8.5). If we imagine the *tie line** DE (Fig. 8.4) to be a lever with its fulcrum at P (which is the composition of the mixture as a whole), from equation (1):

$$(\text{Weight of phase of composition } D).\ DP = (\text{Weight of phase of composition } E).\ PE \quad (2).$$

which is equivalent to 'taking moments' of the weights of the two phases about a fulcrum representing the overall composition of the mixture.

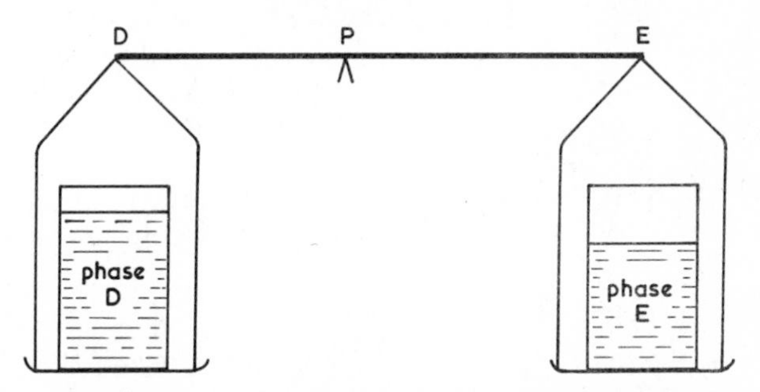

Fig. 8.5—The 'lever rule' as applied to phase equilibrium.

8.6.2 Some general rules for interpreting equilibrium diagrams. For the general interpretation of thermal equilibrium diagrams, particularly those dealing with metallic alloy systems, a number of rules were derived from Gibb's Phase Rule by the French metallurgist, A. Portevin. Some of these rules and definitions may be found helpful:

(*i*) The areas of the diagram are usually referred to as 'phase fields' and on crossing any *sloping* boundary line from one field to the next

* A 'temperature horizontal' terminating at the boundaries of a phase field.

the number of phases will always change by one. Thus in Fig. 8.6 two single-phase fields are separated by a two-phase field containing each of the phases.

(*ii*) In a binary system no phase *field* can contain more than two phases. Three phases can co-exist only at a single point such as a eutectic point (Fig 8.12 and 8.13).

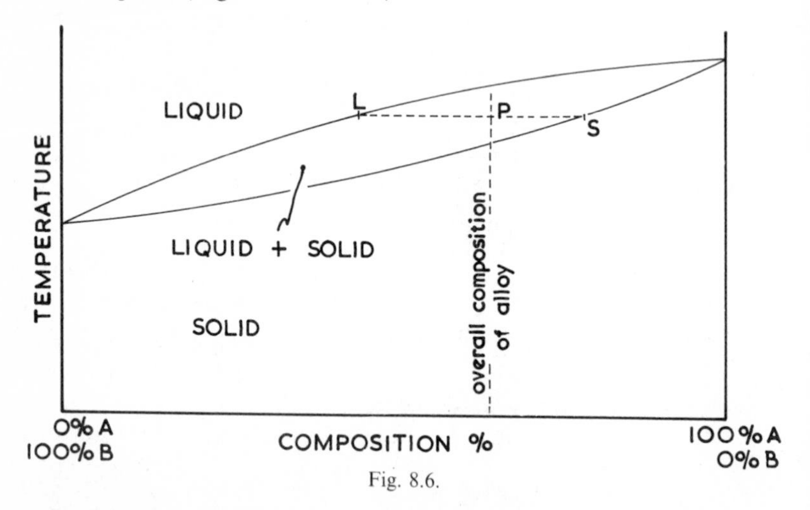

Fig. 8.6.

(*iii*) At point *P* (Fig. 8.6) two phases can exist together. If a temperature horizontal is drawn through *P* the composition of one phase (in this instance a liquid) is given by *L* and the composition of the other phase (in this case a solid) which is in equilibrium with it, by *S*. *P* represents the composition of the alloy mixture as a whole.

(*iv*) The relative amounts of the two phases present at the temperature and composition represented by *P* are given by the lengths *LP* and *PS* respectively:

$$\frac{\text{Weight of phase of composition } L}{\text{Weight of phase of composition } S} = \frac{PS}{LP}$$

This is of course the Lever Rule derived above.

(*v*) A phase which does not occupy a field by itself but appears only in a two-phase field is either a pure metal (Fig. 8.13) or an intermetallic compound of invariable composition (Fig. 8.18). (In each case the 'phase field' is in fact infinitely narrow, being a vertical line representing the appropriate fixed composition.)

(*vi*) If a vertical line representing the overall composition, *C*, of some alloy (Fig. 8.7) crosses a sloping line of the diagram it means that a change in the number of phases will occur at that point (*O*), i.e. a phase will be precipitated or absorbed. Thus in Fig. 8.7 on raising the

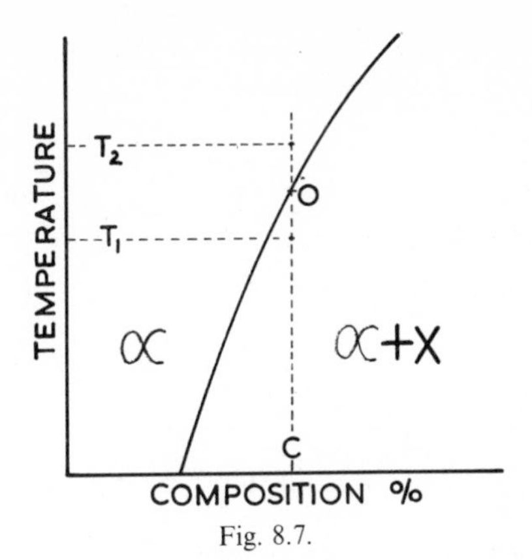

Fig. 8.7.

temperature from T_1 to T_2 the phase X will be absorbed, whilst on cooling from T_2 to T_1 the phase X would be precipitated.

8.6.3 Two components which are soluble in each other in all proportions in both the liquid and solid states. Since this case is of particular importance in a study of metallic alloys we shall deal here with a system in which both components are metals and which form a series of disordered substitutional solid solutions in all proportions (7.2.4). Many pairs of metals form solid solutions which are basically of this type, e.g. Ag/Au; Ag/Pd; Au/Pt; Bi/Sb; Co/Ni; Cu/Ni; Cu/Pt; Fe/Pt; Nb/U; Ni/Pt and Ta/Ti. (In some of these systems the equilibrium diagram may be further complicated by phase changes which occur in the solid state due to allotropic transformations in one of the metals.) Since the Cu/Ni alloys are of commercial interest we shall use this system as our example.

8.6.3.1 Whilst a pure metal freezes and melts at a constant single temperature an alloy, with certain exceptions (8.6.4.1), freezes and melts over a range of temperatures. Suppose we make up a number of copper/nickel alloys of different compositions. We shall find that they solidify over different temperature ranges and by plotting the cooling curves (Fig. 8.8)

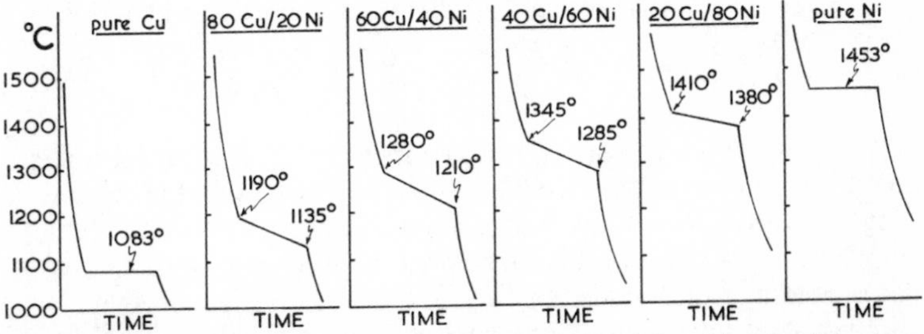

Fig. 8.8—Cooling curves for pure copper, pure nickel and a series of intermediate alloys of the two metals.

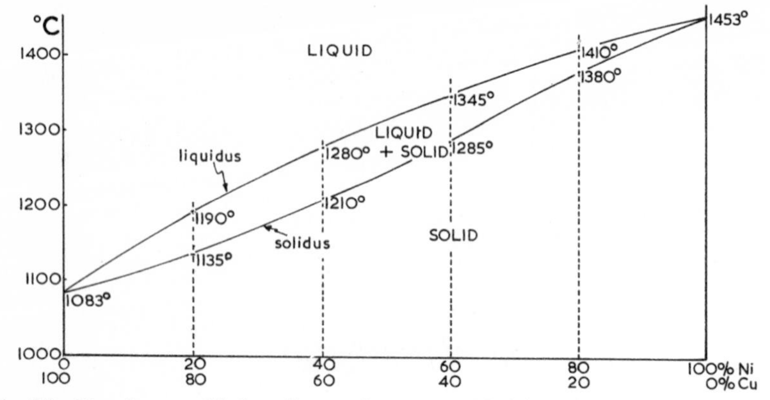

Fig. 8.9—The phase equilibrium diagram for copper and nickel using the salient points of the cooling curves from Fig. 8.8.

we can determine the temperatures at which freezing begins and ends for each alloy. This information can then be used to plot the phase equilibrium diagram for the complete system (Fig. 8.9), assuming, of course, that each alloy cooled slowly enough for it to attain something approaching equilibrium at each stage of the cooling process.

The resultant phase diagram is one of the most simple we encounter in a study of the equilibrium systems of metals. It consists of two lines only, the upper—or *liquidus*—above which any point will represent in temperature and composition a completely liquid phase, and the lower—or *solidus*—below which any point will represent in temperature and composition a completely solid phase. Between these two lines is a phase field in which both liquid and solid phases co-exist.

Applying the Phase Rule ($f + p = n + 1$) to the single-phase fields, both above the liquidus and below the solidus, we have:

$$f + 1 = 2 + 1$$
$$\text{or} \quad f = 2$$

so that the system, being bivariant, both temperature and composition can be varied independently within limits without any alteration in the structure taking place. In the two-phase field, between liquidus and solidus:

$$f + 2 = 2 + 1$$
$$\text{or} \quad f = 1$$

so that the system is univariant and temperature and composition can no longer be altered independently without a loss of equilibrium. Any change in temperature must be accompanied by an appropriate change in composition (and vice versa) or balance between the proportions of the phases present will be upset.

Suppose we have made up an alloy containing 70% nickel and 30% copper (X in Fig. 8.10). At temperatures above T this will exist as a

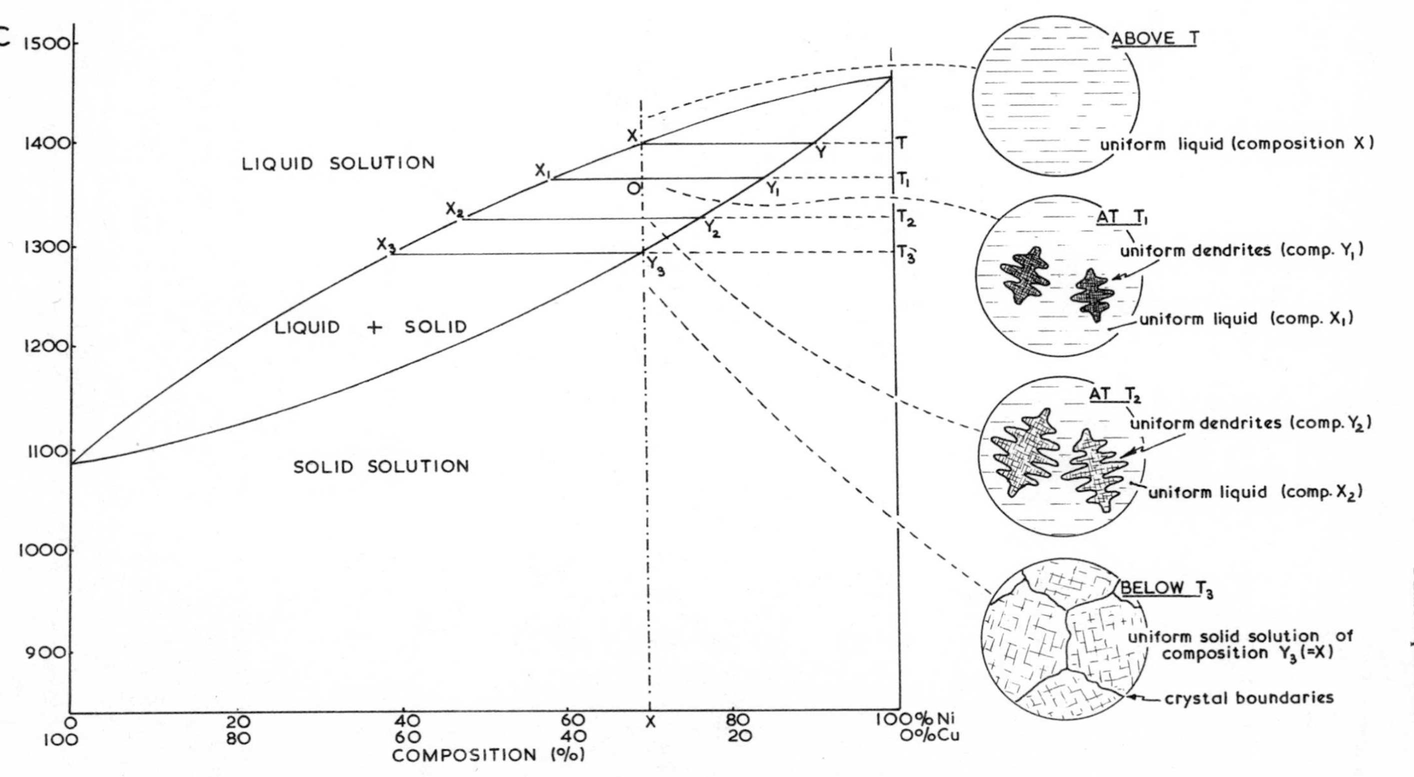

Fig. 8.10—The copper–nickel equilibrium diagram.

homogeneous liquid but as the temperature falls to T, the liquidus temperature of this alloy, solidification will begin. If we draw a temperature horizontal through X and T it cuts the solidus at Y. Y is therefore the composition of the solid phase which can exist in equilibrium with the liquid phase of composition X at the temperature T. Consequently dendrites of composition Y begin to solidify.

Since these dendrites are much richer in nickel (approximately 91% Ni as read from the diagram) than was the original liquid, the remaining liquid will have become depleted in nickel and correspondingly richer in copper as a result. Thus the composition of the remaining liquid moves to the left of the diagram—say, to X_1. No further solidification occurs until the temperature has fallen to T_1 and at this temperature we find that the solid phase which is in equilibrium with the liquid X_1 is now of composition Y_1, i.e. rather less rich in nickel than was the original dendrite. Solid of composition Y_1 therefore forms a coating over the original dendrite.

However, we will assume that the system is *cooling extremely slowly so that phase equilibrium is attained at every stage of solidification*. Clearly, (at T_1), crystals of solid of *uniform* composition Y_1 throughout constitute the equilibrium phase under these conditions and this uniformity is achieved by diffusion which proceeds according to Fick's Laws (7.4.2), nickel atoms moving outwards from the cores of the dendrites, to be replaced by copper atoms which move inwards.

Assuming that a state of equilibrium has been thus achieved at the temperature T_1 we can use the 'lever rule' to obtain complete information about the system at this temperature:

Weight of liquid (composition X_1). X_1O = Weight of solid (composition Y_1). OY_1

$$\text{or} \quad \frac{\text{Weight of liquid (composition } X_1)}{\text{Weight of solid (composition } Y_1)} = \frac{OY_1}{X_1O}$$

In the case of the alloy chosen—70% Ni/30% Cu—we have (reading from Fig. 8.10):

$$\frac{\text{Weight of liquid (59\% Ni)}}{\text{Weight of solid (85\% Ni)}} = \frac{(85-70)}{(70-59)} = \frac{15}{11}$$

We have therefore been able to determine the precise compositions of each phase and the proportions in which they co-exist in equilibrium at 1365° C (T_1).

Solidification continues in this way as the temperature falls slowly, the composition of the liquid changing (due to mixing caused by convection) along $X_1X_2X_3$ whilst the composition of the solid changes (due to diffusion) along $Y_1Y_2Y_3$. Thus at temperature T_2 the composition of the growing dendrites will have changed to Y_2 whilst the composition of the uniform liquid in equilibrium with them will be X_2. Meanwhile the proportion of liquid to solid will also have changed to OY_2/OX_2.

Finally, at temperature T_3 the last trace of liquid (of composition X_3) solidifies at the crystal boundaries and is immediately absorbed by diffusion so that the solid becomes uniform and of composition Y_3 throughout—which is, of course, the same as that of the original liquid (X). It could not be otherwise since matter has been conserved and the initial liquid phase and the final solid phase are of uniform composition.

8.6.3.2 In the above exposition we assumed that the rate of cooling was 'infinitely slow' so that complete equilibrium by means of convection (in the liquid) and diffusion (in the solid) could be achieved at each stage of the process. Under prevailing conditions of industrial casting, however, the rate of solidification outstrips the rate of diffusion and the solid structure is unable to attain uniformity and equilibrium. Thus the composition of the core of each dendrite is in the region of Y, that is it contains more nickel than the equilibrium amount, so that the outer fringes must contain correspondingly more copper than the equilibrium amount. This is the mechanism of coring and, clearly, the greater the rates of solidification and cooling the greater will be the extent of coring in the final structure since there will be less chance for diffusion to take place (Fig. 8.11).

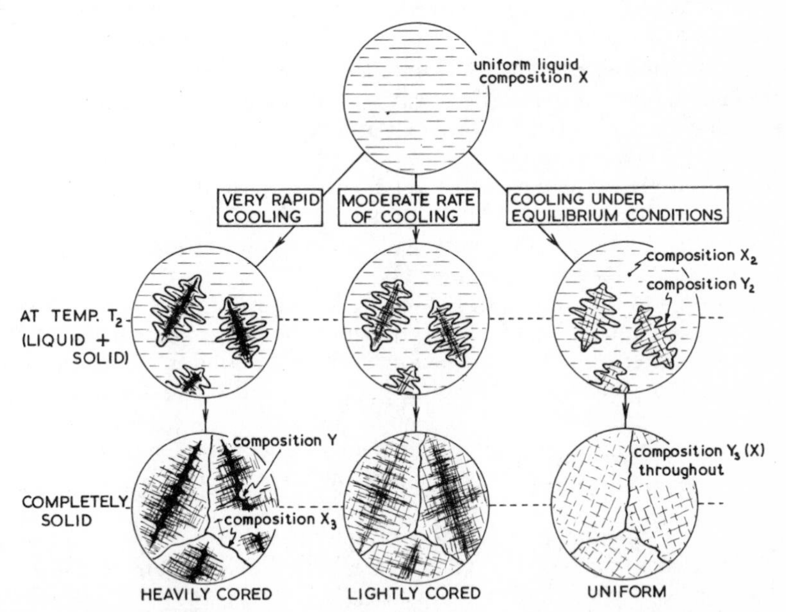

Fig. 8.11—Illustrating the effects of the rate of cooling on the extent of coring in the 70–30 nickel–copper alloy dealt with here.

If the cored structure is heated to a sufficiently high temperature diffusion will take place according to Fick's Laws and after an adequate period of time the concentration gradient will approach zero as nickel atoms diffuse outwards from the core and are replaced by copper atoms moving inwards towards the core. Diffusion ceases when a uniform composition prevails. As the degree of coring approaches zero during annealing, the *rate* of any further diffusion becomes extremely slow so that detectable coring is still present in most industrially annealed materials.

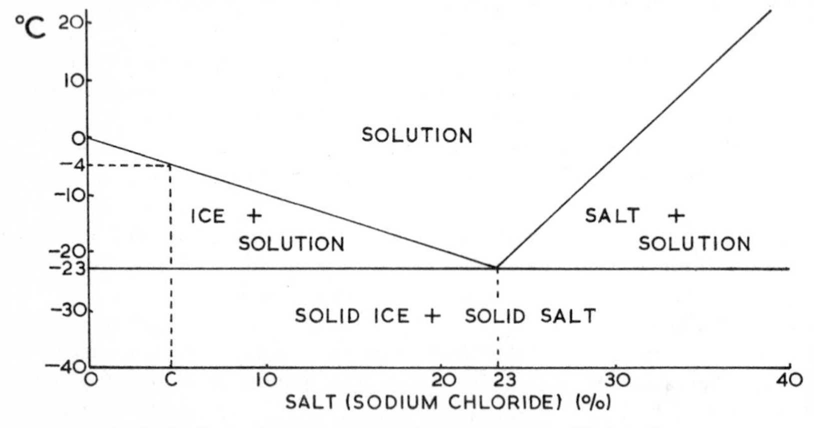

Fig. 8.12—Part of the ice–sodium chloride phase equilibrium diagram.

8.6.4 Two substances, completely miscible as liquids, become totally insoluble in each other as solids. Most readers will know that the easiest way to remove a coating of ice from the garden path is to sprinkle a small amount of 'common salt' (sodium chloride) on it. The ice soon begins to melt and a glance at the relevant part of the phase diagram (Fig. 8.12) explains why this happens. Although we may use only a small proportion of salt (say *C*) we see that both ice and solution are in equilibrium at temperatures between about $-4°$ C and $-23°$ C. Therefore 'slush' will soon form and the slipperiness of the ice be eliminated due to the formation of some liquid. In the unlikely event of the temperature being below $-23°$ C scattering salt would be unfruitful since solid ice and solid salt are then the phases in equilibrium with each other and ice would not melt. When salting our roads the Local Authority uses substances which are cheaper than sodium chloride but equally effective both in de-frosting the tarmac and in causing corrosion of our motor cars.*

8.6.4.1 The diagram is a typical example of what is commonly called a *eutectiferous series*. The term *eutectic* is derived from the Greek,

* Calcium chloride is used principally. It is a low-value by-product of soda manufacture. The eutectic mixture (with water) freezes at $-51°$ C making it effective at very low temperatures.

'eutektos' or 'capable of being melted easily', a reference to the fact that a mixture of precisely eutectic composition (23·6% sodium chloride/76·4% water in the above example) has the lowest freezing temperature range in the series. In a system of this type the components are mutually soluble as liquids but during solidification separate out either partially or completely. In this section we are dealing with cases of complete insolubility in the solid state. The phenomenon of *depression of freezing point* is involved here. Thus the addition of sodium chloride to water progressively depresses its freezing point. Conversely the addition of water to sodium chloride will progressively depress its freezing point.* The two curves meet in a minimum—or eutectic—point at −23° C and 23·6% (by mass) of sodium chloride. Whichever component is present in excess of this eutectic composition will crystallise out first as the mixture cools so that the final liquid will always be of eutectic composition. For this reason any mixture in the series—other than one of eutectic composition—will solidify over a range of temperature. A mixture of exactly eutectic composition, however, will solidify at a single temperature in the same manner as pure water.

8.6.4.2 It is doubtful whether two substances can ever be classed as being *completely* insoluble in each other. For example, we normally speak of, say, silver chloride as being 'insoluble in water' whereas in fact it has a measurable slight solubility. The same is true of solids, and in particular, metals. Total insolubility in the solid state is extremely rare and probably non-existent when those two metals were initially completely soluble in each other as liquids. Thus, although we shall discuss the bismuth/cadmium system here as an example of *complete insolubility* in the solid state we now believe that a small degree of solid solubility exists between the metals. Nevertheless we shall ignore this slight solid solubility for the sake of our argument.

Bismuth, one of the 'less-metallic metals' in a chemical sense, forms a fairly complex type of crystal structure. Whilst this is basically rhombic the metallic bond is partly replaced by covalent bonding with the result that atoms arrange themselves in layers with van der Waal's forces operating between these layers. It is not surprising therefore that bismuth is extremely brittle and that it is reluctant to form mixed crystals with cadmium which crystallises in a normal *CPH* pattern. Again the phase diagram is of simple form consisting of two lines only, the liquidus *AEB* and the solidus *CED*, or, more properly *ACEDB*.

8.6.4.3 We will consider the solidification of a bismuth/cadmium alloy of composition *X* (Fig. 8.13) that is, containing 25% bismuth/75% cadmium. At any point above the liquidus line *AEB* this will consist of a completely homogeneous liquid and solidification will commence when

* The melting point of sodium chloride (801° C) is of course above the boiling point of water but the relationship holds true for water-rich mixtures below 100° C.

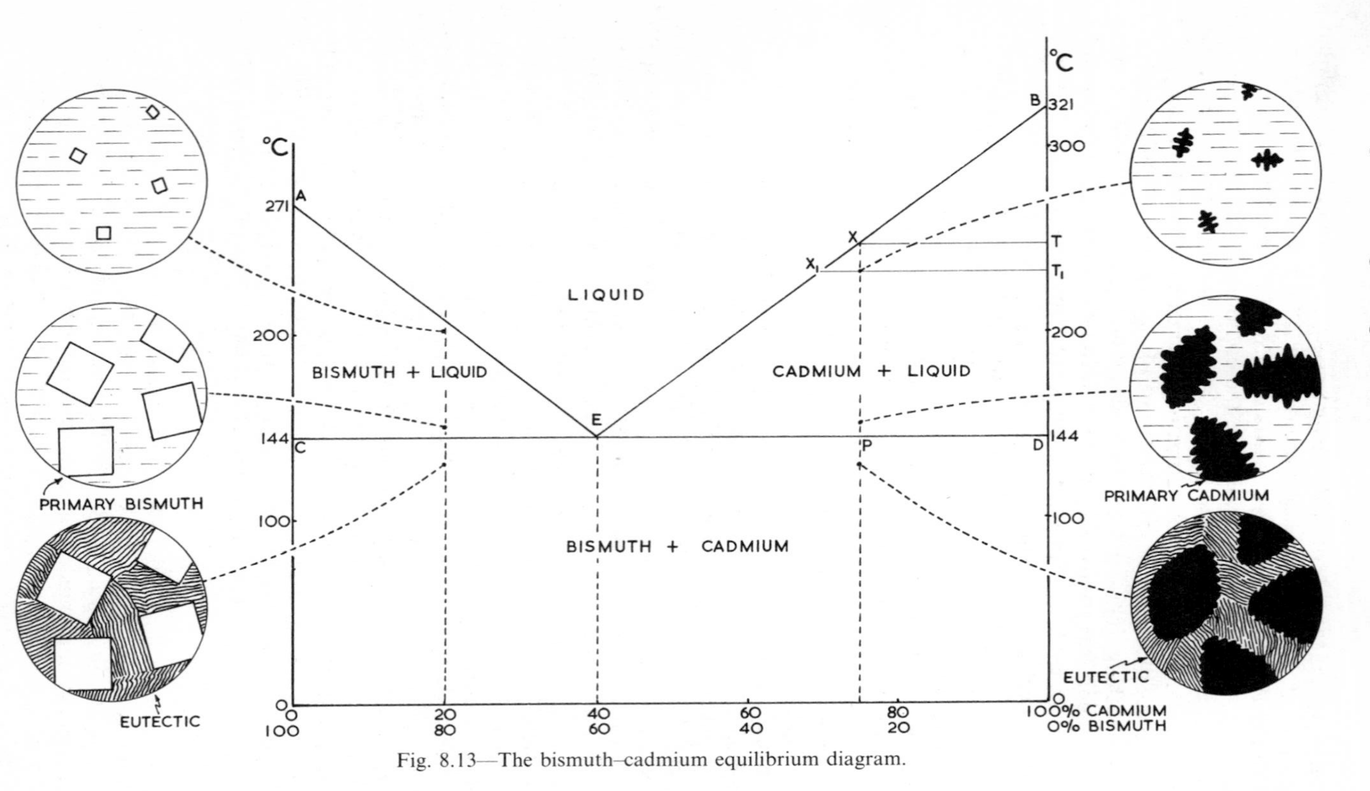

Fig. 8.13—The bismuth–cadmium equilibrium diagram.

the temperature falls to T. At this temperature the solid in equilibrium with the liquid X is pure cadmium since the relevant solidus phase boundary in this case is the 100% cadmium ordinate. Hence dendrites of pure cadmium begin to form. This causes the remaining liquid to become correspondingly richer in bismuth so that its composition moves to the left, say to X_1. Further solidification occurs as the temperature falls to T_1 when more pure cadmium will crystallise. This process continues until the temperature has fallen to 144° C by which time the remaining liquid contains 60%bismuth/40% cadmium, that is, the temperature and composition are represented by the point E—the eutectic point.

Applying the 'lever rule':

$$\text{Weight of liquid (composition } E) \cdot EP = \text{weight of pure cadmium} \cdot DP$$

or

$$\frac{\text{Weight of liquid (composition } E)}{\text{Weight of pure cadmium}} = \frac{DP}{EP}$$

$$(\text{Substituting values from Fig. 8.13}) = \frac{100-75}{75-40}$$

$$= \frac{5}{7}$$

Thus there are 7 parts by mass of solid cadmium to 5 parts by mass of liquid.

At this stage the two metals are in equilibrium with each other in the remaining liquid. It is reasonable to suppose, however, that due to the momentum of crystallisation of the primary cadmium, a little too much cadmium deposits throwing the composition of the liquid slightly to the left of E. Equilibrium is almost immediately restored as a film of bismuth deposits and in this manner the remaining liquid solidifies by depositing alternate layers of cadmium and bismuth as the liquid composition swings to and fro about E. The temperature remains constant at 144° C until the solidification process is complete.

The final structure therefore consists of *primary* crystals of cadmium in a matrix of eutectic which in turn consists of alternate layers of pure cadmium and pure bismuth in a laminated structure which is typical of metallic eutectics. The proportion of primary cadmium to eutectic will obviously be the same as indicated above since the liquid referred to then has now solidified as eutectic. Thus there will be 7 parts by mass of solid cadmium to 5 parts by mass of eutectic which in turn contains 60% bismuth and 40% cadmium.

8.6.4.4 A homogeneous liquid containing, say, 80% bismuth/20% cadmium will begin to solidify by depositing primary dendrites of bismuth. These do not resemble the dendrites of other metals but are roughly cubic in shape (Fig. 8.13). Since bismuth has been rejected from solution the latter has become correspondingly richer in cadmium so that its composition moves to the right towards the eutectic point. Ultimately the liquid will contain 60% bismuth and 40% cadmium at 144° C when it will solidify as eutectic as in the case of the 25% bismuth/75% cadmium alloy. In fact any bismuth/cadmium alloy will contain eutectic which will always be of fixed composition, namely, 60% bismuth/40% cadmium. Any liquid solution of composition either side of the eutectic amount will first deposit which ever metal is in excess of the eutectic composition in the form of primary crystals until the final liquid is of the eutectic composition at 144° C (Fig. 8.14). It must be emphasised that in the final structure there is no question of solid solubility since even in the eutectic portion the alternate layers of the two pure metals are readily resolvable using quite a low-power microscope.

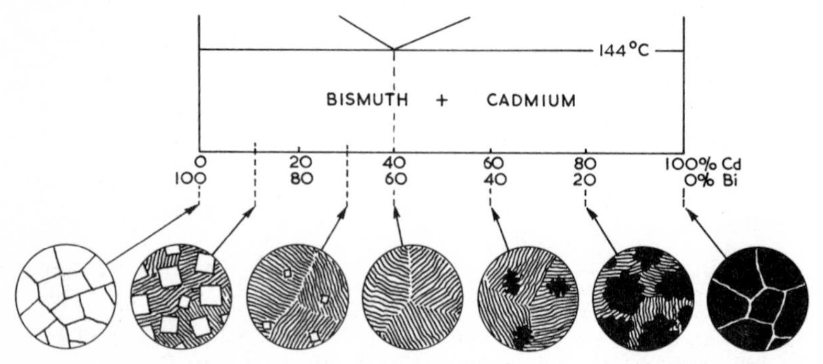

Fig. 8.14—Structures of bismuth–cadmium alloys in relation to the phase diagram.

At the eutectic point three phases can co-exist in equilibrium, i.e. liquid, solid cadmium and solid bismuth. The solid cadmium and solid bismuth of course constitute the eutectic. Hence there are two components (cadmium and bismuth) and three phases. Applying the modified Phase Rule:

$$f + p = n + 1$$
$$f + 3 = 2 + 1$$
$$\therefore f = 0$$

This means that the system is invariant so that if either composition or temperature are altered at least one of the phases will disappear.

8.6.4.5 We can summarise the general properties of the eutectic as follows:

(*i*) In a binary system it consists of a heterogeneous *mixture* of two phases, i.e. it must not be regarded as a single phase itself.

(*ii*) It solidifies at a single fixed temperature—the lowest temperature in the series.

(*iii*) It will be of fixed composition whatever the overall composition of the alloy in which it exists.

(*iv*) It solidifies in typical lamellar manner, though in some cases the lamellae of one of the phases may degenerate to globular form due to surface-tension effects.

Plate 8.1—A typical lamellar eutectoid structure. This micrograph is of a cast 11·8% aluminium-in-copper alloy annealed at 700° C. It shows lamellae of solid solution α and the intermediate phase γ_2. (Etched in Acid ferric chloride solution.) × 500.

8.6.5 Two substances completely miscible as liquids but only partially soluble in each other as solids. In the two preceding sections we have considered cases both of complete solid solubility (Cu/Ni) and total insolubility (Bi/Cd) in the solid state. These were, in a sense, special cases of which this one is the general, since here we shall deal with the case of partial solid solubility. Since the two special cases referred to permit of simpler interpretation they were considered first, so that the reader could progress from the simple to the more complex.

8.6.5.1 The silver–copper system is a typical example of partial solid solubility. The principal features not encountered during our study of previous phase systems are the phase boundaries CRC_1R_1F and DUD_1-U_1G in Fig. 8.15. These boundaries occur *below* the solidus line and consequently separate phase fields containing only solid phases. This means that phase changes can take place in an alloy after solidification is complete. A line such as *CF* or *DG* is generally termed a *solvus*. As the diagram indicates we have in the system two solid solutions α and β* respectively since:

(*i*) silver will dissolve up to a maximum of 8·8% copper at the eutectic temperature forming the solid solution α;

(*ii*) copper will dissolve up to a maximum of 8·0% silver at the eutectic temperature forming the solid solution β.

The slope of these boundary lines *CF* and *DG* indicates that both the solubility of copper in silver (α) and the solubility of silver in copper (β) decrease with fall in temperature. This is a phenomenon mentioned earlier (7.5.2) and is also a common feature of partial solubility in the liquid state as instanced by the zinc/lead system (8.6.1).

8.6.5.2 We will now examine the solidification of two representative alloys in the series. First, we will consider an alloy of composition *X* (Fig. 8.15) that is, containing 35% Ag/65% Cu. This will begin to solidify as the liquidus is reached at temperature *T*, when dendrites of the solid solution β of composition *Y* will commence to form. Since the β dendrites are relatively rich in copper then the remaining liquid will be left correspondingly rich in silver and its composition will move to the left, say to X_1. Further deposition of β, this time of composition Y_1 will occur as the temperature falls to T_1 and, if we assume *slow cooling*, then the original dendrites will also have changed to this composition (Y_1) by means of diffusion. Thus, at temperature T_1, we have:

Weight of uniform liquid (composition X_1) . X_1O

= Weight of uniform solid solution β (composition Y_1) . OY_1

As the temperature continues to fall slowly, β continues to deposit and at the same time change in composition along Y_1D whilst the remaining liquid changes in composition along X_1E. At 779° C, the eutectic temperature, we ultimately have liquid of composition *E* and solid solution β of composition *D* in the ratio:

Weight of uniform liquid (composition *E*) . EO_1

= Weight of uniform solid solution β (composition *D*) . O_1D

As the temperature begins to fall below 779° C the remaining liquid solidifies as a eutectic by depositing alternate layers of α (composition *C*) *and* β (composition *D*) until the liquid is used up. Obviously the ratio:

$$\frac{\text{Weight of primary } \beta}{\text{Weight of eutectic}}$$

will be the same as that indicated above, i.e. EO_1/O_1D.

* In phase diagrams of metallic alloys it is the custom to label phases from left to right using letters of the Greek alphabet as appropriate.

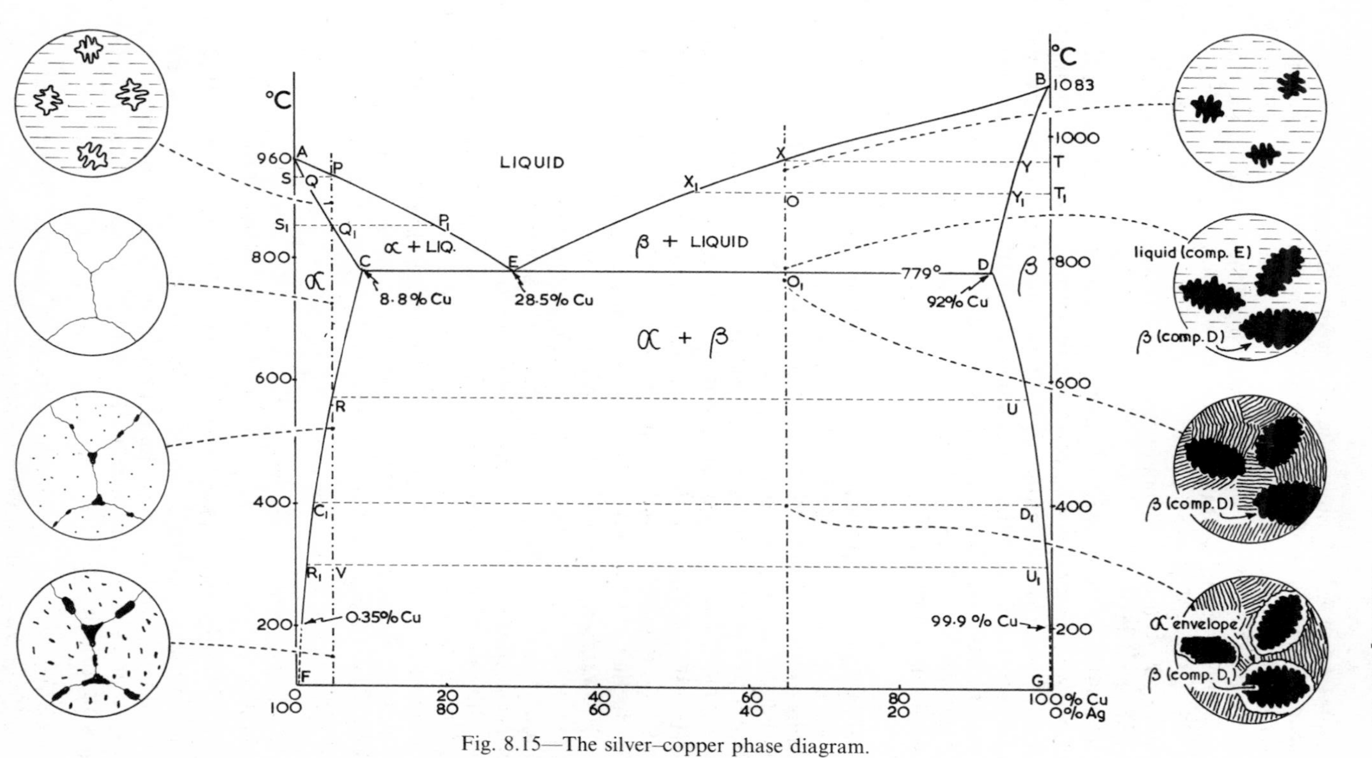

Fig. 8.15—The silver–copper phase diagram.

If we assume a very slow rate of cooling then further changes will occur in the structure of the solid alloy as the temperature falls and the solid solubilities of each metal in the other decrease along CF and DG respectively. Thus copper will be rejected from the α layers of the eutectic in the form of particles of the other solid solution β. These particles will join the layers of β adjacent to those of α. Similarly silver will be precipitated from the β layers in the form of a small amount of α which will join the α layers.

The crystals of primary β will also reject some silver in the form of α and this will sometimes be seen as a film surrounding the primary β crystals (Fig. 8.15) in cases like this. At, say, 400° C the composition of α will be given by C_1 and the composition of β (both eutectic and primary) by D_1. Because of the extremely slow rate of ionic diffusion at low temperatures little change in structure can be expected in this case below approximately 300° C.

8.6.5.3 We will now consider the cooling of an alloy of composition P, that is, containing 95% Ag/5% Cu. This will, of course, begin to solidify at temperature S (Fig. 8.15) by depositing dendrites of α of composition Q. Solidification will be complete at temperature S_1 when the final traces of liquid will be of composition P_1 but if we assume slow cooling such that equilibrium is maintained throughout, then liquid P_1 will be absorbed by diffusion so that the overall composition of the solid solution α is Q_1 (which, of course, is the same as that of the original liquid P).

No further changes take place in the uniform α crystals as they cool slowly towards R. On reaching R, however, the solid solution α becomes fully saturated with copper and as the temperature falls below that corresponding to R, copper will be rejected from solution, not of course as pure copper but as solid solution β of composition U, which is the phase in equilibrium with α at that temperature. As the temperature falls slowly β continues to precipitate from α and both solid solutions in fact change in compositions along RF and UG respectively until at, say, 300° C, α contains little more than 0·5% copper (R_1) and β no more than 0·2% silver (U_1). The relative amounts of α and β at this temperature will be given by:

$$\text{Weight of } \alpha \text{ (composition } R_1) \,.\, R_1V = \text{Weight of } \beta \text{ (composition } U_1) \,.\, VU_1$$

Up to this point we have been dealing with the solidification and cooling of alloys on the assumption that cooling has been slow enough for equilibrium to have been maintained at each stage. In practice this is rarely so and in a system such as this, rapid cooling is likely to have the following principal effects:

(*i*) primary crystals, whether of α or β, will be heavily cored;
(*ii*) α is likely to be supersaturated with copper, and β with silver

since cooling will have been too rapid to permit diffusion and precipitation. This applies both to primary α and β and also to eutectic α and β.

8.6.6 A system in which a peritectic reaction occurs. So far we have been dealing with alloy systems in which solidification is either of the solid solution or eutectic type. A further mode of solidification involves the *peritectic reaction*. Here two phases which are already present together in a heterogeneous mixture react to produce a third phase, one or both of the original phases disappearing in the process. Frequently one of the reacting phases is a liquid though this is not a pre-condition and quite often two solids will take part—albeit very slowly—in a peritectic reaction. When a liquid reacts with a solid in this way it invariably coats it with an envelope of the new phase. This film or envelope effectively insulates the initial solid phase from further reaction, except by the very slow process of diffusion. The term 'peritectic' is derived from the Greek, 'peri' which means 'around' (as in '*peri*meter'), a reference here to this surrounding envelope of the new phase.

8.6.6.1 We will consider the platinum/silver system which is one of the few based almost solely on a peritectic reaction. In this system any alloy containing between 12% and 69% silver will begin to solidify by forming dendrites of the solid solution α but, when the temperature falls to 1185° C (the peritectic temperature), α reacts with the remaining liquid to produce a new solid solution δ, i.e.

$$\alpha + \text{liquid} \rightleftharpoons \delta$$

First let us consider the solidification of an alloy of composition X (75% Pt/25% Ag) (Fig. 8.16). This begins to solidify in the normal manner of a solid solution as the temperature falls to T by depositing dendrites of α of composition A. Assuming that cooling is taking place very slowly so that equilibrium is attained at each stage of the process, then by the time the peritectic temperature of 1185° C has been reached the structure consists of uniform α of composition P and liquid of composition R in the ratio:

$$\text{Weight of solid solution } \alpha \text{ (composition } P) \,.\, PX_1 = \text{Weight of liquid (composition } R) \,.\, X_1R$$

Just below 1185° C the peritectic reaction takes place between the solid solution α and the remaining liquid:

$$\alpha + \text{liquid} \xrightarrow{1185^\circ\text{ C}} \delta$$

The composition of the new solid solution δ at this temperature is given by Q, i.e. 55% Pt/45% Ag. However the overall composition of the alloy is X, i.e. 75% Pt/25% Ag and it follows that not all of the α will be used up since it was obviously in excess of that required to produce a structure solely of δ. Therefore when the reaction is complete some α will remain.

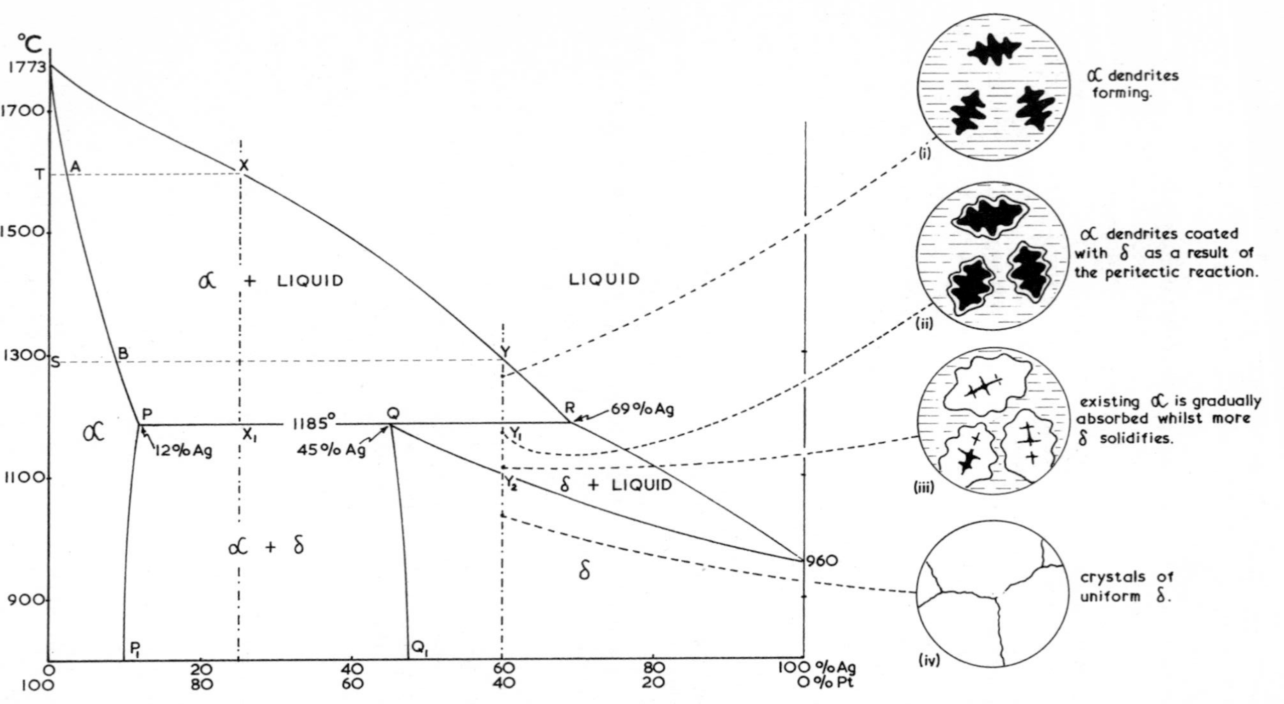

Fig. 8.16—The platinum–silver phase diagram. (Diagrams (ii) and (iii) do NOT represent equilibrium structures but are typical of the sort of situation which would be likely to occur with a moderate rate of cooling.)

In fact the composition of the alloy, when solidification is just complete is given by:

Weight of solid solution α (composition P) . PX_1
= Weight of solid solution δ (composition Q) . X_1Q

Substituting appropriate values in the above equation:

Weight of solid solution α (88% Pt/12% Ag)(25–12)
= Weight of solid solution δ (55% Pt/45% Ag)(45–25)

$$\text{or} \quad \frac{\text{Weight of solid solution } \alpha \text{ (88\% Pt/12\% Ag)}}{\text{Weight of solid solution } \delta \text{ (55\% Pt/45\% Ag)}} = \frac{45\text{–}25}{25\text{–}12}$$

$$= \frac{20}{13}$$

As the alloy continues to cool slowly α will change in composition along PP_1 and δ along QQ_1.

An alloy of composition Y (40% Pt/60% Ag) will also begin to solidify by depositing solid solution α. In this case solidification begins at temperature S as dendrites of α of composition B form. Again assuming very slow cooling we shall have a solid solution α of composition P and remaining liquid of composition R by the time the temperature has fallen to 1185° C. However, the proportions of α and liquid are different from what they were in the previous case since this alloy is richer in silver. Thus:

Weight of solid solution α (composition P) . PY_1
= Weight of liquid (composition R) . Y_1R

Consequently the liquid is present in such excess that when the peritectic reaction takes place all of the α is used up and some liquid remains. Just below the peritectic temperature we have:

Weight of solid solution δ (composition Q) . QY_1
= Weight of liquid (composition R) . Y_1R

As the temperature continues to fall the remaining liquid solidifies as δ, the crystals of which change in composition along QY_2. Solidification is complete at Y_2 when uniform crystals of δ of this composition (which of course is the same as Y) remain.

8.6.6.2 Mention has already been made of the tendency for a phase produced as a result of a peritectic reaction to coat a solid reactant. This will most certainly happen during any but extremely slow rates of cooling. The effect is illustrated in Fig. 8.16. Here the α dendrites (shown in black), on reacting with the remaining liquid, have become coated with the new solid solution δ (light) (ii). This will inevitably 'seal off' the α from the remaining liquid and so stop further direct reaction between the two. As the temperature falls the remaining liquid solidifies as δ.

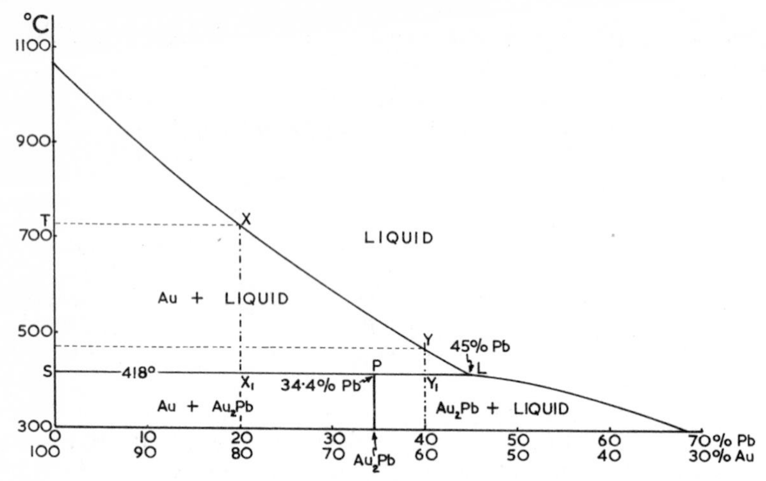

Fig. 8.17—Part of the gold–lead phase diagram.

However, due to diffusion, platinum atoms will migrate outwards from the α cores so that these regions will change progressively in composition to that of δ (iii). If cooling is slow enough, or if the solid alloy is subsequently annealed at a high temperature, diffusion will proceed to completion and uniform δ ultimately remain (iv).

8.6.6.3 A peritectic reaction may also result in the formation of an intermetallic compound of fixed composition. Such a situation is illustrated in Fig. 8.17 which consists of the relevant section of the gold/lead phase diagram. Here an alloy of composition *X* (80% Au/20% Pb) will begin to solidify at temperature *T* by depositing dendrites of pure gold. This process continues until at 418° C, the peritectic temperature, we have crystals of pure gold along with the remaining liquid of composition *L* (55% Au/45% Pb). A peritectic reaction then occurs between the gold dendrites and the remaining liquid to produce the new phase Au_2Pb, an intermetallic compound of fixed composition as is indicated by the fact that its phase 'field' is the ordinate through 65·6% Au/34·4%Pb. When the alloy is completely solid its phase make-up will be given by:

$$\text{Weight of gold} \,.\, SX_1 = \text{Weight of Au}_2\text{Pb} \,.\, X_1P$$

$$\text{or} \quad \frac{\text{Weight of gold}}{\text{Weight of Au}_2\text{Pb}} = \frac{X_1P}{SX_1}$$

$$= \frac{34{\cdot}4 - 20}{20}$$

$$= \frac{14.4}{20}$$

An alloy containing, say, 60% Au/40% Pb (*Y*) will begin to solidify by depositing dendrites of gold as before but when the peritectic reaction occurs (at Y_1) all of the solid gold will be used up in forming Au_2Pb. (Some liquid will remain and as the temperature falls the structure will undergo further changes not indicated in this limited section of the gold/-lead phase diagram.)

8.6.7 A system which contains one or more intermediate phases. So far we have been dealing with alloy systems in which the metals have had little chemical affinity for each other. Consequently, apart from the gold/lead system just mentioned, no intermetallic compounds have been formed in these alloy series and the resultant phase diagrams have been of a relatively simple nature.

If the reader refers to Dr. Max Hansen's comprehensive work Constitution of Binary Alloys,* he will find that many phase diagrams are of complex construction and often contain twenty or more phase fields. This state of affairs is due to the formation of a number of intermediate phases in the series. Indeed the complete gold/lead system referred to above is of this type since it includes other intermetallic compounds as well as Au_2Pb.

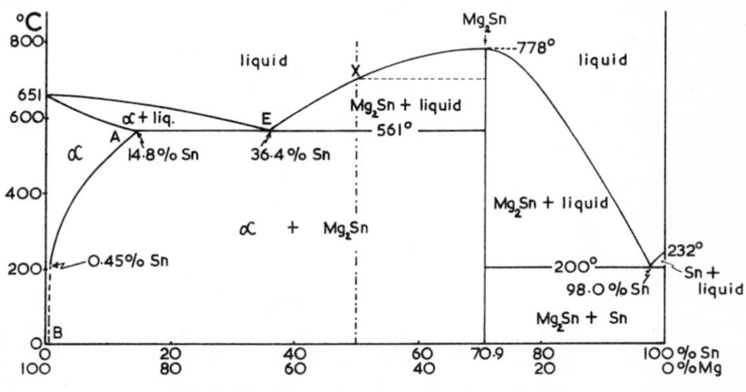

Fig. 8.18—The magnesium–tin phase diagram.

8.6.7.1 The magnesium/tin system (Fig. 8.18) is one which includes a single intermediate phase. This phase is an intermetallic compound of the 'valency' type, Mg_2Sn, and its phase 'field' is represented by the ordinate through 70·9% tin. This indicates that the phase is of invariable composition and that Mg_2Sn dissolves neither magnesium nor tin. It will also be noted that this stable intermetallic compound has a melting point (778° C) which is higher than that of either magnesium or tin.

* Since Dr. Hansen's very useful book was published many years ago and has not recently been revised many of the phase diagrams are now out of date in the light of more recent investigational work.

The system can be interpreted as though we were dealing with two separate eutectic series joined together at the Mg_2Sn ordinate. For example a 50/50 alloy will begin to solidify at X by depositing crystals of pure Mg_2Sn (the temperature horizontal through X cuts the other boundary of the phase field at the ordinate representing pure Mg_2Sn). Mg_2Sn continues to crystallise out until at 561° C, the eutectic temperature of the Mg/Mg_2Sn section of the diagram, the remaining liquid is of composition E. This liquid then solidifies as a eutectic consisting of alternate layers of α (composition A) and Mg_2Sn. On further slow cooling, α changes in composition along AB by rejecting particles of Mg_2Sn (which will join the Mg_2Sn layers in the eutectic). Since Mg_2Sn is of fixed composition no change will take place either in its primary crystals or in its eutectic layers. It will be noted that whilst interpreting the solidification of this alloy X we have not crossed the Mg_2Sn ordinate to the other half of the diagram.

8.6.7.2 Quite often an intermetallic compound will take into solution a limited amount of one or both of its constituent components. Then the intermediate phase field occupies 'width' on the diagram instead of being represented by an ordinate through the composition of the intermetallic compound. Part of the nickel/tantalum phase diagram (Fig. 8.19) is used

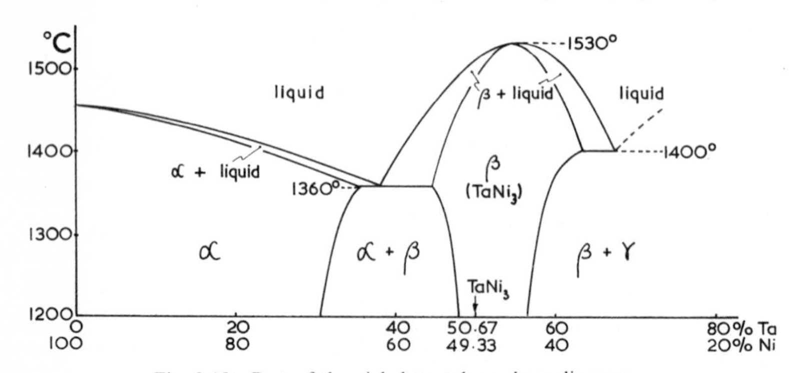

Fig. 8.19—Part of the nickel–tantalum phase diagram.

here to illustrate this case. The intermetallic compound $TaNi_3$ contains 50·67% Ta/49·33% Ni when pure. However, the diagram indicates that $TaNi_3$ will take into solid solution both tantalum and nickel giving rise to the phase field labelled 'β'.

Ternary alloy systems

8.7 So far we have been dealing only with two component systems. Since, in studying metallic systems in particular, we have neglected the effects of pressure variation, only two variables have required consideration. One of these has been temperature and the other, variation of the amount of one component with respect to the other.* Hence only two graphical dimensions are required to represent such a system and flat paper suffices for this purpose.

In the case of a ternary alloy, however, the amounts of any two of the three metals can be varied independently and the extra variable involved requires an extra dimension for its representation, i.e. we must use a three-dimensional phase diagram. The base of this will be in the form of

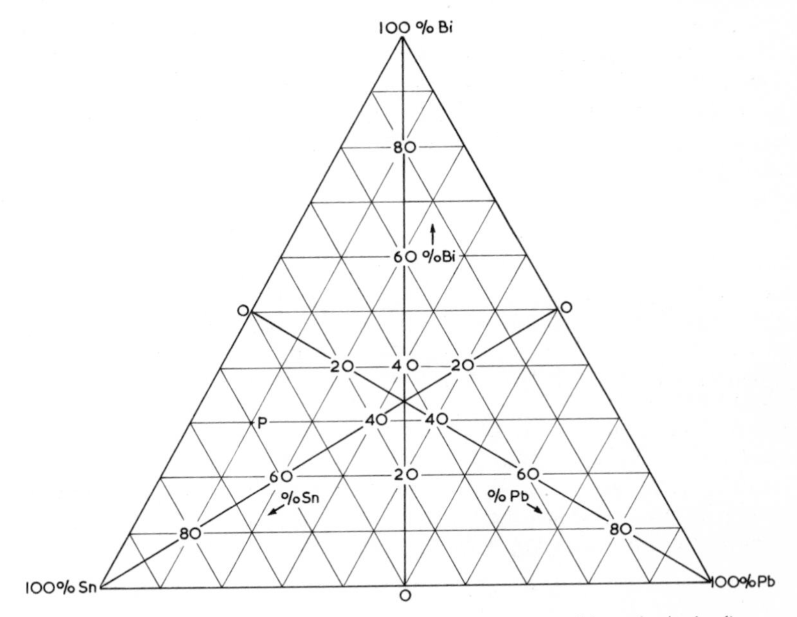

Fig. 8.20—The base plane for a three-component system (bismuth–tin–lead).

an equilateral triangle, each apex of which represents 100% of one of the metals (Fig. 8.20). Thus an alloy containing 30% Bi/60% Sn/10% Pb will be represented by point *P*. (This can be demonstrated mathematically.) Temperature will be represented by ordinates normal to the triangular base.

8.7.1 Fig. 8.21 illustrates (within its limitations) the ternary alloy system for bismuth/tin/lead. The binary systems tin/lead, lead/bismuth and bismuth/tin are fundamentally of the eutectic type and are represented by the 'vertical' end faces of the space model (Fig. 8.21 (ii)). The binary liquidus lines develop to liquidus surfaces in the three-dimensional model and these surfaces intersect in three 'valleys' which drain down to a point

* Clearly, in a binary system the proportions of each metal must add up to 100%. Thus if the amount of one is changed the amount of the other changes in sympathy and there is only one independent variable involved.

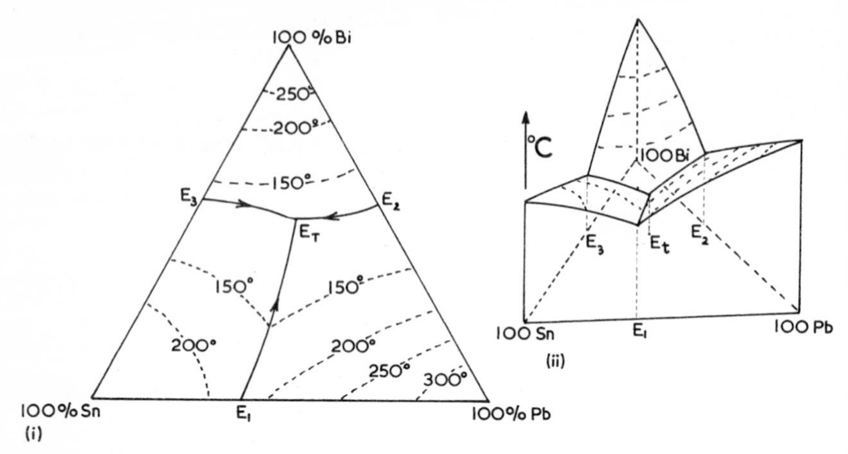

Fig. 8.21—The bismuth–tin–lead phase system. (i) This is a projection of the liquidus surfaces on to a horizontal plane in the manner of a contoured ordnance survey map. (ii) is a space model showing the liquidus surfaces. E_1 is the eutectic point of the Sn–Pb system (183° C); E_2 the eutectic point of the Pb–Bi system at 124° C and E_3 the eutectic point of the Bi–Sn system at 135° C. The *ternary* eutectic point E_t is at 100° C.

of minimum temperature vertically above E_t. This is the ternary eutectic point of the system, its temperature being lower than that of either of the binary eutectic points E_1, E_2, E_3. For this system the ternary eutectic contains 50% Bi/28% Pb/22% Sn and freezes at 100° C.

At this ternary eutectic point four phases can co-exist, i.e. solid bismuth, solid tin, solid lead and liquid. The components of the system are bismuth, tin and lead. Applying the modified Phase Rule:

$$f + \mathrm{p} = n + 1$$
$$f + 4 = 3 + 1$$
$$f = 0$$

Thus at this eutectic point, as at a binary eutectic point, the system is invariant.

8.7.2 In considering this ternary system we have dealt only with the liquidus 'surface'. Generally we shall be more involved with changes which take place in *solid* alloys at lower temperatures, i.e. 'inside' the space model. Clearly such three-dimensional phase 'volumes' are rather inconvenient to illustrate, though for demonstration purposes three-dimensional space models may be constructed using bits of differently coloured electrical wire to indicate phase boundaries. Many years ago such a project used to be a favourite exercise leading to the award of a Ph.D. in many university metallurgy departments. Much painstaking work was involved including the casting of alloys of different compositions within the series, followed by careful chemical analysis, prolonged annealing of samples at different temperatures and then quenching. The phase structure was then examined, generally using the microscope.

It is generally more useful to use either horizontal 'slices' (that is, isothermal sections) at the required temperatures; or vertical slices at a fixed amount of one of the components. Fig. 8.22 represents a vertical

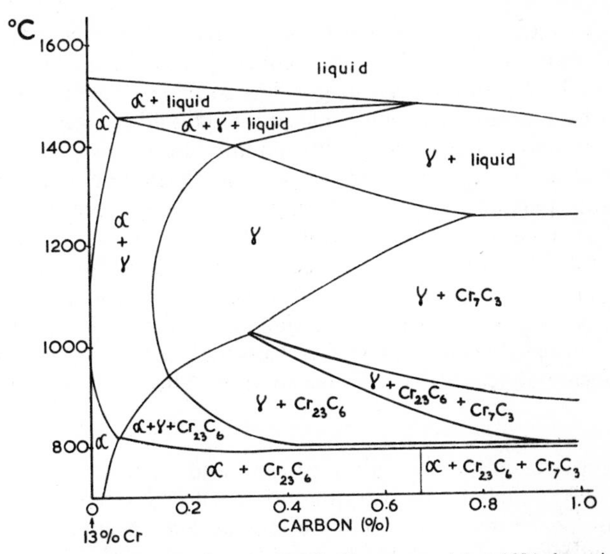

Fig. 8.22—A section through the Fe–Cr–C space model at 13% chromium.

slice from the ternary iron/chromium/carbon system. This has been taken at the 13% Cr base line and the section shown will be parallel to the iron/carbon end face of the space model.

The same sort of arrangement can be used for multi-component alloys. Then, all but one of the components are fixed and a two-dimensional diagram drawn showing temperature and *one* component as variables. These diagrams—usually called 'pseudo-binary' diagrams—are very useful for studying alloy steels which may contain up to six component elements.

Chapter Nine
Steel and its Heat-treatment

9.1 Iron was known to the Ancients as 'Metal from Heaven' or 'Metal from the Stars' from which we may conclude that such iron as was used in those times was of meteoric origin. The deliberate smelting of iron ores seems to have originated some 3500 years ago and by that time the Iron Age can be said to have arrived. Presumably historians would agree that we are still in the Iron Age for, despite the developments of aluminium alloys and plastics materials during the present century, steel undoubtedly remains our most important engineering material. It is unlikely that the present state of technological development could have been achieved in any field without the help of steel as our main structural, constructional and tool material—and all this at a relatively low financial outlay. It is inevitable therefore that in a book of this type we must allocate at least two chapters to those alloys which are based on iron.

9.1.1 Steel is essentially an alloy of iron and carbon containing up to roughly 2·0% carbon. By varying this carbon content and the heat-treatment of the resultant alloy we can obtain an enormous range of mechanical properties, whilst the addition of alloying elements such as nickel, chromium and molybdenum extends the properties still further. Nevertheless it is a sobering thought that, but for the *polymorphic* changes which occur in iron (3.6.4), the heat treatment of steel as we practise it would be impossible and that iron/carbon alloys would then have extremely limited properties. Certainly tool steels and high-strength constructional steels would not be available to the engineer. It is possible that we would still be in the Bronze Age.

9.1.2 The relatively small carbon atoms dissolve interstitially in iron. However, whilst the FCC form of iron will dissolve up to 2·0% C, the maximum solubility of carbon in BCC iron is only 0·02% at 723° C and this falls almost to zero (probably of the order of $2{\cdot}0 \times 10^{-7}$%) at 0° C. This is because the interatomic spaces in BCC iron are very small so that carbon atoms cannot readily be accommodated. In fact carbon atoms are too large for extensive interstitial solid solution but too small to form a stable substitutional solid solution with iron. Whilst the interatomic spaces are larger in FCC iron they are not large enough to accommodate carbon atoms freely. Some crowding therefore occurs producing strain in the structure so that not all spaces can be occupied.

The maximum solubility of carbon in the FCC form of iron is 2·0% and this sets the limit of the carbon content of a steel. As the FCC form changes to BCC any carbon in solution in excess of 0·02% will be precipitated from solid solution assuming that it is cooling slowly enough to permit this. Under the slowest rate of cooling which is possible carbon does not precipitate as graphite but as iron carbide, Fe_3C. Like most metallic carbides this is a hard substance. Consequently as the carbon content of a steel increases so does its hardness even though it has not been heat-treated but is only in the slowly cooled condition.

9.1.3 Since steel is such an important alloy and has been studied since the very early days of metallography* names have been adopted for the three phases mentioned above. The interstitial solid solution of carbon in FCC iron is known as *Austenite*. It can contain up to 2·0% C and is a stable phase between 723 and 1493° C. The interstitial solid solution of carbon in BCC iron is known as *Ferrite*. This is stable below 910° C and can contain a maximum of only 0·02% C (at 723° C). Nevertheless although ferrite contains so little dissolved carbon this carbon exerts considerable influence on the mechanical properties of ferrite by opposing the movement of dislocations (7.5.1.2). For this reason very high purity iron (99·999 9%) is extremely soft and ductile as compared with the low-carbon ferrite of dead-mild steel. Iron carbide, Fe_3C, is generally known as *Cementite*. Other phases, and mixtures of phases, are also given special names as we shall see subsequently.

9.1.4 What is popularly known as the 'iron/carbon diagram' is strictly speaking not a true equilibrium diagram since cementite (iron carbide) is not a completely stable phase. Graphite is in fact more stable than cementite but nucleation of cementite takes phace in advance of precipitation of graphite mainly because at that stage the FCC iron is saturated with carbon. Once cementite has formed it is quite stable and for most practical considerations can be treated as an equilibrium phase. However, whilst Fig. 9.1 represents what is generally known as the 'iron/-carbon equilibrium diagram' we should bear in mind that it is more properly labelled the 'iron/*iron carbide* metastable system'.

9.1.5 If iron did not undergo polymorphic changes at 1400 and 910° C but retained its FCC structure at all temperatures it would presumably form a simple eutectic system with a eutectic point at 4·3% C and 1147° C, and the solid solubility of carbon in austenite falling from 2·0% at 1147° C to a very low value at 0° C. The resultant alloy would have limited use and would not be amenable to the extensive heat-treatments applied to carbon steels as we know them, though some limited precipitation-hardening treatment might be possible.

The very limited solid solubility of carbon in the BCC forms (α and δ)

* Professor Henry Sorby (of Sheffield) was conducting the systematic microscopical metallography of steels in the early 1860s.

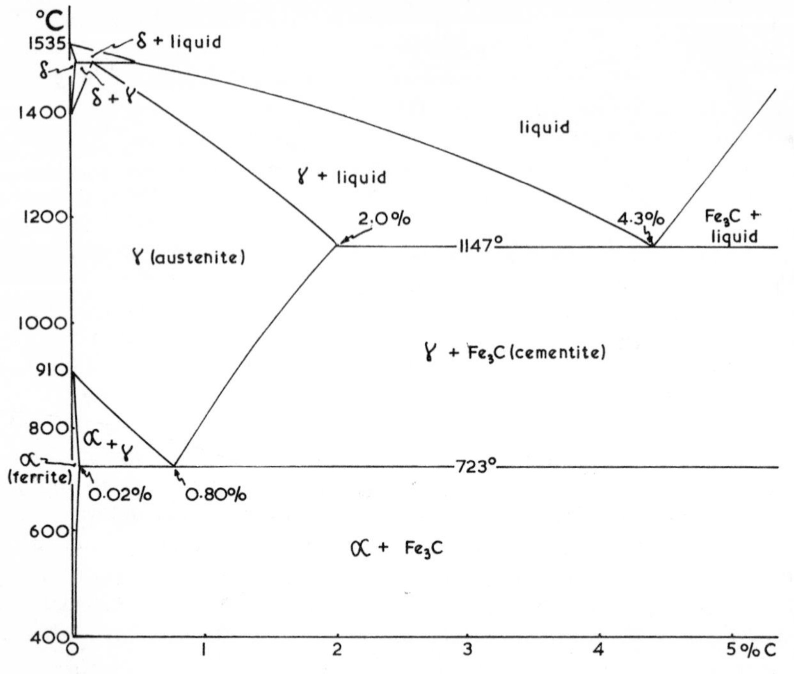

Fig. 9.1—The iron–carbon phase equilibrium diagram (more properly the 'iron–iron carbide metastable system').

introduces modifications to the diagram as indicated. Thus a steel containing less than 0·51% C will begin to solidify as dendrites of δ which, at 1493° C, will undergo a peritectic reaction with the remaining liquid to produce γ (austenite). As the steel cools the FCC structure will ultimately begin to transform to one which is BCC and carbon will be thrown out of solid solution as cementite.

Since carbon dissolves interstitially in iron the diffusion of carbon atoms is relatively rapid and normally there is no coring of carbon with respect to iron. Other alloying elements, and particularly impurities, may, however, be cored especially if they dissolve substitutionally in iron.

The transformation of austenite under near equilibrium conditions

9.2 We will now study the solidification and transformations which occur as a 0·4% C steel cools slowly from the liquid state to room temperature. Such a steel will begin to solidify when its temperature falls to T (Fig. 9.2) and dendrites of BCC δ (of composition D) begin to form. This phase, δ, continues to crystallise and at the same time change in composition along DD_1 until, at 1493° C, its composition is given by D_1 (0·1% C). The remaining liquid in equilibrium with δ at this temperature is of composition C (0·51% C). A peritectic reaction then occurs between δ

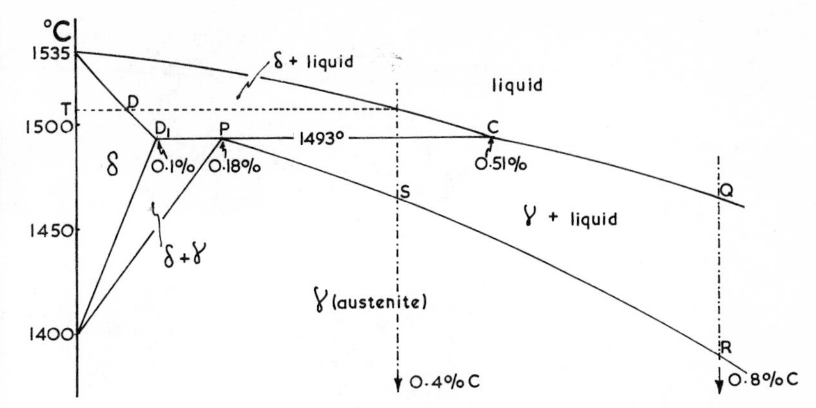

Fig. 9.2—The peritectic region of the iron–carbon diagram.

and this remaining liquid producing a new phase—the solid solution γ (austenite). Since the overall composition of the steel lies between P and C all of the δ will be used up in the reaction and the surplus liquid will then solidify out as γ whilst the temperature falls to S and the γ changes in composition along PS by diffusion of carbon.

No further phase change takes place until the temperature has fallen to T_1 (Fig. 9.3). This is known as the *upper critical*—or A_3—temperature for this particular steel.*

As the temperature falls below T_1 crystals of BCC α (ferrite) of composition X begin to grow within the austenite which of course they gradually replace. Since the ferrite contains less than 0·02% C it follows that the remaining austenite becomes richer in carbon so that its composition moves to the right towards E. By the time the temperature has fallen to 723° C we shall have ferrite of composition L (0·02% C) and remnant austenite of composition E (0·80% C) in the ratio:

$$\frac{\text{Weight of ferrite (0·02\% C)}}{\text{Weight of austenite (0·8\% C)}} = \frac{L_1E}{LL_1} \qquad (1)$$

i.e. there are approximately equal masses of each phase present at this instant.

9.2.1 The *lower critical*—or A_1—temperature is given by the line LO (Fig. 9.3) and is of course constant at 723° C for all carbon steels. The point E will be recognised as being associated with a eutectic type of transformation. Previously we have been dealing with the formation of eutectics from liquid solutions. Here, however, the transformation of a *solid* solution (austenite) is involved and so the resultant structure in this case is termed a *eutectoid*.

* The *upper critical temperature* of a steel is given by the appropriate point corresponding with its composition on the line UEV (Fig. 9.3).

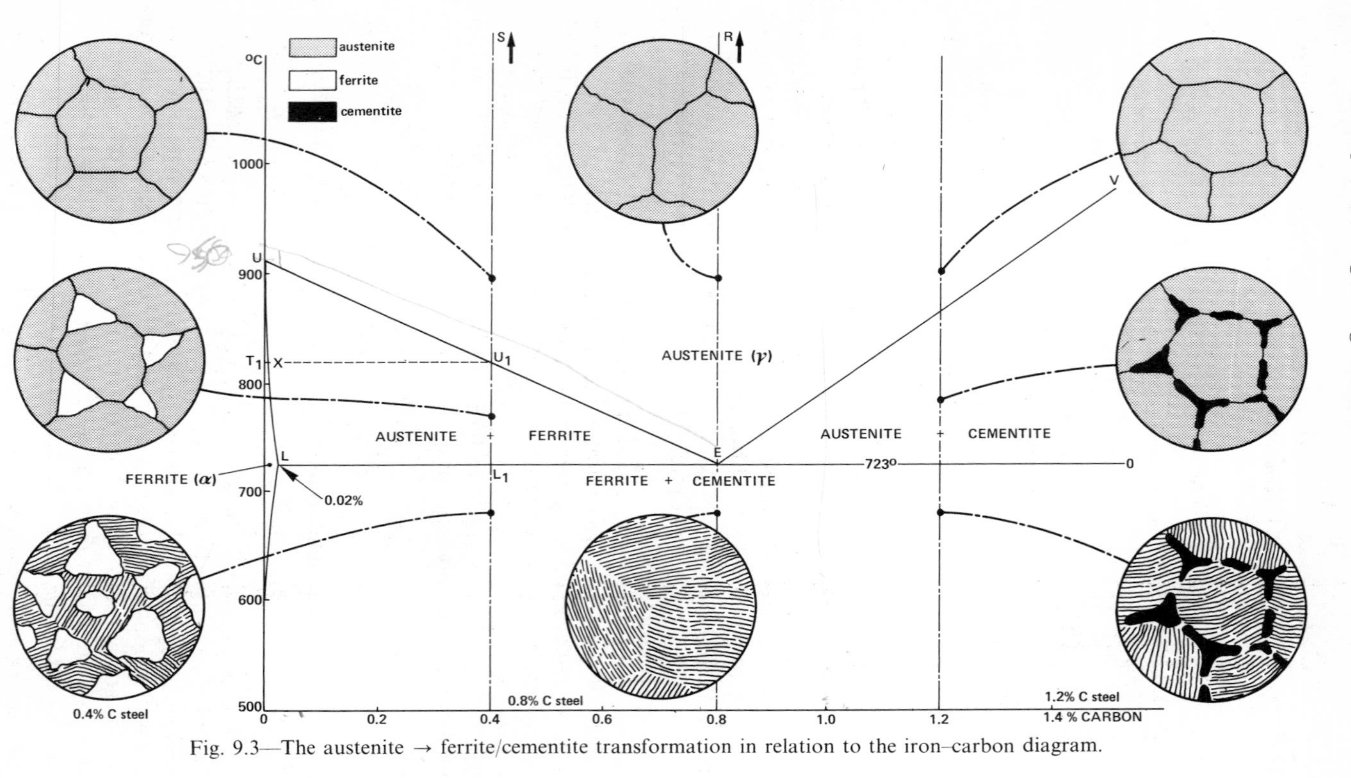

Fig. 9.3—The austenite → ferrite/cementite transformation in relation to the iron–carbon diagram.

As the temperature falls just below 723° C the remaining austenite, now of composition *E*, transforms to eutectoid by precipitating alternate layers of ferrite (composition *L*) and cementite (Fe_3C). This particular eutectoid is known as *pearlite*, a reference to the fact that, like mother-of-pearl, it behaves as a diffraction grating in dispersing white light. It is important to realise that pearlite is *not* a phase but a mixture of two phases, ferrite and cementite. Since the austenite present in the structure at just above 723° C has now transformed to pearlite the ratio:

$$\frac{\text{Weight of primary ferrite}}{\text{Weight of pearlite}} = \frac{L_1E}{LL_1}$$

i.e. as in (1) above. All *hypo*-eutectoid steels, that is, those containing *less* than 0·8% C consist of varying quantities of pearlite and primary ferrite, assuming that they are in an equilibrium state.

Further fall in temperature will have no significant effects on the microstructure. The carbon contents of both primary and eutectoid ferrite will decrease from the already low value of 0·02% to virtually zero. This carbon will precipitate as tiny particles of cementite mainly at the ferrite grain boundaries.

9.2.2 A steel containing exactly 0·8% C will solidify directly as austenite (between *Q* and *R*—Fig. 9.2) and will undergo no further changes until it reaches the eutectoid point *E* when the whole of the structure will transform from austenite to pearlite. *Hyper*-eutectoid steels, i.e. those containing *more* than 0·8% C, will also solidify directly as austenite. On reaching the upper critical temperature (a point on *EV*), primary cementite will begin to separate out—generally as embrittling films at the austenite grain boundaries—so that the composition of the remaining austenite moves to the left towards *E*. At 723° C this remaining austenite contains exactly 0·8% C and will transform to pearlite as before.

The formation of pearlite

9.3 When austenite transforms isothermally just below 723° C the reaction product is pearlite, the eutectoid of ferrite and cementite indicated by the iron/carbon diagram. As emphasised already pearlite is not a phase but a mixture of two phases. These two phases, ferrite and cementite, are present in a definite ratio as indicated by the lever rule:

$$\text{Weight of ferrite} \,.\, (0{\cdot}8) = \text{Weight of cementite} \,.\, (6{\cdot}67\text{–}0{\cdot}8)$$

(Since cementite, Fe_3C, contains 6·67% by weight of carbon)

$$\therefore \frac{\text{Weight of ferrite}}{\text{Weight of cementite}} = \frac{(6.67\text{–}0.8)}{0.8}$$

$$= \frac{5.87}{0.8}$$

$$7.3/1$$

As the relative densities of ferrite and cementite are roughly equal the lamellae of ferrite are approximately seven times the width of those of cementite. As a result ferrite forms a continuous phase with the plates of cementite surrounded by this ferrite matrix. Since ferrite is tough and ductile this makes pearlite tougher and stronger than it would be if the brittle cementite were continuous.

9.3.1 The mechanism of the austenite → pearlite transformation is typical of many reactions within solid alloys in that it begins at grain boundaries in the main and continues into the existing grains. This is predictable since the atoms at grain boundaries have higher energy states than atoms within the grains. Hence these atoms at grain boundaries require less extra energy in order to break away from their neighbours to form a new structure.

It is thought that carbon atoms collect together and so form a cementite nucleus (Fig. 9.4 (ii)). As this nucleus grows inwards from the boundary the surrounding austenite becomes successively depleted in carbon and, since its composition moves to the left of the eutectoid point, so layers of ferrite separate on either side of the cementite nucleus (iii). Beyond these ferrite layers an increase of carbon occurs so that more cementite nucleates (iv) and so on, the structure building up as alternate layers of cementite and ferrite (vi).

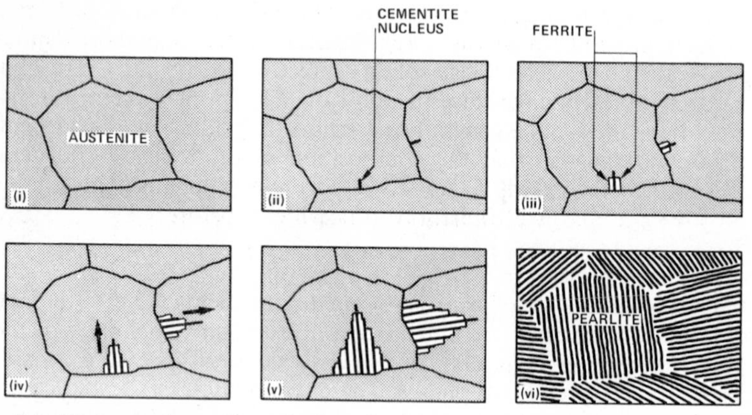

Fig. 9.4—The nucleation and growth of pearlite (it is assumed that the steel in this instance contains 0·8% carbon).

Uses of steel

9.4 Ordinary carbon steels can be classified conveniently into four main groups in terms of their specific uses:

9.4.1 *'Dead-mild' steels* containing up to 0·15% C which are used for general presswork and other applications where high ductility is necessary in forming, e.g. motor-car bodies, tin cans, nails and wire.

	Relevant specifications	Typical compositions (%)	Heat-treatment	Typical mechanical properties Y.P. (N/mm²)	T.S. (N/mm²)	Elong. (%)	Impact (J)	Hardness (Brinell)	Uses
Mild steel	B.S. 970:040A10	0·10 C 0·40 Mn	No heat treatment—except process annealing (9.5.1) to remove the effects of cold-work	—	300	28	—	—	Lightly stressed parts produced by cold-forming processes, e.g. deep-drawing and pressing
Structural steels	B.S. 15	0·20 C	No heat treatment	240	450	25	—	—	General structural steel
	B.S. 968	0·20 C 1·50 Mn	No heat treatment	350	525	20	—	—	High tensile structural steel for bridges and general building construction—fusion welding quality
Casting steel	B.S. 1504/161B	0·30 C	No heat treatment other than 'annealing' (9.5.4) to refine grain	265	500	18	20	150	Castings for a wide range of engineering purposes where medium strength and good machinability are required
Constructional steels	B.S. 970:080M40	0·40 C 0·80 Mn	Harden by quenching from 830/860° C. Temper at a suitable temperature between 550 and 660° C	500	700	20	55	200	Axles, crankshafts, spindles, etc., under medium stress
	B.S. 970:070M55	0·55 C 0·70 Mn	Harden by quenching from 810/840° C. Temper at a suitable temperature between 550 and 660° C	550	750	14	—	250	Gears, cylinders and machine-tool parts requiring resistance to wear
Tool steels	—	0·70 C 0·35 Mn	Heat slowly to 790/810° C and quench in water or brine. Temper at 150/300° C	—	—	—	—	780	Hand chisels, cold sates, mason's tools, smith's tools, screwdriver blades, stamping dies, keys, cropping blades, miner's drills, paper knives
	B.S. 4659:BW1A	0·90 C 0·35 Mn	Heat slowly to 760/780° C and quench in water or brine. Temper at 200/300° C	—	—	—	—	800	Press tools; punches; dies; cold-heading, minting and embossing dies; shear blades; woodworking tools; lathe centres; draw plates
	B.S. 4659:BW1B	1·00 C 0·35 Mn	Heat slowly to 770/790° C and quench in water or brine. Temper at 150/350° C	—	—	—	—	800	Taps; screwing dies; twist drills; reamers; counter sinks; blanking tools; embossing, engraving, minting, drawing, needle and paper dies; shear blades; knives; press tools; centre punches; woodworking cutters; straight edges; gouges; pneumatic chisels; wedges
	B.S. 4659: BW1C	1·20 C 0·35 Mn	Heat slowly to 760/780° C and quench in water or brine. Temper at 180/350 °C	—	—	—	—	800	Engraving tools; files; surgical instruments; taps; screwing tools

9.4.2 *Mild steels* containing 0·15–0·3% C. Wrought forms are used as RSJ and other structural members, shafting, levers and various forgings. Steels used for sand castings usually contain 0·3–0·35% C.

9.4.3 *Medium-carbon constructional steels* containing 0·4–0·6% C which are widely used for components in the engineering industries, e.g. axles, connecting rods, gears, wire ropes, rails, etc.

9.4.4 *Tool steels* containing 0·6–1·5% C.

9.4.5 Steels in the first two of these groups, i.e. containing less than 0·3% C, do not harden sufficiently when heat treated to make this generally worth while. This is not due specifically to lack of carbon as we shall see later (9.6.6). Such steels are therefore used in the cast, hot-rolled, cold-worked

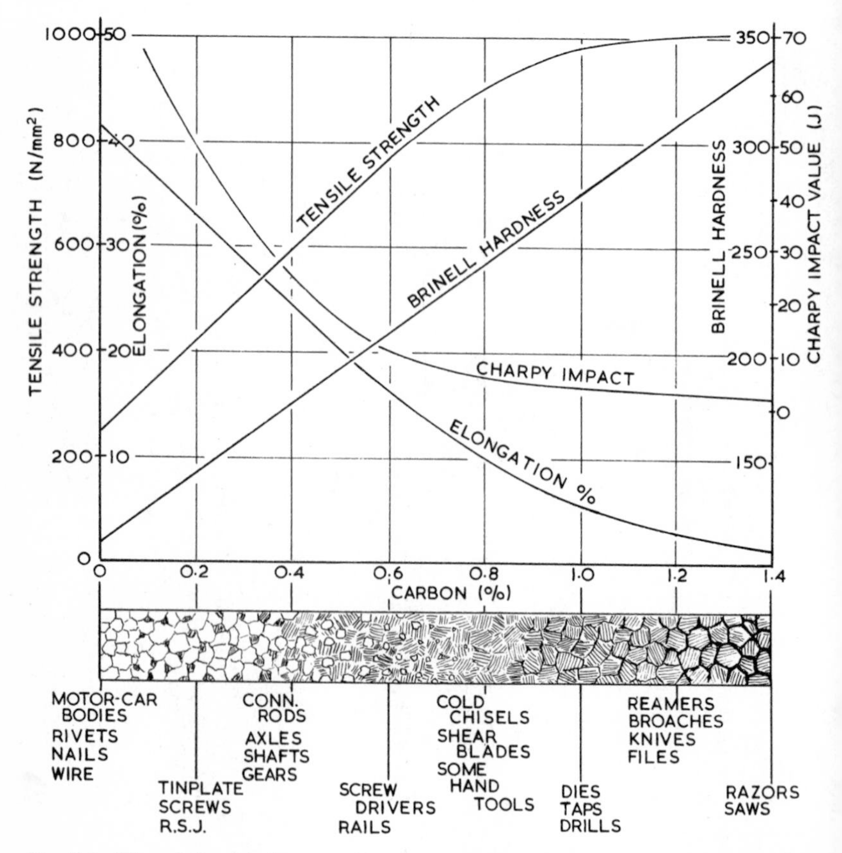

Fig. 9.5—The relationship between % carbon, uses and mechanical properties of carbon steels in the hot-rolled condition.

or normalised forms. The third and fourth groups contain steels with more than 0·4% C and can be hardened or strengthened by an appropriate heat treatment programme.

In Britain the motor-car industry is the largest single user of steel, followed by various branches of the building and plant construction industries. Thus, despite the attractive properties offered by steel in its heat-treated forms we must not forget that the bulk of ferrous material is of the low-carbon type which is not heat-treated. Some of the many uses of carbon steels related to carbon content and mechanical properties are shown in Table 9.1 and in Fig. 9.5.

Heat-treatment processes which produce equilibrium structures in steels

9.5 Several heat-treatment processes applied to steels result in the formation of pearlitic* structures. In the first two of these processes mentioned here no phase change occurs since the treatment is carried out *below* the A_1—or lower critical—temperature.

9.5.1 *Process annealing.* This is an inter-stage annealing operation such as might be applied to any heavily cold-worked metal or alloy in order to achieve recrystallisation (5.8.2) and lead to the recovery of malleability and ductility so that further cold-work may be applied. In this context it is used mainly to soften mild and dead-mild steels which have become work-hardened by cold-rolling or drawing and which must undergo further cold-forming. A typical example is the drawing of low-carbon steel wire.

The presence of small amounts of carbon in the form of scattered patches of pearlite is not significant here. We are involved mainly with the work-hardening of ferrite. The recrystallisation temperature of ferrite is of the order of 500° C but on an industrial scale annealing at about 650° C is generally employed so that recrystallisation will proceed more quickly and a treatment time of about one hour be ample. The structure remains one of ferrite/pearlite throughout and since no phase change occurs the equilibrium diagram is not involved here.

9.5.2 *Spheroidising anneals.* The machining of high-carbon steels even in the normalised condition is difficult since the layers of pearlitic cementite provide hard barriers to cutting. The structure can be suitably modified by a prolonged annealing treatment—upwards of twelve hours—at a temperature between 650° C and 700° C, that is, just *below* A_1. Again, no phase change will occur but, due to surface energy effects, the cementite layers of the pearlite gradually break up and assume a globular form (Fig. 9.6).

Not only is machinability improved but cold-forming processes are

* In this sense the term 'pearlitic' means that the steel has cooled slowly enough for pearlite to form, not that the structure will necessarily be wholly pearlitic, which of course it will not be if less—or more—than 0·8% C is present.

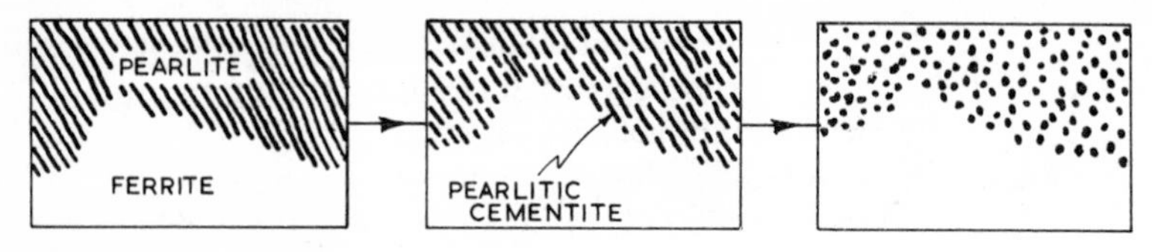

Fig. 9.6—The spheroidisation of pearlitic cementite on prolonged annealing just below A_1.

possible when the steel is in this condition. After shaping is completed the steel can be hardened or the original pearlitic structure restored by normalising.

9.5.3 *Normalising*. The general aim of this process is to refine the grain of a steel which, for some reason, has become coarse-grained. Such refinement of grain improves the mechanical properties particularly in respect of ductility and toughness. Normalising is applied to hypoeutectoid steels and involves heating the material to a temperature not more than 50° C above the A_3—or upper critical temperature. Between A_1 and A_3 the initial structure recrystallises to austenite but, and this is the important point, to austenite of *very fine grain.* After a short period of 'soaking' the steel is removed from the furnace and cooled in still air when the fine-grained austenite will transform to fine-grained ferrite and pearlite.

It is important that neither the temperature nor the time of treatment be exceeded or grain growth of the newly formed austenite will take place and give rise to coarse-grained ferrite and pearlite. Normalising also improves the machinability of hypo-eutectoid steels. The resultant surface finish is also superior to that obtained by machining an annealed component, since high ductility in the latter leads to a tendency to local tearing of the surface.

9.5.4 *Annealing of castings*. Sand castings in steel commonly contain about 0·35% C and, because of the slow rate of solidification and cooling in the sand mould, considerable grain growth takes place in the austenite whilst it is at temperatures above 1000° C. Consequently such a steel is very coarse grained by the time the temperature has fallen to A_3 and the primary ferrite tends to separate out as elongated plates at the austenite grain boundaries and also along the close-packed {111} planes *within* the coarse austenite grains (Plate 9.3). This is termed a Widmanstätten* structure and is characterised by brittleness since fracture can propagate easily along the boundaries of the coarse primary ferrite plates.

The structure can be refined by heat treatment in a process closely resembling normalising. The casting is heated to some 50° C above the

* A structure of this type was first observed on the surface of a meteorite by Aloys Beck von Widmanstätten in 1804—possibly one of the first excursions into practical metallography.

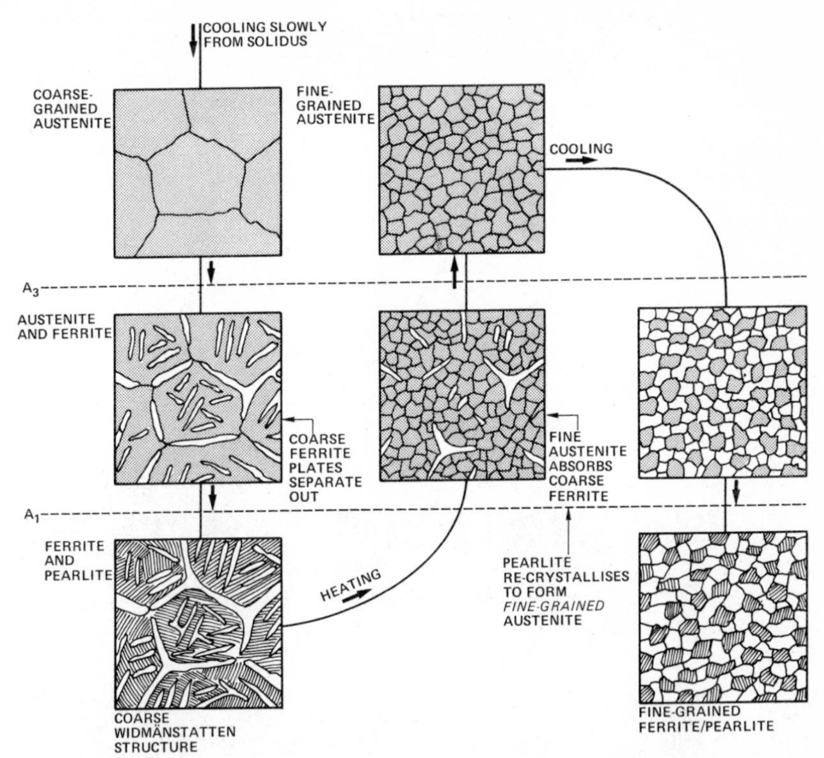

Fig. 9.7—The replacement of the coarse Widmanstätten pattern by a fine-grained structure during the annealing of a casting.

A_3 temperature, soaked for a period depending upon the section of the casting and then cooled, usually in the furnace. Air cooling would be more likely to lead to the formation of internal stresses particularly in castings of variable section which would therefore cool unevenly.

As the casting is heated to the A_1 temperature small austenite grains replace the coarse pearlite and as the temperature rises to A_3, the coarse Widmanstätten ferrite plates gradually transform to *fine-grained* austenite also. On cooling again this fine-grained austenite is replaced by fine-grained ferrite/pearlite. Whilst the tensile strength is little affected by this treatment, ductility is almost doubled and impact toughness increased considerably.

The non-equilibrium transformation of austenite

9.6 Phase transformations in solid alloys do not occur instantaneously but are very time-dependent. Many such transformations rely on diffusion which in substitutional solid solutions is inevitably a slow process (7.4). For this reason, particularly in many industrial processes, the rate of temperature fall far outstrips the ability of the alloy to maintain phase

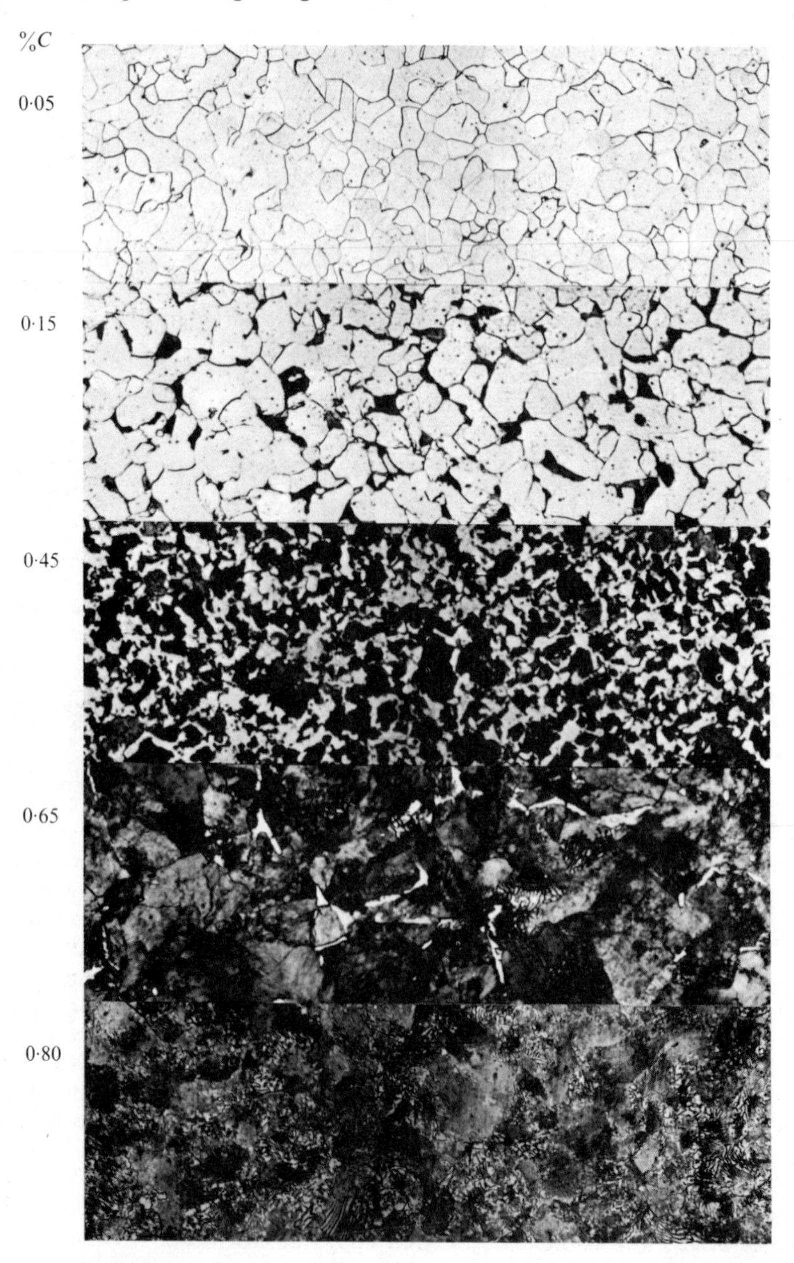

Plate 9.1—Illustrates the effect of increase in carbon content on the structures of normalised plain carbon steels. Ferrite (light) with increasing amounts of pearlite (dark). The pearlite is not generally resolved at this magnification. (Etched in 2% Nital.) × 150.

Plate 9.2—1·2% carbon tool steel, annealed. Pearlite and primary cementite (Fe_3C). Here the high magnification is sufficient to resolve the pearlite showing lamellae of cementite in a ferrite matrix. (Etched in 2% Nital.) × 1000.

equilibrium during cooling. Often when ambient temperature is reached the thermal energy of the system is too low to allow further transformation, and a non-equilibrium structure is trapped, sometimes permanently unless further heat-treatment is applied.

9.6.1 Since carbon dissolves interstitially in γ-iron it is able to diffuse comparatively quickly so that, for example, the resultant coring of carbon with respect to iron is negligible. Nevertheless when austenite is cooled rapidly below the A_1 temperature—to about 200° C or so—the driving force tending to cause transformation from FCC to BCC is so great that it occurs *before* carbon has an opportunity to form separate crystals of carbide. Consequently supersaturation of the new BCC structure with carbon is bound to result. This causes considerable distortion of the structure which in turn makes slip virtually impossible, a fact which manifests itself in the great hardness produced by quenching a steel in cold water. Such quenching of a steel from the austenitic condition gives rise to a solid solution of carbon in BCC iron which is of the order of at least one thousand times supersaturated. It appears under the microscope as acicular (needle-shaped) crystals though what is being observed is in reality a cross-section through roughly discus-shaped plates. This structure, known as *Martensite*, is single-phased since all of the carbon has been retained in solid solution.

If the degree of undercooling is less severe allowing transformation to

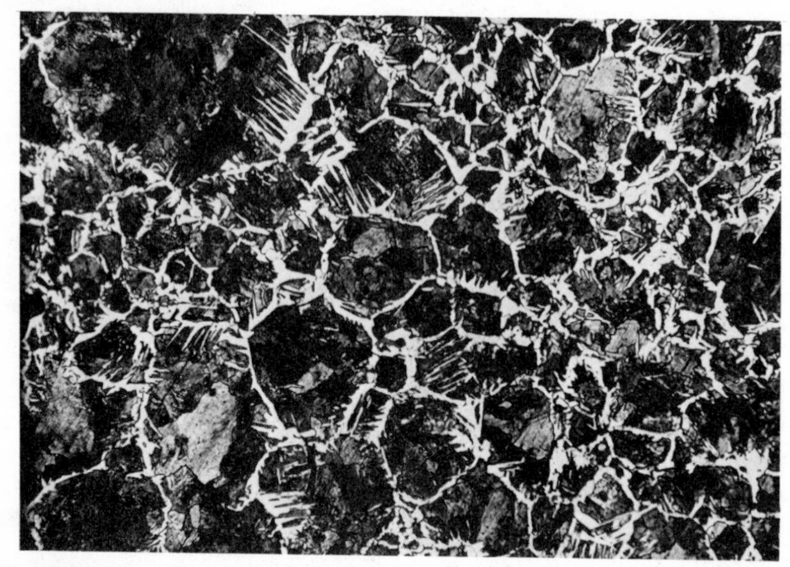

Plate 9.3—0·5% carbon steel overheated to 1100° C and furnace cooled. Ferrite (light) has precipitated along the original austenite grain boundaries and also as Widmanstätten plates. (Etched in 2% Nital.) × 150.

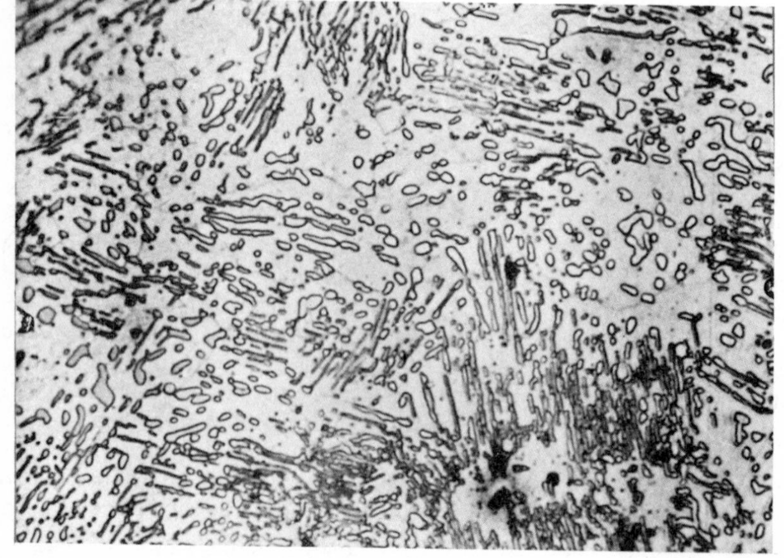

Plate 9.4—0·5% carbon steel annealed at 700° C for 12 hours to give spheroidisation of the pearlitic cementite lamellae. (Etched in 2% Nital.) × 1000.

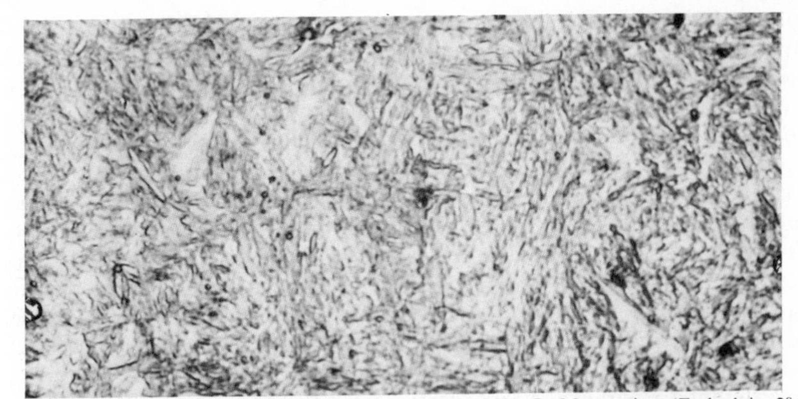

Plate 9.5—0·5% carbon steel water quenched from 850° C. Martensite. (Etched in 2% Nital.) × 1000.

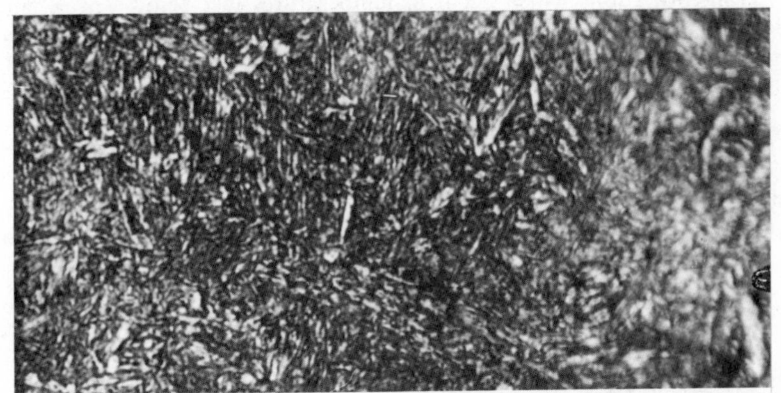

Plate 9.6—0·5% carbon steel, water quenched from 850° C and tempered at 400° C. Tempered martensite. (Etched in 2% Nital.) × 1000.

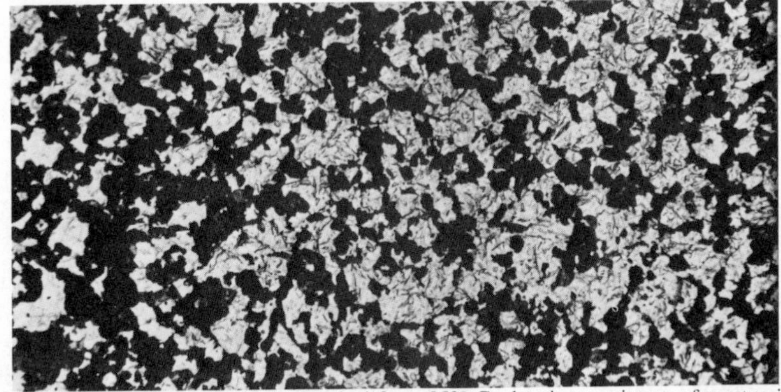

Plate 9.7—0·5% carbon steel oil quenched from 850° C, showing a mixture of martensite (light) and bainite (dark). (Etched in 2% Nital.) × 150.

proceed isothermally at some temperature between 250° C and 550° C then carbide will precipitate though not in the lamellar form found in pearlite. In this case the carbide tends to form as small particles too small to be visible except at high magnifications, and the structure is known as *Bainite*.

Whereas pearlite formation is initiated by the nucleation of carbide, bainite formation appears to be initiated by ferrite growth which in turn is followed by the precipitation of carbide particles adjacent to the ferrite. Transformation which occurs at the lower end of the range (around 250° C) gives rise to very small carbide particles and the structure, known as *lower bainite*, resembles most closely that of *martensite* in shape. Transformations occurring in the region of 500° C on the other hand, allow much larger carbide particles to form and the structure, which has a different appearance to that formed around 250° C, is usually known as *upper bainite*.

Transformations which occur at temperatures just below 723° C and down to about 550° C will give rise to pearlite. The lower the transformation temperature within these limits the finer the structure of pearlite produced.

9.6.2 The relationship between the temperature at which transformation of austenite occurs and the structure and properties produced can be studied by reference to a time–temperature–transformation (TTT) diagram. Such a diagram (Fig. 9.10) relates the temperature at which the

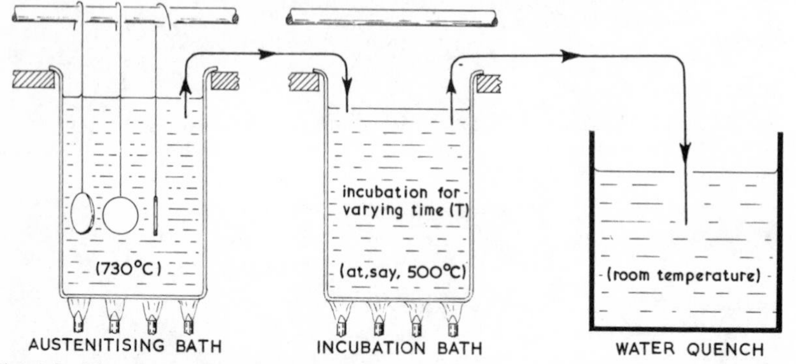

Fig. 9.8—The thermal treatment sequence for the derivation of a TTT diagram. The specimens are of roughly the dimensions of a 1p piece.

transformation of austenite is taking place with the time necessary for this to occur and the type of structure produced as outlined above. This form of diagram is produced for a steel of single composition and we shall consider one of eutectoid (0·8% C) composition.

As Fig. 9.3 indicates this steel becomes completely austenitic when heated just above 723° C so a sufficient number of specimens, all identical in size and composition (0·8% C) are heated in a bath of molten salt at 730° C so that they attain this state. About ten of these specimens are then removed simultaneously from the austenitising bath and quickly transferred to another salt bath which will be maintained at some predetermined temperature *below* 723° C. Individual specimens are then removed from this second—or *incubation*—bath after increasing time intervals and immediately quenched in cold water. This final quench effectively halts any transformation which was taking place at the incubation temperature in the second bath and any austenite which remained will be converted to martensite by the final quench.

Microscopical examination will reveal the extent to which transformation had occurred during incubation in the second salt bath (Fig. 9.9). Thus a specimen which was almost completely martensitic (A) had only just begun to transform after being held for one second at the incubation

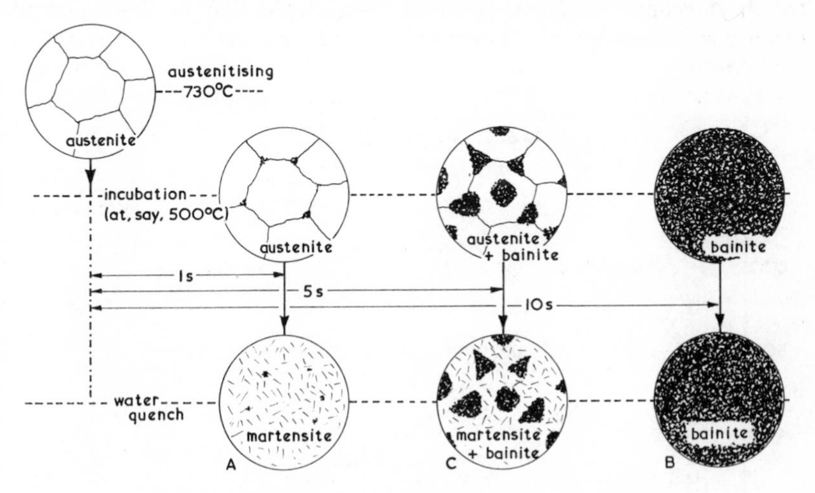

Fig. 9.9—The extent of transformation during incubation.

temperature (500° C in this case); whilst one which contained no martensite but which was almost completely bainite (B) had already transformed to bainite after ten seconds at the incubation temperature. The final water quench could then have no further effect on the structure since the transformation of austenite was already complete. Any specimen incubated at 500° C for a time intermediate between these two extremes would, on quenching in the final water bath, contain a mixture of bainite and martensite (C) indicating that transformation was incomplete after that holding time (in this case five seconds).

The same sequence of operations is then repeated with other batches of the same austenitised steel but using different incubation temperatures. At each incubation temperature the time for transformation to begin and to be completed can then be ascertained and the complete TTT curves plotted.

In practice identical small *thin* specimens of approximately the dimensions of a 1p piece are used in order that they shall attain very quickly the temperature of the incubation bath to which they are transferred. A length of temperature-resistant wire is generally fixed to the specimen to facilitate rapid transfer and quenching.

9.6.3 The resultant TTT curves plotted from information obtained by studying the *isothermal* transformations of a 0·8% C steel are shown in Fig. 9.10. The horizontal broken line representing the temperature of 723° C is of course the A_1 temperature above which the structure of the 0·8% C steel in question consists entirely of stable austenite. At temperatures below this line austenite becomes increasingly unstable and the two approximately C-shaped curves indicate the time intervals necessary for the austenite → ferrite + carbide transformation to begin and to reach completion when a specimen is cooled rapidly to the incubation temperature.

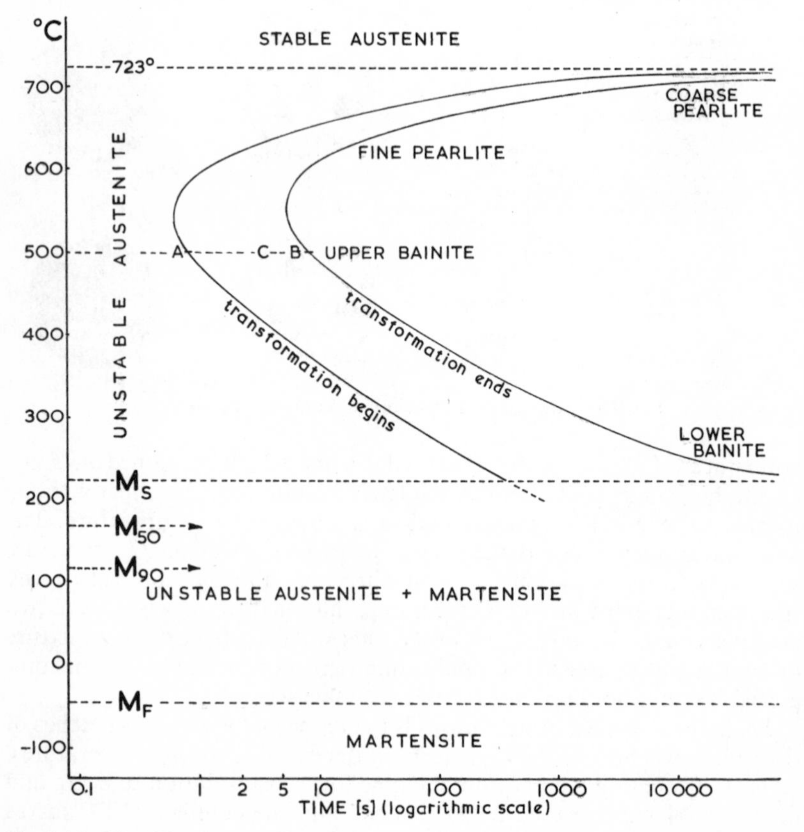

Fig. 9.10—Time Temperature Transformation (TTT) curves for a 0·8% plain carbon steel.

As might be expected transformation is sluggish at temperatures immediately below 723° C since austenite is then only slightly short of its lowest equilibrium temperature, but as the temperature falls towards 550° C both the delay in starting and the time required for completion of transformation decrease rapidly. Again this is to be expected since the greater the degree of undercooling of the austenite the greater will be its urge to transform. Both the interval before transformation begins and the period of transformation itself are at a minimum at about 550° C.* Any transformation which takes place between 723° C and 550° C is initiated by cementite precipitation since in this range the diffusion of carbon in FCC iron is rapid and the transformation product is pearlite (9·3)—coarse pearlite just below 723° C where transformation is sluggish and broad lamallae have time to form; but fine pearlite when rapid transformation occurs near 550° C.

At temperatures between 550° C and 220° C transformation becomes increasingly sluggish with decrease in temperature, for, although the FCC structure becomes increasingly unstable the further the temperature falls below 723° C, the thermal diffusion of carbon at lower temperatures becomes much slower and influences transformation. In this temperature range transformation is initiated by *ferrite* precipitation (instead of cementite as in pearlite formation) and the product, bainite, varies in appearance from a dark feathery mass for that formed in the region of 550° C (upper bainite) to an acicular structure resembling martensite for that formed in the region of 220° C (lower bainite).

9.6.4 The horizontal lines at the foot of the diagram are, strictly speaking, not part of the TTT diagram but represent the temperature range over which the transformation: austenite → martensite begins (M_s) and is about 90% complete (M_{90}). The M_f line at which transformation is 100% complete is at approximately −50° C so that most steels quenched to room temperature exhibit some 'retained austenite' in the microstructure.

It must be emphasised that these TTT curves represent *isothermal* transformations, that is, transformations which have taken place at a series of single incubation temperatures, and it would be inadmissible to superimpose on them cooling curves which represent more or less continuous rates of cooling. However, it is possible to derive modified TTT curves which are related to continuous rates of cooling. These are similar in shape to true TTT curves but are displaced somewhat to the right as indicated in Fig. 9.11.

9.6.5 On this diagram curves *X*, *Y* and *Z*, representing different continuous cooling rates, are superimposed. Curve *X* indicates a rate of cooling such as might prevail during a normalising process. Transformation of

* Note that a logarithmic scale is used.

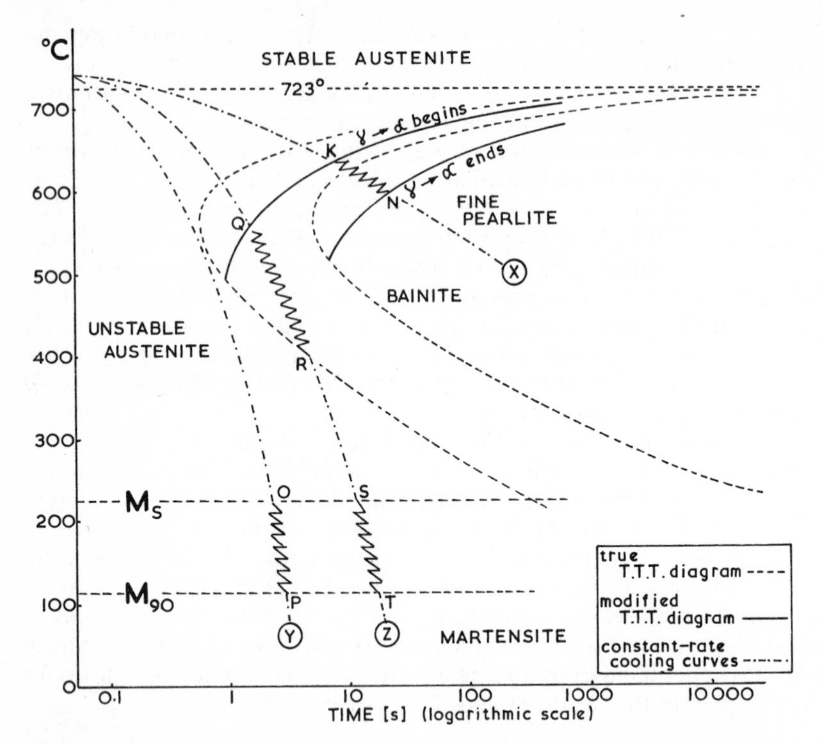

Fig. 9.11—The relationship between true TTT curves and those representing continuous-cooling conditions.

the unstable austenite begins at *K* and is complete at *N* and the product is of course fine pearlite. A very rapid fall in temperature is represented by curve *Y*. Here the 'nose' of the modified transformation-begins curve is missed so that unstable austenite persists until at *O* (on the M_s line) transformation direct to martensite begins. The rapid cooling necessary in order to avoid the nose of the transformation-begins curve explains why it is not possible to harden uniformly very heavy sections in ordinary carbon steel. Whilst the *surface* of a thick bar might cool sufficiently rapidly to be represented by curve *Y* and so become martensitic, the core of the same bar will cool more slowly and be represented, say, by curve *Z*. Thus, in the core, transformation to bainite will begin at *Q*, be arrested again somewhere in the region of *R* and the austenite which remains will begin to transform to martensite at *S*. Consequently whilst the outer skin of the bar will be totally martensitic the core will contain a mixture of martensite and softer bainite.

9.6.6 The TTT curves we have been considering are those for a 0·8% C steel. If the carbon content is either above or below this amount the curves are displaced to the left so that the critical cooling rate* required to produce a totally martensitic structure will be even greater. Hence the need for drastic water quenches in order to harden plain carbon steels becomes apparent. With a steel of less than 0·3% C the transformation-begins curve has moved so far to the left (Fig. 9.12 (i)) that it is impossible to cool a steel such that the formation of considerable amounts of ferrite in the upper temperature ranges is avoided. The resultant structure therefore contains martensite interspersed with soft ferrite.

9.6.7 An unfortunate result of this need to cool plain carbon steels so rapidly, if they are to be completely hardened, is that internal quenching strains are set up. These strains lead to distortion and, particularly with components of complex shape, to quench cracking. With rapid cooling it

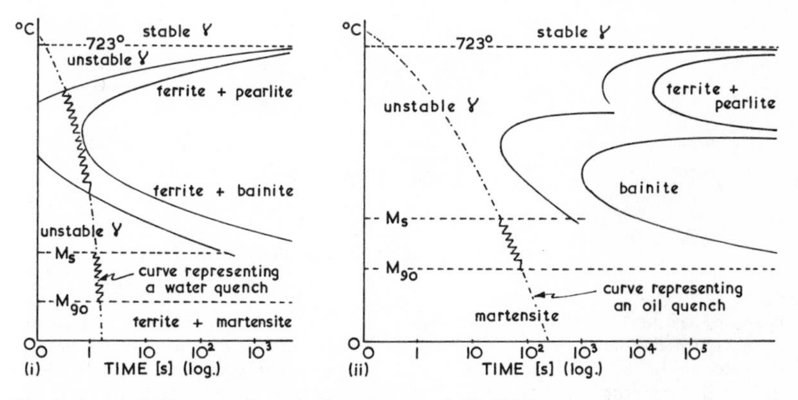

Fig. 9.12—(i) TTT curves for a 0·3% carbon steel. (ii) TTT curves for a low-alloy steel.

is inevitable that the outer skin and the inner core of a piece of steel will pass through the FCC → BCC transformation after *different time intervals*. Since a volume change accompanies this transformation (3.6.3) it follows that the skin, as it changes to BCC, will tend to expand away from the core which at that instant is still FCC in structure. Internal strain is set up and this is very likely to cause cracking of the skin which has just transformed to hard, brittle martensite.

Fortunately the addition of alloying elements like nickel, chromium or molybdenum to a steel reduces considerably the rate of cooling necessary to produce a martenistic structure. Thus the TTT curves are displaced to the right (Fig. 9.12 (ii)). This makes it possible to oil quench—or even air harden—a suitable steel and still obtain a hard martensitic structure throughout.

* Represented by a cooling curve which just grazes the nose of the relevant transformation-begins curve.

Mass effects and hardenability

9.7 It will have become obvious from the foregoing account that there is a maximum cross-section for carbon steels which, if exceeded, will not be uniformly martensitic even when quenched at the greatest possible rate from the austenitic state. Thus whilst a small-section bar (Fig. 9.13) will have cooled rapidly enough for martensite to have been formed throughout, thicker bars will contain bainite or even fine pearlite in those regions where austenite did not cool fast enough to avoid cutting into the nose of the appropriate transformation-begins curve. The displacement of TTT

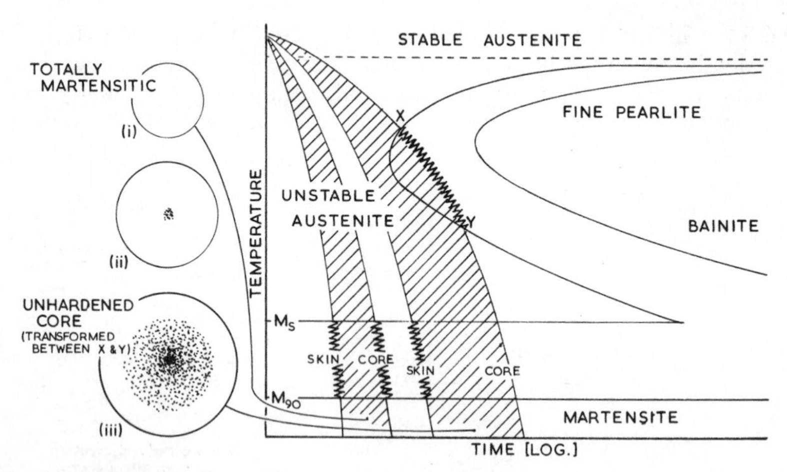

Fig. 9.13—The mass effects of heat-treatment. The heavier sections (ii) and (iii) will cool too slowly for martensite to form throughout; (ii) is of slightly greater diameter than the ruling section, but (i) has hardened completely throughout. The core of (iii) has cooled so slowly that transformation to bainite and pearlite has taken place between *X* and *Y*.

curves to the right due to the addition of alloying elements to a carbon steel makes it possible to oil- or air-harden quite heavy sections of suitably alloyed steels. However, to prevent the misuse of their steels by over-optimistic customers who may assume incorrectly that any alloy steel can be oil-hardened in heavy sections, manufacturers generally specify a *ruling section* for a particular alloy steel. This ruling section is the maximum diameter which can be hardened (using the recommended heat-treatment programme) if the stated mechanical properties are to be obtained. If this diameter is exceeded then the mechanical properties may vary across the section since the core will not have cooled rapidly enough to have transformed to martensite during quenching. Any subsequent tempering will not produce the desired structure or properties.

9.7.1 The *Jominy end-quench test* is of practical value in assessing the 'depth of hardening' of a steel. Here a standard test piece (Fig. 9.14) is heated to its austenitic range, dropped into position in a suitable holding

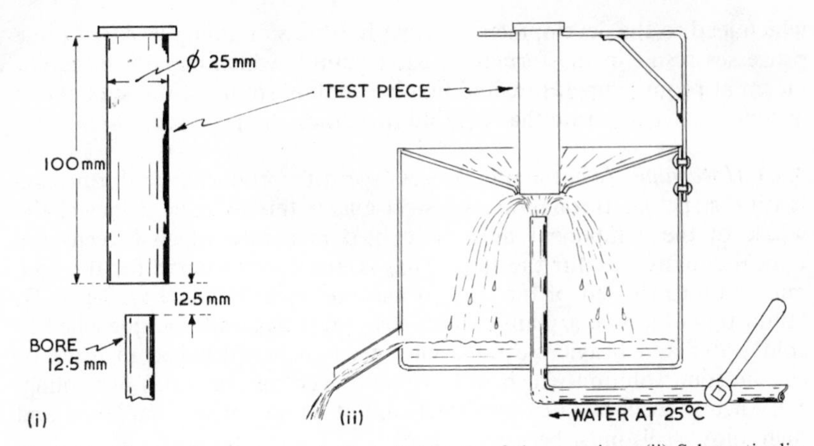

Fig. 9.14—The Jominy End-quench Test. (i) Details of the test-piece. (ii) Schematic diagram of the test apparatus.

jig as shown and then quenched at its end by a jet of water at standard temperature and pressure. In this way different rates of cooling are achieved along the length of the test piece. When it is cool, a 'flat' about 0·4 mm deep is ground along the side of the test piece and hardness determinations made every millimetre along the length from the quenched end. The results are then plotted (Fig. 9.15). Whilst there is no simple relationship between the results of this test and the ruling section of a steel, it is of value in making a preliminary estimation of the latter.

Heat-treatments based on the non-equilibrium transformation of austenite

9.8 Heat-treatments most popularly associated with steels are those

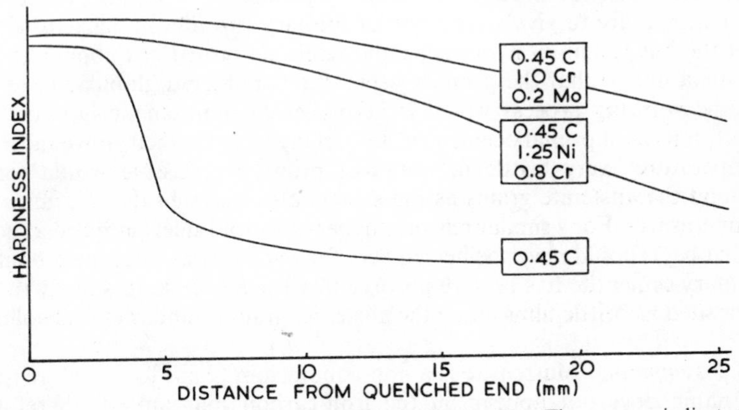

Fig. 9.15—Depth of hardening as indicated by the Jominy Test. These curves indicate that low-alloy steels harden to a greater depth than do plain carbon steels of similar carbon content.

which lead to the development of great hardness or strength. All of these processes result in the formation of structures which are not in equilibrium at room temperature and involve cooling from the range of stable austenite at such a rate that pearlite formation is prevented.

9.8.1 *Hardening*. A hardening process seeks to produce a uniform martensitic structure throughout the steel and if this is to be achieved the whole of the component must be cooled at a rate which exceeds the critical cooling rate for the steel. That is, the cooling curve for the core must pass to the left of the transformation-begins curve (Fig. 9.13). To attain this objective a plain carbon steel must generally be quenched in cold water and heavier sections may need to be quenched in brine or caustic soda solution which will promote even quicker rates of cooling. Very heavy sections in carbon steel cannot be completely hardened and then alloy steels must be used instead.

A major difficulty associated with water quenching plain carbon steels is that distortion and cracking are always likely to occur (9.6.7). Less drastic quenching in oil can be used for thin sections in carbon steel and for heavier sections in low-alloy steels and a martensitic structure still obtained. In all quenched steels varying amounts of austenite may be retained since the M_f line is at sub-zero temperatures. Such retained austenite will give rise to soft spots but can be made to transform to martensite by further treatment in solid carbon dioxide or liquid oxygen as required.

A hypo-eutectoid steel is generally hardened by heating it to a temperature just above A_3 so that it becomes completely austenitic, and then quenching it in the appropriate medium. Symmetrical components like axles should be quenched 'end-on' and agitated vigorously so that they cool as uniformly as is possible thus minimising distortion. Hyper-eutectoid steels are usually hardened by quenching from just above the A_1 temperature to give a structure of primary carbides in martensite. In fact the hot-rolling of hyper-eutectoid steels is finished just above the A_1 temperature so that the primary cementite is rendered globular in form by the working process which is taking place simultaneously with the precipitation of primary cementite. If working were finished above the A_{cm} temperature* very brittle networks of primary cementite would form around the austenite grains as the steel cooled between the A_{c_m} and A_1 temperatures. For a similar reason a hyper-eutectoid steel (supplied as hot-rolled bar) should never be heated very far above A_1 as once the globular primary cementite has been re-absorbed by the austenite it is likely to be deposited as brittle films along the austenite grain boundaries on cooling.

9.8.2 *Tempering*. Martensite is a non-equilibrium phase for which reason its name does not appear on the iron-carbon diagram. If, therefore,

* The temperature at which primary cementite begins to separate from the austenite, in a hyper-eutectoid steel.

martensite is heated the structure is enabled to proceed in some degree towards equilibrium depending upon the temperature used. Essentially tempering causes the dispersed dissolved carbon atoms to precipitate as carbide particles which increase in size as the temperature increases.

At low temperatures (between 100° C and 200° C) extremely thin platelets of ϵ-carbide, which is of different composition to Fe_3C, are formed and the carbon content of the remaining martensite falls to about 0·3% as a result. At first the hardness increases slightly (Fig. 9.16) as carbide precipitation begins but then begins to fall again as the amount of precipitated carbide increases. At about 400° C the ϵ-carbide begins to change to ordinary Fe_3C and low-carbon martensite starts to revert to ferrite. These changes are accompanied by a fall in hardness but a corresponding increase in ductility and toughness.

During low-temperature tempering the original martensite becomes dark in appearance due to the precipitation of the ϵ-carbide particles. These particles are too small to be visible using an optical microscope. When higher tempering temperatures are used the structure loses its original martensitic appearance and globular carbide particles in a ferrite matrix can be seen at high magnifications using an optical instrument. The former terms of 'troostite' and 'sorbite' to describe these respective structures are now generally obsolete and metallurgists prefer to speak in terms of 'tempered martensite'.

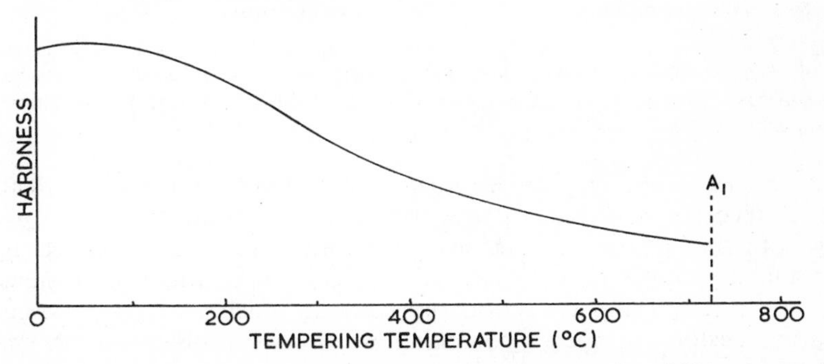

Fig. 9.16—The variation of hardness with tempering temperature for a 0·8% carbon steel.

At tempering temperatures approaching A_1 precipitation of Fe_3C is such that the structure resembles that produced by a spheroidising anneal (9.5.2). In fact this treatment is often used to spheroidise a steel following a suitable quenching operation since it is quicker and gives a more regular dispersion of cementite.

Plain-carbon (and also low-alloy) steels are usually tempered below 250° C where hardness and wear-resistance are of paramount importance. When greater ductility is required, as in stress-bearing parts in aircraft and automobile engineering, then higher tempering temperatures up to 650° C are employed.

9.8.3 *Isothermal treatments.* These treatments are based on information obtained about the isothermal transformations of steels as embodied in the TTT curves (9.6.2), though it must be admitted that many old industrial processes such as the 'patenting' of steel wire predated the *theoretical* consideration of such matters by many years. The principal advantage of isothermal transformation is that it reduces the possibility of internal stresses being generated due to rapid cooling through the M_s–M_f (or more properly M_s–M_{90}) range.

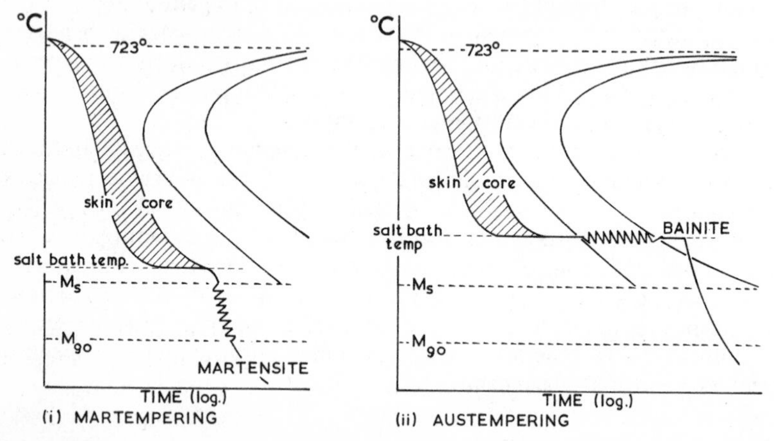

Fig. 9.17—Isothermal treatments (i) martempering and (ii) austempering. In each case the steel remains in the salt bath for long enough for a *uniform temperature to be attained throughout* before transformation of the unstable austenite begins. This is the most important feature of these processes.

9.8.3.1 *Martempering.* This is rather an unfortunate title for this treatment since it is, in no sense, a tempering process. For that matter it is not an isothermal treatment either since transformation to martensite occurs on a falling temperature gradient. The object of the treatment is of course to minimise warping, distortion and cracking which are likely to occur during customary quench-hardening which might otherwise be employed.

The treatment involves heating the steel to its austenitising temperature and then quenching it *as rapidly as possible* in a molten salt bath which is maintained just above the M_s temperature. The steel is held in the salt bath for long enough to permit its structure to attain the same temperature throughout (Fig. 9.17 (i)). It is then withdrawn from the salt bath and allowed to cool *quite slowly* to room temperature (remember the TTT diagram employs a logarithmic time axis). As the diagram indicates both the skin and the core of the component will then pass through the M_s–M_{90} range more or less simultaneously so that distortion arising from the generation of internal stresses will be minimised. A uniform martensitic structure will have been produced.

9.8.3.2 *Austempering.* The object of this treatment is again to eliminate the need for a drastic water quench and the 'thermal shock' associated with it. In this case a true isothermal transformation to bainite is involved. In so far as *mechanical* properties are concerned bainite produced by isothermal transformation is as satisfactory as the parallel structure produced by tempering martensite (9.8.2). The former is initiated by ferrite separation, the latter by carbide precipitation. Thus, by using isothermal transformation the need for an extra tempering process is eliminated but, more important still, the customary severe quench needed to precipitate the birth of martensite becomes unnecessary. Moreover the toughness and ductility of austempered steels is often greater than than of similar steels hardened and tempered in the orthodox ways to similar hardness values.

Again, as Fig. 9.17 (ii) shows, it is necessary to cool the steel rapidly past the nose of the transformation-begins curve. However, the quenching-bath temperature, which is always of course above M_s, is adjusted to give bainite of the desired form. The steel is allowed to remain in the quenching bath long enough for transformation to bainite to be complete, when it can be cooled to room temperature at any convenient rate.

In view of the obvious advantages of these isothermal treatments the reader may well ask: 'Why are the old methods of quenching/tempering still used?' The answer lies in the fact that successful isothermal treatments are dependent upon cooling the steel rapidly enough to miss the 'nose' of the transformation-begins curve, otherwise fine pearlite will form in the range 500°–600° C. Thus, as far as *plain-carbon* steels are concerned these processes are limited to the treatment of materials not more than 10 mm thick which can be cooled sufficiently rapidly. For example, austempering is a useful treatment for springs, pins, needles, screws, garden spades and similar thin components. High-carbon steel wire too is similarly treated.

With alloy steels the scope is greater because of the displacement of the TTT curves which indicate the lower critical cooling rates prevailing. In such cases an oil quench will be very roughly equivalent to martempering since the cooling curve representing an oil quench levels out near to the M_s line because of the higher boiling point of oil as compared with that of water.

Surface hardening of steels

9.9 Frequently components like gears and axles need to be tough and ductile but, at the same time, have a hard, wear-resistant surface. Ductile materials are usually soft whilst hard materials are generally brittle, so the most effective way to achieve this difficult combination of properties is to use a steel which has a surface of different composition from that of the core; as is contrived by carburising or nitriding. Alternatively a steel of uniform carbon content can be used and its surface heat-treated in a different manner from the core.

9.9.1 *Case-hardening.* This is a title generally given to those processes which involve carburising the surface of a tough, ductile low-carbon steel. Since carbon dissolves interstitially in FCC iron it follows that the low-carbon steel employed must be heated to above A_3 (about 900°–950° C) in an atmosphere which will release free carbon atoms. Treatment time must be sufficient to produce a 'case' of the required depth.

Solid, liquid and gaseous carburising media are used. Solid media contain some form of carbon such as charcoal along with an activator like barium carbonate. The components are packed into steel boxes with this medium and heated for up to five hours. Liquid media are based on molten sodium cyanide/sodium carbonate mixtures. The poisonous nature of these mixtures and the disposal of process waste poses problems in the modern world. Most large-scale industrial carburising now employs gaseous media, such as town gas, natural gas or other hydrocarbons like propane. Gaseous methods are clean, rapid and efficient.

Following carburising heat-treatment of the component is necessary. In many instances quenching direct from the carburising medium is employed but the counsel of perfection is to use a double heat-treatment process. The core structure is first refined by normalising the component just above the A_3 temperature *for the core.* The component is then reheated to about 760° C which is above the A_3 temperature for the *case*, and then quenched.

9.9.2 *Nitriding.* This process involves the interstitial solid solution of nitrogen atoms to form a very hard skin on the surface of the component. Special steels containing either 1·0% aluminium or small amounts of chromium, molybdenum and vanadium are used so that hard nitrides of these metals will form in the surface layers. The charge is heated at 500° C for upwards of forty hours in an atmosphere of ammonia. This dissociates:

$$NH_3 \rightleftharpoons 3H + N$$

and the *atomic* nitrogen released dissolves interstitially in the surface of the BCC iron.

The main advantages of nitriding as compared with carburising are:

(1) Since no quenching is required after nitriding, cracking and distortion are unlikely.

(2) High surface hardness up to 1150 *VPN* is obtainable with some of the special 'Nitralloy' steels.

(3) Resistance to corrosion and resistance to fatigue are good.

(4) Hardness is retained up to 500° C whereas a hardened carburised component begins to soften at 200° C due to normal effects of tempering.

The principal disadvantage of nitriding is that the plant outlay is high whereas components can be carburised in solid media quite cheaply. As with many processes the question of unit cost is dependent largely upon the scale of 'throughput'.

9.9.3 *Induction and flame hardening*. In these processes a steel of uniform carbon content—usually about 0·4% C—is used. The component is first normalised to give a tough, ductile, pearlitic core. Its surface is then heated, either by a gas torch or by high-frequency induction methods, and, almost immediately, quenched by means of water jets. This produces a hard martensitic skin on top of the tough pearlitic core. The two zones are generally separated by a layer of bainite which reduces the probability that the case will flake away. Symmetrical components such as gears and axles can be spun between centres during heating and quenching, thus ensuring a more uniform structure.

Chapter Ten
Alloy Steels and Engineering Cast Irons

10.1 Man's early ferrous implements were of wrought origin, forged from meteoric iron and, later, from the spongy mass produced by the low-temperature reduction of iron ore. Development of furnaces which could reduce and *melt* iron and so produce *cast* iron began in about 1000 B.C. in China whence the technology spread to India before the start of the Christian era. The process was independently, if somewhat belatedly, discovered in Europe in the fourteenth century A.D. The World's first metal bridge—that at Ironbridge in Shropshire—was built of cast iron in 1779. The use of cast iron was widespread in Victorian times for ornamental (?) as well as for engineering purposes, whereas it is now used solely as a sophisticated engineering material with a highly developed technology.

10.1.1 Though the first alloy steel was a high-manganese alloy originated by Hadfield in 1882, commercial alloy steels are of this century. Thus, high-speed steel was first produced in the USA by Taylor and White in 1900 and the first stainless steel by Brearley of England in 1913, though it did not attain commercial status until much later.

In the foregoing chapter the structures, properties and heat-treatments of ordinary carbon steels were outlined. Here we shall deal with some of the more complex iron-base alloys which depend for their specific properties on the presence of other elements in addition to carbon.

Alloy steels

10.2 Alloying elements are added to carbon steels to improve their existing properties but, in some cases, also to introduce new properties such as corrosion resistance. Small amounts of manganese (0·3–0·8%) are always present in so-called plain-carbon steels as a residue from deoxidation and desulphurisation. Such manganese has a measurable effect in reducing the critical cooling rate for a steel. However, it is not regarded as an alloying element unless it exceeds 2·0%, whilst 0·1% molybdenum or vanadium; 0·3% tungsten or cobalt, and 0·5% chromium or nickel are quantities of these elements which must be exceeded before they qualify as alloying elements.

The principal effects which the main alloying elements have on the structure and properties of steels can be classified as follows:

10.2.1 *Increase in tensile strength.* Most of the elements used in alloy steels dissolve substitutionally to some extent and, predictably, increase the tensile strength as a result. This of course is a general effect of alloying (7.5). In low-alloy steels the ferrite is strengthened in this way whilst high ductility and toughness are maintained. Though metallurgically unspectacular, this is one of the most important features of alloying in steels.

10.2.2 *Changes in the polymorphic transformation temperatures.* The polymorphic transformation temperatures involved here are the A_3, at 910° C in pure iron, where the $\alpha \rightleftharpoons \gamma$ transformation occurs, and the A_4, at 1400° C, where the $\gamma \rightleftharpoons \delta$ change occurs.

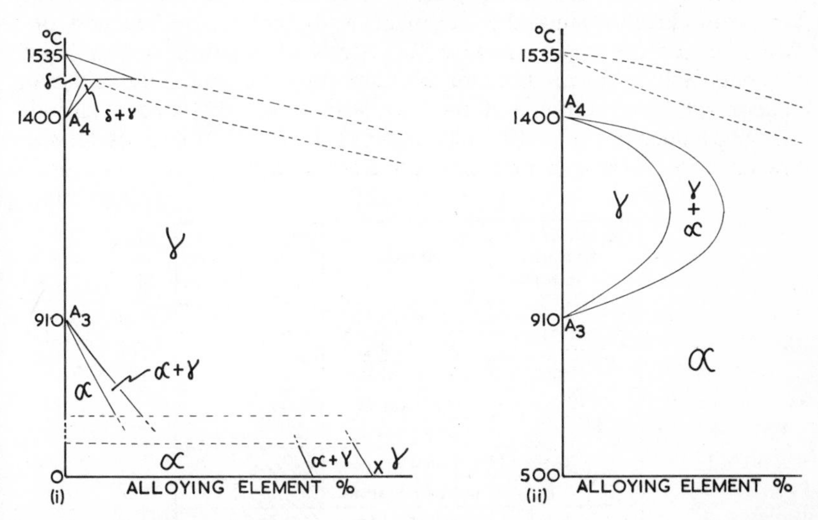

Fig. 10.1—The effects of alloying additions on the A_3 and A_4 temperatures in iron. (i) Elements which tend to stabilise γ, i.e. Ni, Mn, Cu and Co. (ii) Elements which tend to stabilise α, i.e. Cr, W, Mo, V, Al and Si.

Some elements, notably nickel, manganese, copper and cobalt, raise the A_4 temperature and depress the A_3 as indicated in Fig. 10.1 (i). Consequently, when added to a carbon steel these elements will tend to stabilise austenite (γ) by increasing the temperature range over which it can persist as a stable phase. Many of these elements are FCC in structure at room temperature thus favouring the retention of the FCC austenite in which they are substitutionally dissolved and retarding its transformation to BCC ferrite. Moreover since most of these elements do not form carbides, the carbon tends to remain in solid solution in the austenite. When austenite is stabilised at room temperature by amounts of the alloying element in excess of X (Fig. 10.1 (i)) then the steel will also lose its ferromagnetism.

Other elements, the most important of which are chromium, molybdenum, tungsten, vanadium, aluminium and silicon, have the opposite effect in that they stabilise ferrite (α) by raising the A_3 temperature and depressing the A_4. These elements restrict the field over which austenite exists, forming what is generally termed a 'γ-loop' (Fig. 10.1 (ii)) in the appropriate diagram. Most of these elements are BCC in structure at room temperature so it is not surprising that they stabilise BCC ferrite. Since some of them (the Cr–Mo–W group of the transition metals) form very stable carbides they favour the precipitation of carbon in this form and further aid the $\gamma \rightarrow \alpha$ transformation.

10.2.3 *Relative stability of carbides.* Some elements form very stable carbides when added to a carbon steel. Many of these carbides are harder than iron carbide. Since these elements will increase the hardness of a steel they are generally used in tool steels of superior quality. Such elements include chromium, molybdenum, tungsten, vanadium and manganese; whilst titanium and niobium also have very strong carbide-forming tendencies (Fig. 10.2). Complex carbides are often present when one or more of these elements are added to a steel.

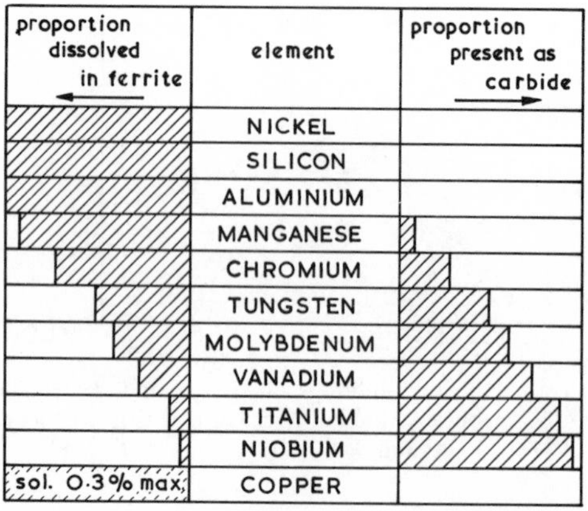

proportion dissolved in ferrite ←	element	proportion present as carbide →
	NICKEL	
	SILICON	
	ALUMINIUM	
	MANGANESE	
	CHROMIUM	
	TUNGSTEN	
	MOLYBDENUM	
	VANADIUM	
	TITANIUM	
	NIOBIUM	
sol. 0·3% max.	COPPER	

Fig. 10.2—The relative carbide-forming tendencies of alloying elements in steel.

Not all of the alloying elements used in steel have a chemical affinity for carbon when in the presence of iron. Thus nickel, aluminium and silicon do not form carbides in the presence of iron and in fact cause instability of Fe_3C so that carbon is precipitated in the form of graphite. Therefore if it is necessary to add appreciable amounts of these elements to steel it can only be done in the presence of one or more of the carbide-stabilising elements already mentioned. For this reason there are no high-carbon steels containing nickel as the sole alloying element.

10.2.4 *Influence on grain growth.* During heat-treatment, particularly at high temperatures, grain growth of austenite will inevitably take place. The presence of some elements will accelerate this grain growth and so increase the resultant brittleness of the component. The most important element which produces this unwelcome effect is chromium. Consequently care must be taken in heat-treating chromium steels to avoid undue over-heating or using excessive treatment times.

Fortunately grain growth is rendered more sluggish by the presence of some elements, notably nickel and vanadium. Hence these elements are generally classed as 'grain refiners'.

10.2.5 *Displacement of the eutectoid point.* The addition of any of the important alloying elements to a carbon steel will cause displacement of the eutectoid point such that the structure will be totally pearlitic even

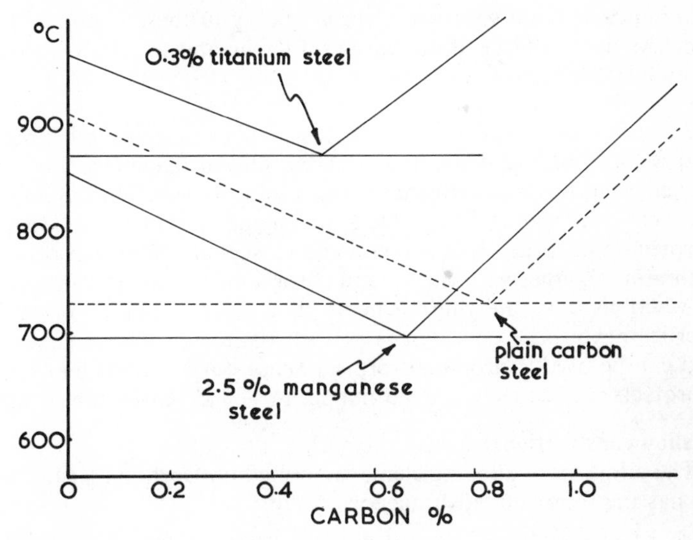

Fig. 10.3—The effects of manganese and titanium on the displacement of the eutectoid point in steel.

though the steel contains less than 0·8% carbon. For example, the addition of 2·5% manganese to a steel containing 0·65% carbon will produce a completely eutectoid structure (Fig. 10.3), whilst a stainless cutlery steel containing 13% chromium and 0·35% carbon is also eutectoid. The eutectoid temperature may be raised or lowered (Fig. 10.3), the A_1 temperature being displaced in sympathy with A_3 as mentioned above (10.2.2). Hence alloy steels generally contain less carbon than plain carbon steels used for similar purposes.

10.2.6 *Retardation of transformation rates.* We have seen (9.6.7) that the addition of alloying elements tends to reduce the critical cooling rates necessary to obtain a completely martensitic structure in steels. This is indicated by the displacement of the TTT curves to the right in Fig. 9.12(ii).

All alloying elements, with the exception of cobalt, have this effect and it is one of the most useful features imparted by alloying since it renders possible oil quenching or even air hardening as alternatives to a drastic water quench. For example, a steel containing 4·25% nickel and 1·25% chromium can be air hardened in thin sections or oil hardened in sections up to 150 mm diameter. Atoms of these elements which are dissolved substitutionally impede the $\gamma \rightarrow \alpha$ transformation.

Extremely small amounts of boron (up to 0·005%) may be added to low-carbon/low-alloy steels in order to improve depth of hardening. The atomic size of boron renders it too large for easy interstitial solid solution but too small for ready substitutional solid solution. It is thought that the boron atoms therefore migrate to lattice imperfections which exist at grain boundaries thus decreasing strain energy in these regions. This in turn delays the incidence of nucleation of ferrite and so retards the transformation of austenite.

10.2.7 *Improvements in corrosion resistance.* Those metals which become coated with a thin but dense and adherent film of oxide (15.2) are often protected from further corrosion by this useful barrier. Thus aluminium, a relatively reactive metal, is stable as an engineering material because of the protection afforded by the dense and closely adherent film of Al_2O_3. Aluminium, together with silicon and chromium, impart this property to steel when used as alloying elements, but when resultant mechanical properties have to be considered, only chromium of the elements mentioned can be used. Chromium forms a dense but adherent film on steel and protects it effectively when a minimum of 13% chromium is used.

Low-alloy constructional steels

10.3 The addition of alloying elements to medium-carbon constructional steels has the following advantageous results:

(1) Mechanical properties, but in particular yield strength, tensile strength, toughness and ductility are improved.

(2) Critical transformation rates are considerably reduced so that a steel can be hardened by oil quenching or air cooling.

10.3.1 Nickel is useful in promoting these properties since it strengthens ferrite by solid solution and improves toughness by its tendency to impede grain growth. Unfortunately nickel also favours graphitisation of any cementite present so that, in a medium-carbon steel, it is generally accompanied by a carbide stabiliser such as chromium, which will also assist it in strengthening ferrite and reducing transformation rates. Such steels contain rather more than two parts of nickel to one of chromium, maximum quantities being 4·25% Ni and 1·25% Cr, giving an air-hardening steel. These nickel–chromium steels regrettably suffer a fault known as 'temper brittleness'. This manifests itself as a very low impact toughness value in a nickel–chromium steel which has been tempered in

Table 10.1—Relative effects of alloying elements on steel.

Element	*Influence on the properties of steel*	*Uses in steel*
Nickel	Stabilises γ by raising A_4 and depressing A_3. It is the universal grain refiner in alloy steels (and many non-ferrous alloys). Strengthens ferrite by solid solution. Unfortunately a powerful graphitiser	In amounts up to 5% as a grain refiner in case-hardening steels. Along with chromium and molybdenum in low-alloy constructional steels. In larger amounts in stainless and heat-resisting steels
Manganese	Like nickel it stabilises γ but unlike nickel it forms stable carbides	Low-manganese steels are not widely used though in recent years it has been used to replace small amounts of more expensive alloying elements, e.g. nickel. The high-manganese (Hadfield) steel contains 12·5% Mn and is austenitic but hardens on abrasion
Chromium	Stabilises α by raising A_3 and depressing A_4. Forms hard stable carbides. Strengthens ferrite by solid solution. In amounts above 13% imparts stainless properties. Unfortunately increases grain growth	In small amounts in constructional and tool steels. Also in ball bearings. In larger amounts in stainless and heat-resisting steels.
Molybdenum	Strong carbide-stabilising influence. Raises the high-temperature creep strength of suitable alloys. Imparts some sluggishness to tempering influences	Reduces 'temper brittleness' in nickel–chromium steels. Increases red-hardness of tool steels. Now used to replace some tungsten in high-speed steels
Vanadium	Strong carbide-forming tendency. Stabilises martensite and increases hardenability. Like nickel it restrains grain growth. Induces resistance to softening at high temperatures once the steel is hardened	Used in steels required to retain hardness at high temperatures, e.g. hot-forging dies, extrusion dies, die-casting dies. Also increasingly in high-speed steels
Tungsten	Has similar effects to chromium in stabilising α. Also forms very hard carbides. Renders transformations very sluggish—hence, once hardened, a steel resists tempering influences	Used mainly in high-speed steels and other tool and die steels, particularly for use at high temperatures
Cobalt	Induces sluggishness to the transformation of martensite. Hence increases 'red hardness'	Super high-speed steels and 'marageing' steels. Permanent magnet steels and alloys
Silicon	A strong graphitising influence—hence not used in high-carbon steels. Imparts casting fluidity. Improves oxidation resistance at high temperatures	Up to 0·3% in steels for sandcastings where it improves fluidity. In some heat-resisting steels (up to 1·0%)

the range 250–400° C. Tensile properties, however, are not substantially impaired. Fortunately this defect can be minimised by adding about 0·3% molybdenum, thus giving rise to the popular range of 'nickel–chrome–moly' steels. Constructional steels of this type are used for the manufacture of axles, shafts, gears, connecting rods, bolts, brackets, etc.

10.3.2 Of recent years greater use has been made of small additions of the relatively cheap alloying element manganese to replace some nickel and chromium in the cheaper grades of low-alloy steels. These steels are used quite widely in the automobile industry. Manganese promotes increased depth of hardening by reducing critical cooling rates. In low-carbon steels it effectively reduces the temperature at which brittle fracture (16.3) will take place. Compositions, heat-treatments, properties and uses of some popular low-alloy constructional steels are given in Table 10.2.

10.3.3 Although most of the constructional alloy steels are in the low-alloy class, a few high-strength/high-alloy steels are produced. These are the 'marageing' steels, typical of which is one containing 18% Ni, 8% Co, 4% Mo and up to 0·8% Ti. Since such a high nickel content will have a considerable graphitising influence carbon must be kept low—about 0·03%. Such a steel is first solution treated (11.3.2) to absorb precipitated compounds. On air-cooling the structure will become martensitic because of the reduction in the transformation rates incurred by the high-alloy content. Unlike ordinary tetragonal martensite this is of BCC form and hence much softer and tougher. However, the steel can now be *precipitation hardened* (11.3.2) at 500° C for two hours so that coherent precipitates of compounds such as $TiNi_3$ are formed and tensile strengths of over 2000 N/mm^2 developed. The term 'marageing' refers of course to the 'age-hardening'—or 'ageing'—of martensite. Such steels are naturally very expensive and their use is generally limited to sophisticated technology such as rocketry.

Alloy tool steels

10.4 Alloying elements are added to tool steels:

(1) To increase hardness and wear-resistance by the formation of carbides which are much harder than Fe_3C.

(2) To impart resistance to the tempering effects associated with high-temperature working. This applies not only to exposure to high temperatures when these steels are used in hot-working dies, but also to frictional heat generated in cutting tools during high-speed machining processes. A correctly hardened high-speed steel will resist tempering up to dull-red heat.

The alloying elements involved here are the refractory metals chromium, tungsten, molybdenum and vanadium, all of which form very hard stable carbides. Moreover since they are all BCC metals they limit

Table 10.2—Low-alloy constructional steels.

Type of steel	Relevant Spec'n; B.S. 970:	Composition (%)	Condition	Mechanical properties Yield stress (N/mm^2)	Tensile stress (N/mm^2)	Elongation (%)	Izod (J)	Heat-treatment	Uses
Low manganese	150M28	0·28 C 1·50 Mn	Normalised	355	587	20	—	Oil-quench from 860° C (water-quench for sections over 38 mm diameter). Temper as required	Automobile axles, crankshafts, connecting rods, etc., where a relatively cheap steel is required
Nickel–manganese	503M40	0·40 C 0·90 Mn 1·00 Ni	Quenched and tempered at 600° C	494	695	25	91	Oil-quench from 850° C; temper between 550° and 660° C and cool in oil or air	Crankshafts, axles, connecting rods; other parts in the automobile industry and in general engineering
Manganese–molybdenum	608M38	0·38 C 1·50 Mn 0·50 Mo	28·5 mm bar, o.q. and tempered at 600° C	1000	1130	19	70	Oil-quench from 830°–850° C; temper between 550° and 650° C and cool in oil or air	A substitute for the more highly alloyed nickel–chrome–molybdenum steels
Nickel–chromium	653M31	0·31 C 0·60 Mn 3·00 Ni 1·00 Cr	28·5 mm bar, o.q. and tempered at 600° C	819	927	23	104	Oil-quench from 820°–840° C; temper between 550° and 650° C. Cool in oil to avoid 'temper brittleness'	Highly stressed parts in automobile and general engineering, e.g. differential shafts, stub axles, connecting rods, high-tensile studs, pinion shafts
Nickel–chromium–molybdenum	817M40	0·40 C 0·55 Mn 1·50 Ni 1·20 Cr 0·30 Mo	O.q. and tempered at 200° C O.q. and tempered at 600° C	— 988	2010 1080	14 22	27 69	Oil-quench from 830°–850° C; 'light temper' 180°–200° C; 'full temper' 550°–650° C—cool in oil or air	Differential shafts, crankshafts and other highly-stressed parts where fatigue and shock resistance are important. In the 'light tempered' condition it is suitable for automobile gears. Can be surface hardened by nitriding
	835M30	0·30 C 0·55 Mn 4·25 Ni 1·25 Cr 0·30 Mo	Air-hardened and tempered at 200° C	1470	1700	14	35	Air-harden from 820°–840° C; temper at 150°–200° C and cool in air	An air-hardening steel for aero-engine connecting rods, valve mechanisms, gears, differential shafts and other highly-stressed parts. Suitable for surface hardening by cyanide or carburising
Manganese–nickel–chromium–molybdenum	945M38	0·38 C 1·40 Mn 0·75 Ni 0·50 Cr 0·20 Mo	28·5 mm bar, o.q. from 850° C and tempered at 600° C	958	1040	21	85	Oil-quench from 830°–850° C; temper at 550°–660° C, and cool in air	Automobile and general engineering components requiring a tensile strength of 700 to 1000 N/mm^2

Table 10.3—Alloy tool and die steels.

Type of steel	*Relevant Specification*	*Composition (%)*	*Heat-treatment*	*Uses*
'60'-carbon–chromium	B.S. 970:526M60	0·60 C 0·65 Mn 0·65 Cr	Oil-quench from 800°–850° C. Temper: (i) for cold-working tools at 200°–300° C; (ii) for hot-working tools at 400°–600° C	Blacksmith's and boilermaker's chisels and other tools. Mason's and miner's tools. Vice jaws. Hot-stamping and forging dies
1% carbon–chromium	B.S. 970:534A99	1·00 C 0·45 Mn 1·40 Cr	Oil-quench from 810° C; temper at 150° C	Ball- and roller-bearings. Instrument pivots. Cams. Small rolls
High carbon–high chromium	B.S. 4659:BD3	2·10 C 0·30 Mn 12·50 Cr	Heat slowly to 750°–800° C and then raise to 960°–990° C. Oil-quench (small sections can be air cooled). Temper at 150°–400° C for 30–60 minutes	Blanking punches, dies and shear blades for hard, thin materials. Dies for moulding abrasive powders, e.g. ceramics. Master gauges. Thread-rolling dies
$\frac{1}{4}$% vanadium	—	1·00 C 0·25 Mn 0·20 V	Water-quench from 850° C; temper as required	Cold-drawing dies; etc.
4% vanadium	—	1·40 C 0·40 Mn 0·40 Cr 0·40 Mo 3·60 V	Water-quench from 770° C; temper at 150°–300° C	Cold-heading dies; etc.

Hot-working die steel	B.S. 4659:BH12	0·35 C 1·00 Si 5·00 Cr 1·50 Mo 0·40 V 1·35 W	Pre-heat to 800° C, soak and then heat quickly to 1020° C and air cool. Temper at 540°–620° C for 1½ hours	Extrusion dies, mandrels and noses for aluminium and copper alloys. Hot-forming, piercing, gripping and heading tools. Brass-forging and hot-pressing dies
High-speed steels *18% Tungsten*	B.S. 4659:BT1	0·75 C 4·25 Cr 18·00 W 1·20 V	Quench in oil or air blast from 1290°–1310° C. Double temper at 565° C for 1 hour	Lathe, planar and shaping tools; millers and gear cutters. Reamers; broaches; taps; dies; drills and hacksaw blades. Bandsaws. Roller-bearings at high temperatures (gas turbines)
12% Cobalt	B.S. 4659:BT6	0·80 C 4·75 Cr 22·00 W 1·50 V 0·50 Mo 12·00 Co	Quench in oil or air blast from 1300°–1320° C. Double temper at 565° C for 1 hour	Lathe, planar and shaping tools, milling cutters, twist drills, etc. for exceptionally hard materials. Has maximum red-hardness and toughness. Suitable for severest machining duties, e.g. manganese steels and high-tensile steels, close-grained cast irons
Molybdenum '562'	B.S. 4659:BM2	0·83 C 4·25 Cr 6·50 W 1·90 V 5·00 Mo	Quench in oil or air blast from 1250° C. Double temper at 565° C for 1 hour	Roughly equivalent to the standard 18–4–1 tungsten high-speed steel but tougher. Drills, reamers, taps, milling cutters, punches, threading dies, cold-forging dies
9% Mo–8% Co	B.S. 4659:BM42	1·00 C 3·75 Cr 1·65 W 1·10 V 9·50 Mo 8·25 Co	Quench in oil or air blast from 1180°–1210° C. Triple temper at 530° C for 1 hour	Similar uses to the 12% Co–22% W high-speed steel

the range over which austenite may exist (as indicated by the γ-loop formed in the binary diagram when each is added to iron) and so stabilise ferrite and martensite.

10.4.1 Low-alloy tool steels generally contain small amounts of chromium and are oil-hardened from temperatures slightly above those used for equivalent plain carbon steels. Such steels are used for ball-bearings, chisels and hot-working dies.

The high-alloy tool steels, however, contain considerable amounts of tungsten, molybdenum and vanadium as well as chromium (Table 10.3). Of recent years the relative scarcity of tungsten has led to the wider use of molybdenum and vanadium, particularly in high-speed steels.

In order that these steels shall not lose their hardness at high working temperatures the maximum amount of tungsten (or molybdenum) must be present in *solid solution* in the original austenite before it is quenched. Fig. 10.4—a pseudo-binary diagram (8.7.2) representing a high-speed steel—indicates that at a temperature of say 900° C, little more than 0·2 % C *along with its associated tungsten* is in solid solution in the austenite and that even at the solidus temperature of about 1300° C only an amount, *S*, is in solid solution. For this reason, in order to retain as much carbon and tungsten in solid solution in the quenched structure as is

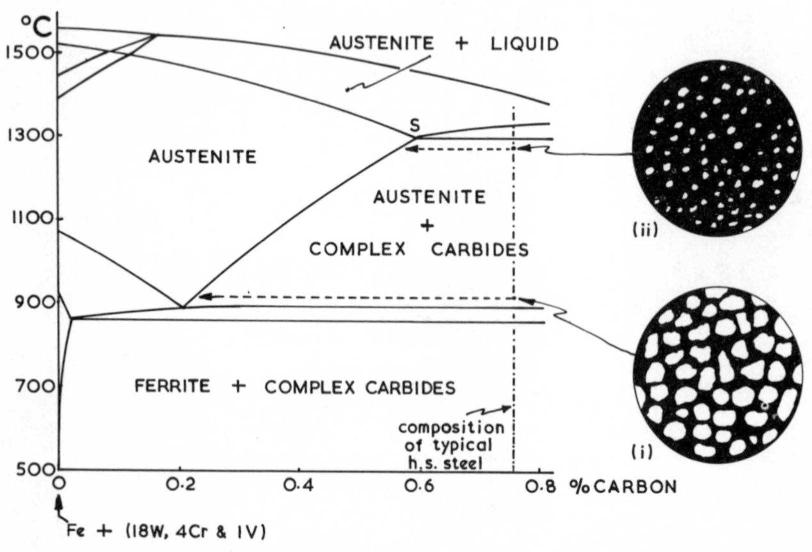

Fig. 10.4—A pseudo-binary diagram representing an 18–4–1 high-speed steel. If such a steel were quenched from 900° C the structure (i) would consist of mixed carbides (light) in a matrix of martensite (dark) which would be deficient in both tungsten and carbon (0·2%). When quenched from almost 1300° C much of the carbide has been dissolved (ii) and the martensite matrix now contains much more dissolved carbon (0·6%) along with its previously combined tungsten.

possible and so obtain both hardness and *temper-resistance*, very high quenching temperatures are necessary. At these high temperatures decarburisation and grain growth are rapid unless special precautions are adopted. Thus a two-chamber furnace enables preliminary pre-heating to be used and so limit grain growth by reducing the time of contact with the high quenching temperature; whilst the use of inert furnace atmospheres prevents decarburisation of the tool surface.

10.4.2 The large amounts of alloying elements present slow down transformation considerably and so air-hardening (or oil-quenching of heavy sections) can generally be used. Considerable amounts of austenite are retained on quenching and a 'secondary hardening' treatment is necessary. This involves heating the tool to about 550° C to promote the transformation of retained austenite to martensite. Two or three such treatments are often necessary before the austenite → martensite transformation approaches completion. Once formed, the martensite is similarly very sluggish to tempering at temperatures up to 700° C, wherein lies the most important feature of high-speed steels.

Stainless and heat-resisting steels

10.5 The role of chromium in imparting corrosion resistance to steels has been mentioned (10.2.7) and amounts between 13% and 21% are commonly used for this purpose. Those steels containing only chromium as an alloying element are of course ferritic since chromium stabilises BCC iron and quenching those chromium steels which contain sufficient carbon will produce a hard martensitic structure. Thus, high-chromium steels containing 0·3% C are used for the manufacture of table knives and other rust-resistant edged tools; whilst those containing little or no carbon are used for press-work and deep-drawing since they can be hardened only by cold-work.

10.5.1 Chromium steels tend to become brittle during heat treatment because chromium promotes grain growth. Stainless steels containing both nickel and chromium are therefore less likely to suffer from brittleness because of the grain-refining influence of nickel. The well-known '18–8' stainless steels are in this class. These are non-hardening alloys in which the carbon content is normally less than 0·1%. Even so the final treatment is usually a quench from 1050° C in order that carbon is retained in solid solution and to prevent it from precipitating as particles of chromium carbide which would impair the corrosion resistance of the alloy (15.7.1). Alternatively the addition of up to 1·0% of either niobium or titanium (which have high affinities for carbon) effectively 'ties up' any carbon present so that it is no longer available to precipitate as chromium carbides. This is necessary for stainless steels which are to be welded and would be prone to 'weld decay' (17.8.2) unless so 'proofed'.

Table 10.4—Stainless and heat-resisting steels.

Type of steel	Relevant specification	Composition (%)	Condition	Typical mechanical properties: Yield stress (N/mm²)	Tensile strength (N/mm²)	Elongation (%)	Hardness (Brinell)	Heat-treatment	Uses
Stainless iron	B.S. 970:403S17	0·04 C 0·45 Mn 14·00 Cr	Soft	340	510	31	—	Non-hardenable except by cold work	Wide range of domestic articles—forks, spoons. Can be spun, drawn and pressed
Cutlery steel	B.S. 970:420S45	0·30 C 0·50 Mn 13·00 Cr	Cutlery temper	—	1670	—	534	Water- or oil-quench (or air-cool) from 950°–1000° C. Temper: (for cutlery)—at 150°–180° C; (for springs)—at 400°–450° C	Cutlery and sharp-edged tools requiring corrosion resistance. Circlips, etc. Approximately pearlitic in structure when normalised
			Spring temper	—	1470	—	450		
18/8 stainless	B.S. 970:302S25	0·05 C 0·80 Mn 8·50 Ni 18·00 Cr	Softened	278	618	50	170	Non-hardening except by cold-work. (Cool quickly from 1050° C to retain carbon in solid solution)	Particularly suitable for domestic and decorative purposes. An austenitic steel
			Cold-rolled	803	896	30	—		
18/8 stainless (weld-decay) proofed)	B.S. 970:321S20	0·05 C 0·80 Mn 8·50 Ni 18·00 Cr 1·60 Ti	Softened	278	649	45	180	Non-hardening except by cold-work. (Cool quickly from 1050° C to retain carbon in solid solution)	A weld-decay proofed steel (fabrication by welding can be safely employed). Used extensively in nitric acid plant and similar chemical processes
			Cold-rolled	402	803	30	225		

Type of steel	Relevant specification	Composition (%)	Condition	Testing temperature (°C)	0·1% P.S. (N/mm²)	Tensile strength (N/mm²)	Elongation (%)	Maximum working temperature (°C)	Uses
Heat resisting steel	AISI Series 311	0·15 C 20·00 Cr 25·00 Ni	Forged or rolled	—	—	—	—	820	Conveyor chairs and skids, heat-treatment boxes, recuperator valves, valves and other furnace parts
Heat resisting steel	AISI Series 302B	0·10 C 1·50 Si 1·00 Mn 19·00 Cr	Air-cooled from 1100° C	20 400 600 800	347 248 217 139	698 527 465 248	55 40 37 45	1000 (air) 950 (flue gases)	A fairly cheap grade of heat-resisting steel with a good combination of

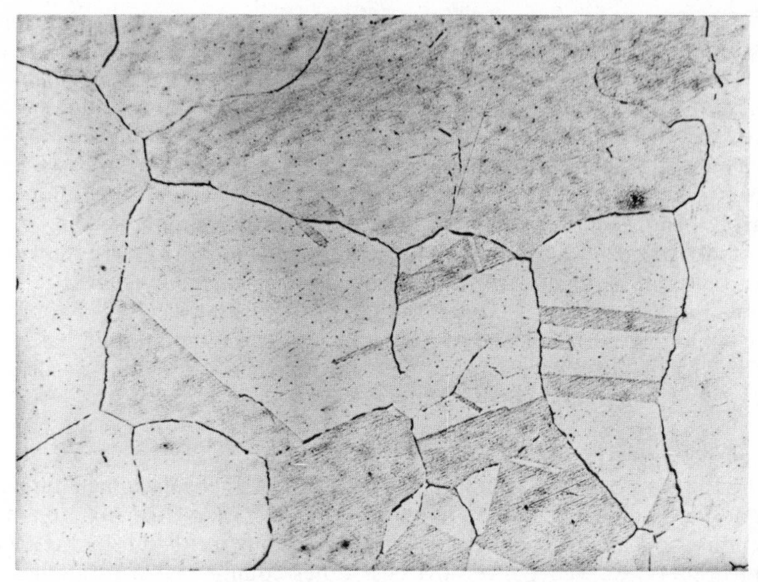

Plate 10.1—18–8 stainless steel, furnace-cooled from 1100° C. This treatment resulted in the precipitation of some chromium carbide as dark films along the austenite grain boundaries. Some austenite crystals show annealing twins in common with many other FCC structures. (Etched in 'Mixed acids with glycerol'.) × 100.

Plate 10.2—18–4–1 high-speed steel, annealed at 900° C. Rounded particles of carbide (light) in a ferritic-type matrix. (Etched in 2% Nital.) × 600.

10.5.2 Heat-resisting steels must resist oxidation at high temperatures and, in some cases, attack by sulphurous gases (15.2.2). For this reason they may contain as much as 30% chromium together with small amounts of silicon. The general tendency for chromium to induce grain growth, with its attendant brittleness, is offset by adding nickel.

These steels must be reasonably strong at high temperatures so they are 'stiffened' by adding small amounts of two or more of the following: carbon, tungsten, molybdenum, titanium and aluminium, so that small particles of carbides or intermetallic compounds are formed. In this manner the limiting creep stress is raised due to *particle hardening* (14.1.5). Metallurgical principles involved in the heat-resisting steels also apply to the 'Nimonic' series of alloys. These are not steels but contain basically 75% nickel–20% chromium, stiffened with small amounts of carbon, titanium, aluminium, cobalt, molybdenum or niobium.

Engineering cast irons

10.6 In terms of cost per unit mass cast iron is the cheapest metallurgical material available to the engineer. It is in fact re-melted pig iron, the composition of which has undergone some adjustments during the melting process. Apart from its low cost other commendable properties of cast iron include good rigidity and *compressive* strength; excellent fluidity so that it makes good casting impressions; and good machinability. Whilst the ductility and *tensile* strength of ordinary grey cast iron are not very high both of these properties can be considerably improved by treatments which modify the microstructures of suitable irons.

During the nineteenth century cast iron was used both for engineering and domestic purposes. Today the Victorian fireplaces, lamp standards and water fountains have largely disappeared—except as collector's items—and cast iron is now used solely as a reasonably sophisticated engineering material. Alloy—and other special cast irons are now used for the manufacture of such components as crankshafts, axles and connecting rods, hitherto the province of forged steels.

10.6.1 *The relationships between composition and structure of cast iron.* Most engineering cast irons contain 2·0–3·5% C. This carbon may exist in the structure either as the carbide, Fe_3C (cementite)—referred to by the foundryman as 'combined carbon'—or as free carbon in the form of graphite. The form in which carbon exists depends to some extent on the amounts of other elements present. We have seen (10.2.3) that the presence of some elements in steel renders carbides unstable so that they tend to decompose forming graphite. Silicon is one of these elements and its presence in varying amounts in cast iron has a similar influence on the state in which carbon exists. Thus a high-silicon iron—containing as much as 2·5% Si—will be almost completely graphitic and will contain little or no cementite. The presence of large amounts of graphite in the structure will lead to a fractured surface having a grey appearance. That is, it will be a 'grey iron'.

If, on the other hand, the silicon content is low—in the region of 0·5%—then cementite will tend to be stable and little or no graphite will form. Since cementite is extremely brittle such an iron will be mechanically weak and a fractured surface will be silvery white since cementite is a white compound. For this reason it is termed a 'white iron'.

Quite small amounts of sulphur will have the opposite effect to that of silicon, that is, sulphur will tend to stabilise cementite and so favour the formation of a white iron. However, the direct effect of sulphur in iron, as in steel, is to increase the brittleness excessively so the foundryman does *not* use sulphur as a means of controlling the structure of cast iron. Generally he will attempt to keep the sulphur content to a minimum, and instead use a varying silicon content as a means of controlling the state of the carbon present.

10.6.2 *The influence of cooling rates on the structure of cast iron.* Even though silicon is present in a cast iron in sufficient quantities to graphitise any cementite present such decomposition does not occur instantaneously but is time dependent. Consequently if a cast iron solidifies rapidly its fracture may well be white because cementite has had insufficient opportunity to decompose, under the influence of silicon, to form graphite.

Clearly the sectional thickness of a casting influences its rate of solidification and cooling and this, in turn, will control the condition of the carbon. An iron which, in heavy sections, will produce a grey iron may, when 'chilled' by casting it as a thin section, give a brittle white structure. The extent to which a particular iron will 'chill' in this way is generally examined by casting a sample in the form of a wedge (Fig. 10.5).

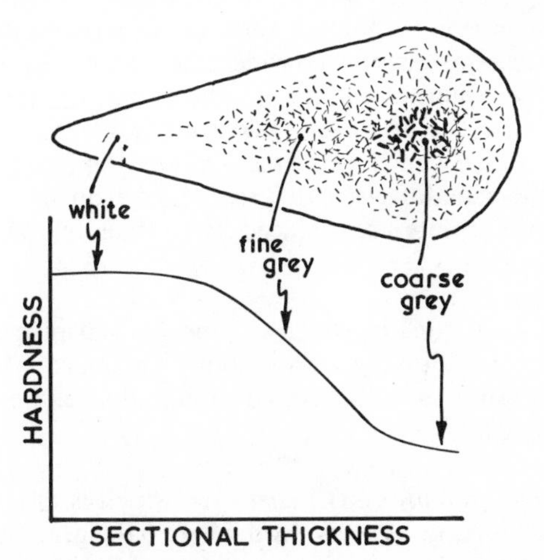

Fig. 10.5—The relationship between section thickness—and hence rate of cooling—and structure in a cast iron.

The structure and properties of cast iron

10.7 A white cast iron contains both *primary* and *pearlitic* cementite. The addition of silicon to the composition causes first the primary cementite and then the pearlitic cementite to decompose to graphite and ferrite. The best combination of mechanical properties in an ordinary grey iron is obtained when the structure consists of small flakes of graphite in a pearlitic matrix. Such a structure will be achieved by the foundryman by adjusting the silicon content of the iron to suit the sectional thickness of the casting and hence the rate of cooling which will prevail during its solidification.

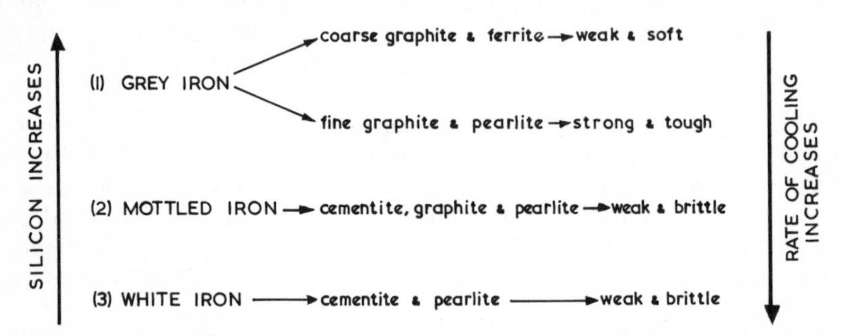

Fig. 10.6—The influence of silicon and cooling rate on the structure and properties of cast iron.

Under the microscope graphite flakes in cast iron appear as irregular strands with pointed ends. In fact what we see are sections through graphite flakes which, in reality, are shaped something like the corn flakes sold as a breakfast cereal. Nevertheless the sharp rims, which appear in cross-section as pointed ends, act as stress-raisers in the structure and are responsible for considerable reductions in strength and toughness of the resultant casting. In the design of stress-bearing components generally, the engineer makes every attempt to avoid in-cut sharp corners which can act as stress-raisers. Similarly, on the microscopic scale, the metallurgist can provide an iron in which the stress-raising flakes of graphite are replaced by rounded globules of carbon.

Special cast irons

10.8 These are cast irons in which the mechanical properties have been improved either by alloying or by some treatment which has caused graphite to precipitate in a spheroidal form and so eliminate the weakening graphite flakes.

10.8.1 *Spheroidal-graphite* (*S.G.*) *cast iron*, which is also known as 'ductile iron' in the USA, is one such iron where flake graphite is replaced by spherical particles of graphite. This is achieved by adding small amounts of magnesium—usually as a magnesium/nickel alloy—to give a resultant

Table 10.5—Compositions, properties and uses of some cast irons.

Type of iron	Composition (%)	Representative mechanical properties: Tensile strength (N/mm²)	Hardness (Brinell)	Uses
Grey iron	3·30 C 1·90 Si 0·65 Mn 0·10 S 0·15 P	Strengths vary with sectional thickness but are generally in the range 150–350 N/mm²	—	Motor brake drums
Grey iron	3·25 C 2·25 Si 0·65 Mn 0·10 S 0·15 P		—	Motor cylinders and pistons
Grey iron	3·25 C 1·25 Si 0·50 Mn 0·10 S 0·35 P		—	Heavy machine castings
Phosphoric grey iron	3·60 C 1·75 Si 0·50 Mn 0·10 S 0·80 P		—	Light and medium water pipes
'Chromidium'	3·20 C 2·10 Si 0·80 Mn 0·05 S 0·17 P 0·32 Cr	275	230	Cylinder blocks, brake drums, clutch casings, etc.
Wear- and shock-resistant	2·90 C 2·10 Si 0·70 Mn 0·05 S 0·10 P 1·75 Ni 0·10 Cr 0·80 Mo 0·15 Cu	450	300	Crankshafts for diesel and petrol engines. (Good strength, shock-resistance and vibration-damping capacity)
'Ni-resist'	2·90 C 2·10 Si 1·00 Mn 0·05 S 0·10 P 15·00 Ni 2·00 Cr 6·00 Cu	215	130	Pump castings handling concentrated chloride solutions. An austenitic corrosion-resistant alloy

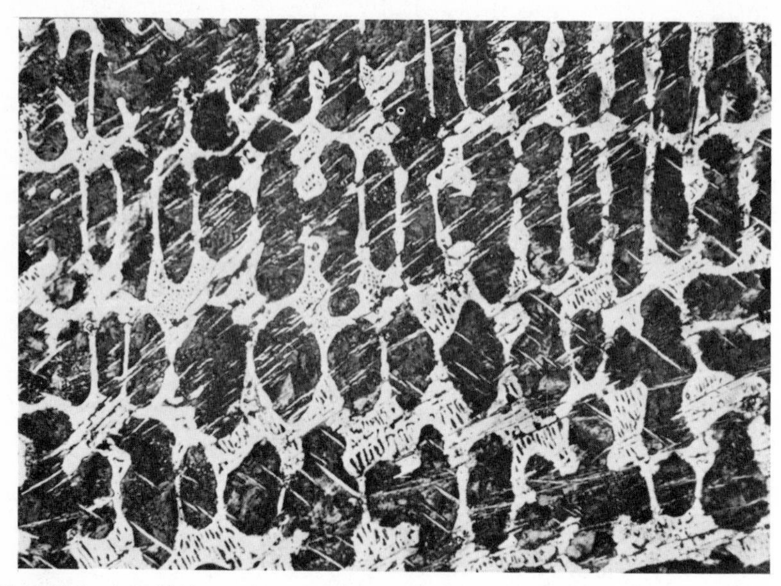

Plate 10.3—A 'white' cast iron consisting essentially of pearlite (dark) and cementite (white). The pearlite 'dendrites' were derived from primary austenite dendrites. On close examination the light areas are seen to be lamellae of cementite and pearlite, arising from a eutectic of cementite and austenite formed at 1147° C. (Si—0·35%.) (Etched in 2% Nital.) × 150.

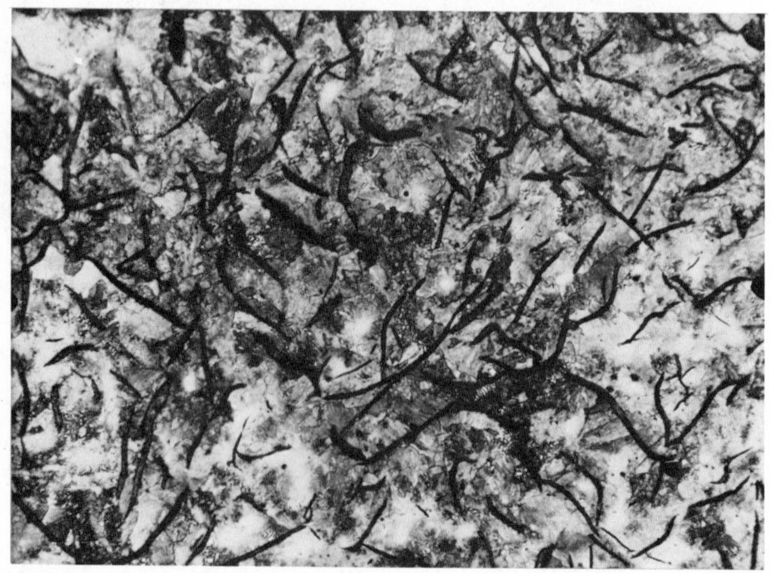

Plate 10.4—A 'grey' cast iron. The structure consists of graphite flakes (black) in a matrix of pearlite. Typical of a good-quality engineering cast iron. (Si—1·85%.) (Etched in 2% Nital.) × 150.

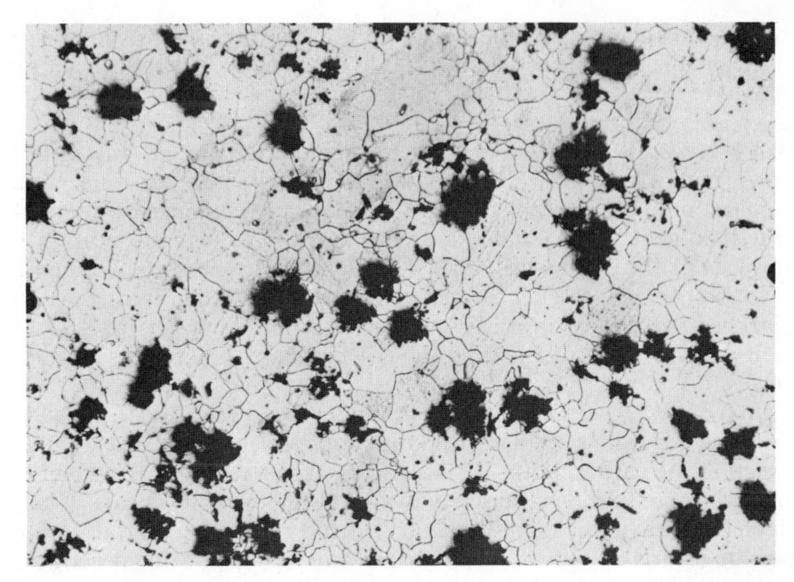

Plate 10.5—Blackheart malleable iron, produced by annealing the appropriate white iron at 900° C for 70 hours. The structure consists of 'rosettes' of graphite in a matrix of ferrite. (Etched in 2% Nital.) × 90.

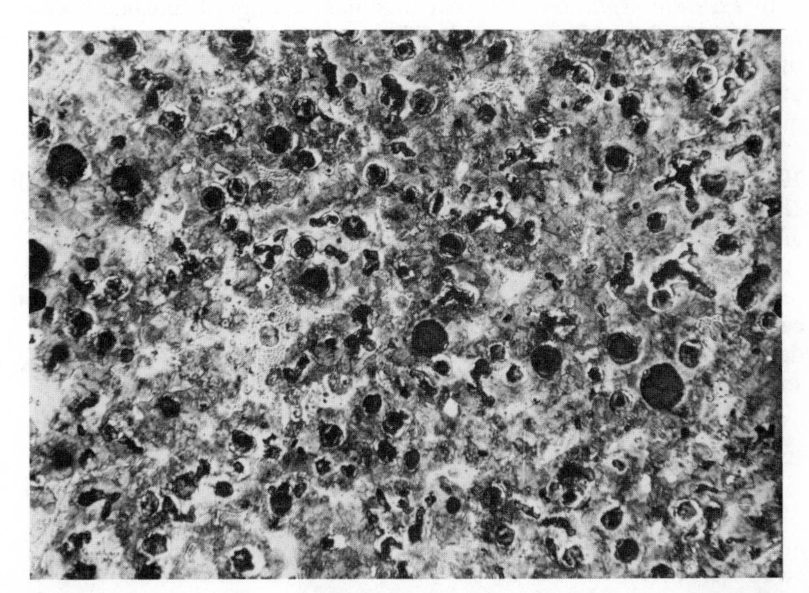

Plate 10.6—Spheroidal-graphite iron, in the cast condition. The structure consists of spheroidal graphite (black) in a matrix of pearlite and cementite (white). Annealing would lead to decomposition of the brittle cementite giving a matrix of pearlite and ferrite. (Etched in 2% Nital.) × 90.

magnesium content of no more than 0·1% in the iron. This is sufficient to 'nucleate' the graphite so that it forms as spherical globules. Such an iron may have a tensile strength of 775 N/mm^2.

10.8.2 *Malleable cast irons* are cast to shape in the normal way using a low-silicon iron which will produce a brittle white structure. Both primary and pearlitic cementite are subsequently eliminated by suitable heat treatment. In the *Blackheart Process* the castings are heated for upwards of forty hours at about 900° C so that cementite decomposes to form spherical 'rosettes' of 'temper carbon' in a tough ferritic matrix. Blackheart malleable iron is used widely in the motor-car industry for such components as rear-axle housings, wheel hubs, brake shoes and door hinges.

In the *Whiteheart Process* the heating process is carried out under oxidising conditions so that carbon is progressively eliminated from the structure. As carbon in the surface layers oxidises and is lost as carbon dioxide, more carbon diffuses outwards from the core and is in turn eliminated leaving a totally ferritic structure though a central core of pearlite invariably remains. The surface often exhibits oxide penetration and malleable iron produced by this method is widely used in fittings for gas, water and steam pipes; bicycle and motor-cycle frame fittings and switch-gear equipment.

In *Pearlitic Malleable Iron* heat treatment is applied to cause decomposition of all primary cementite. The work is then heated to 900° C so that some of the temper carbon is dissolved in the austenite then present. Subsequent cooling leads to the formation of a strong, tough pearlitic matrix.

10.8.3 *Alloy cast irons* make use of the same elements as are employed in alloy steels. *Nickel* is the most common element used. It promotes graphitisation and has a useful grain-refining—and hence, toughening—effect. *Chromium* on the other hand increases hardness and wear-resistance by stabilising carbides. *Vanadium* promotes heat-resistance in cast iron by rendering carbides very stable so that they are less likely to decompose at high temperatures. *Copper*, which is only very sparingly soluble, improves the resistance to atmospheric corrosion.

Chapter Eleven
Non-ferrous Metals and Alloys

11.1 The metals iron, aluminium and copper can be regarded as the 'Big Three' in terms of the annual World output of these metals. Iron is, of course, produced in much greater quantities than all other metallic elements combined, having gradually attained this position as the Iron Age replaced the Bronze Age. Until fairly recently copper continued to occupy second place in terms of World output but increasing scarcity of copper ore, leading to increases in cost of the metal, has caused aluminium to move into second place.

Other metals like lead, zinc, tin, magnesium and nickel have been in use in the pure or alloyed form for considerable periods of time. More recently metals such as titanium have assumed engineering importance but whilst the ores of this metal are more plentiful than those of most other non-ferrous metals, titanium remains expensive because of the difficulty of its extraction.

Copper and its alloys

11.2 A large proportion of the World output of copper is used in the unalloyed form, mainly in the electrical industries. The specific conductivity of copper is second only to that of silver to which it is but little inferior. Since copper is still very much less expensive than silver it is therefore used extensively where high electrical conductivity is required. Its main competitor is pure aluminium which, though having only half the specific conductivity of copper, is much less expensive. Not only is the cost of aluminium per unit mass much less than that of copper, but the *conductivity per unit mass* is much higher, i.e. taking conductors of equal length and equal mass, aluminium will provide the better conduction of electricity.

The presence of alloying elements—or impurities—reduces the specific conductivity of copper. Some impurities have a devastating effect. Thus 0·04% phosphorus will reduce conductivity by as much as 25% (Fig. 18.5). Consequently phosphorus-deoxidised copper is most unsuitable for use as a conductor, and for electrical purposes either electrolytically refined (99·97% pure) copper or fire-refined ('tough-pitch') copper are used. The latter contains small globules of cuprous oxide, Cu_2O the main impurity and whilst these have little or no effect on conductivity they render the product unsuitable for gas-welding since interstitially dissolved hydrogen reacts with these globules:

$$Cu_2O + 2H \rightarrow 2Cu + H_2O$$

Steam, thus released, is insoluble and is rejected at the crystal boundaries where it forms minute fissures which make the structure extremely brittle.

11.2.1 Pure copper is a very ductile metal, being FCC in structure. In the 'soft' condition it has a % elongation of 60 but is relatively weak with a tensile strength of no more than 215 N/mm^2. Cold-drawn copper has a strength of only 340 N/mm^2 and for use as telephone wires the addition of 1% cadmium will increase the strength to 460 N/mm^2 whilst reducing specific conductivity by only 5% (Fig. 18.5). Because of its high ductility and good corrosion resistance much copper is also used in domestic plumbing, particularly since the availability of standard pipe connectors (containing pre-set solder) has made joining a simple procedure.

11.2.2 Copper-base alloys are now important for specific purposes involving such properties as high ductility, good corrosion resistance or electrical conductivity coupled with reasonable strength and stiffness. In the more important copper-base alloys the degree of solid solubility of the solute metal is generally great, as indicated in the appropriate diagram (Fig. 11.1). In each case the α-solid solution formed is a tough ductile phase, though in brasses, bronzes and aluminium bronzes some of the duplex structures formed when primary solubility is exceeded are also useful.

11.2.3 *Brasses* are copper–zinc alloys containing up to 45% zinc and sometimes small amounts of other elements. The α-phase brasses containing up to approximately 37% zinc are noted mainly for high ductility which reaches a maximum at 30% zinc. 70–30 brass is used for the deep-drawing of components requiring good resistance to corrosion. This group of brasses is shaped almost entirely by cold-working processes in the finishing stages.

Brasses containing between 40 and 45% zinc are of duplex structure. The β^1-phase (an ordered structure) is hard and somewhat brittle whilst β (a disordered solid solution) is very malleable. As the diagram (Fig. 11.1 (i)) indicates a 60–40 brass when heated to about 700° C becomes completely β in structure as α is absorbed. Hence, 60–40-type brasses are hot-worked by forging, rolling and extrusion.

Additions of up to 2% lead, which is insoluble in both liquid and solid brass, improves the machinability of brass by initiating chip-cracks in advance of the cutting tool; whilst 1·0% tin will increase corrosion resistance.

11.2.4 *Tin bronzes* also fall into two groups. The single-phase α-alloys are tough, ductile, strong and corrosion resistant. They are more heavily cored in the cast condition than are the α brasses and this is indicated by the wide liquidus/solidus range in the copper–tin diagram (Fig. 11.1 (ii)) as compared with the narrow liquidus/solidus range in the copper–zinc diagram. Moreover diffusion of tin in copper is slow so that cast copper–tin alloys with as little as 6% tin may contain some of the brittle intermetallic compound δ ($Cu_{31}Sn_8$). In industrial practice cooling is never slow enough to permit transformation to ϵ (Cu_3Sn) to occur; for which

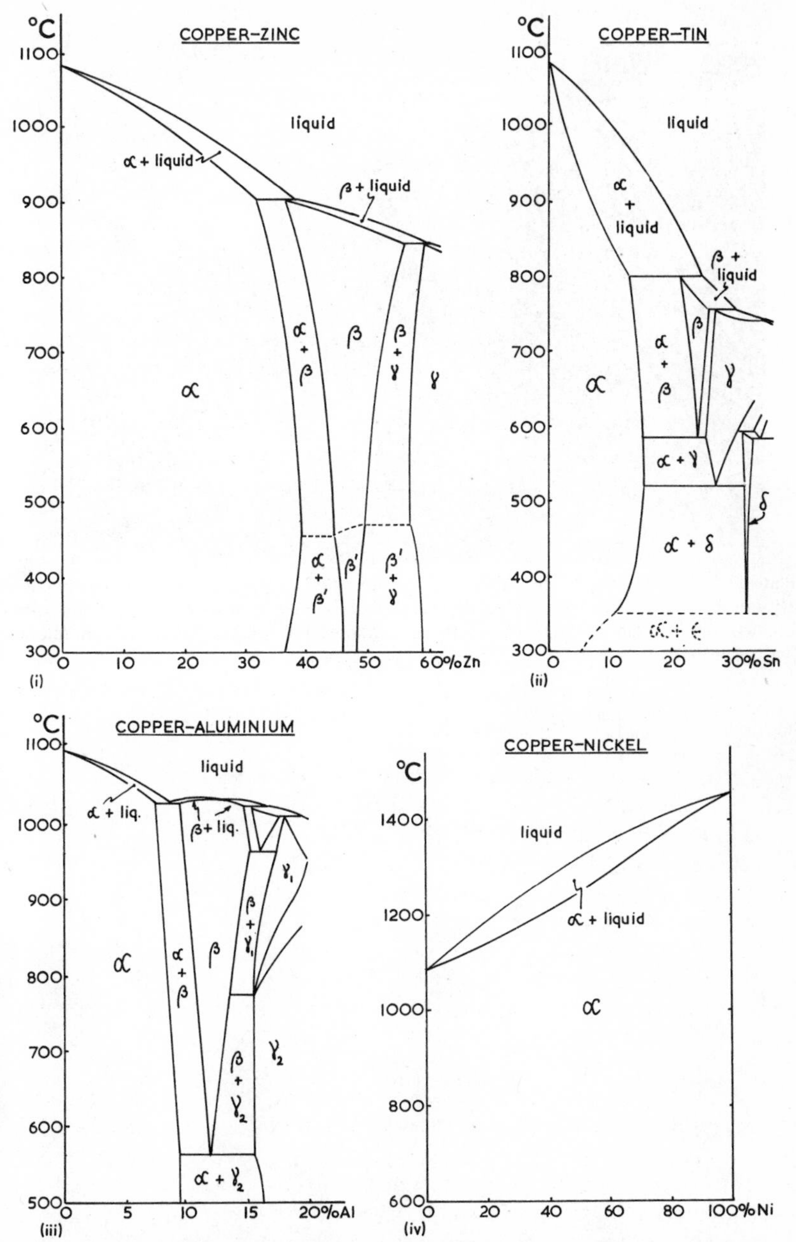

Fig. 11.1—Phase diagrams for the more important copper-base alloy systems. In systems (i), (ii) and (iii) further phase changes occur below 400° C when alloys are cooled *very slowly* under laboratory conditions. In industrial practice cooling processes are too rapid to permit such changes to occur. Hence they need not concern us at this stage.

Plate 11.1—70–30 brass in the chill-cast condition. The structure consists of heavily cored crystals of the α phase only. Neighbouring crystals have etched to different intensities because of different orientations of the crystallographic planes. (Etched in ammonia/hydrogen peroxide.) $\times$ 60.

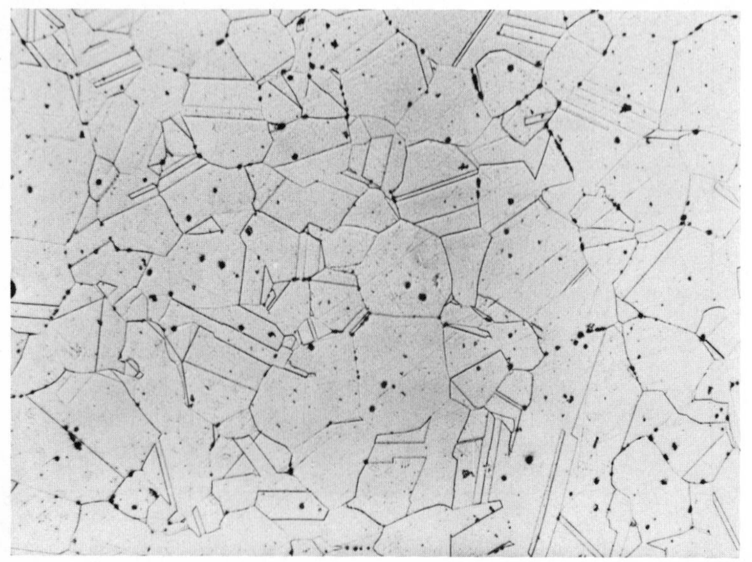

Plate 11.2—70–30 brass, cold-worked and then annealed at 650° C. Uniform α solid solution with annealing twins produced on recrystallisation of the cold-worked structure. (Etched in ammonia/hydrogen peroxide.) $\times$ 60.

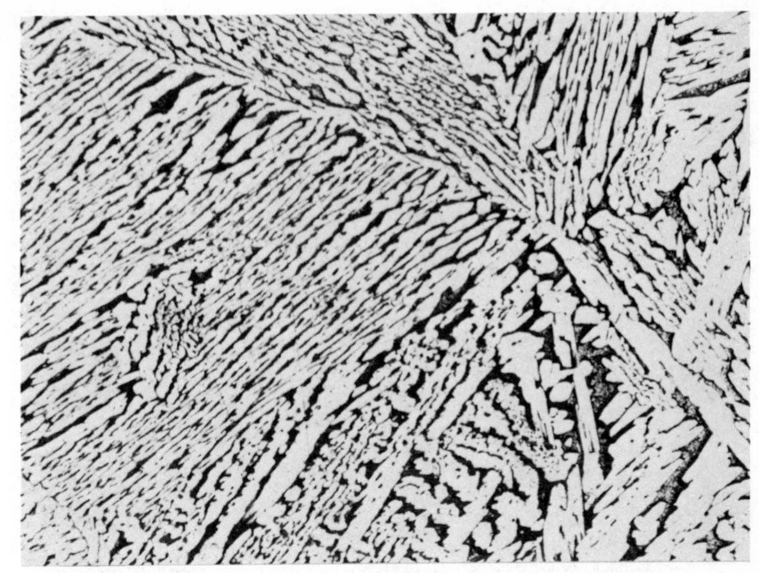

Plate 11.3—60–40 brass in the chill-cast condition. This shows a Widmanstätten distribution of α (light) in a matrix of β' (dark). (Etched in ammonia/hydrogen peroxide.) × 150.

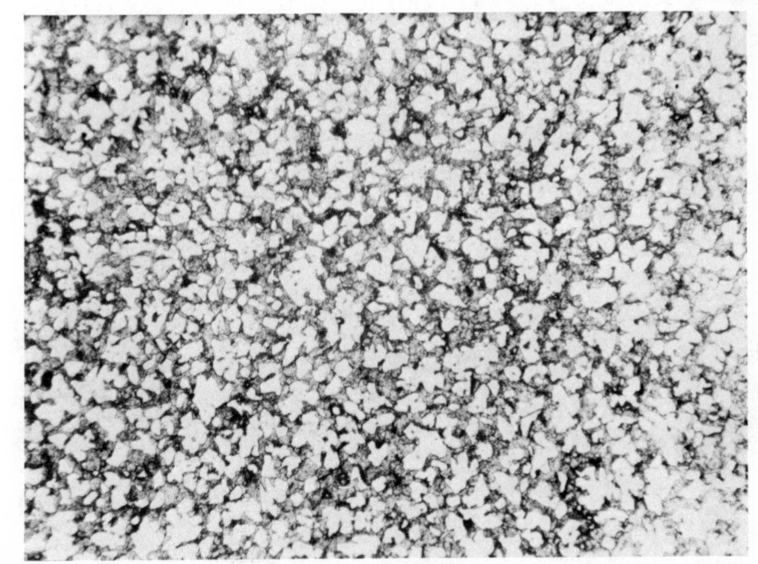

Plate 11.4—60–40 brass in the extruded condition. Extrusion in a temperature range in which α was being deposited from β has resulted in the former being broken up and the weak Widmanstätten structure prevented from re-forming. (Etched in ammonia/hydrogen peroxide.) × 150.

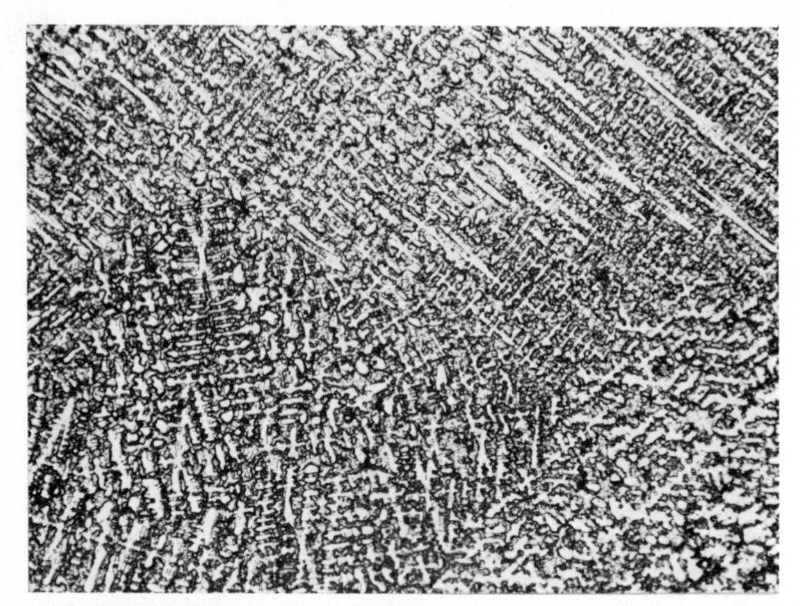

Plate 11.5—95–5 tin bronze, chill cast. Cored dendritic structure of α solid solution. (Etched in ammonia/hydrogen peroxide.) $\times$ 100.

Plate 11.6—90–10 tin bronze, chill cast. Cored dendrites of α (dark) with an infilling of $\alpha + \delta$ eutectoid. The δ has coagulated so that the lamellar nature of the eutectoid has disappeared. The large dark areas are shrinkage cavities (they follow the outline of the dendrites). (Etched in ammonia/hydrogen peroxide.) $\times$ 150.

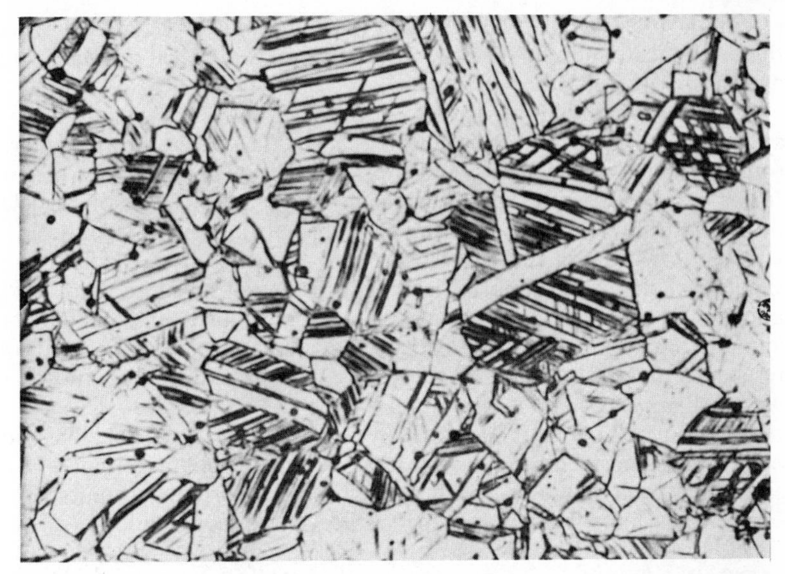

Plate 11.7—95–5 tin bronze, cold-worked, annealed and cold-worked again. Uniform α solid solution exhibiting annealing twins from the previous annealing operation and large numbers of strain bands caused by the last cold-working operation. (Etched in ammonia/hydrogen peroxide.) × 250.

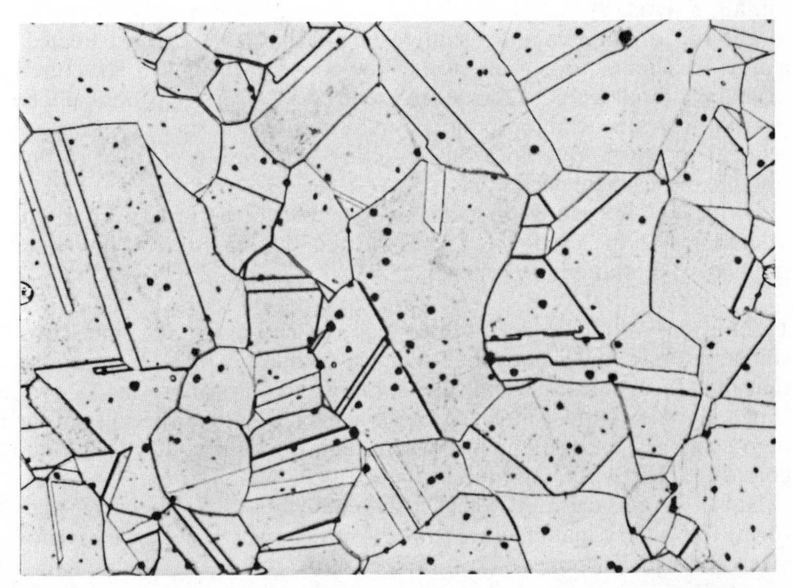

Plate 11.8—95–5 tin bronze, cold-worked and annealed. Uniform α solid solution, many of the crystals showing annealing twins. The strain bands are now absent because the alloy has recrystallised, and is now soft. (Etched in ammonia/hydrogen peroxide.) × 250.

reason the lower (broken-line) part of the diagram is of theoretical rather than practical interest. These α bronzes are cold-rolled to sheet and drawn to wire and other sections.

Bronzes containing in excess of 10% tin may show considerable amounts of δ due to excessive coring of the α. They are too brittle therefore to permit cold-working and are used for the manufacture of corrosion-resistant castings and bearings in which the presence of particles of the hard δ phase, cushioned in a tough matrix of α provides a hard-wearing, low-friction surface. Tin bronzes are generally deoxidised with small amounts of phosphorus, leaving a residue of no more than 0·04% phosphorus in the alloy. The addition of phosphorus in excess of this amount and up to 1·0% is made when increased strength, corrosion resistance and casting fluidity are necessary. These alloys are known as *phosphor bronzes*.

In 'gunmetal' some of the expensive tin is replaced by zinc. This not only reduces cost but makes deoxidation with phosphorus unnecessary since zinc itself has a high affinity for oxygen forming insoluble zinc oxide which joins the dross on the surface of the molten alloy.

11.2.5 *Aluminium bronzes* can also be classified into two main groups, viz. single-phase and duplex-phase alloys. The α alloys are of course ductile and are also corrosion resistant. Since a 5% aluminium alloy has a colour similar to that of 18 ct. gold it is used largely for the manufacture of imitation jewellery.

The 10% aluminium alloy is interesting in that it can be heat-treated in a manner similar to carbon steel. Slow cooling will give a structure of primary α (analogous to ferrite) in a eutectoid of $\alpha + \gamma_2$ (analogous to pearlite) and quenching produces a hard martensitic structure (β^1). Subsequent tempering modifies this structure with results parallel to those obtained in carbon steel.

Difficulties in casting these alloys arise due to the rapid oxidation of aluminium at the casting temperature and this has always limited the popularity of aluminium bronzes.

11.2.6 *Cupro-nickels.* Since both copper and nickel are FCC metals with similar atomic dimensions and chemical properties it is not surprising that they form substitutional solid solutions in all proportions. They are tough, ductile and reasonably strong when cold-worked. Because of their high cost they are used mainly where high ductility coupled with good corrosion resistance are required.

Nickel-silvers are copper–nickel–zinc alloys used (in the silver-plated form) for cutlery manufacture, though now largely replaced by stainless steels for such purposes.

11.2.7 *Beryllium bronze* contains up to 2% beryllium and can be strengthened by precipitation treatment to give tensile strengths in the region of

1400 N/mm^2. It is used as a tool material in the gas industries and explosives factories where sparks, generated from steel tools, might have disastrous results.

Details of some of the more important copper-base alloys are given in Table 11.1.

Aluminium and its alloys

11.3 Because of its high affinity for oxygen, aluminium could not be extracted on a commercial scale until the electrolytic decomposition of aluminium oxide was introduced in 1886. Hence the technology of aluminium has been developed almost entirely in the present century. Despite its great affinity for oxygen the corrosion-resistance of aluminium is high. This is because a thin but impervious film of oxide forms on the surface of the metal and protects it from further oxidation. The film can be further thickened by 'anodising' and since aluminium oxide is extremely hard, wear-resistance is also increased by the process. The high affinity of aluminium for oxygen makes it useful as a de-oxidant in steels.

The usefulness of aluminium as an electrical conductor has been mentioned (11.2). *Weight for weight* it is a better conductor than copper and, where space is available as in power-grid cables, it is widely used. However, pure aluminium is relatively soft and weak—it has a tensile strength of no more than 90 N/mm^2 in the annealed condition—and for most engineering purposes is used in the alloyed form. The strengths of many aluminium-base alloys can be further increased by *precipitation hardening* (7.5.2) to produce a strength/mass ratio of the same order as for high-tensile steels. The greater relative volume of aluminium alloy involved for a specific force-bearing capacity means that greater flexibility in design is possible. Aluminium alloys used in both cast and wrought forms may be precipitation hardened if of suitable composition.

11.3.1 *Aluminium alloys which are not heat-treated.* Although these alloys are not metallurgically exotic as are those which can be precipitation hardened, they none the less account for the greater proportion of the industrial output. The most widely used casting alloys are those containing 10–13% silicon which imparts a considerable increase in casting fluidity. These alloys are of approximately eutectic composition (Fig. 11.2) so that their freezing range is very small making them particularly suitable for die casting where rapid ejection from the die is desirable in the interests of output. The rather coarse eutectic structure can be refined by a process known as *modification*. This involves adding a small amount of sodium (about 0·01% of the mass of the charge) to the melt just before casting. This delays precipitation of silicon when the normal eutectic temperature is reached so that when nucleation begins a fine grain results from rapid crystallisation due to under-cooling. A shift of the eutectic composition to the right is also caused. Hence a 12% silicon alloy when modified will contain an excess of primary α instead of embrittling

Table 11.1—Some important copper-base alloys.

	Relevant specifications	Composition (%) (Balance Cu)	Condition	*Typical mechanical properties* 0·1% P.S. N/mm²	Tensile strength N/mm²	Elongation (%)	Hardness (VPN)	Characteristics and uses
Brasses	B.S. 2870/5:CZ106	30 Zn	Annealed Hard	77 510	325 695	70 5	65 185	*Cartridge brass*: deep-drawing brass, having maximum ductility of the copper–zinc alloys
	B.S. 2870/5:CZ108	37 Zn	Annealed Hard	95 540	340 725	55 4	65 185	'*Common brass*': a general purpose alloy suitable for limited forming operations by cold-work
	B.S. 2870/5:CZ109	40 Zn	Hot-rolled	110	370	40	75	Hot-rolled plate used for tube plates of condensers. Also as extruded rods and tubes. Limited capacity for cold-work
	B.S. 2870/5:CZ114	Mn, Al, Fe, Sn } up to 7% total 37 Zn	Grade A Grade B	230 280	465 540	20 15	— —	*High-tensile brass*: wrought sections for pump rods, etc. Cast alloys: marine propellors, water turbine runners, rudders. Locomotive axle boxes
Tin bronzes	B.S. 2870:PB101	3·75 Sn 0·10 P	Annealed Hard	110 620	340 740	65 5	60 210	*Low-tin bronze*: good elastic properties combined with corrosion resistance. Springs and instrument parts
	B.S. 1400:PB1/c	10 Sn 0·5 P	Sand-cast	125	280	15	90	*Cast phosphor bronze*: mainly bearings—cast as sticks for machining of small bearing bushes
	B.S. 1400:G1/c	10 Sn 2 Zn	Sand-cast	125	295	16	85	'*Admiralty gunmetal*': pumps, valves and miscellaneous castings, particularly for marine purposes because of good corrosion resistance. Also statuary because of good casting properties
Aluminium bronzes	B.S. 2870/5:CA101	5 Al Ni, Mn } up to 4·0% total	Annealed Hard	125 590	385 775	70 4	80 220	Imitation jewellery, etc. Excellent resistance to corrosion and to oxidation on heating, hence used in engineering particularly in tube form
	B.S. 1400:AB1/c	9·5 Al 2·5 Fe Ni, Mn } up to 1·0% each	Cast	185	525	30	115	The best-known aluminium bronze for both sand- and die-casting. Corrosion-resistant castings
Cupro-nickels	B.S. 2870/5:CN105	25 Ni 0·25 Mn	Annealed Hard	— —	355 600	45 5	80 170	Mainly for coinage, e.g. the current British 'silver' coinage
	B.S. 3073/6:NA13	68 Ni 1·25 Fe	Annealed Hard	215 570	540 725	45 20	120 220	*Monel Metal*: Combines good mechanical properties with excellent corrosion resistance. Mainly in chemical

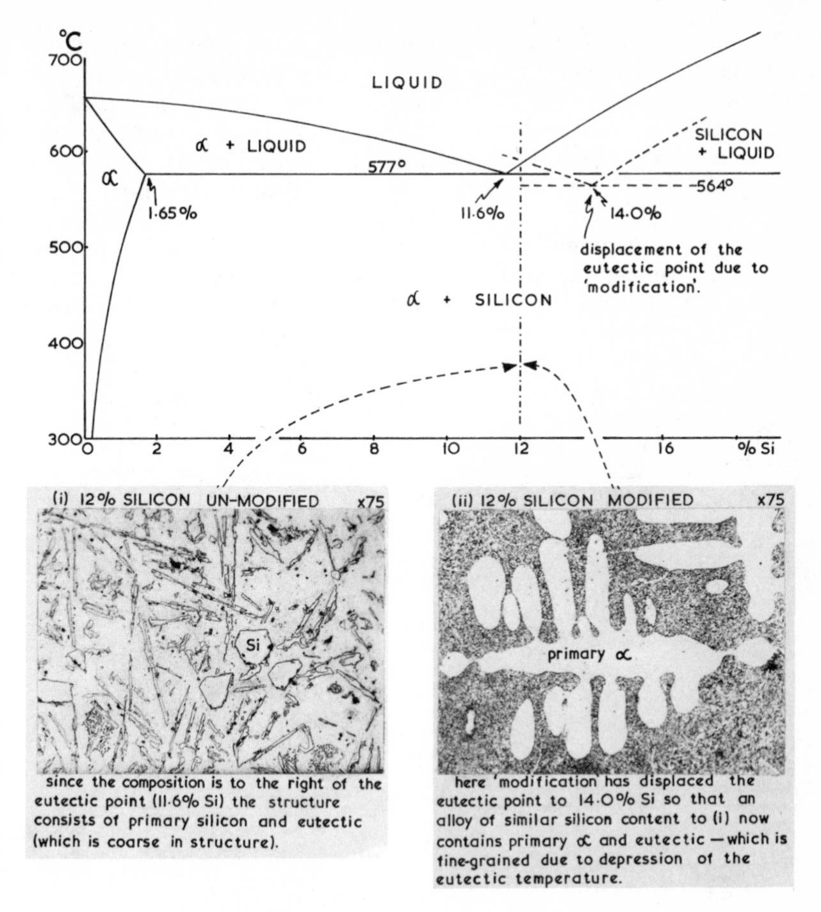

Fig. 11.2—The 'modification' of aluminium–silicon alloys.

primary silicon which appears in the structure of the unmodified 12% alloy (Fig. 11.2). Modification raises the tensile strength from 120 to 200 N/mm^2 and the % elongation from 5 to 15.

The cast aluminium–magnesium alloys include the well-known 'Birmabright' series notable mainly for a good combination of corrosion resistance and rigidity, making them particularly suitable for moderately stressed parts working in marine atmospheres.

Those wrought alloys which are not heat-treated rely for their strength on the formation of solid solutions which are subsequently work-hardened by cold-working processes. A mere 1% manganese is all that is necessary to strengthen aluminium sufficiently to make it suitable for the manufacture of milk-bottle caps and vehicle panelling, but where good corrosion resistance is also necessary then up to 5% magnesium may be added.

11.3.2 *Aluminium alloys which are heat-treated.* Although precipitation hardening can be applied to many alloys, both ferrous and non-ferrous, it is most commonly used to strengthen suitable aluminium-base alloys. 'Age-hardening', as it was then called, was in fact discovered in some aluminium-base alloys at the beginning of the century and subsequently developed for use in military aircraft during the First World War.

A number of wrought 'duralumin-type' alloys are in use but possibly the best known is that based on 4% copper. At temperatures above that indicated by *A* (Fig. 11.3) the structure will consist entirely of α-solid

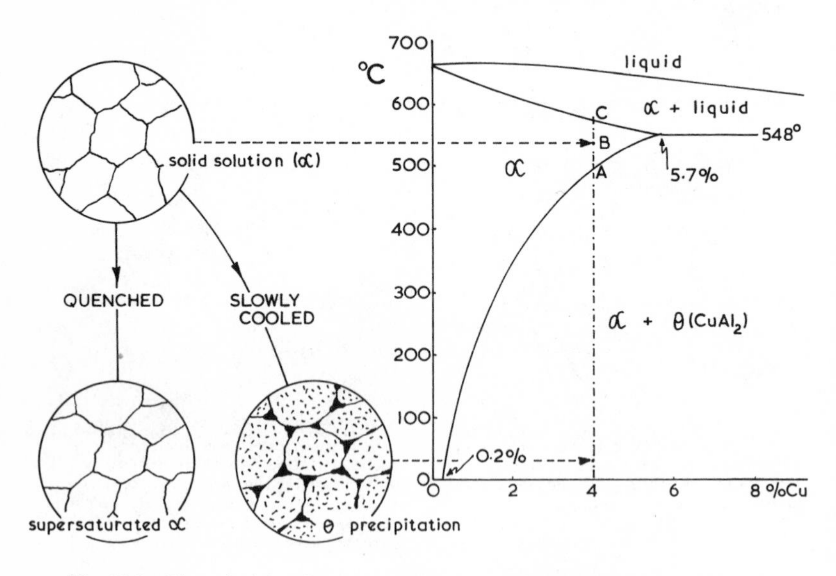

Fig. 11.3—The principle of 'solution treatment' of an aluminium–copper alloy.

solution. If the alloy is allowed to cool slowly to room temperature then the phase θ ($CuAl_2$) will be rejected as a non-coherent precipitate in accordance with the equilibrium diagram which indicates that the solid solubility of copper in aluminium falls from 5·7% at 548° C to 0·2% at 0° C. The resultant structure lacks strength because the α matrix contains so little copper in solid solution, whilst it will be brittle because of the presence of brittle particles of θ.

If the alloy is slowly reheated θ is gradually re-dissolved until at *A* the structure once more consists entirely of the uniform solid solution α. In industrial practice a solution-treatment temperature given by *B* will be used to ensure maximum solution of θ. Care must of course be taken that the temperature does not exceed that given by *C* or the alloy will begin to melt.

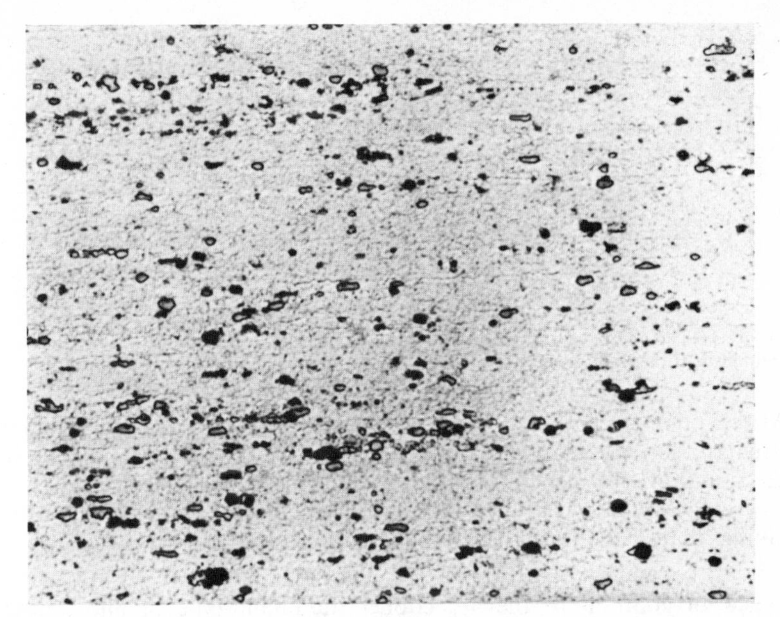

Plate 11.9—A duralumin-type alloy (Cu 4·5; Si 0·75; Mg 0·7; Mn 0·6; Fe 0·4) in the extruded condition. Particles of various compounds (mainly $CuAl_2$ and an Al–Mn–Fe complex) elongated along the direction of extrusion. (Unetched.) × 250.

The alloy is quenched from *B* so that copper is retained in supersaturated solid solution at room temperature. This increases strength because the whole of the 4% copper is now in solid solution and brittleness decreases because the brittle particles of θ ($CuAl_2$) are now absent. However, since the structure is not in equilibrium subsequent migration of copper atoms occurs to the extent that they form coherent precipitates (7.5.2.1) within the α lattice:

Cu	+	Al	→	θ'
(Al lattice)		(Al lattice)		(Intermediate *coherent* precipitate)

Since the coherent precipitates impede the movement of dislocations the strength of the alloy increases.

11.3.3 The extent of the formation of coherent precipitates at ordinary temperatures is limited so that strength attains a fairly low maximum value in a few days and this process used to be called 'age-hardening'. At higher temperatures the formation of coherent precipitates proceeds further and so the strength continues to increase (Fig. 11.4). However, a point is reached where the thermal activation is such that tiny *non-coherent* particles of θ ($CuAl_2$) begin to form in accordance with phase equilibrium:

θ'	→	$\theta(CuAl_2)$
(coherent)		(non-coherent)

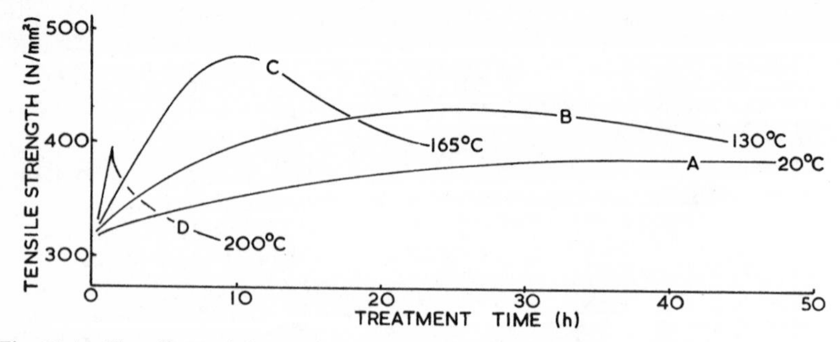

Fig. 11.4—The effects of time and temperature of precipitation treatment on the structure and tensile strength of a suitable alloy.

At this point strength and hardness begin to fall rapidly. Thus optimum strength, for the alloy relevant to Fig. 11.4, would be obtained by precipitation treatment at 165° C for 10 hours. Treatment at 200° C (curve *D*) would lead to the formation of non-coherent θ virtually 'over-taking' that of the formation of coherent θ'.

11.3.4 In addition to the 4% copper duralumin-type of alloy, other wrought aluminium alloys which can be precipitation hardened are those containing small amounts of magnesium and silicon. These form the compound Mg_2Si the solubility of which, like that of $CuAl_2$, increases considerably with temperature.

A number of aluminium-base casting alloys can also be precipitation hardened with advantage. Possibly the best-known of these is the aircraft piston alloy known as '*Y* alloy'.* This contains small amounts of copper, nickel, manganese and titanium and is notable for its tolerance of relatively high working temperatures.

Compositions and properties of some of the more important aluminium-base alloys are described in Table 11.2.

Magnesium-base alloys

11.4 Despite its high affinity for oxygen magnesium, like aluminium, is a stable metal with a high resistance to corrosion at ordinary temperatures. Its chief attribute is its very low relative density (1.7) which makes it useful in the aircraft and aerospace industries. It has a low tensile strength but its alloys containing suitable amounts of aluminium, zinc or thorium can be strengthened by precipitation hardening. The solid solubility of each of these elements decreases as the temperature falls (Fig. 11.5) thus making the precipitation treatment possible. Both cast and wrought alloys are used, particularly as castings and forgings in the aircraft industries—landing wheels, crank cases, petrol tanks and many

* The reference code used during the development of this alloy at the NPL during the First World War.

	Relevant specifications	Composition (%) (balance Al)	Condition	Typical mechanical properties: 0·1% P.S. (N/mm²)	Tensile strength (N/mm²)	Elongation (%)	Characteristics and uses
Wrought alloys— not-heat-treated	B.S. 1470/7:N3	1·2 Mn	Soft Hard	45 170	110 200	34 4	Metal boxes, milk bottle caps, food containers, cooking utensils, roofing sheets, panelling of transport vehicles
	B.S. 1470/7:N4	2·5 Mg	Soft $\frac{3}{4}$ hard	75 215	185 265	24 4	Marine superstructures, life boats, panelling for marine atmospheres, chemical plant, panelling for land-transport vehicles
	B.S. 1470/7:N6	5·0 Mg	Soft $\frac{1}{4}$ hard	125 215	265 295	18 8	Ship-building and applications requiring high strength and corrosion resistance
Cast alloys— not-heat-treated	B.S. 1490:LM4	5·0 Si 3·0 Cu	Sand cast Chill cast	70 80	150 170	2 3	Sand castings; gravity and pressure die-castings. General purpose alloy where mechanical properties are of secondary importance.
	B.S. 1490:LM6	11·5 Si	Sand cast Pressure die cast	55 85	170 215	7 4	Sand castings; gravity and pressure die-castings. Excellent foundry properties. One of the most widely useful aluminium alloys ('modified'). Radiators, sumps, gear boxes and large castings
Wrought alloys— heat-treated	B.S. 1470/7:HT14	4·0 Cu 0·8 Mg 0·5 Si 0·7 Mn	Solution treated at 480° C, quenched and aged at room temperature for 4 days	280	400	10	General purposes—stressed parts in aircraft and other structures. The original 'duralumin'
	B.S. 1470/7:HT30	1·0 Mg 1·0 Si 0·7 Mn	Solution treated at 510° C; quenched and precipitation hardened at 175° C for 10 hours	150	250	20	Structural members for road, rail and sea transport vehicles; architectural work; ladders and scaffold tubes. High electrical conductivity —hence used in overhead lines
	DTD. 5074	1·6 Cu 2·5 Mg 6·2 Zn 0·3 Ti	Solution treated at 465° C; quenched and precipitation hardened at 120° C for 24 hours	590	650	11	Highly stressed aircraft parts such as booms. Other military equipment requiring a high strength/mass ratio. The strongest aluminium alloy produced commercially
Cast alloys— heat-treated	B.S. 1490:LM8	4·5 Si 0·5 Mg 0·5 Mn 0·15 Ti	Solution treated at 540° C for 8 hours; quenched in oil, precipitation hardened at 165° C for 10 hours	—	280	2	Good casting properties and corrosion resistance. Mechanical properties can be varied by heat-treatment
	B.S. 1490:LM14	4·0 Cu 0·3 Si 1·5 Mg 2·0 Ni 0·2 Ti	Solution treated at 510° C; precipitation hardened in boiling water for 2 hours or aged at room temperature for 5 days	215	280	—	Pistons and cylinder heads for liquid and air-cooled engines. General purposes. The original 'Y alloy'

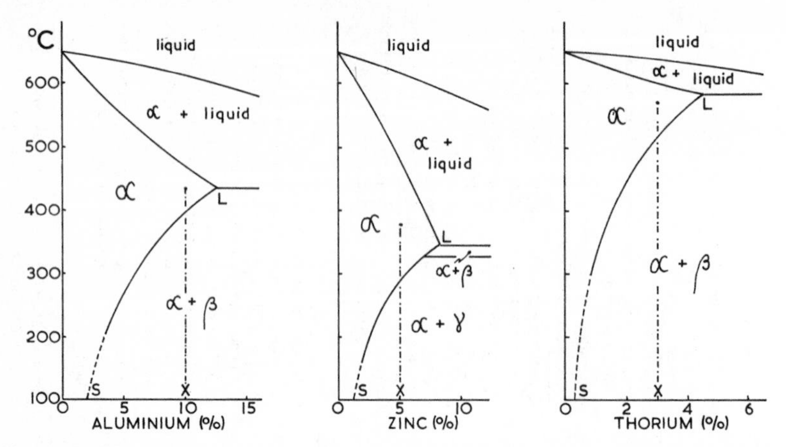

Fig. 11.5—The rapid changes in solid solubility with temperature of Al, Zn and Th in magnesium (as shown by the sloping solvus line SL in each case) make possible the precipitation treatment of alloys of composition in the region of *X*.

engine parts in both piston and jet engines. Some of the more important magnesium-base alloys are given in Table 11.3.

Nickel-base alloys

11.5 Nickel is a valuable constituent both in corrosion-resistant alloys and, since it restrains grain-growth at high temperatures, in high-temperature alloys.

11.5.1 *Corrosion-resistant alloys.* Nickel is a principal constituent of many corrosion-resistant alloys used in the chemical industries. Many of these alloys are tough and ductile as well as being corrosion resistant. Thus 'Corronel B' (66 Ni; 28 Mo; 6 Fe) is particularly resistant to mineral acids and acid chloride solutions and can be produced in the form of tubes and other sections for use in the chemical and petro-chemical industries. 'Hastelloy D' (85 Ni; 10 Si; 3 Cu; 1 Al) on the other hand is a tough, very hard casting alloy which is particularly resistant to hot concentrated sulphuric acid.

11.5.2 *High-temperature alloys.* Nickel–chromium alloys are particularly suitable for high-temperature working. Chromium imparts a low rate of oxidation (15.2) at all temperatures whilst nickel limits grain growth and the brittleness arising therefrom at high temperatures. Since these alloys also have a high specific resistance they are particularly suitable for resistance wire used in heater elements of electric fires, immersion heaters, toasters, etc. A grade of 'Brightray' containing 80 Ni–20 Cr has a maximum working temperature of 1150° C.

The best-known of the nickel–chromium base high-temperature alloys are those of the 'Nimonic' series which played a leading part in the

Table 11.3—Some magnesium-base alloys.

			Typical mechanical properties		
	Composition (%) (balance Mg)	*Condition*	*0·1% Proof stress* (N/mm²)	*Tensile strength* (N/mm²)	*Elongation* (%)
Cast alloys	10·0 Al 0·3 Mn 0·7 Zn	Chill-cast	115	200	2
	4·0 Zn 0·7 Zr 1·2 Rare earths	As-cast Heat-treated	95 130	170 215	5 4
	0·7 Zr 3·0 Th	Heat-treated	100	210	8
Wrought alloys	1·5 Mn	Rolled	95	200	5
	6·0 Al 0·3 Mn 1·0 Zn	Forged Extruded	155 140	280 215	8 8
	3·0 Zn 0·7 Zr	Rolled Extruded	170 215	265 310	8 8
	1·0 Mn 3·0 Th	Rolled	215	280	10

development of the jet engine, but are now used for a large number of applications where a high creep strength at elevated temperatures is required. Most 'Nimonic' alloys contain approximately 75 Ni and 20 Cr to form a tough oxidation-resistant matrix at temperatures sometimes in excess of 1000° C. The alloy is stiffened for use at high temperatures by small additions of titanium, cobalt, zirconium, aluminium and carbon in suitable associations. These elements help to raise the limiting creep stress at high temperatures by the formation of dispersed particles of carbides or compounds such as $NiAl_3$.

Zinc-base alloys

11.6 These are exclusively die-casting alloys sold under such trade names as 'Mazak'. They contain 4% aluminium and up to 2·7% copper, the balance being zinc, and are used for the manufacture of a large number of domestic and engineering components. Door handles for motor cars, bodies for windscreen wiper motors, frames for electric fires, accurately scaled children's toys and parts for washing machines and refrigerators are typical examples.

Such alloys have quite good strength (320 N/mm^2) and rigidity, but are sensitive to the presence of small amounts of impurities like cadmium, lead and tin which cause intercrystalline brittleness and some dimensional instability which manifests itself as swelling of the casting during use. These faults are largely due to intercrystalline corrosion for which reason good quality zinc-base die-casting alloys are generally made from 'four nines' zinc, i.e. zinc which is 99·99% pure.

Bearing metals

11.7 The mechanical requirements of a bearing metal can only be met by the intelligent use of alloying. A bearing must be hard and wear-resistant with a low coefficient of friction but at the same time be tough, shock resistant and sufficiently ductile to allow for 'running in'. These properties of hardness, toughness and ductility cannot be found to the required degree in a single-phase alloy. Thus, intermetallic compounds are hard and have a low coefficient of friction but are extremely brittle, whilst pure metals and solid solutions though ductile are usually soft and with a relatively high coefficient of friction. A suitable combination of mechanical properties can, however, be obtained by using an alloy in which particles of a hard intermetallic compound are embedded in a matrix of ductile solid solution—or, in some cases, a eutectic of two solid solutions. During the 'running in' process the soft matrix tends to wear leaving the hard particles standing proud. This not only reduces the overall coefficient of friction of the bearing surface but also provides channels through which lubricant can flow.

11.7.1 '*White*'-*bearing metals* are either tin-base or lead-base. The former, the high-quality Babbitt metals, contain as much as 90% tin and up to

Plate 11.10—A white-bearing metal containing 10% Sb; 3% Cu; 40% Sn; balance-Pb. A copper–tin compound (Cu_6Sn_5) crystallises first (light needles) forming a network which prevents the very light SbSn cuboids from segregating to the surface when they begin to form. The matrix is a eutectic of tin–antimony and tin–lead solid solutions. (Etched in 2% Nital.) × 70.

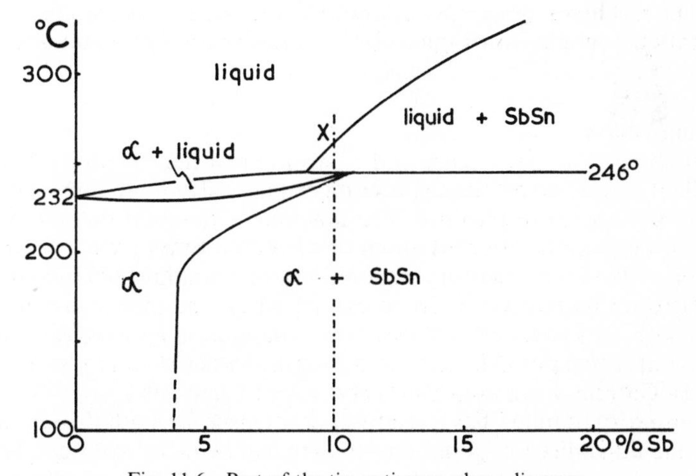

Fig. 11.6—Part of the tin–antimony phase diagram.

10% antimony. These two metals form 'cuboid' crystals of the intermetallic compound SbSn which are easily identified in the microstructure and which constitute the hard, low-friction phase. In lead-free white metals the cuboids are held in a matrix of the solid solution α (Fig. 11.6) whilst in white metals to which lead has been added in amounts up to 80% in order to reduce cost, the SbSn cuboids will be in a matrix consisting of a eutectic of two solid solutions—one tin-rich and the other lead-rich.

Cuboids of SbSn separate from the liquid phase between X and 246° C and since the relative density of SbSn is less than that of the liquid, the cuboids float to the surface as they form. To prevent this type of segregation up to 3% copper is added to white-bearing metals. This forms the intermetallic compound Cu_6Sn_5 which separates out as a network *before* the formation of SbSn cuboids. When these latter subsequently begin to form they become trapped by the network of Cu_6Sn_5 crystals and so prevented from segregating at the surface of the casting.

11.7.2 *Copper-base-bearing alloys* include the plain tin bronzes (10–15% Sn) and phosphor bronzes (10–13% Sn; 0·3–1·0% P). As the phase diagram (Fig. 11.1 (ii)) indicates these alloys will be of $\alpha + \delta$ structure, and though nominally the hard compound δ ($Cu_{31}Sn_8$) will be present as a eutectoid with α, in practice the layers of δ coagulate to form hard particles in a tough matrix of the solid solution α thus fulfilling the general requirements of a bearing metal.

11.7.3 Sintered bronze bearings produced by powder metallurgy (6.5.1) retain sufficient porosity to allow them to be impregnated with lubricating oil. These 'oil-less' bearings are widely used in vacuum cleaners, washing machines, gramophone motors and other applications where lubrication might be either uncertain or a lubrication system unjustifiably expensive.

Titanium alloys

11.8 In terms of its occurrence in the Earth's crust titanium is a relatively abundant metal and, of the engineering metals, only aluminium, iron and magnesium are more plentiful. The amount of titanium ore within the range of mining operations is about one hundred times greater than that of copper. However, the very high affinity of titanium for both oxygen and nitrogen makes it difficult to extract whilst the molten metal itself reacts with all known refractories. Thus titanium is an expensive metal because of its cost of extraction and forming rather than any scarcity of its ores. Currently titanium billets cost about £2.50 per kg.

High-purity titanium has a relatively low tensile strength (216 N/mm^2) and a high ductility (50%) but the strength can be raised considerably by alloying. It is a polymorphic element. The α phase (CPH) transforms on heating to 882·5° C to β (BCC) (Fig. 11.7 (i)) and this change-point is affected by alloying as is the A_3 point in iron. Thus alloying elements

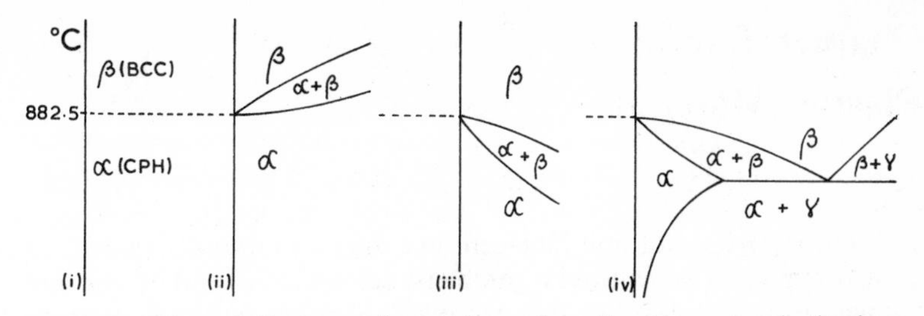

Fig. 11.7—Effects of alloying on the $\alpha \rightleftharpoons \beta$ transformation temperature in titanium.

which have a greater solubility in α than in β tend to stabilise α over a wider range of temperatures (Fig. 11.7 (ii)). These elements include aluminium, which dissolves substitutionally, and oxygen, nitrogen and carbon all of which dissolve interstitially. Elements which dissolve more readily in β stabilise it in preference to α (Fig. 11.7 (iii) and (iv)). They include the transition metals iron, chromium, molybdenum, etc. Alloys represented by a phase diagram of the type (iv) can be precipitation hardened as might be expected with a diagram of this form (7.5.2.1). A typical titanium-base alloy, IMI 700 (6 Al; 5 Zr; 3 Mo; 1 Cu) attains a tensile strength of 1540 N/mm^2.

11.8.1 The relative density of titanium is only 4·5 and suitable alloys based on it have a high specific strength. Moreover creep properties up to 500° C are very satisfactory whilst the fatigue limit is also high. Titanium alloys therefore find application in the compressors of jet engines. In turbine engineering generally titanium alloys, being of low relative density, impose much lower centrifugal stresses on rotors and discs for a given blade size. Titanium alloys are also very resistant to erosion by wet steam and other media so that steam turbo-generators, gas turbines and condenser tubing all make use of these alloys.

Because of the high specific strength, high temperature resistance and good corrosion resistance generally, many titanium alloys find use in supersonic aircraft—including 'Concorde'—as structural forgings. Such an alloy is IMI Ti680 (11 Sn; 2·25 Al; 4 Mo; 0·2 Si) which develops a strength of 1300 N/mm^2.

Chapter Twelve
Plastics Materials

12.1 The townships of mid-nineteenth-century Colorado attracted a motley assortment of adventurers, gamblers, saloon ladies and others bent on the acquisition of easy wealth. Amongst the more respectable establishments of these towns were billiard saloons and in an attempt to provide them with enough billiard balls, John Hyatt sought a substitute for scarce and expensive ivory. His early products consisted of a core of ivory dust bonded with shellac and an exterior coating of the somewhat unstable compound collodion. Unfortunately violent impact of the balls often caused the collodion to explode so that every gunman in the saloon instinctively 'pulled a gun'.*

However, in 1868 Hyatt successfully adopted a substance developed some fourteen years earlier by Alexander Parkes of Birmingham, England. It consisted of a mixture of cellulose nitrate and camphor and had been known as 'Parkesene'. This, the first of the man-made plastics, was renamed 'celluloid' and was used widely until a few years ago for the manufacture of many articles, in particular toys such as dolls. Its dangerous inflammability finally led to its obsolescence so that comparatively little of it is now used. When in 1907 Baekeland patented his process for the manufacture of 'bakelite' the plastics industry could be said to have 'arrived', for this material was quickly adopted by the rapidly expanding electrical trades as an insulation material.

12.1.1 The term 'plastics' is not one which is easy to define precisely. Many familiar 'plastic' articles like buttons, switchgear parts and egg-cups are in fact no longer plastic after being moulded to shape since they become hard and brittle and can no longer be deformed on heating. Plastics can be defined as organic materials containing molecules of high molecular weight (i.e. between 10^4 and 10^7) and which can be moulded to shape by the application of pressure at moderately high temperatures. Once moulded they may retain their plasticity in the manner of polythene or nylon or they may become permanently hard and brittle like bakelite. Some plastics, however, are compounded as liquids and proceed to 'set' without the application of either heat or pressure.

12.1.2 Until the beginning of the present century Man's activities in the field of the organic (carbon) compounds were involved mainly in the destruction of these substances. Thus the complex compounds present in coal were destructively distilled to produce a larger number of simpler

* From *The First Century of Plastics—Celluloid and its Sequel* by M. Kaufman (The Plastics Institute, London).

materials the more important of which is coal gas (a mixture of methane, carbon monoxide and hydrogen). Much earlier than this Man had learned to ferment substances containing starches and sugars in order to produce alcohol. Fermentation of the juice of the grape was well-known in Biblical days, whilst the fact that the England of former days earned the title of 'Merrie' is not unconnected with the fermentation of honey to produce mead. All over the world various fruits, vegetables and cereals are fermented for the production of all kinds of national drinks from beer and vodka to okolehao. Fermentation depends upon the action of yeast which produces both alcohol and carbon dioxide from the more complex sugar molecules:

$$\underset{\text{starch}}{(C_6H_{10}O_5)_n} \xrightarrow{H_2O} \underset{\text{glucose}}{C_6G_{12}O_6} \xrightarrow{\text{yeast}} \underset{\text{Ethanol (alcohol)}}{C_2H_5 . OH + CO_2}$$

12.1.3 It is thermodynamically more difficult to build up than to break down but in the present technological age the organic chemist has learned to reassemble some of the bits produced by these destructive processes. More important still he has built up complex compounds from these simple units to produce substances such as have never been known in Nature. Foremost amongst these substances are the *super-polymers*—or *plastics*—so much a part of life in this second half of the twentieth century.

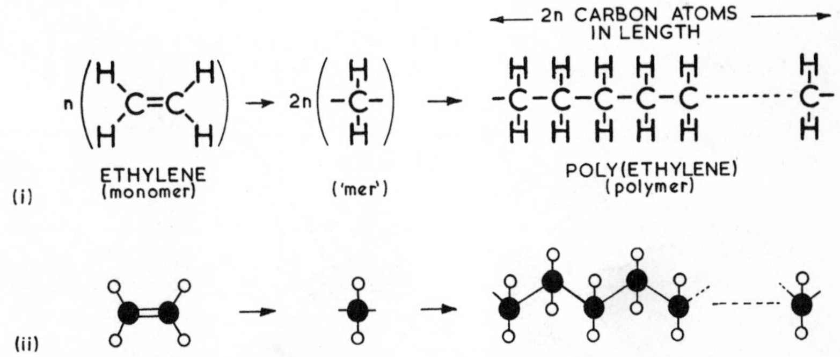

Fig. 12.1—The polymerisation of ethylene. (i) Using orthodox chemical notation ('n' is approximately 600). (ii) Illustrated diagrammatically where ● = carbon and ○ = hydrogen.

The synthesis of polythene from ethylene—a by-product of petroleum 'cracking'—has already been mentioned (2.3.3). In this instance ethylene, C_2H_4, is known as the *monomer* whilst polythene $(CH_2)_n$ is of course the *polymer*. The single repeating unit, $-CH_2-$ (which cannot exist by itself as a molecule) is often referred to as the *mer*. Such a process is known as *addition polymerisation*.

Thermoplastic polymers

12.2 Compounds based on ethylene in which one or more of the hydrogen atoms have been replaced by a different atom or group of atoms are known collectively as *vinyl* compounds (Table 12.1). Some of these vinyl compounds will polymerise spontaneously at ordinary temperatures (Fig. 12.2) whilst others polymerise when heated in the presence of a catalyst. A typical vinyl compound may exist as a very mobile liquid (Fig. 12.2 (i)). The liquid flows as easily as water because only weak van der Waal's forces operate between the relatively small molecules. As polymerisation proceeds (Fig. 12.2 (ii)) the molecules reach a size where the van der Waal's forces become effective so that the liquid becomes relatively viscous. As polymerisation nears completion (Fig. 12.2 (iii)) the molecules become very long so that the sum of van der Waals forces acting between adjacent molecules becomes considerable and the material so viscous as to be regarded as a solid. It no longer flows unless it is heated. This increases the energy of the molecules so that they are able to overcome the van der Waals forces acting between them. Such a substance is said to be *thermoplastic* since it can be softened repeatedly by the application of heat.

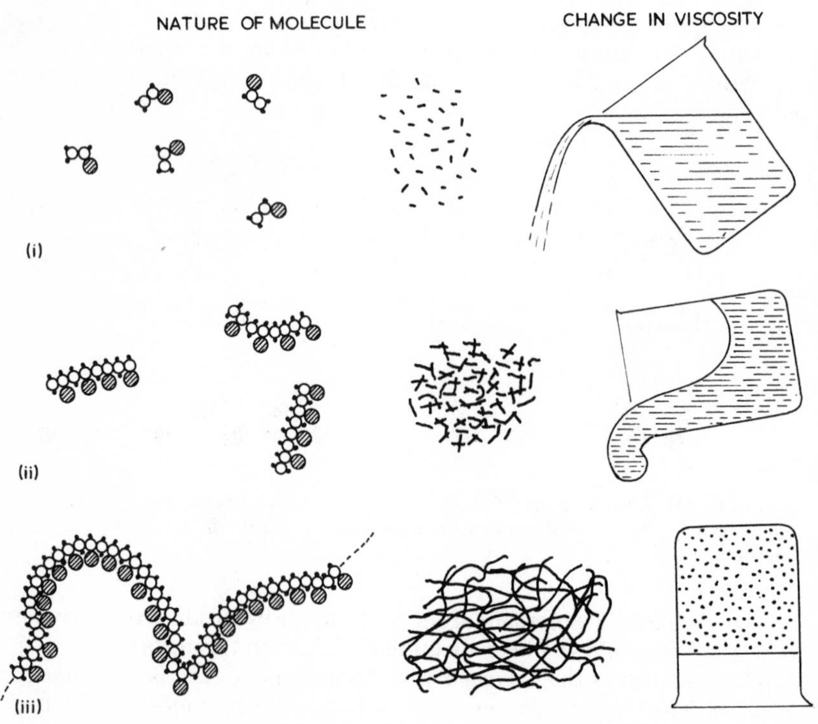

Fig. 12.2—Stages in the polymerisation of a vinyl compound and the changes in viscosity which ensue.

Table 12.1—Common vinyl polymers.

Monomer	*Polymer*
Ethylene $H_2C{=}CH_2$	Polyethylene (polythene) $-CH_2-CH_2-CH_2-CH_2-CH_2-CH_2-CH_2-CH_2-CH_2-\ldots\ldots$
Vinyl chloride $H_2C{=}CHCl$	Polyvinyl chloride (PVC) $-CH_2-CHCl-CH_2-CHCl-CH_2-CHCl-CH_2-CHCl-\ldots\ldots$
Vinyl acetate $H_2C{=}CH-O-C{=}O\,(-CH_3)$	Polyvinyl acetate $-CH_2-CH(O-C{=}O-CH_3)-CH_2-CH(O-C{=}O-CH_3)-CH_2-CH(O-C{=}O-CH_3)-CH_2-CH(O-C{=}O-CH_3)\ldots\ldots$
Methyl Methacrylate $H_2C{=}C(CH_3)-C{=}O\,(-O-CH_3)$	Polymethyl Methacrylate ('Perspex') $-CH_2-C(CH_3)(C{=}O-O-CH_3)-CH_2-C(CH_3)(C{=}O-O-CH_3)-CH_2-C(CH_3)(C{=}O-O-CH_3)-CH_2\ldots\ldots$
Styrene $H_2C{=}CH-C_6H_5$ (hexagon) = benzene ring C_6H_5 with a circle in the centre (ring carbons C; H on all but one of them)	Polystyrene $-CH_2-CH(C_6H_5)-CH_2-CH(C_6H_5)-CH_2-CH(C_6H_5)-CH_2\ldots\ldots$
Vinylidene chloride $H_2C{=}CCl_2$	Polyvinylidene chloride $-CH_2-CCl_2-CH_2-CCl_2-CH_2-CCl_2-CH_2-CCl_2\ldots\ldots$
Tetrafluoro ethylene $F_2C{=}CF_2$	Polytetrafluoroethylene (PTFE or 'Teflon') $-CF_2-CF_2-CF_2-CF_2-CF_2-CF_2-CF_2-CF_2-CF_2\ldots\ldots$

12.2.1 Sometimes two different monomers are used in the synthesis of the large molecule. The latter is then known as a *copolymer*. For example a mixture of vinyl chloride and vinyl acetate can be polymerised to produce the copolymer, polyvinyl chloride–acetate:

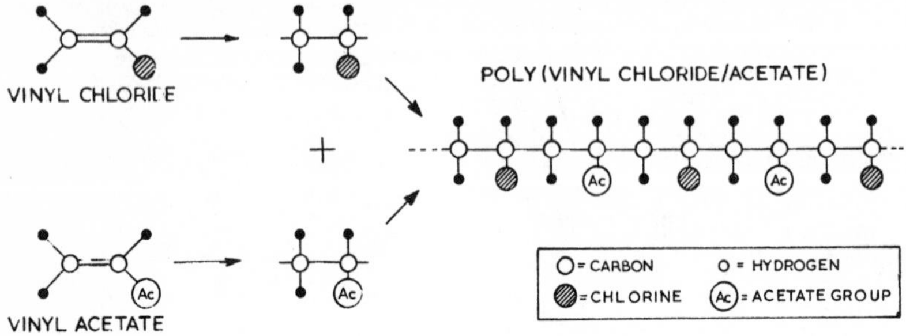

Of course, as is indicated in Table 12.2 not all thermoplastics are vinyl compounds.

Thermosetting polymers

12.3 The polymers mentioned so far have all been of the *thermoplastic* type. That is they can be softened and remoulded repeatedly by the application of heat. A number of polymeric substances, however, are of the *thermosetting* variety. These are plastic in the primary stages of manufacture but once moulded to shape they 'set' and cannot subsequently be softened by re-heating. Thermosetting is due to the formation of covalent bonds between chain molecules.

12.3.1 In the early days of this century a Belgian chemist Dr. Leo Baekeland developed the first synthetic plastics material. Recognising the importance of his discovery he patented it and named the product 'bakelite'. This is a thermosetting material made from the two organic chemicals phenol, $C_6H_5 . OH$, and formaldehyde, $H . CHO$. Under suitable conditions these two substances can be made to react with each other (Fig. 12.3). As a result of this reaction a chain-type of molecule is

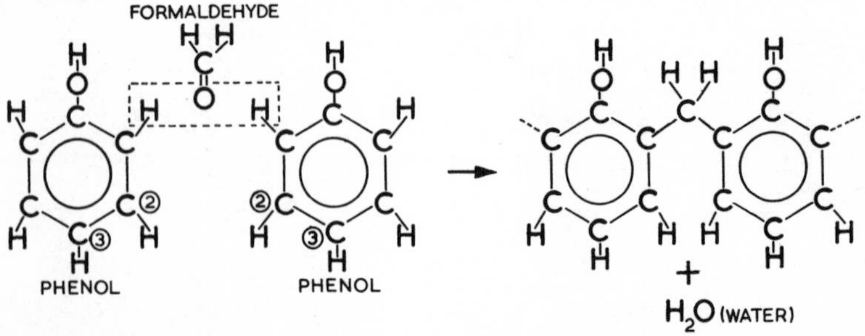

Fig. 12.3—Formation of the 'novolak'. The phenol molecules could equally well be linked at positions ② and ③.

Table 12.2—Some important thermoplastic polymers.

Type	Structure and/or composition		Relative density	Typical mechanical properties: Tensile strength (N/mm²)	Elongation (%)	Impact (J)	Max. service temperature (° C)	T_g (° C)
Polyethylene ('polythene')	$\left[\begin{array}{cc} H & H \\ \mid & \mid \\ -C- & C- \\ \mid & \mid \\ H & H \end{array}\right]_n$	Homopolymer—with various degrees of chain branching, which governs both flexibility and density	Low dens. —0·93	11·0	90–650	No fracture	85	−120
			High dens. —0·96	31·0	50–800	2·0–18·0	125	
Polypropylene	$\left[\begin{array}{cc} H & H \\ \mid & \mid \\ -C- & C- \\ \mid & \mid \\ H & CH_3 \end{array}\right]_n$	Homopolymer	0·91	30·0–35·0	50–600	1·0–10·0	150	−27
Polystyrene	$\left[\begin{array}{cc} H & H \\ \mid & \mid \\ -C- & C- \\ \mid & \mid \\ H & C_6H_5 \end{array}\right]_n$	$C_6H_5 = \begin{array}{ccc} H & C & H \\ C & \bigcirc & C \\ C & & C \\ H & C & H \\ & \mid & \\ & H & \end{array}$	1·07	Properties depend largely upon structure 28·0–53·0	1·0–35	0·25–2·5	65–85	100
Polyvinyl alcohol	$\left[\begin{array}{cc} H & H \\ \mid & \mid \\ -C & C- \\ \mid & \mid \\ H & OH \end{array}\right]_n$	Some acetate groups may be present, replacing the hydroxyl groups	1·26	120·0	400	—	100	85
Polyvinyl chloride (PVC)	$\left[\begin{array}{cc} H & H \\ \mid & \mid \\ -C- & C- \\ \mid & \mid \\ H & Cl \end{array}\right]_n$	The pure polymer is rigid but is often plasticised (12.6) to produce a more flexible material	Rigid form—1·37	49·0	10–130	1·5–18·0	70	87
			Flexible form —1·35	7·0–25·0	240–380	—	60–105	
Polyvinyl acetate (PVA)	$\left[\begin{array}{cc} H & H \\ \mid & \mid \\ -C- & C- \\ \mid & \mid \\ H & COO.CH_3 \end{array}\right]_n$		1·19	Too soft (showing excessive cold flow) for use in moulded plastics generally				29
Polyvinyl acetate/chloride		A co-polymer of the above two monomers	1·35	30·0–35·0	200–400	0·7–1·5	—	—

Table 12.2—continued

Type	*Structure and/or composition*	*Relative density*	*Typical mechanical properties* Tensile strength (N/mm²)	Elongation (%)	Impact (J)	*Max. service temperature* (° C)	T_g (° C)
Polytetrafluoroethylene (*PTFE or 'Teflon'*)	$[-\mathrm{C(F)(F)-C(F)(F)}-]_n$	2·17	17·0–25·0	200–600	3·0–5·0	260	126
Polyvinylidene chloride	$[-\mathrm{C(H)(H)-C(Cl)(Cl)}-]_n$ Homopolymer, or, generally, a copolymer with with about 20% vinyl chloride	1·68	25·0–35·0	up to 200	0·4–1·3	60	−17
(*PMM or 'Perspex'*)	$[-\mathrm{C(H)(H)-C(CH_3)(COO.CH_3)}---]_n$	1·18	50·0–70·0	3·0–8·0	0·5–0·7	95	0
Acrylonitrile-butadiene-styrene (ABS)	Copolymers containing varying proportions of the monomers: $-\mathrm{C(H)(H)-C(H)(CN)}--$ (acrylonitrile); $-\mathrm{C(H)(H)-C(H)=C(H)-C(H)(H)}-$ (butadiene); $-\mathrm{C(H)(H)-C(H)(C_6H_5)}--$ (styrene)	1·05	30·0–35·0	10·0–140	7·0–12·0	100	−55
Polyamides (nylons)	Polymers in which various aliphatic groups separate recurring amide groups $[-\mathrm{N(H)-C(=O)}-(\mathrm{C(H)})-\mathrm{C(=O)-N(H)}-(\mathrm{C(H)})-]$	'Nylon 66' 1·14	50·0–87·5	60·0–300·0	1·5–15·0	120	50

	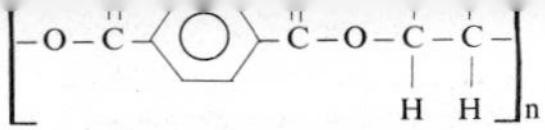 $[-O-CO-C_6H_4-CO-O-CH_2-CH_2-]_n$ poly(ethylene terephthalate)						
Acetals (*polyformaldehyde or polyoxymethylene*)	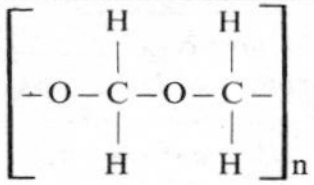 $[-O-CH_2-O-CH_2-]_n$ Homopolymer derived from formaldehyde 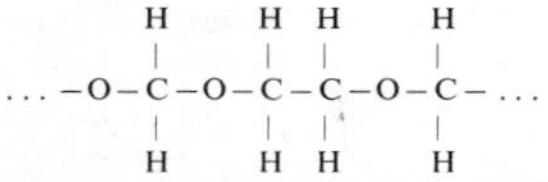 $\ldots-O-CH_2-O-CH_2-CH_2-O-CH_2-\ldots$ Copolymer- extra CH_2 mers increase stability at high temperatures	1·41	50·0–70·0	15·0–75·0	0·5–2·0	105	−73
Polycarbonates	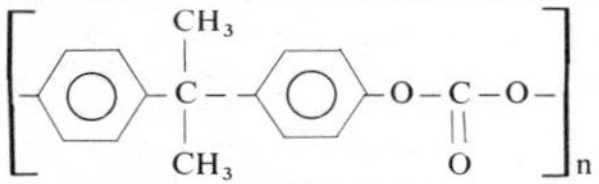 $[-C_6H_4-C(CH_3)_2-C_6H_4-O-CO-O-]_n$	1·2 (unfilled)	60·0–70·0	60–100	10·0–20·0	130	150
Polyurethanes (*thermoplastic types*)	Polymers prepared from di-isicyanates and polyethers or polyesters 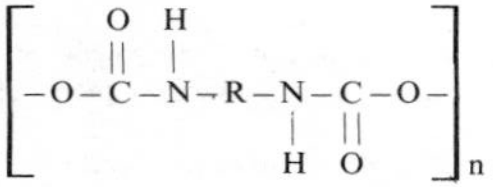 $[-O-CO-NH-R-NH-CO-O-]_n$	1·21	up to 55·0	up to 700	—	—	—
Cellulose nitrate	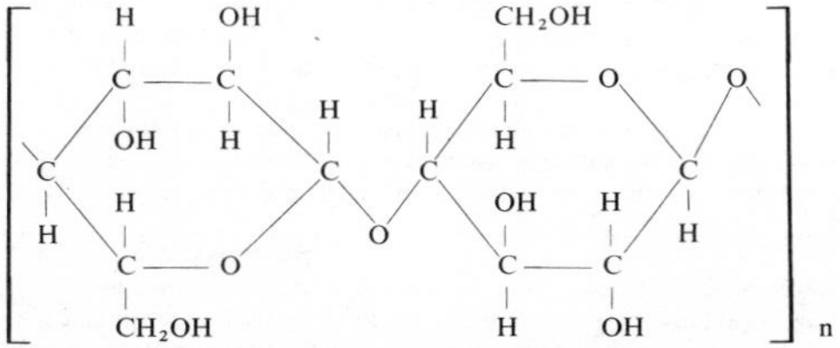 [cellulose monomer ring structure]$_n$ Some, or all, of the hydroxyl groups (OH) of the cellulose monomer (above) are nitrated: $\geq C-OH \xrightarrow{HNO_3} \geq C-O.NO_2$	1·40	35·0–70·0	10–40	3·0–11·0	—	53
Cellulose acetate	Various degrees of acetylation of the cellulose monomer (above) are possible—about 80% of the available hydroxyl groups are generally acetylated	1·28	24·0–65·0	5–55	0·7–7·0	70	120
Cellulose acetate/butyrate	Esters with usually 20 40% butyrate groups	1·20	17·0–50·0	8–80	0·7–6·5	70	122
Ethyl celluloses	About 85% of the hydroxyl groups of cellulose (above) are etherified	1·15	40·0–60·0	10–40	4·5–8·0	60	43

Table 12.2—*continued*

Type	*Other properties and characteristics*	*Uses*
Polyethylene ('polythene')	One of the most versatile thermoplastics. Tough and flexible over a wide temperature range. Good dimensional stability. Easily moulded. Good resistance to most common solvents. Good weathering properties but some deterioration on long exposure to light	Used largely in packaging as bags and squeeze-bottles. Also a wide range of household goods such as buckets, bowls and other containers. Piping, chemical equipment, coating for cables and wires
Polypropylene	Similar properties to polythene but has better heat resistance. Good mechanical properties. Excellent resistance to chemical attack by acids, alkalis and salts even at high temperatures. Tough, rigid and light in weight	Mouldings for hospital and laboratory equipment; chemical plant and domestic hardware. Sheets used in packaging. Monofilaments used as ropes, fishing nets and filter cloths
Polystyrene	A tough, dense plastic—hard and rigid. Good dimensional stability. Moulds with high surface gloss. Good strength but inclined to be brittle due to high T_g. Retains properties at low temperatures. Attacked by petrol and other organic solvents. Can be foamed to produce rigid but extremely light cellular material	Refrigerator trays, boxes and many articles of household holloware. Toys, display figures, etc, etc. Rigid foams used in building and refrigeration as insulation against heat and sound, e.g. ceiling tiles, and in packaging for insulation against shock. Buoyancy equipment
Polyvinyl alcohol	Tough and strong. Flexible. Resistant to many common organic solvents but dissolves slowly in boiling water. Softens above 100° C. Processed by moulding, extrusion and solution coating	In water solution can be used for sizing papers and fabrics. Also in adhesives. High solubility in water limits its use
Polyvinyl chloride (PVC)	In the unplasticised form is tough and hard. If plasticised it becomes soft, flexible and rubbery. Good dimensional stability. Good resistance to water, acids, alkalis and most common solvents. Plasticised material stiffens with age	The rigid form is used in mouldings of all kinds. The flexible form in imitation leather cloth, table cloths, raincoats; etc. Also for piping, guttering and electrical cable covering. Safety helmets; chemical tank linings
Polyvinyl acetate (PVA)	Soft, with poor dimensional stability. Good adhesive properties. Resistant to water but dissolves in many common organic solvents	Since it is soluble in many organic solvents it is used in lacquers, paints, coatings and adhesives
Polyvinyl acetate/chloride	Tough, more flexible than PVC. Good dimensional stability. Good resistance to acids, alkalis and most common solvents. Good electrical insulation	The unplasticised sheet is used for packaging. Other uses include hand luggage, upholstery, covering for wires and cables, long-playing gramophone records, chemical equipment
Polytetrafluoroethylene (PTFE or 'Teflon')	Tough and flexible. Excellent heat-resistance. Does not burn. *Not attacked by any known reagent and no known solvent*. Excellent electrical insulation. Waxy surface and low coefficient of friction. Relatively expensive	Bearings, fuel hoses, gaskets and tapes. Non-stick coatings for fry-pans and other cooking utensils. Also in chemical industry because of its resistance to attack
Polyvinylidene chloride	Very strong and tough. Flexible. Good resistance to attack by acids and alkalis and most organic solvents. Excellent weather-resisting properties	Food wrapping; outdoor furniture fabrics; public transport seat coverings; chemical equipment; screens filters, wire coating, filaments for doll's hair
Polymethyl Methacrylate (PMM or 'Perspex')	Excellent transmission of light. Strong and rigid but easily scratched. Softens in boiling water but unattacked by most household chemicals. Attacked by petrol and many organic solvents. Good electrical insulation	Lenses; dentures; telephones; knobs and handles. Aircraft windows; sinks; building panels; lighting fittings; baths; protective shields; dials; corrugated roof lights. Fancy goods; advertising displays; ornamental lighting. Adhesives, surface coatings

Acrylonitrile-butadiene-styrene (*ABS*)	High impact value along with high strength (resists impact by flying stones). High dimensional stability over widely differing conditions. Remains tough at sub-zero temperatures. Resists attack by acids, alkalis and many petroleum-type liquids	Pipes, protective [illegible], valves, refrigerator parts, battery cases, [illegible], tool handles, textile bobbins, radio cabinets. Now widely used in the automobile industry for radiator grills and other fittings because of its high impact value and the fact that it can be chromium plated
Polyamides (*nylons*)	Very strong and tough. Flexible. Good resistance to abrasion. Good dimensional stability. Absorbs water but has good resistance to most common solvents. Deteriorates with long exposure outdoors. Good electrical insulation	A number of different varieties are in use. Gears, valves, electrical equipment, handles, knobs, bearings, cams, shock absorbers, combs. Surgical and pharmaceutical packaging. Raincoats. Filaments for bristles, climbing ropes, fishing lines and textiles
Polyesters (*thermoplastic types*)	Highly crystalline but if quenched from a melt to below T_g an amorphous structure results. This crystallises if annealed at above T_g. Used as fibre or film—strong and stable. Good dimensional stability and resistance to many organic solvents but attacked by strong acids and alkalis. Excellent electrical insulation	Widely used as a textile fibre ('Terylene', 'Dacron'). In the electrical industries as insulation. Drums and loudspeaker cones; recording tapes; drafting materials; photographic films for special purposes; book covering; packaging. Gaskets; mould release foils; insulating tapes
Acetals (*polyformaldehyde or polyoxymethylene*)	Very strong and stiff—creep resistant. Good fatigue endurance. High crystallinity and high melting point make it suitable for replacing some metals. Good electrical properties. Resistant to alkalis and many solvents but attacked by mineral acids. Slight water absorption. Adversely affected by prolonged exposure to ultra-violet light—not recommended for outdoor use in strong sunlight	Impellors and other parts for water pumps, housings, washing machines and extractor fans. Instrument panels and housings for automobiles. Plumbing fittings. Pipe clips (replacing the well-known Jubilee clips). Bearings, cams and gears. Hinges and window catches. Parts for aerosol containers
Polycarbonates	Good impact strength and heat resistance. Good dimensional stability. Resistant to oils and most solvents. Surface takes a high gloss.	Electronic equipment; aircraft parts; fittings for automobiles
Polyurethanes (*thermoplastic types*)	Physical properties of the thermoplastic varieties are similar to nylon. Good dimensional stability. Strong and tough. Low water absorption and resistant to most common solvents	Packaging film, filaments for bristles, etc. Textile coatings, tank linings, surgical gloves, gears, hosing. Elastic thread for foundation garments
Cellulose nitrate	Tough and flexible. Easily worked. Softens in boiling water. Good resistance to water but dissolves in some organic solvents such as alcohol, ketones and esters. Not suitable for exposure outdoors. Its *extreme inflammability* limits its use now that many more suitable compounds are available	Solutions are used for lacquers and leather cloth. Combs, toys (though inadvisable due to high inflammability), handles, spectacle frames, brushes, windows, piano keys, drawing instruments, pens and pencils, handle grips.
Cellulose acetate	Tough and durable. Good impact strength. High absorption of water which affects dimensions. Not suitable outdoors. Softens in boiling water. Resistant to most household chemicals but is attacked by acids, alkalis and alcohol	Toys, containers, lampshades, machine guards, brush handles, spectacle frames. Photographic film base, book covers, packaging. Beading, insulation, handle covers. Combs, brushes, steering wheels, oil pumps, buttons, door handles, radio cabinets
Cellulose acetate/butyrate	Tough and flexible. Better resistance to water than the acetate. Better dimensional stability. Softens in boiling water. Resistant to most household chemicals but not organic solvents	Machine guards; advertising displays. Beading special purpose hoses. Handles for brushes and cutlery. Pens and pencils. Furniture
Ethyl celluloses	Tough and flexible. Good impact strength at low temperatures. Good dimensional stability. Low absorption of water and good resistance to acids and alkalis, but dissolves in many organic solvents. Not suitable for outdoor use. Good electrical insulation	Packaging, watch 'glasses'. Handle and grip covers, special hoses. Telephones, furniture, automobile fittings, tool handles, vacuum cleaner parts, strippable protective coatings

produced. However, this method of polymerisation differs from that involved in addition polymerisation as typified by the poly-vinyls. Here parts of the reacting molecules are eliminated (in this particular instance as molecules of water, H_2O) and this type of reaction is termed *condensation polymerisation.*

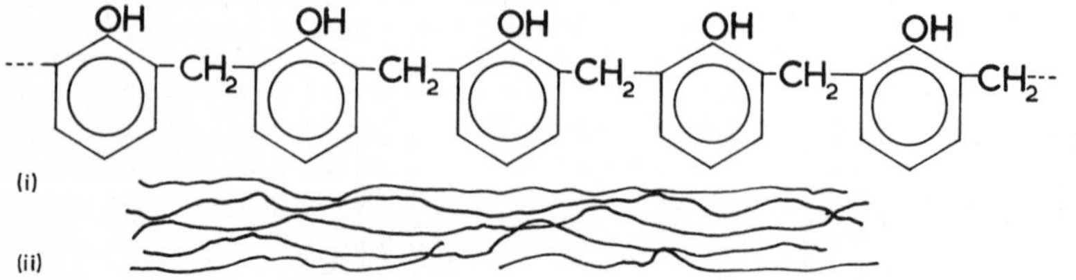

Fig. 12.4—Chain molecule at the 'novolak' stage. (i) The phenol groups covalently linked by $-CH_2-$ groups. (ii) Only van der Waal's forces will operate between the resultant chain molecules.

In the above reaction only sufficient formaldehyde is used to link the phenol molecules in a simple linear manner (Fig. 12.4). Consequently at this stage the material, known in the trade as a 'first-stage resin' or 'novolak', is still thermoplastic. It is allowed to cool and is then powdered and mixed with hexamethylene tetramine. During the subsequent

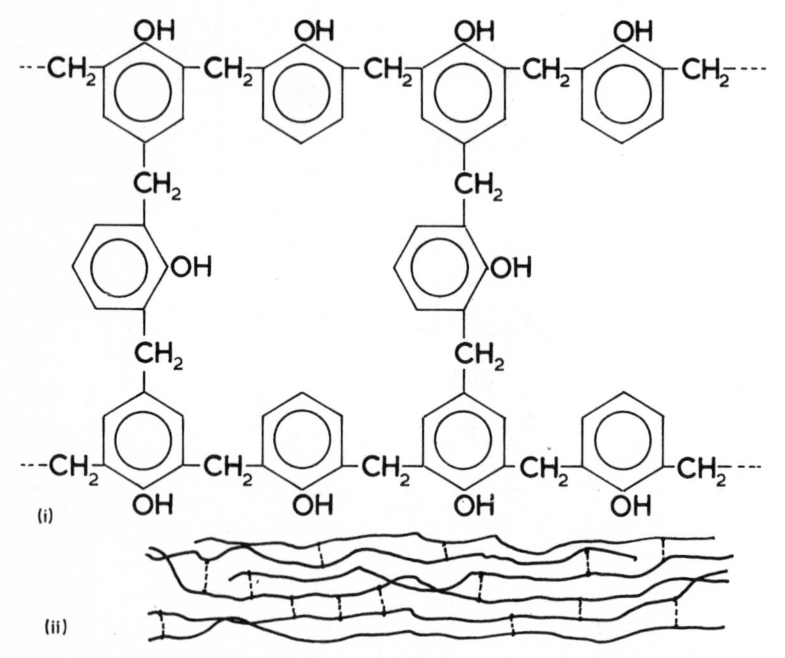

Fig. 12.5—The type of cross-linking which results in thermosetting. (i) Covalent linking between chains by means of other phenol and $-CH_2-$ groups. (ii) Covalent linking now holds chain molecules rigidly together.

moulding processes the hexamethylene tetramine decomposes, releasing more formaldehyde which forms cross links between the chain-molecules of the novolak (Fig. 12.5). The formation of these cross links means that the chain-molecules are now joined by covalent bonds which are very strong when compared with the relatively weak van der Waals forces which hold together the chain molecules in a thermoplastic material. For this reason thermosetting materials are generally stronger at elevated temperatures than are thermoplastic materials in which van der Waals forces are readily overcome by thermal vibrations within the molecules.

12.3.2 *Cold-setting plastics.* Similar in principle to the thermosetting materials are the cold-setting plastics. These are formed by reagents which will undergo polymerisation when they are mixed at ordinary temperatures. In addition to moulding materials some varnishes and adhesives are of this type.

The physical nature of solid polymers

12.4 So far we have been considering the form of single polymer molecules. With the exception of 'three-dimensional' condensation polymers of the thermosetting type these molecules can be visualised as thread-like entities bonded to their neighbours by van der Waals forces. Some molecules like those of polyethylene, are 'symmetrical' in section whilst others may carry relatively bulky side groups, non-symmetrically placed. Thus, in polystyrene these appendages consist of phenyl groups ⬡, and in polypropylene the group, $-CH_3$. Moreover, any sample of a pure polymer will contain molecules of varying length. Thus, whilst a polyethylene molecule consists *on average* of a chain some 1200 carbon atoms in length, individual molecules may contain many more—or less—than the average number.

Likewise varying degrees of 'side-branching' can occur in polyethylene molecules (Fig. 12.6), producing branches of varying lengths. The

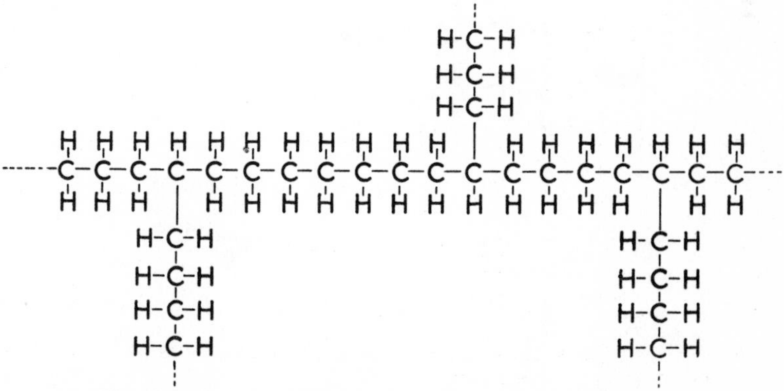

```
                                        :
                                      H-C-H
                                      H-C-H
                                      H-C-H
                                        |
 H  H  H  H  H  H  H  H  H  H  H  |  H  H  H  H  H  H  H
-C--C--C--C--C--C--C--C--C--C--C--C--C--C--C--C--C--C--C-
 H  H  H  |  H  H  H  H  H  H  H  H  H  H  H  H  |  H  H
       H-C-H                                  H-C-H
       H-C-H                                  H-C-H
       H-C-H                                  H-C-H
       H-C-H                                  H-C-H
          :                                      :
```

Fig. 12.6—Side branching in polyethylene (polythene) molecules.

Table 12.3—Some important thermosetting plastics.

Type	*Structure and/or composition*	*Relative density*	*Typical mechanical properties* Tensile strength (N/mm²)	*Elongation* (%)	*Impact* (J)	*Max. service temperature* (° C)
Casein plastics	Natural protein (skimmed milk) cross-linked with formaldehyde (thermoplastic when not cross-linked in this way)	1·34	55–70	2·5–4	1·5–2	150
Phenol formaldehyde (bakelite)	See Fig. 12.5 A condensation product of either phenol or cresol with formaldehyde	1·35 ('filled' with wood flour)	35–55	1	0·3–1·5	75
Urea formaldehyde	A condensation product of urea and formaldehyde	1·5 (cellulose filled)	50–75	1·0	0·3–0·5	75
Melamine formaldehyde	A condensation product of melamine and formaldehyde	1·5 (cellulose filled)	56–80	0·7	0·2–0·45	100

```
       H     C=O    H
       |      |
 ··· -N-C----N----C-N- ···
    |  |          |  |
    |  H          H  |
  O=C                C=O
    |  H          H  |
    |  |          |  |
 ··· -N-C----N----C-N- ···
       |     |     |
       H     .     H
             .
             .
```

```
              |
             N-H
              |
              C
            //  \
           N     N
      H    |     ||    H
      |    |     ||    |
 ··· -N-C        C-N- ···
         \\     /
            N
```

Epoxides		$\cdots-CH_2-CH(-)-C_6H_4-C(CH_3)_2-C_6H_4-O-CH_2-CH(-)-\cdots$ (H H / C–C / H; CH₃ / C / CH₃; H H / C–C / H, with dotted cross-links) Polymers generally cross-linked by amines	1·15 (rigid and unfilled)	35–80	5–10	0·5–1·5	200
Polyesters	(i) *unsaturated polyesters*	Unsaturated linear polymers, cross-linked by various vinyl monomers	1·12	55	2	0·7	220
	(ii) *alkyd resins*	Condensation products of, for example, glycol and phthalic anhydride	2·0	25	—	0·25	150
Polyurethanes (*thermosetting variety*)		They contain the group: $\cdots-N(H)-C(=O)-O-\cdots$ Foams are produced due to the release of carbon dioxide simultaneously with the cross-linking process	1·2 (solid polymer)	up to 55	up to 500	—	Rigid foams up to 150
Silicones		$\cdots-O-Si(CH_3)-O-Si(CH_3)-\cdots$ with each Si linked through O to $\cdots-Si-O-Si-\cdots$ (cross-linked network)	1·88	40	—	0·4	450

Table 12.3—*continued*

Type	*Properties and characteristics*	*Uses*
Casein plastics	Tough and rigid. Takes good polish. Absorbs water but is not affected by most common solvents. Brittle below O° C. Not recommended for exposure to weather. A poor electrical insulator	Imitation tortiseshell and 'mother o'pearl-buckles, brush backs, buttons, ornament pens and pencils, knitting pins, umbrella handles
Phenol formaldehyde (bakelite)	In the 'raw' state phenolics are brittle—hence 'filled' with fibrous materials to increase strength. Wide variation of properties depending upon filler used. Brittle in thin sections. Absorbs water but resistant to alcohol, oils and most common solvents. Does not soften—decomposes above 200 °C	Electrical equipment, handles, buttons, radio cabinets, furniture, vacuum cleaner parts, cameras, ash trays, engine ignition equipment. Cheap jewellery, ornaments. Laminated materials—electrical equipment, gears, bearings, aircraft parts, chemical equipment, clutch and brake linings
Urea formaldehyde	Properties vary considerably with fillers or laminating materials. Hard and rigid. Scratch resistant. Tough if a suitable filler is used. Absorbs water with some loss in dimensional stability. Resistant to most solvents and common reagents. Unaffected by detergents. Good electrical resistance	*As a syrup or resin*: a binder for foundry core production, finishes for textiles and papers, adhesives, laminates. *Mouldings*: electrical equipment (plugs and switches), buttons, buckles, bottle tops, knobs, cups, saucers and plates, toys, clock cases, radio cabinets, kitchen ware. *Laminates*: building panels
Melamine formaldehyde	Properties are generally similar to those of urea formaldehyde but with improved resistance to heat. Also harder than UF and absorbs less water. Excellent electrical properties	*As syrups and resins*: surface coating of papers and textiles, laminates. *Mouldings*: electrical equipment, table ware (particularly cups and saucers), knobs and handles. *Laminates*: electrical equipment, panels
Epoxides	Commonly used in conjunction with glass fibre (available to D.I.Y. experts) for increased strength. Good adhesion to metals. Good resistance to water and most solvents and common reagents. Excellent heat resistance. Good electrical insulation	Sold as resins and syrups. Used as adhesives for gluing metal, low-pressure laminations, surface coatings, casting and 'potting', as a putty for repairing castings. *Laminates* (*with glass fibre*): boat hulls, table tops, laboratory furniture
Polyesters (i) *unsaturated polyesters*	Mechanical properties variable, depending upon polymer used. Generally good impact resistance and surface hardness. Good resistance to water and most common solvents but attacked by conc. acids and alkalis. Resistant to weather	Adhesives, surface coatings. Lampshades, radio grilles, refrigerator parts, laminates for boat hulls, car bodies and wheel barrows. Helmets, fishing rods and archery bows
(ii) *alkyd resins*	Excellent heat resistance, retaining good dimensional stability. Good resistance to water and many organic solvents. Excellent electrical properties	*Liquid resins* are used as paints and enamels for cars, refrigerators, washing machines, etc. *Mouldings* for electrical equipment of cars, fuses, switches, TV components, etc.
Polyurethanes (*thermosetting variety*)	Strong, shock-resistant and similar in many ways to nylons. Good adhesive properties for bonding synthetic rubbers and fibres in tyres. Good dimensional stability. Good resistance to wear. Resistant to most common solvents, oils and chemicals	Gears and bearings. Electrical equipment. Knobs, handles, cable covering, packaging film. Footwear, paints and lacquers. *Foams*: upholstery, toilet sponges, insulation. *Rigid foams*: reinforcement for aircraft wings
Silicones	Available as oils, greases, rubbers, thermoplastic and thermosetting polymers, depending upon the monomeric material and polymerising conditions. Water-repellent and very resistant to weather. Not attacked by mineral acids. Excellent dielectric properties even when exposed to moisture	Water-proof coatings for fabrics. Anti-foaming agents. Hydraulic fluids. *Mouldings*: electrical equipment, e.g. coil formers, switch parts, induction heating equipment, insulation for motors and generator coils. Gaskets; seals in chemical plant

presence of these side branches has considerable effect on the mechanical properties, generally increasing the strength and hardness.

The forces acting between individual atoms within a polymer molecule are of the primary valency type—strong covalent bonds (2.2)—whilst forces acting between adjacent molecules are of a secondary nature known collectively as van der Waals forces (2.4). Naturally the magnitude of these relatively weak secondary bonding forces govern the ease with which molecules may be separated and so affect such properties as solubility, viscosity and frictional properties. Whilst strength depends largely upon the covalent bonding within the molecule chains it is also affected by van der Waals forces operating between adjacent chains.

12.4.1 Many solid substances exist in purely crystalline form. This is usually so when the particles of which they are composed, whether ions or molecules, are small and are able to fall easily into a predetermined regular pattern. The long polymer chains in plastics materials, however, are likely to become entangled, particularly if large side appendages are present. Consequently only a limited degree of crystalline arrangement is generally possible in the case of these substances and limited regions occur in which the linear chains are arranged in an ordered pattern (Fig. 12.7). These ordered regions are known as *crystallites*. They can be detected by the use of X-ray analysis (3.4) which produces appropriate diffraction patterns. Those polymers which are completely amorphous of course produce no such clear patterns.

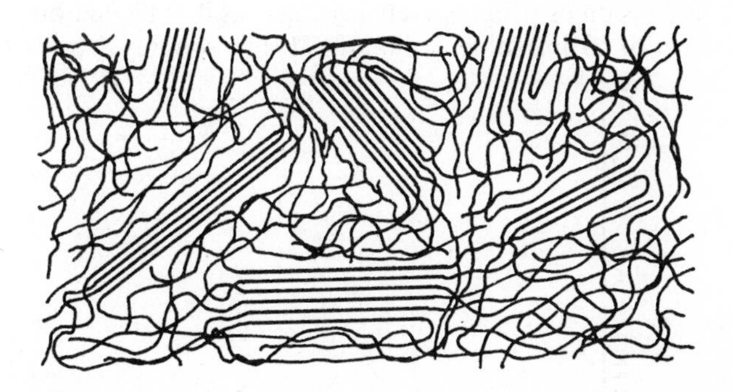

Fig. 12.7—The existence of 'crystallites' in a solid polymer.

In the solid state, therefore, polymers can be regarded as consisting of both crystalline and amorphous regions in relative amounts which vary from one polymer to another and which depend upon the molecular structure as well as upon the previous thermal history. Those polymers in which the molecular structure is regular and the section highly symmetrical, show a high proportion of crystallites. Such polymers include polythene— $\left[\begin{array}{c} H \\ | \\ -C- \\ | \\ H \end{array} \right]$ and polytetrafluoroethylene $\left[\begin{array}{c} F \\ | \\ -C- \\ | \\ F \end{array} \right]$ where short repetitive units of high symmetry allow chains to align themselves. Bulky side chains, such as are present in polyvinyl acetate, on the other hand, impede the formation of crystalline regions and such substances are largely amorphous. Crystalline regions vary in length from as much as 50 nm for strongly crystalline materials like polythene to as little as 2 nm for poorly crystalline polymers like polyvinyl chloride. Highly crystalline polymers have a crystallinity of 80–90%.

When a polymer crystallises freely in the absence of any external forces the crystallites so formed are randomly orientated, but if an external stress is applied in a single direction the crystallites orientate themselves so that their longitudinal axes are parallel to the direction of the applied stress. During the process of cold stretching the degree of crystallinity of a partially crystalline polymer does not increase appreciably but in the case of amorphous polymers or those of very low crystallinity, some increase in crystallinity may occur. As the degree of crystallinity and the extent of orientation increase so do the tensile strength and stiffness.

12.4.2 *Melting points of crystalline polymers.* A pure and completely crystalline solid, for example a metal or an inorganic salt, melts at a single definite temperature called the melting point. An amorphous solid, however, merely becomes progressively less rigid as it is heated but shows no sharp transition from solid to liquid. This is roughly what one would

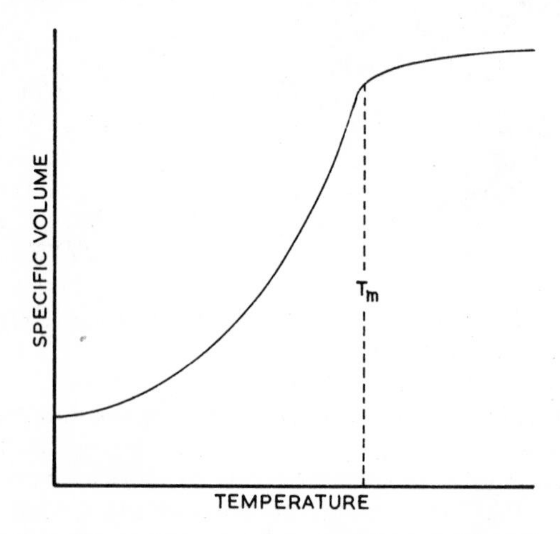

Fig. 12.8—The melting curve for a partially crystalline polymer.

expect since, as with a liquid (which is amorphous), viscosity decreases progressively as van der Waals forces decrease with rise in temperature. For those polymers which are particularly crystalline, however, a melting curve similar to that shown in Fig. 12.8 is obtained. Here the change in specific volume is used as an indication of the change in structure. Melting begins as the smaller, less perfect, crystallites change to an amorphous structure and this is indicated by the gradual change in the shape of the curve. At a high temperature there is a fairly sudden change in the specific volume with temperature, and this is regarded as the melting point. The melting point, T_m, of a crystalline polymer is defined as the temperature at which the crystallinity disappears and the structure

Table 12.4—Melting points of some crystalline polymers.

Polymer	*Melting point, T_m. (° C)*
Polyethylene (50% crystalline)	120
Polyethylene (80% crystalline)	135
Polypropylene	176
Natural rubber	28
Polyvinyl chloride (PVC)	212
Polyvinylidene chloride	198
Polytetrafluoroethylene (PTFE)	327

becomes completely amorphous. Melting point is, to some extent, a measure of the van der Waals forces acting between molecules. Thus, increase in crystallinity of polythene leads to an increase in melting point presumably because inter-molecular forces are greater in the crystalline regions where chains are more closely aligned.

12.4.3 *Glass transition temperature.* Under normal conditions polymers may be either soft, flexible substances or hard brittle glassy materials. If a soft, flexible polymer is cooled sufficiently a temperature is reached where it becomes hard and glassy. Thus, a soft rubber ball when cooled in liquid air becomes hard and brittle and will shatter to fragments if an attempt is made to bounce it whilst it is still at a low temperature. The temperature at which this change in properties takes place is known as the *glass transition temperature*. For some polymers the glass transition temperature is above room temperature so that these materials are normally hard but will become soft and pliable when heated. Polystyrene (Table 12.2) is a polymer of this type.

The glass transition temperature, T_g, can be assessed by measuring the change in some property such as specific volume or specific heat with change in temperature and is indicated by a change in slope of the curve produced (Fig. 12.9). The glass transition temperature is less well defined than is the crystalline melting point, T_m.

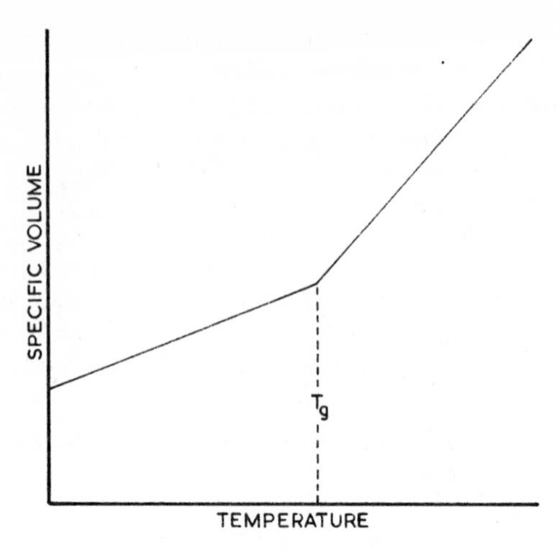

Fig. 12.9—Derivation of the glass-transition temperature, T_g.

12.4.4 The glass transition temperature is a property associated with the amorphous regions in a polymer and is usually described as the temperature at which, on heating, the polymer chains acquire sufficient thermal energy to vibrate in a co-ordinated manner. Below T_g the atoms, apart from restricted vibrations about equilibrium positions, are in a rigid frozen state and the polymer is hard and glassy as a result. Above T_g the co-ordinated vibrations referred to above impart to the polymer its rubber-like properties. The presence of plasticisers and other additives which decrease the magnitude of the van der Waals forces give rise to a reduction in T_g.

Table 12.5—Glass transition temperatures of some polymers.

Polymer	*Glass transition temperature,* T_g (°C)
Polyethylene	−120
Rubber	−73
Polypropylene	−27
Polyvinylidene chloride	−17
Polymethyl methacrylate	0
Polyvinyl acetate	27
Cellulose nitrate	53
Polyethylene terephthalate	69
Polyvinyl chloride	87
Polystyrene	100
Cellulose acetate	120
Polytetrafluoroethylene	126

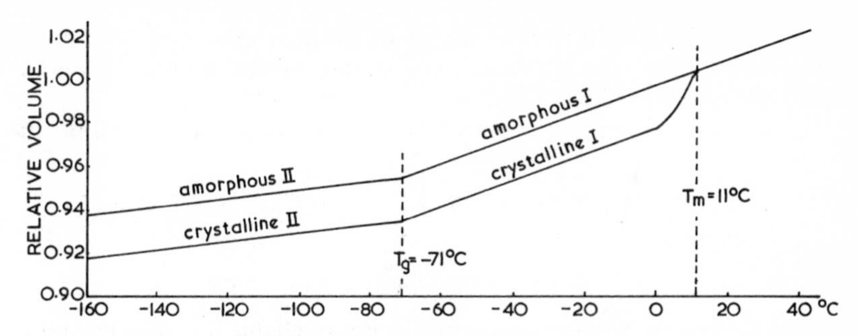

Fig. 12.10—The volume–temperature relationship of a high polymer. The upper curve is for a completely amorphous polymer whilst the lower is for a partly crystalline one.

The mechanical properties of polymers

12.5 Engineers are familiar with the elastic properties of metals and with the extent to which they are reasonably constant under normal conditions. Even with metals, of course, large deviations occur when high temperatures or large deformations are involved, whilst the effects of creep over long periods of time are also well known. With plastics, however, such deviations occur when relatively normal conditions prevail making it impossible to relate deformation in plastics in terms of simple elastic constants. As one might expect from a study of the structure of a polymer, deformation is very dependent upon *time*. Molecular response is slow to reach equilibrium with the external forces and so the material continues to deform, or creep, almost indefinitely. When the applied forces are removed some recovery may occur but a degree of permanent deformation may also persist.

12.5.1 Deformation in plastics is also very dependent upon temperature and this must be very seriously considered in design to a far greater extent than is necessary with metals. As temperature increases the weak van der Waals forces are further diminished and slip occurs more easily between adjacent chains. When a crystalline material, such as a metal, is stressed atoms are progressively displaced from their equilibrium positions until new internal forces set up balance the externally applied forces. Assuming that the distortion produced is not too great, the atoms will return to their original positions when the stress is removed. Larger stresses will produce permanent deformation but provided that the stress does not reach these values the change in shape of the material is instantaneous and elastic.

This mode of deformation may also occur in a plastic upon the application of stress. When two atoms within a chain are held in fixed positions relative to each other by strong covalent bonds, some displacement of the atoms relative to each other occurs instantaneously on the application of stress (Fig. 12.11). On the removal of stress the atoms immediately return to their original positions. In highly cross-linked polymers and in thermoplastics below their glass-transition point, atoms can be regarded as being fixed in this manner so that this is the only type of deformation possible.

In thermoplastics above T_g this type of elastic deformation will be due

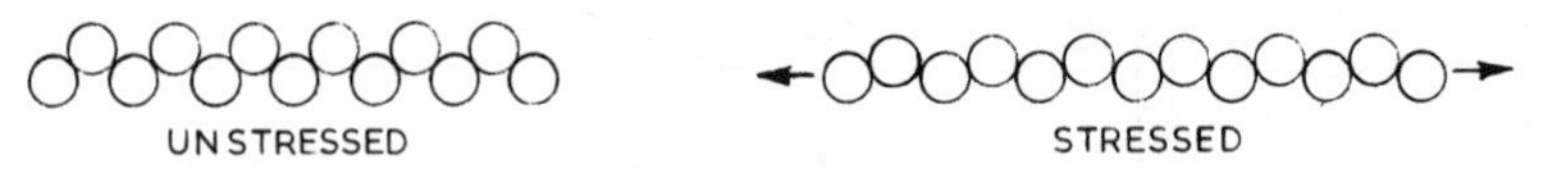

Fig. 12.11—Purely elastic deformation in chain molecules.

not only to bonds between individual atoms in chains but also, if crystallites are present, to inter-molecular forces as well. In the amorphous regions of such thermoplastics, however, the molecules are only attracted to each other by relatively weak van der Waals forces, so that they are able to 'slip' into new permanent positions relative to each other (Fig. 12.12). This type of movement is not instantaneous but is dependent upon the viscosity of the material. *Viscoelastic* deformation thus occurs.

Fig. 12.12—The slip of chain molecules into new positions.

12.5.2 In addition to elastic deformation which is dependent upon small movements of atoms relative to each other, and plastic deformation caused by wholesale slipping of chain molecules past one another into permanent new positions, further deformation occurs due to straightening of coiled and folded chain molecules themselves in the direction of the applied stress (Fig. 12.13). Essentially this type of distortion is elastic

Fig. 12.13—Deformation in chain polymers due to chain straightening.

since it is due to chain-straightening which will be a reversible process when stress is relaxed. It is also a time-dependent change so that the resultant deformation is viscoelastic. Straightening of molecules tends to take place more quickly than slipping so that much of the initial deformation is elastic.

12.5.3 'Spring and dash-pot' models are widely used to represent the modes of deformation of a completely amorphous thermoplastic above T_g (Fig. 12.14). Here it is assumed that the springs obey Hooke's Law and that the dash-pots contain a Newtonian fluid in which the deformation/time relationship is linear for a given stress.

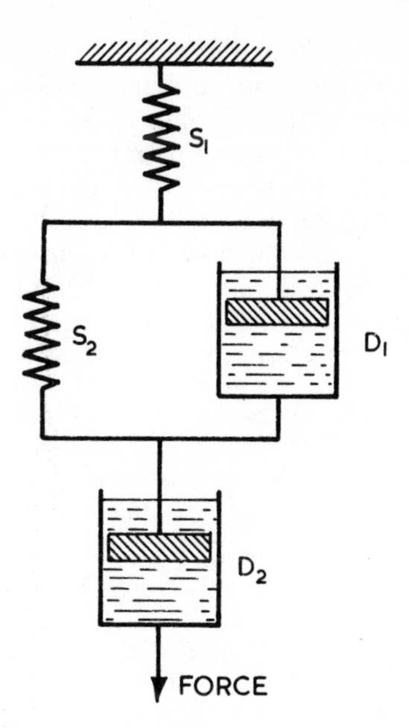

Fig. 12.14—A 'spring and dash-pot' model to illustrate the method of deformation of a stressed thermoplastics material.

The spring S_1 represents the instantaneous elastic deformation of the atomic bonds between individual atoms, whilst the spring S_2 represents the elastic deformation arising from chain straightening, the dash-pot D_1 introducing the time-dependence of that deformation. Dash-pot D_2 simulates the non-elastic, non-reversible time-dependent deformation due to molecules slipping relative to each other into new positions.

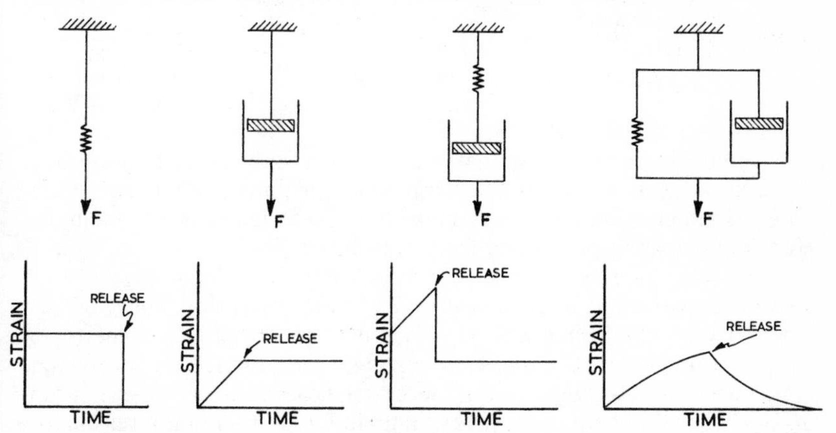

Fig. 12.15—Strain-time curves for the various components of the system shown in Fig. 12.14.

Under the action of a force the various components of the system shown in Fig. 12.14 will give strain/time diagrams as indicated in Fig. 12.15. If we now consider the sudden application of a force, F, to the system as a whole we obtain a strain/time relationship like that indicated in Fig. 12.16. The sudden application of force results in an elastic deformation S_1 (ii) which is followed by a viscoelastic response S_2 in the S_2/D_1 unit coupled with viscous flow in the dash-pot D_2 (iii) in time, t_1. When the displacing force is removed the elastic element (S_1) relaxes immediately (iv) and the viscoelastic element slowly (v) during time (t_2–t_1), but the viscous flow D_2 is never recovered (v).

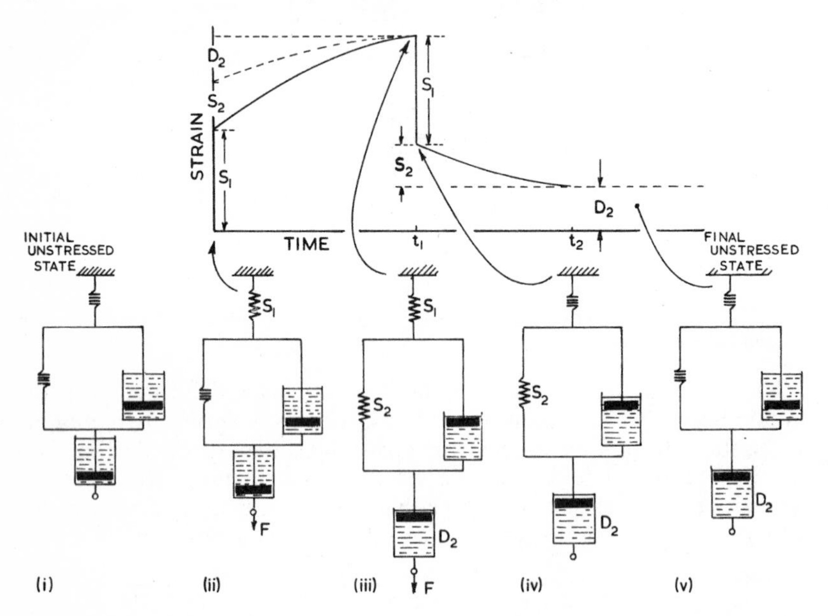

Fig. 12.16—Strain-time relationship for the 'spring and dash-pot' model shown in Fig. 12.14.

It should be appreciated that these models only exhibit the *general* characteristics of the viscoelastic behaviour of polymers and are necessarily oversimplified. There is considerable deviation both from the Hookean elastic response and from Newtonian flow.

12.5.4 *Stress–strain measurements.* The force-extension type of test already well established with metals was soon applied as a matter of course to polymer materials. Unfortunately tests carried out at a single temperature and a single rate of force application are relatively meaningless when applied to polymers and in order to obtain reasonably useful data a series of tensile tests carried out over a range of temperatures and application rates are necessary. The most successful method of testing is to stretch the test piece at a constant rate and measure the tensile force produced in it by means of strain gauges or by some electronic device.

Stress–strain curves for different types of polymer were classified some years ago by Carswell and Nason into five main groups (Fig. 12.17).

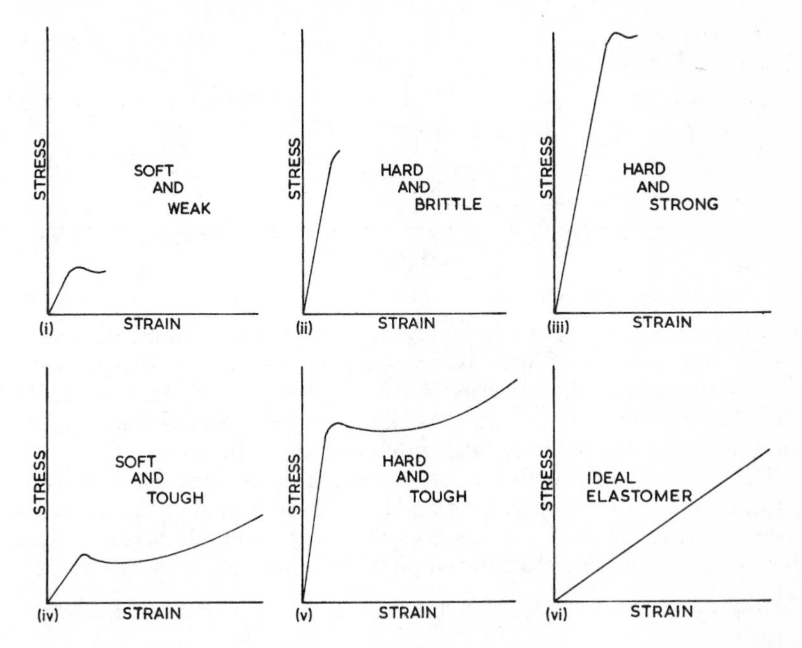

Fig. 12.17—Types of stress–strain diagram for different polymer groups (after Carswell and Nason).

Weak, soft polymers such as would be of little use practically have not only a low tensile strength and modulus of elasticity but also a relatively low elongation (Fig. 12.17 (i)). Hard and brittle materials, such as polystyrene, polymethyl methacrylate and some of the phenolic resins at room temperature, have fairly high tensile strengths and quite high moduli of elasticity, but they often break with an elongation of no more than 1%(Fig. 12.17 (ii)). Hard and strong polymers (Fig. 12.17 (iii))—some rigid polyvinyl chlorides and modified polystyrene—have a relatively high tensile strength and modulus of elasticity coupled with a moderate elongation of up to 10%; whilst many soft, tough materials such as plasticised polyvinyl chlorides and some rubbers have low yield values and elastic moduli but high elongations of up to 500% or more, coupled with moderately high tensile strengths (Fig. 12.17 (iv)). Tough, hard polymers including nylons and cellulosics are characterised by high yields, high tensile strengths and high elastic moduli. They also elongate considerably, generally by 'necking' (Fig. 12.17 (v)).

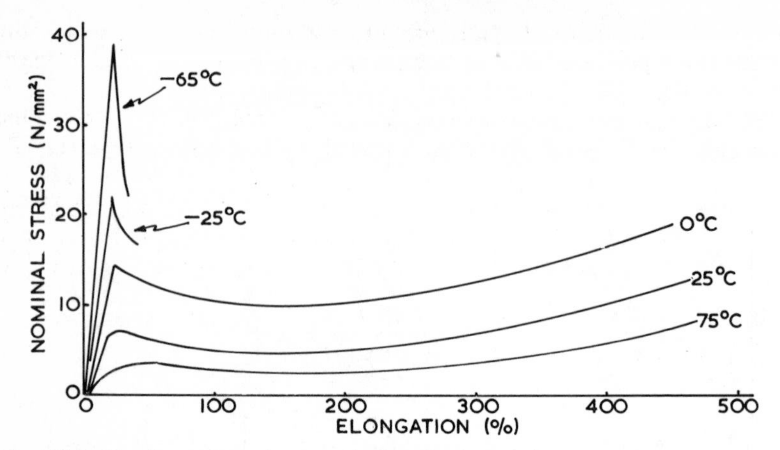

Fig. 12.18—Tensile stress–strain curves for low-density polythene at various temperatures.

12.5.5 Tensile properties are highly dependent upon temperature (Fig. 12.18) and a series of tests is necessary in order to give worthwhile information about the material. For rigid polymers elongation usually increases as temperature increases though for rubbers elongation may increase as the temperature falls within certain limits.

The rate at which deformation is produced also has a considerable effect on the tensile properties of a polymer as indicated in Fig. 12.19. As the rate of elongation is increased so the tensile strength increases. Thus an increase in the rate of deformation has a similar effect to a decrease in temperature

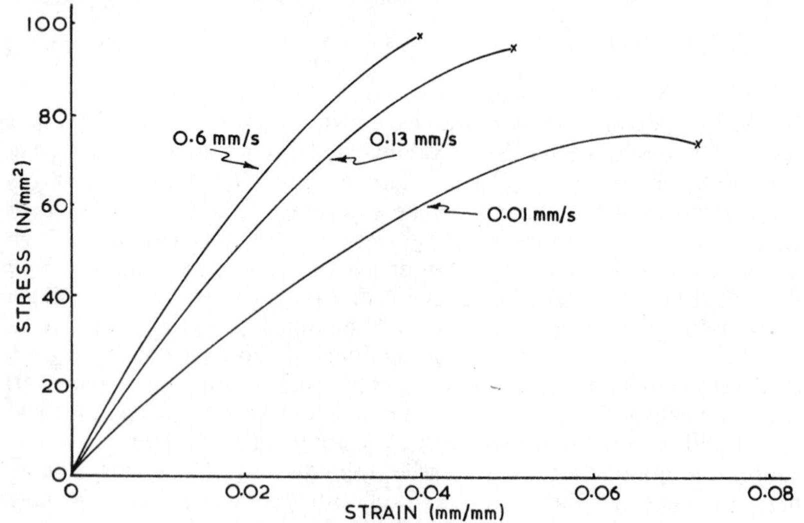

Fig. 12.19—The effect of rate of deformation on the stress–strain characteristics of polymethyl methacrylate.

12.5.6 *Creep*, or gradual extension under a constant force, is a phenomenon which must be considered in the case of metals, particularly if they are destined to work at high temperatures (16.4). Polymers, however, are much more prone to this type of deformation as has been suggested in the foregoing section. In the form of an engineering component a rigid polymer must be able to withstand reasonable tensile or compressive forces over long periods of time without alteration in shape. Even a soft plastic raincoat must be capable of supporting its own weight so that it can hang in the wardrobe for long periods without gradually flowing on to the floor.

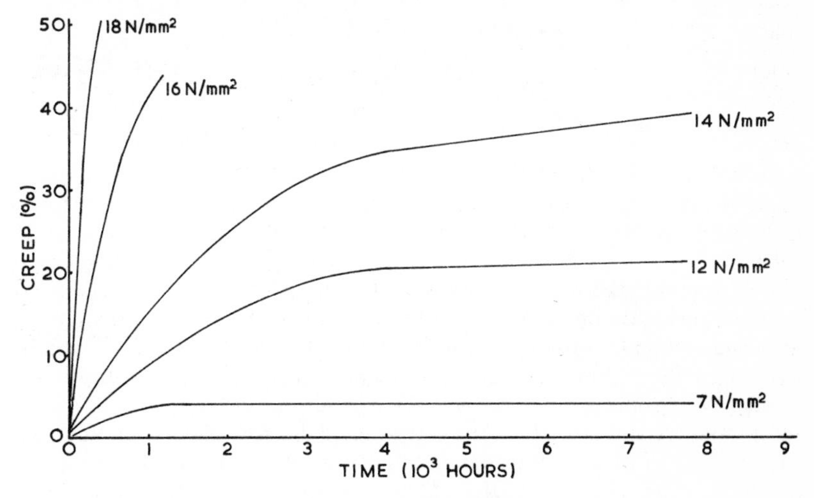

Fig. 12.20—Creep curves for cellulose acetate at 25° C.

The series of creep curves shown in Fig. 12.20 are typical of most thermoplastic polymers. The test piece undergoes an initial elongation as soon as the specified force is applied and this is followed by a period during which there is a rapid rate of creep. This creep rate then decreases to a constant value, which approaches zero for small values of stress. As would be expected the greater the applied force the greater the creep rate and elongation.

Creep properties of a polymer are very dependent upon temperature. At temperatures well below the glass-transition temperature, T_g, a polymer tends to be rigid and have a low creep rate for a given stress. As the temperature is increased the creep rate and the amount of elongation increase. At T_g a given load gives a much greater elongation than the same load at lower temperatures. Even more significant is the great increase in the rate of elongation for, whilst at temperatures well above T_g greater *elongations* will be obtained, the actual *creep rate* decreases.

Plasticisation

12.6 Plasticisers are substances which are added to plastics materials in order to improve their flow properties and in some cases to reduce the brittleness of the product. Two principal groups of plasticisers are in use. Some substances—called 'primary plasticisers'—contain polar groups which neutralise the fields of force of the polymer polar groups, thus reducing the van der Waals forces between adjacent polymer chains. In many cases, however, the plasticiser is a compatible and inert material without polar groups but which exists dispersed throughout the polymer providing mechanical 'spacers' which separate the polymer chains and so reduce the van der Waals forces of attraction between them. These are termed 'secondary plasticisers' (Fig. 12.21 (ii)).

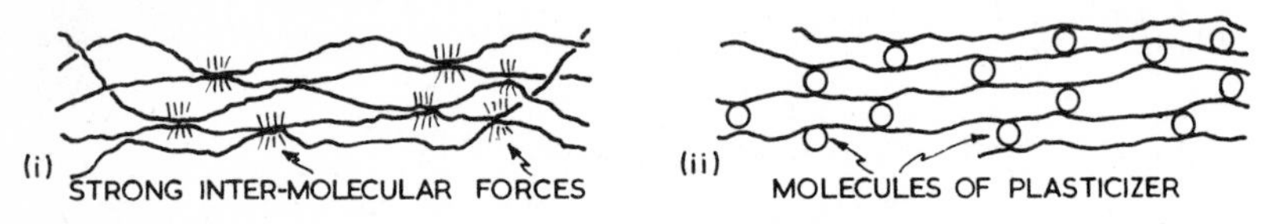

Fig. 12.21—The function of a secondary plasticizer in separating polymer chains (ii) and so reducing intermolecular attraction (van der Waal's forces) (i).

12.6.1 *Internal plasticisation* involves modifying the polymer chain by introducing small amounts of another monomer unit during polymerisation. This second monomer requires bulky side groups. Thus PVC can be plasticised by adding 15% of vinyl acetate during polymerisation. The presence of the occasional bulky acetate groups in the polymer so formed act as spacers and so reduce the inter-molecular forces between adjacent chains. This in turn increases the amount of viscous movement possible. In this case increase in plasticity depends upon the formation of a new copolymer rather than upon the use of a plasticiser and is less often employed.

12.6.2 *External plasticisation* is more commonly used. Here the plasticiser is added to the polymer *after* polymerisation is complete. It is often a compatible non-volatile solvent liquid which will disperse in molecular form throughout the polymer virtually forming a very viscous solution with the latter. Only the amorphous regions are so affected and a highly crystalline polymer cannot successfully be plasticised in this way since insufficient solvent will be absorbed.

An alternative form of external plasticisation involves the use of a non-solvent oil. This disperses as tiny droplets throughout the plastics material. These droplets act as spacers between the polymer chains decreasing the van der Waals forces in the usual way.

Fillers and other additives

12.7 To a large extent fillers are used in plastics materials in the interests of economy. Often up to 80% of the additive is used. In general fillers are finely powdered naturally occurring non-polymeric materials which are readily available at low cost. Phenolic (bakelite) type resins are nearly always compounded, or filled with substances such as wood flour, short cellulose fibres from wood pulp, cotton flock, mica and asbestos. In addition to the economic advantages obtained from the increase in bulk so produced, these materials greatly improve the strength, toughness and dimensional stability of the product. Mica improves electrical resistance whilst graphite reduces friction. Asbestos, of course, improves temperature resistance.

Some soft thermoplastics are blended with up to 80% mineral solids such as crushed quartz, clay or limestone. In materials of this type the plastics material functions as an inter-particle adhesive in much the same way as cement in concrete. On the whole, however, fillers are more often used with the thermosetting moulding compounds of the phenol-formaldehyde type.

In some cases the polymer is the vehicle for a filling material with special properties. Thus, some magnetic materials can be bonded in particle form with a suitable polymer so that better control of eddy currents is possible and more complex shapes more easily produced.

In addition to such materials as colourants, other additions may also be made to plastics materials. Thus *antioxidants* are sometimes used to prevent or inhibit the oxidation of polymers. The antioxidants are themselves substances which readily oxidise and in this way protect the polymer itself. *Stabilisers* are also sometimes used, for example, to prevent damage to a polymer by long exposure to ultra-violet light which often causes photo-decomposition as evidenced by a yellowish tinge and a reduction in transparency as well as a serious deterioration in mechanical properties. The addition of carbon black to the polymer renders it opaque to light and in this way protects it.

Foamed or 'expanded' plastics materials

12.8 Foamed rubber has been in use for many years but more recently a number of polymers have become available in this form. Of the thermoplastic polymers 'expanded' polystyrene is well-known as an extremely lightweight packing material, also as a thermal insulator and for the manufacture of ceiling tiles for the enthusiastic home decorator. Up to 97% of its volume consists of air bubbles making it an excellent flotation material, since it has only about one-tenth the relative density of cork and does not become waterlogged.

Chief of the thermosetting polymers available as foams are the polyurethanes. These are available in the flexible form for the manufacture of upholstery and sponges whilst rigid polyurethane foams are used for the internal reinforcement of aircraft wings.

Two varieties of foamed polymers are manufactured—*open* (interconnecting) or *closed* cellular structures—though in practice foams are likely to contain some of each type of structure with one predominating. As suggested above, thermoplastics, thermosetting polymers and elastomers can all be foam expanded and a number of methods are available for introducing the gas pockets into the polymer:

(1) By mechanically mixing air with the polymer whilst it is still in the molten state or dissolved in some solvent which will evaporate.

(2) By dissolving a gas such as carbon dioxide in the mix by the use of pressure. During the hardening process pressure is reduced so that the gas is released from solution—rather like the bubbles of carbon dioxide in mineral waters. Since hardening is in progress the gas bubbles are trapped producing a foam.

(3) Similar in principle is the addition of a volatile component to the mix. During the hardening process this vaporises forming bubbles.

(4) Alternatively a gas may be evolved during a cross-linking process, e.g. as a result of condensation polymerisation (12.3.1). This gas or vapour can be utilised to produce the foaming process during hardening.

(5) A 'blowing agent' may be added which will release gas in the mix during heating. This is useful for foaming thermoplastics. A common reagent is sodium bicarbonate which reacts in the same way as it does in baking powder, that is, by releasing bubbles of carbon dioxide on heating thus causing the cake to 'rise':

$$2\,NaHCO_3 \xrightarrow{\text{Heat}} Na_2CO_3 + H_2O + CO_2$$

Chapter Thirteen
Ceramics and Rubbers

13.1 The atoms or molecules of any crystalline solid are permanently located with respect to their neighbours. Atoms vibrate continuously about equilibrium positions because of the energy they have received in the form of heat. The total internal energy of a solid increases as its temperature rises and for this reason both the frequency and amplitude of atomic vibrations increase so that atoms or molecules are forced further away from their equilibrium positions. This movement away from original positions manifests itself as *thermal expansion* and also explains the storing of energy by a body in terms of its heat capacity (*specific heat*).

Atomic vibrations also furnish one method by which heat can travel through a substance, that is, by *conduction*. Metals are particularly good conductors of heat since their valency electrons, constituting a common 'cloud', are able to move freely under the influence of a temperature gradient just as they will under the action of an electrical potential difference.

Materials such as glasses, rubbers, cellulose and many of the plastics materials are usually considered as solids in the sense that they are of definite and permanent shape. On the other hand they are not crystalline since their molecules are not arranged in a regular pattern. For this reason, as well as for others such as the lack of a definite melting point, they are often regarded as super-cooled liquids. When heated to a sufficiently high temperature range they become increasingly plastic as the temperature rises. In this respect they resemble viscous liquids which become more mobile at higher temperatures. Pure crystalline solids on the other hand pass from a strictly geometrical lattice structure to a liquid state at a single fixed temperature.

Ceramics

13.2 The term 'ceramics' is applied to a range of inorganic materials of widely varying uses. Generally these materials are non-metallic and in most cases have been treated at a high temperature at some stage during manufacture. The word 'ceramic' is derived from the Greek, 'keramos'— or 'potter's clay', though the group of materials now so described includes glass products; cements and plasters; some abrasives and cutting tool materials; building materials such as bricks, tiles and drain pipes; various electrical insulation materials; refractory linings for furnaces; porcelain and other refractory coatings for metals; as well as the more traditional uses in pottery, crockery and sanitary ware.

From the point of view of composition and structure ceramics can be classified conveniently into four main groups:

(1) *Amorphous ceramics:* these are substances referred to generally as 'glasses'. They include those such as 'obsidian' which occur naturally, and man-made glasses used for the manufacture of bottles, windows and lenses.

(2) *Crystalline ceramics:* these may be single-phase materials like magnesium oxide or aluminium oxide, or various mixtures of materials such as these. In addition some carbides and nitrides belong to this group.

(3) *Bonded ceramics:* materials in which individual crystals are bonded together by a glassy matrix as in a large number of products derived from clay.

(4) *Cements:* a number of these are crystalline but some may contain both crystalline amorphous phases.

Structures of ceramics

13.3 Ceramics are far less ductile than metals and tend to fracture immediately any attempt is made to deform them by mechanical work. They are often of complex chemical composition and their structures may also be relatively complex. These structures fall into two main groups:

13.3.1 *Simple crystal structures.* These include structures in which the bonding is either ionic or covalent. Thus magnesium oxide, MgO, is an ionic compound of cubic structure (Fig. 13.1) similar to that of sodium chloride. Magnesium oxide is a refractory material used in the linings of furnaces for the manufacture of steel by basic processes and can withstand temperatures in excess of 2000° C.

Silicon carbide is also a ceramic material of simple composition and structure. Here the bonding is covalent and the structure tetrahedral such

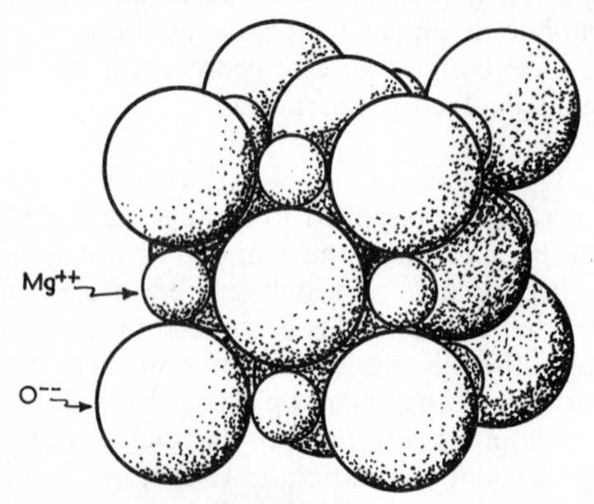

Fig. 13.1—The crystal structure of magnesium oxide in which each O^{--} ion is surrounded by six Mg^{++} ions and, similarly, each Mg^{++} ion is surrounded by six O^{--} ions.

that each silicon atom is surrounded by four carbon atoms and, conversely, each carbon atom by four silicon atoms. As a result the structure is very like that of diamond. This will come as no surprise when it is realised that both elements are quadrivalent and members of the same family of elements (Group IV). Due to the stable covalent bonds which operate and also the symmetry of the structure, silicon carbide is a very hard material. Though less hard than diamond which it resembles in structure it is nevertheless a well-known abrasive material.

13.3.2 *Complex silicate structures.* The majority of ceramic materials, and particularly those derived from clay, sand or cement, contain the element silicon in the form of silicates. The similarity between silicon and carbon in respect of many chemical properties has already been mentioned. Just as carbon polymer molecules are built up from a succession of repeating 'mers' (12.1.3) so are many silicates formed as chain, sheet or framework structures. The most simple silicon–oxygen 'mer' is the 'SiO_4' group in which a small silicon atom is surrounded tetrahedrally by four oxygen

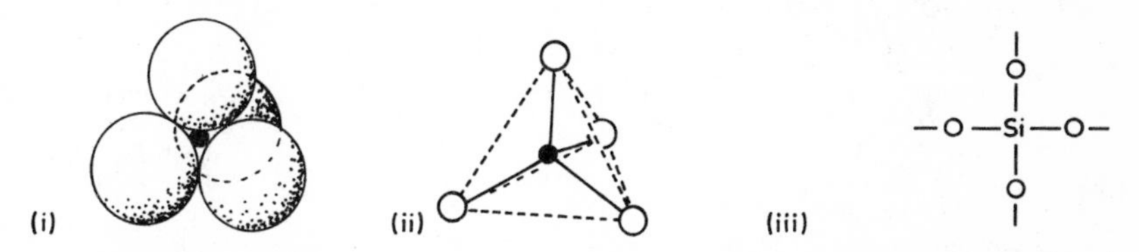

Fig. 13.2—The tetrahedral arrangement of the $(SiO_4)^{4-}$ ion. (iii) Indicates the unsatisfied valencies of the oxygen atoms.

atoms (Fig. 13.2). In this way the quadrivalency of the silicon is satisfied but the bivalency of the oxygen atoms is not. Hence to satisfy the valency of the oxygen atoms, SiO_4 groups are frequently ionically bonded to metallic atoms thus producing negatively charged silicate ions $(SiO_4)^{4-}$ and positively charged metallic ions, e.g. Ca^{2+}; Na^+; Mg^{2+} and Al^{3+}. The positive and negative ions then form a crystal structure.

Alternatively the electron deficiency of the oxygen atom can be made up by combination with another silicon atom thus making it part of an adjacent tetrahedron (Fig. 13.3). The resultant unit $(Si_2O_7)^{6-}$ is present in only a few minerals.

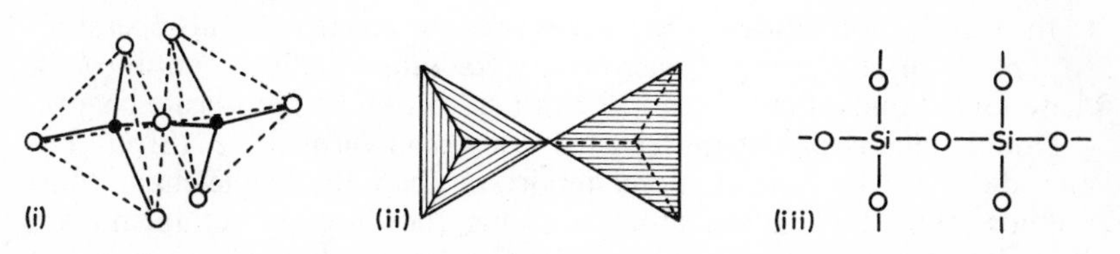

Fig. 13.3—The structure of the $(Si_2O_7)^{6-}$ group. (i) 'Picture view'. (ii) 'Plan view'. (iii) Valency diagram.

An example of a mineral based on the $(SiO_4)^{4-}$ ion is magnesium silicate or *forsterite*. The empirical formula, that is, that formula which indicates the ratio of atoms present in a compound, is Mg_2SiO_4, but it must be remembered that, as in the case of sodium chloride (3.2.3), the unit involved here is the crystal and not the molecule. *Forsterite* is useful as a refractory material since its melting point is in the region of 1900° C.

13.3.3 The structure depicted in Fig. 13.4 is fundamentally that of a chain based on silicate ions $(SiO_3)^{2-}$, in which adjacent chains are ionically bonded by bridging metallic ions such as Ca^{2+}. Herein lies the difference in mechanical properties between an organic polymer and a silicate ceramic. In the case of the organic polymer adjacent chain molecules are

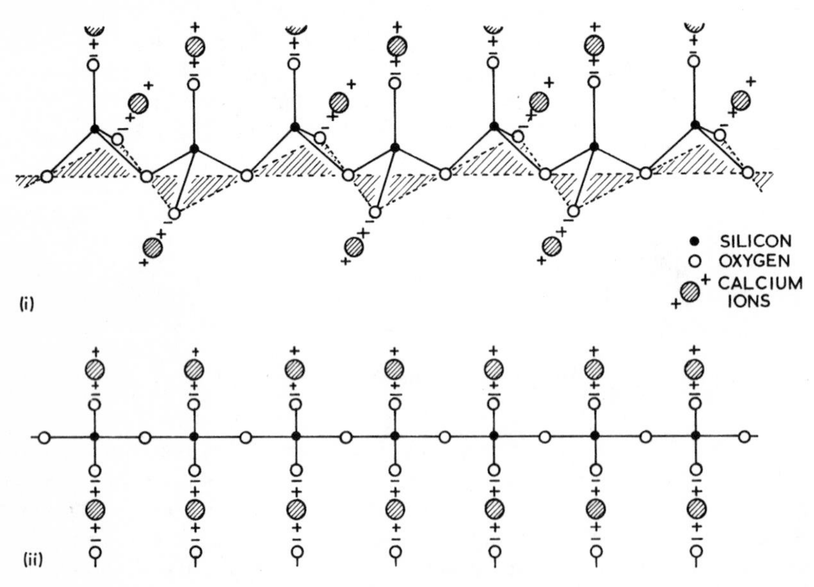

Fig. 13.4—A silicate single-chain structure based on $(SiO_3)^{3-}$. The chains are ionically linked by Ca^{++} ions.

held together by weak van der Waals forces but in the ceramic these forces are replaced by stronger ionic bonds making ceramics hard and relatively strong. Nevertheless the ionic bonds linking the silicate chains to the metallic ion 'bridges' are weaker than the covalent bonds operating within the silicate group. Consequently fracture usually occurs through these ionic bonds, that is, parallel to the direction of the silicate chains.

Double chains of silicate tetrahedra are also formed, i.e. $(Si_4O_{11})^{6-}$ and these are in general more important than the single-chain compounds. Here the oxygen atoms forming the bases of tetrahedra are shared with adjacent units (Fig. 13.5). The double-chain units are linked by metallic ions forming characteristically rigid crystalline structures. In

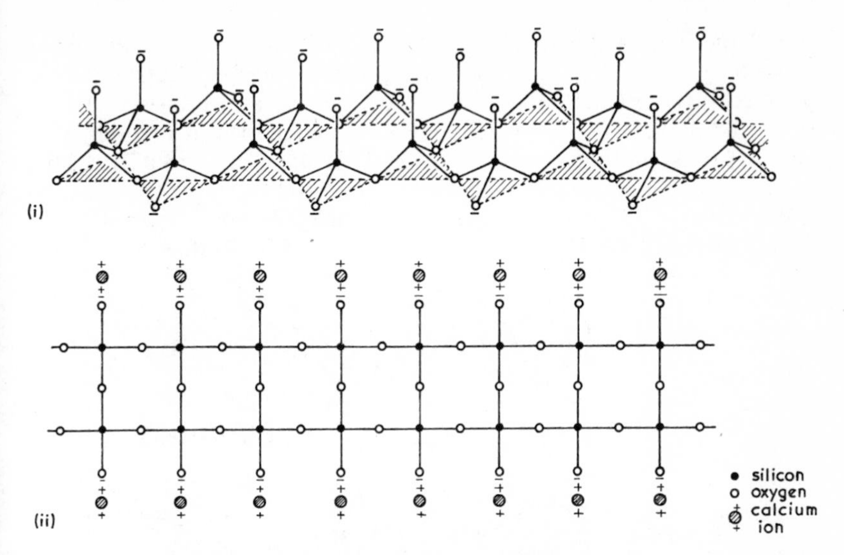

Fig. 13.5—A double-chain structure of silicate tetrahedra.

some cases the chains are held together by van der Waals forces as well as by ionic bonds thus allowing the material to be separated into fibres. Asbestos is an example of this class of mineral which have empirical formulae of the type $3MgO . 2SiO_2 . H_2O$.

13.3.4 In many minerals of the group which include clays, talc and micas the silicate tetrahedra are linked in sheet form instead of in chain form. Fig. 13.6 represents a 'plan' view of this system in which the positions of the oxygen atoms at the apices of the tetrahedral units are shown. The empirical formula of the silicate unit is $(Si_2O_5)^{2-}$. On the basal plane of the sheet all oxygen valencies are satisfied, i.e. each oxygen atom has eight available electrons. The mineral *kaolinite*, the simplest type of clay, has aluminium ions (Al^{3+}) which satisfy the valencies of the upper oxygen

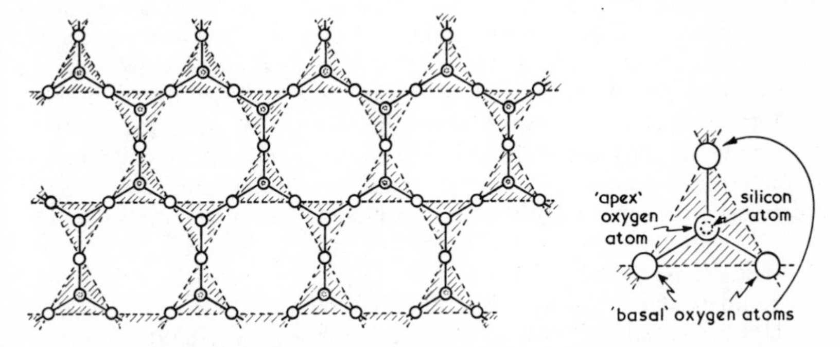

Fig. 13.6—A sheet arrangement of silicate tetrahedra.

atoms; the remaining two valencies of the Al^{3+} ions then in turn being satisfied by $(OH)^-$ ions which are also present. Since all valencies within the sheet structure are now satisfied adjacent silicate sheets can attract each other only by weak polarisation forces of the van der Waals type. This explains the plasticity of clay and the fact that mica is readily separable into extremely thin sheets.

When water is added to clay the H_2O molecules are attracted to the surface layers of the sheets by polarisation forces (2.4) thus providing a

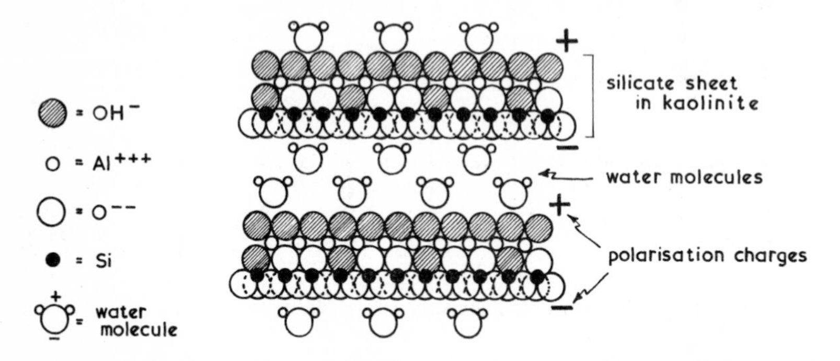

Fig. 13.7—The absorption of water at the surfaces of the silicate sheets in clay (kaolinite). This absorption is due to the attraction between polarisation forces on the surfaces of the kaolinite sheets and the hydrogen bond of the water molecules (2.4).

boundary lubricant between adjacent sheets (Fig. 13.7). When water is dried out the plasticity of the clay is correspondingly reduced and when the clay is baked cross-linking between adjacent sheets produces a strong brittle structure. This is a non-reversible reaction comparable with the thermosetting of bakelite and the clay has now permanently lost its plasticity.

13.3.5 The development of the silicate tetrahedral structure into three dimensions produces a framework type of arrangement which is in effect a giant molecule (3.8.1) of empirical formula $(SiO_2)_n$. Each silicon atom is surrounded by four oxygen atoms and each oxygen atom is jointly shared by an adjacent unit. In this way all valency bonds are used up and a hard, rigid framework structure is produced. The simplest of these structures is that of *cristobalite* whilst another is that of *quartz*. Both of these materials are hard and fracture in an irregular manner since there is no preferred direction of cleavage, bonding forces being equal in all directions. Other framework structures include various aluminosilicates.

Properties of ceramics

13.4 Until recent years most ceramic materials consisted of crystalline particles cemented together by glass. These materials, mainly alumino-silicates of varying compositions, were used as furnace linings; for the manufacture of pottery and electrical insulators and for many building materials such as bricks and tiles. However, modern technologies of sintering have produced structures which are wholly crystalline. Some of these materials maintain high strength at elevated temperatures since low melting point glasses are absent.

13.4.1 *Strength.* Although ceramics are hard and reasonably strong they lack ductility and this, in turn, prevents any stress concentrations which may be present from being relieved by the onset of plastic flow. Since ceramics tend to suffer from the presence of micro-cracks which act as potential stress-raisers it follows that overall stresses must be kept low if sudden failure is to be prevented. To explain this lack of ductility in terms of structure we will consider a part of the lattice structure of a metal and a part of the lattice structure of an ionic crystal such as magnesium oxide or sodium chloride (Fig. 5.2), both being under the action of shearing forces. In Fig. 5.2 (i) it is obvious that when the metallic lattice has slipped by one atom spacing the fundamental lattice pattern has not been altered and positively charged ions remain under the 'cementing' influence of the negatively charged electron cloud. In Fig. 5.2 (ii), however, a movement of one lattice spacing of one layer with respect to the other has brought like-charged ions close together so that a strong force of repulsion will occur between the layers forcing them apart. Sudden cleavage will therefore occur.

The compressive strengths of ceramics are high compared with their tensile strengths. Retention of strength at high temperatures is one of the useful features of such materials. For example, titanium diboride will maintain a compressive strength of 250 N/mm^2 at 2000° C making it one of the strongest materials known at that temperature.

13.4.2 *Hardness.* Ceramic materials generally are hard when compared with most other engineering materials. This hardness makes certain ceramics useful as abrasives or, suitably bonded, as cutting tool materials. Thus, cubic boron nitride is almost as hard as diamond whilst materials like silicon carbide, alumina (aluminium oxide) and beryllium oxide are already well known. Some hard ceramics are listed in Table 13.1, hardened tool steel being included for comparison.

Table 13.1—Hardnesses of some ceramics.

Material	*Knoop hardness number*
Diamond	7000
Cubic boron nitride	7000
Boron carbide	2900
Silicon carbide	2600
Alumina	2000
Beryllium oxide	1220
Hardened steel	700

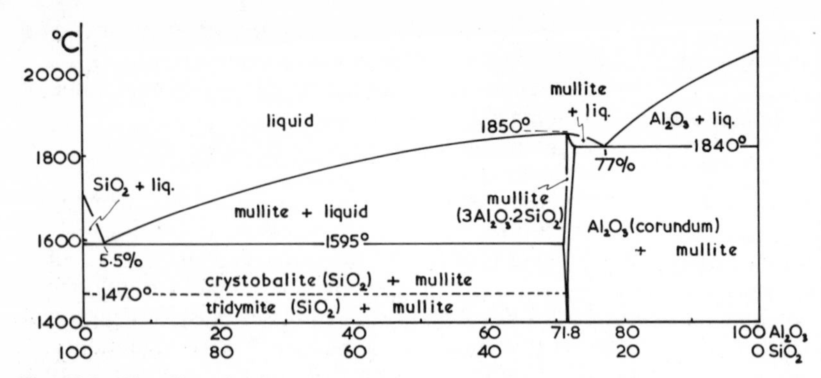

Fig. 13.8—The silica–alumina phase diagram. (Crystobalite and tridymite are allotropic forms of silica.)

13.4.3 *Refractoriness* is generally defined as the ability of a material to withstand high temperatures without appreciable deformation under service conditions. Some refractories like fireclay and high-alumina brick soften gradually over a range of temperature and may ultimately collapse under load at a temperature well below the normal fusion temperature. Materials most widely used as refractories in furnaces are clays containing various proportions of silica (SiO_2) and alumina (Al_2O_3). As the thermal equilibrium diagram (Fig. 13.8) indicates mixtures containing in the region of 4–8% alumina are best avoided for use at high temperatures since they will melt completely at a little above the eutectic temperature of 1595° C. As the amount of alumina increases above 5·5% the quantity of liquid (produced at 1595° C) will decrease and so the bricks will be stronger at that temperature. For severe conditions where temperatures reach 1800° C the alumina content must exceed 71·8%.

Certain ceramic materials have very high melting points indeed and for special purposes these may be used, assuming that their relative high cost is justified. A few of these high temperature ceramics are included in Table 13.2.

Table 13.2—Melting points of some high-temperature refractory materials.

Material	*Melting point* (°C)
Hafnium carbide	3900
Tantalum carbide	3890
Thorium oxide	3315
Magnesium oxide	2800
Zirconium oxide	2600
Beryllium oxide	2550
Aluminium oxide	2050
Silicon nitride	1900
Fused silica glass	1680 (approx.)
Soda glass	340–430 (approx.)

13.4.4 *Electrical properties.* Until a short while ago the use of ceramics in the electrical industries was restricted to insulation. Today ceramic materials include many substances varying in properties from low-loss high-frequency insulators to semi-conductors (18.6), conductors and ferro-magnets. Consequently they are used as thermistors, thermoelectrics, transistors and storage cells in memory systems.

Glasses

13.5 Glass is popularly regarded as a hard, brittle, transparent substance with a fairly high softening temperature, and as being relatively insoluble in water and other common solvents. From a scientific standpoint, however, it may be defined as a product of the fusion of inorganic materials which has been cooled to a hard condition without crystallisation having taken place. Certain organic polymers also fit this description and behave in some ways like glasses since they too are amorphous and have no definite melting points.

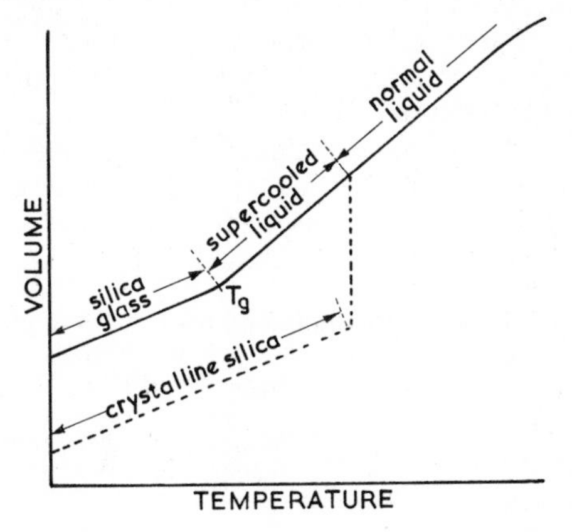

Fig. 13.9—The glass-transition (or *fictive*) temperature for silica glass.

It is perhaps reasonable to consider glasses as supercooled liquids in as much as they are non-crystalline. At high temperatures glasses form normal liquids and since their atoms are free to move they respond readily to shear stresses. When a liquid glass is supercooled normal thermal contraction takes place due to some rearrangement of the atoms leading to their more efficient packing. This type of contraction is common to all liquid glasses (Fig. 13.9). With further reduction in temperature a point is reached where a sudden change in the coefficient of expansion occurs. Below this temperature, called the *glass transition temperature* (or *fictive temperature*), T_g (12.4.3) there is no further rearrangement of atom packing and such smaller amounts of contraction as do take place are due solely to a reduction in the thermal vibrations of the atoms.

This lower coefficient of expansion is roughly the same as that for *crystals* of similar composition to the glass (Fig. 13.9). In rigid crystalline materials contraction can only result from decreased thermal vibrations. Hence the term *glass* could well be defined as a substance having expansion characteristics similar to those indicated in Fig. 13.9. Other physical properties which are affected in a similar manner to the coefficient of expansion include density, electrical resistivity, optical properties and viscosity.

A glass will sometimes crystallise if given sufficient time—perhaps several hundred years. This process, known as *devitrification*, sometimes occurs with very old stained-glass windows and is generally manifested as a considerable increase in brittleness as crystallisation takes place. Glass does not always devitrify, of course, and some glass at least 3000 years old has retained its vitreous—or 'glassy'—properties.

13.5.1 Glass was one of the earliest man-made materials, though, prior to the present technological age, its composition fell within a limited range. Commercial glasses are produced from various inorganic oxides of which silica (in the form of sand) is generally the most important constituent. The principal ingredients of common soda glass are silica sand, lime (obtained from limestone) and soda ash (crude sodium carbonate). These materials are heated together in a 'tank' furnace where they react to form complex silicates we know as glass. Many special glasses are also manufactured of which a few are included in Table 13.3.

Table 13.3—Compositions and uses of some glasses.

Type of glass	*Approx. Composition (%)*	*Properties and Uses*
Soda glass	72 SiO_2 15 Na_2O 9 CaO 4 MgO	Windows, plate glass, bottles and other containers
Lead glasses	60 SiO_2 15 $Na_2O + K_2O$ 25 PbO	Has a high electrical resistance—used for lamps, valves, etc
	40 SiO_2 7·5 $Na_2O + K_2O$ 47·5 PbO 5 Al_2O_3	High refractive index and dispersive power—used for lenses and other optics. Also for 'crystal' glass tableware
Borosilicate glass	70 SiO_2 7 $Na_2O + K_2O$ 3 Al_2O_3 20 B_2O_3	Low expansion and good resistance to chemicals—used for heat-resistant ware ('Pyrex') and laboratory apparatus. Can 'seal' to a number of metals
Aluminosilicate glasses	35 SiO_2 30 CaO 25 Al_2O_3 5 B_2O_3 5 MgO	Has a high softening range with T_g as high as 800 °C

13.5.2 As we have seen glasses differ from solid crystalline substances in respect of a number of properties. They are plastic at high temperatures and rigid at low temperatures but under normal manufacturing conditions they do not crystallise. Thus glass is in a condition which is continuous with, and analogous to, the liquid state of the substance but which, as a result of having been cooled from a fused state, has acquired a high degree of viscosity so as to be, for most practical purposes, rigid.

13.5.3 The structure of *crystalline* silica, quartz, is fundamentally tetrahedral (Fig. 13.2) as determined by X-ray crystallographic methods. In this three-dimensional structure each silicon atom is chemically bonded to four oxygen atoms placed at the corners of the tetrahedron, each oxygen atom being shared with an adjacent silicon atom from another tetrahedron. However, for the sake of clarity we will consider a two-dimensional

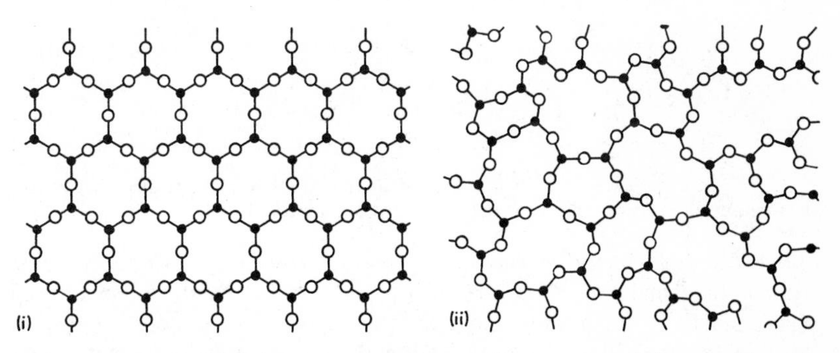

Fig. 13.10—A two-dimensional representation of the structure of (i) crystalline silica (quartz) and (ii) silica glass. Since the diagram is two-dimensional the silicon atoms are shown bonded to only three oxygen atoms instead of four as in Fig. 13.2. In (i) there is long-range order whilst in (ii) only short-range order prevails, i.e. individual silica tetrahedra themselves.

adaptation of this structure as shown in Fig. 13.10 (i). Here the structure is one of long-range order, that is, a regular lattice structure is apparent. If, however, fused silica is cooled at such a rate as to make normal crystallisation impossible, a structure such as is suggested in Fig. 13.10 (ii) is obtained for silica glass. Here the long-range order is replaced by a short-range one in which individual tetrahedra still persist as they do in crystalline quartz, but the random orientation of individual tetrahedra, resulting from local distortions, destroys the periodicity of the lattice network and the structure is no longer crystalline.

When metallic oxides such as soda and lime are added to the silica the ratio of oxygen atoms to silicon atoms is increased. The extra oxygen atoms cannot be incorporated into each of two tetrahedra and as a consequence the network loses its continuity in regions where these ions are present. The structure is therefore less rigid at high temperatures and soda glass is more easily worked than pure silica glass. The sodium and calcium ions take up suitable available spaces within the structure (Fig. 13.11). Since sodium ions are relatively small they are able to migrate through the structure with relative ease even at room temperatures. This influences some of the properties of soda glass.

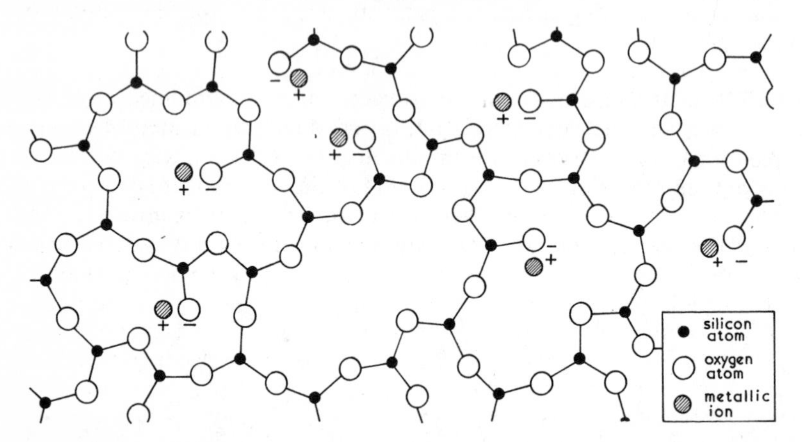

Fig. 13.11—A two-dimensional representation of the structure of soda glass. Again the silicon atoms are shown bonded to only three oxygen atoms.

13.5.4 *The mechanical properties of glasses.* Since only very short-range order is present in the structure of a glass it lacks the slip planes along which plastic deformation can occur as it does in a metal. Thus, in contrast to crystalline materials which deform plastically by slip, glass deforms plastically by a process of *viscous flow*. The rate of viscous flow is dependent mainly upon the prevailing temperature. It is also dependent upon the composition and structure of the glass.

Small applied stresses cause the more highly strained bonds within the structure to be ruptured. The bond rearrangement which follows this process leads to a small degree of permanent strain. As this process continues more bonds become strained in turn to the extent where they will break and so allow viscous flow to continue. As would be expected the stresses necessary to cause a prescribed degree of viscous flow are less at elevated temperatures since bond-strength becomes smaller at high temperatures.

The amount of viscous flow in glass at ordinary temperatures is very small. However, glass will flow very slowly under its own weight and a number of cases have been reported where large plate-glass windows have become measurably thicker at the lower edge over a period of years.

There is little in common between the mechanical properties of glasses and metals when under the action of applied forces. Such differences can be classified as follows:

(1) Under short-time testing methods, glasses are brittle at ambient temperatures. They are elastic right up to the point of fracture and fail without any previous yield or plastic deformation. Even the most brittle of metals show some plastic flow.

(2) Although an external load may be applied in compression, failure in glass always results from a tensile component of the stress. The strength of such materials can therefore best be described in terms of tensile strength.

(3) The time for which a static load is applied has a great influence on the strength of glass in the long term. Thus the extrapolated infinite-time modulus of rupture for glasses is usually between a third and a half of that for short-time loading.

13.5.5 *Glass fibres.* When in the form of fibre glass is considerably stronger than it is in any other form and as such enters the ranks of engineering materials. Silica glass fibre in the freshly-drawn condition, that is if tested within ten minutes of being drawn and 'untouched by human hand' so to speak, may have a tensile strength in the region of 15 kN/mm^2. The most minute of scratches, even such as would be produced by drawing a feather lightly across the surface of the fibre, will cause fracture at considerably lower stress. Such minute cracks or scratches act as stress-raisers so that at the tip of the crack the stress is well above that necessary to cause fracture, whereas the overall average stress in the fibre is much lower than this. Consequently, the crack will extend and, as long as the stress concentration is maintained at the new tip of the crack, fracture will continue. (Fig. 13.12 (i)). A ductile material on the other hand can adjust to the situation by means of plastic strain so that the high concentration of stress is automatically reduced (Fig. 13.12 (ii)).

Thus the high strength of glass fibre is due partly to the fact that, unlike ordinary glass, these fibres are relatively free from such surface defects as are likely to lead to crack propagation and consequent failure at

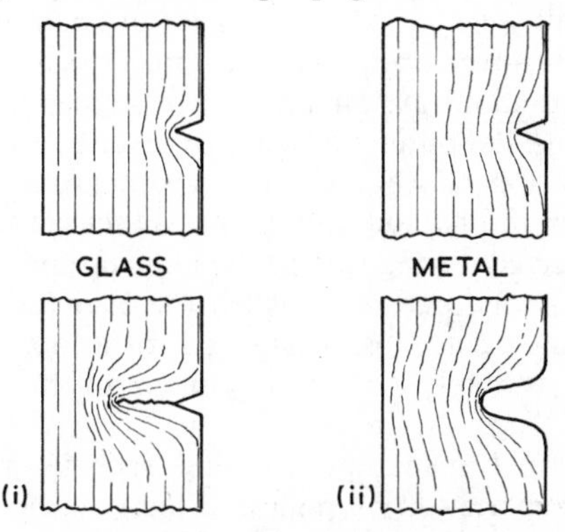

Fig. 13.12—The propagation of a crack in a brittle material (i) as compared with a ductile material (ii).

low stresses. Moreover, in the absence of the possibility of deformation by slip, fairly strong chemical bonds operate along the length of the fibre. Glass fibre is of course widely used to produce composite materials in combination with various polymers (14.2.3).

13.5.6 *The heat-treatment of glass.* Glass responds to suitable heat-treatment immediately following the shaping process.

13.5.6.1 *Annealing* is applied to glass in order to relieve stresses set up during the later stages of shaping as a result of rapid cooling. Failure to anneal the glass in a suitable manner would leave it weak and brittle. Annealing is usually carried out immediately after the shaping process. The glass is reheated in a special furnace to the necessary annealing temperature and then allowed to cool very slowly in the furnace. The small amount of viscous flow which takes place during annealing leads to stress relaxation within the glass which is thus rendered relatively stress-free. This treatment bears some similarity to the low-temperature stress-relief annealing given to some cold-worked metals and alloys such as 70–30 brass in that atoms move short distances into positions nearer to equilibrium. At the same time it contrasts sharply with those annealing processes which lead to complete recrystallisation in metals.

13.5.6.2 *Tempering.* Glass is sometimes toughened by a tempering process which seeks to reduce the formation of surface cracks by putting the surface in a state of compression. The principle involved is similar to that applied in the pre-stressing of concrete in order to prevent cracks from developing in its surface (14.6.6). Tempering of glass is achieved by heating it to much the same temperature as that used for annealing it, and then cooling the surface rapidly by means of air jets. In common with other substances glass contracts as it cools. Initially the outside surface cools, contracts and hardens more quickly than the inside. The latter, being still soft, will yield in the early stages of cooling. During the final stages of cooling, however, both inside and outside behave in an elastic manner but their contractions are already out of step. Hence, when the glass has finished cooling the outer surface is in a state of compression whilst the inside is in tension. Since the surface layers are in compression these compressive forces will balance any moderate tensile forces to which the glass may be subjected during service. Glass invariably fails as a result of tensile components of forces even when these forces are, overall, compressive. Glass thus treated has been used for a long time in motor-car windscreens and is likely to be developed in the future for many industrial uses.

Strain patterns in transparent materials such as ‘perspex’ and glass become visible when viewed by polarised light and to a lesser extent by white light which becomes partly polarised when reflected from non-metallic surfaces. For this reason it is often possible to see the pattern of air-jet impingements used for cooling the glass when a car windscreen is inspected at an oblique angle or, preferably, by using polarising spectacles.

Carbon fibres

13.6 The traditional engineering materials have been exploited to the limit, particularly in the aerospace industries where weight considerations are of paramount importance. In these industries the design engineer is interested in obtaining the maximum in terms of mechanical properties for the expenditure of a certain mass of material. What was loosely termed the 'strength weight ratio' became the criterion by which to judge the usefulness of a material. When *stiffness* is the prime consideration, however, Young's Modulus of Elasticity is a better guide to desirable properties and a value termed the *specific modulus* is now generally accepted as being representative. The Specific Modulus is the ratio of Young's Modulus to the relative density of the material:

$$\text{Specific Modulus} = \frac{\text{Young's Modulus of Elasticity}}{\text{Relative Density}}$$

For most of the structural materials generally available to the engineer, viz. steel, titanium, aluminium alloys, magnesium alloys, glass and wood, the specific modulus is approximately 25 MN/mm^2. For this reason wood has remained an important structural material in cases where space is not limited and was in fact used as a material in aircraft construction until late in the Second World War.

It follows that to make lighter but stiffer and stronger structures new materials of high Young's Modulus but low relative density are needed. Carbon fibres in the form of yarn and cloth have been available for some years but their mechanical properties were poor so that these materials were used mainly for thermal insulation and filters. The first suggestion of a breakthrough came when, in the USA, graphite whiskers were produced with a tensile strength of 20 kN/mm^2 and Young's Modulus of 700 kN/mm^2.

13.6.1 The graphite crystal is *anisotropic** in respect of many of its properties and this in turn is due to its layer structure (Fig. 3.25). For example Young's Modulus measured in a direction parallel to the layers is 1030 kN/mm^2 whilst perpendicular to the layers it is only 35 kN/mm^2. This is to be expected since individual carbon atoms in a layer are held to their neighbours by strong covalent bonds only 0·142 nm long, whereas only weak van der Waals forces operate between adjacent layers which are comparatively far apart (0·335 nm). Thus the bond energy due to the covalent bonds operating within the layers is about 600 kJ/gram-atom, whereas the interlayer binding energy is only 5·4 kJ/gram-

* Anisotropy is the characteristic of exhibiting different physical properties in different directions in a body of material.

atom. The relatively high binding energy within the layers gives rise to the high value of Young's Modulus since high elasticity is dependent on a large restoring force acting between atoms.

13.6.2 In practice carbon fibres are polycrystalline and consist of a large number of small crystallites. These are about 10^{-8} m thick and $2{\cdot}5 \times 10^{-8}$ m in diameter and consist of parallel planes of carbon atoms grouped together. The main difference between the structure of these crystallites and the ideal structure shown in Fig. 3.25 is that there is no regularity of the orientation of the parallel planes and the structure is said to be *turbostatic*. The structure of perfect graphite has been likened to a pack of playing cards which has been tapped on a table to square up individual cards (layer planes) whilst the structure in the turbostatic graphite of carbon fibres has been likened to the pack of cards collected together into a pile but not squared at the edges so that the orientation of indvidual cards (layer planes) is random. Examination by the electron microscope shows that the crystallites are grouped together in units known as *fibrils*, which are sub-units of the fibre. The fibre itself some 8×10^{-6} m in diameter consists of a network of closely packed fibrils each about $2{\cdot}5 \times 10^{-8}$ m in diameter.

13.6.3 At present carbon fibres are produced by the 'pyrolysis' of polyacrylonitrile filaments. The polymer material is heated at a high temperature in an inert atmosphere so that decomposition takes place and polymer chains are, so to speak, stripped of all other atoms and groups of atoms leaving a 'skeleton' of carbon atoms in the structure of graphite. The fibres are kept in tension during pyrolysis in order to prevent curling and loss of desirable properties. The final treatment temperature, which may be in excess of 2000° C, to some extent governs the mechanical properties of the resultant carbon fibre. Two varieties of fibre are obtained by different heat-treatments—a high-*strength* fibre and a high-*modulus* fibre (Table 13.4).

Table 13.4—Some mechanical properties of carbon fibres compared with other strong materials.

Property	*Relative density*	*Tensile strength* (N/mm²)	*Specific strength* (N/mm²)	*Young's modulus* (E) (N/mm²)	*Specific modulus* (E/ρ) (N/mm²)
Heat-treated medium C steel	7·9	1 000	130	210 000	27 000
Aluminium	2·8	400	140	70 000	25 000
Glass-reinforced plastic	2·0	1 100	550	40 000	20 000
Carbon-fibre reinforced plastics					
high-strength fibre	1·5	1 400	930	130 000	87 000
high-modulus fibre	1·6	1 000	600	190 000	110 000

It is fairly obvious that carbon fibre by itself is not a particularly useful material to the engineer, but when used as a component of a composite material, i.e. as a strengthening or stiffening agent for plastics substances and later, it is hoped, for metals, its properties can be utilised.

Carbon-fibre reinforcement of plastics has been successfully used in the manufacture of such diverse products as the fan blades of aero-engine compressors and the handles of cricket bats. Other uses include pressure bottles, racing kayaks, moulded gear wheels in carbon-fibre reinforced nylon, and racing-car bodies. Since it costs several thousands of pounds

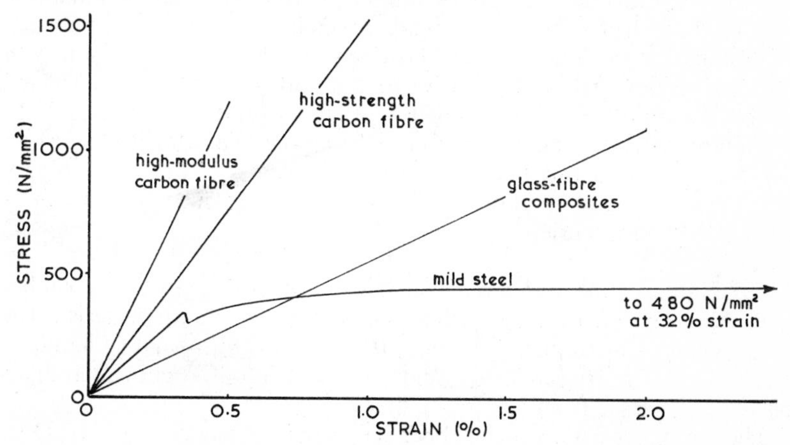

Fig. 13.13—The tensile properties of carbon fibres as compared with mild steel and also glass-fibre composites.

sterling to lift one kilogram mass through the Earth's atmosphere, carbon fibre has a guaranteed market in aero-space projects. Even with ordinary aircraft, one kilogram mass saved means a saving of some £200 over the life of the aircraft so that, even though expensive, carbon fibre is an attractive proposition in these directions. As production methods are developed and demand increases the cost of carbon fibre will doubtless fall. Then many other applications will be possible such as helicopter blades, many parts of aircraft, hovercraft propellors and sports equipment such as skis, dinghy masts and boat hulls.

Rubbers

13.7 Rubber is a very important *elastomer*. This is a substance in which the arrangement of the polymer molecules allows considerable reversible extension to take place at normal temperatures. Elastomers exist as long chain molecules which are irregularly coiled, bent and generally entangled when in the unstressed state. Since the coils are not arranged in a regular pattern the structure is amorphous. When a tensile force sufficient to stretch the material is applied, the molecules partially disentangle and straighten out so that they become orientated in the general direction in which stretching occurs. Since the molecules become more closely aligned along the direction of stretching, the forces of attraction between adjacent molecules also increase so that the material becomes stronger and stiffer.

As temperature is reduced polymers become less elastic and pass into a glassy state. This occurs at the glass transition temperature, T_g, and this varies from one polymer to another. T_g for most useful elastomers is well below room temperature. This can be demonstrated by cooling a rubber ball to a very low temperature in liquid air. When an attempt is made to bounce the ball it shatters in a manner similar to glass. Under normal conditions the molecules of an elastomer are in constant motion with adjacent sections vibrating relative to each other. As T_g is reached the molecular structure becomes rigid so that movement is restricted. In crystalline polymers the presence of multiple and hence fairly strong van der Waals forces acting between adjacent molecules will raise T_g to very near the softening temperature of the polymer. Consequently such materials are not elastomeric.

13.7.1 Natural rubber is a polymer derived from the sap of the rubber tree, *Hevea brasiliensis*. Since this tree is native to South America it is believed that Christopher Columbus was the first European to handle a piece of rubber. It seems that some tribes of South American Indians played ball games from very early times.

The monomer from which rubber polymerises is isoprene, C_5H_8. During the natural polymerisation process, double bond positions in the molecule are rearranged (Fig. 13.14 (i)) so that isoprene units link up in the usual manner of polymers to form a chain molecule some 20 000 carbon atoms—or approximately 0·002 5 mm—in length. This type of chain molecule can display the property of *stereoisomerism*. For our purposes this can be explained in a simplified manner as indicated in Fig. 13.14 (ii) and (iii). Here it will be noticed that the CH_3 groups can be arranged on different 'sides' of the chain. The arrangement indicated in Fig. 13.14 (ii) is termed the 'cis'-configuration and is the structure present in rubber, whilst that shown in Fig. 13.14 (iii) is known as the 'trans'-

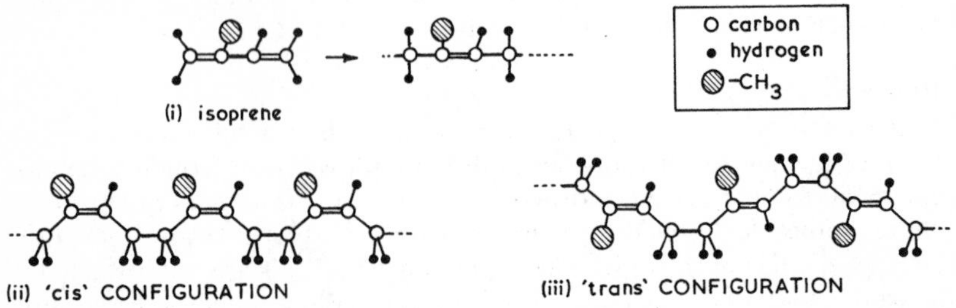

Fig. 13.14

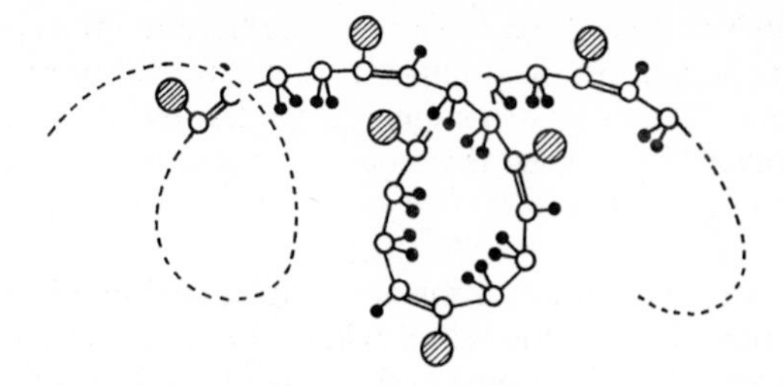

Fig. 13.15—The ability of *cis*-polyisoprene (rubber) to form a coiled molecule.

configuration and is the molecular pattern of *gutta-percha*. Rubber is an elastomer whilst gutta-percha is inflexible and hard by comparison. This could well be due to the fact that a chain molecule of the 'cis' type (rubber) is able to bend and coil easily in one direction, i.e. 'away' from the CH_3 groups (Fig. 13.15) whilst the relatively even distribution of CH_3 groups on both 'sides' of the molecule in the 'trans' configuration (gutta-percha) limits flexibility of the molecule in either direction. Since the double bond cannot rotate, the CH_3 groups are in fixed positions. Hence both the cis- and trans-configurations are stable and one form cannot convert to the other.

13.7.2 In its natural state rubber is both elastomeric and thermoplastic. Being thermoplastic it becomes *permanently* deformed when it is

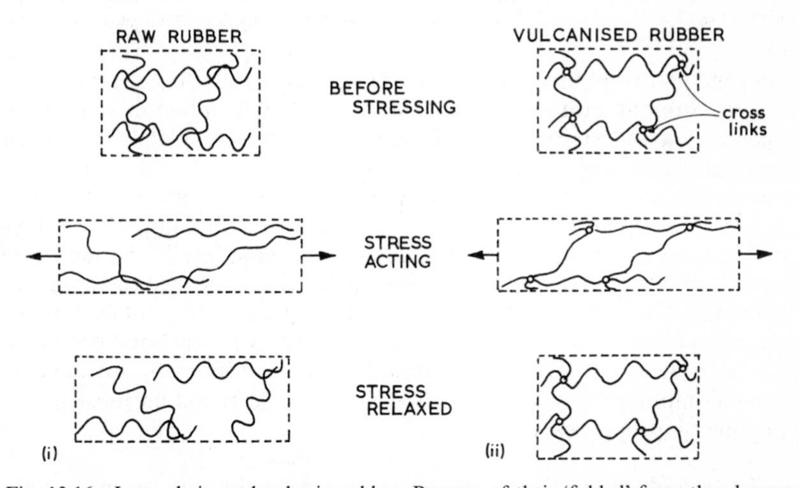

Fig. 13.16—Long-chain molecules in rubber. Because of their 'folded' form they become extended in tension, but return to their original shapes when the stress is removed. In 'raw' rubber (i) a steady tensile force will cause individual molecules to slide slowly past each other into new positions, so that when the force is removed some plastic deformation remains, although the elastic deformation has disappeared. By 'vulcanising' the raw rubber (ii) the chain molecules are chemically bonded so that no permanent plastic deformation can occur and only elastic deformation is possible.

stretched since van der Waals forces operate between the chain molecules only at a limited number of points. In this state rubber is useless as an engineering material, but when rubber is 'vulcanised' these van der Waals forces are replaced by strong permanent covalent bonds and the material becomes thermosetting (Fig. 13.16).

13.7.3 The process of vulcanisation was introduced by Goodyear in 1839 and involves the use of sulphur to provide cross links between the polyisoprene molecules. The raw rubber is first 'masticated' or kneaded between rolls so that a degree of depolymerisation occurs and the material becomes more plastic as a result. It can then be mixed with sufficient sulphur—or other vulcanising agents—before being heated to 200° C to cause cross-linking between the polyisoprene molecules (Fig. 13.17).

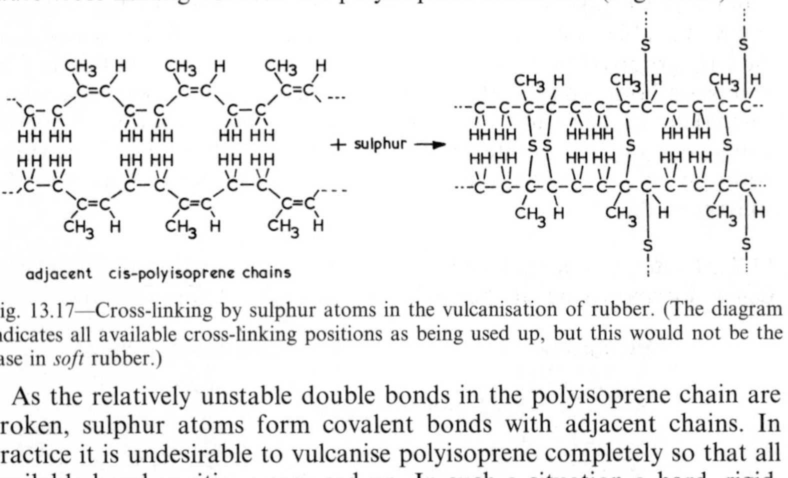

adjacent cis-polyisoprene chains

Fig. 13.17—Cross-linking by sulphur atoms in the vulcanisation of rubber. (The diagram indicates all available cross-linking positions as being used up, but this would not be the case in *soft* rubber.)

As the relatively unstable double bonds in the polyisoprene chain are broken, sulphur atoms form covalent bonds with adjacent chains. In practice it is undesirable to vulcanise polyisoprene completely so that all available bond positions are used up. In such a situation a hard, rigid, non-elastomeric material would be produced since little movement of one chain relative to another would then be possible. In fact in the early days of the plastics industry materials such as 'vulcanite' and 'ebonite' were made in this way but have since been replaced by cheaper rigid plastics materials. The rubber used in the manufacture of automobile tyres is only vulcanised to the extent of about 5% of the available bond positions. This is sufficient to reduce the plasticity of the rubber by introducing enough sulphur cross links but at the same time retaining most of the elastomeric properties.

13.7.4 In addition to the vulcanising agent raw rubber is generally compounded with other ingredients or 'fillers'. Carbon is widely used to increase both strength and abrasion resistance. The unsaturated double bonds still remaining in rubber after vulcanisation are susceptible to oxidation, particularly by ozone. This effectively reduces elasticity by forming cross links in those remaining positions not already linked by sulphur atoms. As a result the rubber becomes brittle or 'perished'. To reduce this tendency various anti-oxidants are incorporated in the mixture.

The mechanical properties of rubber

13.8 The resistance of rubber to deformation in tension at first decreases (as with metals) but then increases as indicated in the stress/strain diagram (Fig. 13.18). Rubber does not obey Hooke's Law when in tension. This is due to the fact that in materials with high elasticity the considerable extension is accompanied by a big reduction in cross-section whereas in metals, which obey Hooke's Law, this reduction is negligible. The ratio of lateral contraction to longitudinal extension (Poisson's Ratio) is therefore important in considering the tensile extension of rubbers.

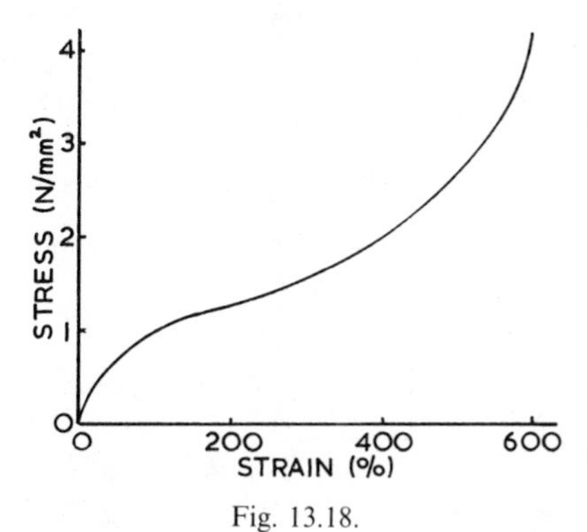

Fig. 13.18.

As might be expected, Young's Modulus of Elasticity for rubber is relatively low since a large amount of strain is produced by a small stress, e.g. for hard rubber E is about 35 N/mm^2.

The term *drift* as applied to rubbers is analogous to creep in metals. It refers to permanent distortion of the material under long-time loading. As in the case of metals the rate of drift increases with increase of temperature. Live loads produce greater drift than do static loads.

Synthetic rubbers

13.9 The Japanese invasion of Britain's Far Eastern rubber plantations in 1941 led to a serious crisis in the war potential of the Allies. The situation was saved by the development of the American synthetic rubber known as GR–S. This is a co-polymer of butadiene and styrene and, except in some of the mechanical and physical properties where it is inferior, it is very similar to natural rubber. GR–S is cheaper than natural rubber and is used in the manufacture of tyres. It is also used, alone or blended, in footwear, hosepipe, conveyor belts and cable insulation.

13.9.1 *Neoprene*, another synthetic rubber, is closely related chemically to natural rubber as its structural formula shows (Table 13.5). The substitution of a chlorine atom for a $-CH_3$ group leads to a superior resistance to

Table 13.5—Some synthetic rubbers.

TYPE	MONOMER(S)	POLYMER UNIT
natural rubber	isoprene	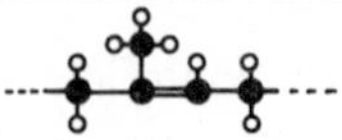 or $\left[-CH_2-\overset{CH_3}{\overset{\mid}{C}}=CH-CH_2-\right]_n$
neoprene	chloroprene	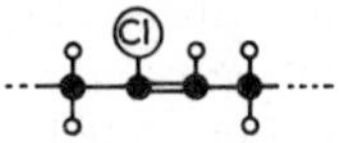 or $\left[-CH_2-\overset{Cl}{\overset{\mid}{C}}=CH-CH_2-\right]_n$
styrene rubber	75% butadiene & 25% styrene	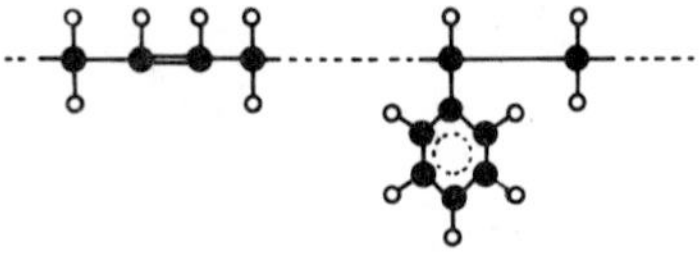 or $\left[-(CH_2-CH=CH-CH_2)_x-\cdots-\underset{C_6H_5}{\underset{\mid}{CH}}-CH_2-\right]_n$
butyl rubber	98% isobutylene & 2% isoprene	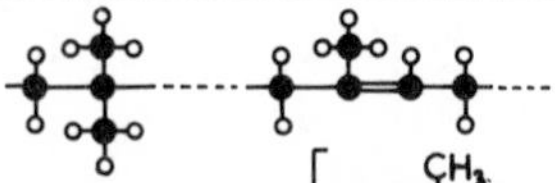 or $\left[(-CH_2-\overset{CH_3}{\underset{CH_3}{C}}-)_x-CH_2-\overset{CH_3}{\overset{\mid}{C}}=CH-CH_2-\right]_n$
nitrile rubber	butadiene & acrylonitrile	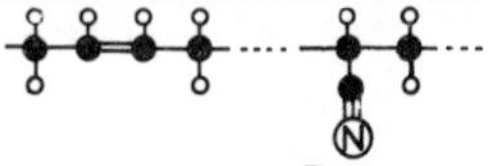 or $\left[(-CH_2-CH=CH-CH_2)_x-\underset{CN}{\underset{\mid}{CH}}-CH_2-\right]_n$

vegetable and mineral oils and also to high temperatures. Unfortunately solubility in some electrolytes is increased. Neoprene is best vulcanised with metallic oxides such as MgO and ZnO. It is an expensive material generally used for hoses, conveyor belts, gaskets and cable sheathing.

13.9.2 *Butyl rubber* is produced by co-polymerising isobutylene with small amounts of isoprene. Since the product has a low degree of 'unsaturation' (few double bonds) it is chemically very stable and has a good resistance to heat and to many chemicals. It is vulcanised in the same way as natural rubber. Butyl rubber is cheaper than natural rubber and since it is resistant to ozone because of its low unsaturation, it is less liable to 'perish'. It is also impermeable to air and other gases making it useful for inner tubes and tubeless tyres, air bags, sporting equipment and moulded diaphragms. Being impervious to many chemicals it is also used for hoses, tank linings and conveyor belts.

13.9.3 *Nitrile rubber* (GR–A) is a co-polymer of butadiene and acrylonitrile. It is very resistant to oil but is attacked by alkalis more easily than are natural and styrene rubbers. Vulcanised nitrile rubber is more resistant to high temperature than is natural rubber, but, like neoprene, it is expensive. It is used for hoses, conveyor belts, gaskets and cable sheathing.

Chapter Fourteen
Composite Materials

14.1 New jargon frequently appears in the field of technological terminology. For example the vocabulary of the materials scientist was enriched by the term 'composite materials' a few years ago. Yet the principles of these composite materials are by no means new. The poet may sing: 'I think that I shall never see a poem lovely as a tree . . .' but to the materials technologist wood is one of the oldest composite materials and, incidentally, still one of the most useful especially when 'specific cost' is taken into account.

The term 'composite material' covers a very wide range of substances and is now used to describe a simple but old idea of putting dissimilar materials in service together so as to achieve a new complex whose overall properties are different in type and magnitude from those of the separate constituents. As suggested above the principle of composite materials had its origin in nature. Thus the concept of reinforcement of

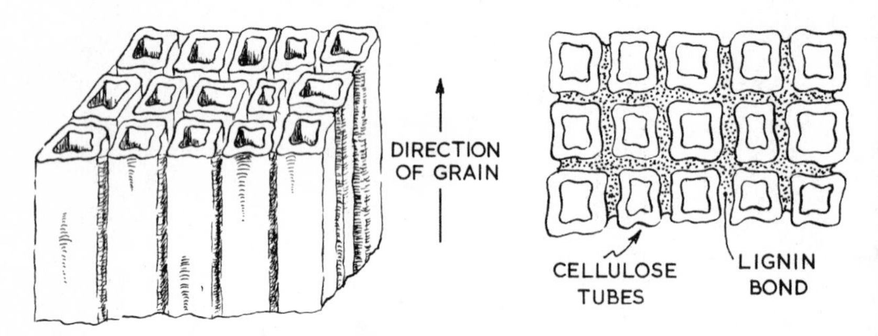

Fig. 14.1—The composite nature of wood (white pine).

plastics by using glass fibre in the 1940s was based on the natural structure of bamboo. Ordinary wood is in fact a composite material in which tubes of cellulose fibre are bonded by the natural plastic, lignin (Fig. 14.1). Man-made composites such as reinforced concrete have been with us for a long time and even the Ancients reinforced their air-dried bricks with straw—a practice which may still be seen even in some European countries.

14.1.1 Other composite materials depending upon fibre-reinforcement include:

(1) Plastics materials reinforced with carbon fibres (13.6);
(2) Plastics materials reinforced with laminations of woven textiles (e.g. 'Tufnol');

(3) Vehicle tyres in which vulcanised rubber is strengthened by woven textile cords.

In the examples quoted so far strength has been produced by employing the reinforcement material in fibre form. Other composites however depend for their strength upon a degree of particle hardening (14.3). Such materials include 'cermets' which are a mixture of metal and ceramic substances and are generally compounded with the object of producing a combination of hardness and toughness such as would be required in a tool material. Yet a further group of composite materials relies upon dispersion hardening. Here the movement of dislocations is impeded by strong particles of microscopical dimensions only.

14.1.2 Regarded in this light many metallic alloys would seem to qualify as composite materials. For example, the structure of a tool steel, which has been correctly heat-treated, contains particles of carbide in a matrix of a different composition. Alloys, however, are normally derived from a single liquid phase and though the resultant solid structure may be of a duplex character, they are *not* classed as composite materials. These latter may be defined as a mixture of two or more separate substances which have been compounded mechanically and in which at least one constituent remains in a specific manufactured form such as a fibre, fabric or particle of predetermined size. Enamelled, plated or carburised steel; 'expanded' polymers; asphalt and concrete, all fall under the classification of 'composite materials' but first we shall deal with some of the more sophisticated cermets and fibre-reinforced materials.

14.1.3 If a material contains particles these suffer elastic strains when the material is stressed. In this way the particles contribute to the load-carrying capacity of the material as well as providing obstacles to the movement of dislocations, assuming that the particles themselves are strong. Thus if the volume of such strong particles in the composite is proportionally large they will provide a high strength by producing a high load-carrying capacity. The most satisfactory method of increasing tensile strength is to use particles in the form of long fibres. When stressed the matrix may begin to flow but in doing so will cause a force to be set up at the surface of the fibre. If the fibre is long enough this transmitted force will ultimately lead to its fracture and the fibre will therefore have contributed fully to the strength of the composite material. Obviously strength will be directional, i.e. at a maximum value parallel to the direction of the fibres.

14.1.4 The nature of the interface between the particles or fibres and the matrix has a bearing on the extent to which the load will be transferred from the matrix to the strengthening material. Cohesion at the interface may be achieved by one or more of the following methods:

(1) *Mechanical bonding*. This involves a high enough coefficient of friction acting between the surfaces, e.g. fibres in a simple hemp rope;
(2) *Physical bonding* depending upon van der Waals forces acting between surface molecules;
(3) *Chemical bonding* at the interface—this can, however, give rise to weak brittle compounds in some instances;
(4) Multiple bonds formed by solid solution effects.

14.1.5 Generally speaking sophisticated engineering composites fall into three main groups:

(*i*) Those in which the matrix is relatively weak and brittle but is reinforced by fibres or platelets of high tensile strength. Such materials are useful when high tensile strength and stiffness are required as in a fishing rod or the blades of a turbine rotor.
(*ii*) Materials in which a tough and ductile matrix carries the second component in the form of hard spherical particles. These composites are characterised by great hardness and high compressive strength and are represented by tool and die materials.
(*iii*) Dispersion-hardened materials in which the strength of the matrix is increased by the action of microscopical particles in impeding the movement of a 'dislocation front'. Since such dislocations cannot pass *through* such particles—assuming that they are harder and stronger than the matrix—'loops' are formed (Fig. 14.2). These dislocation loops then act as further barriers to the progress of any dislocation fronts which may follow.

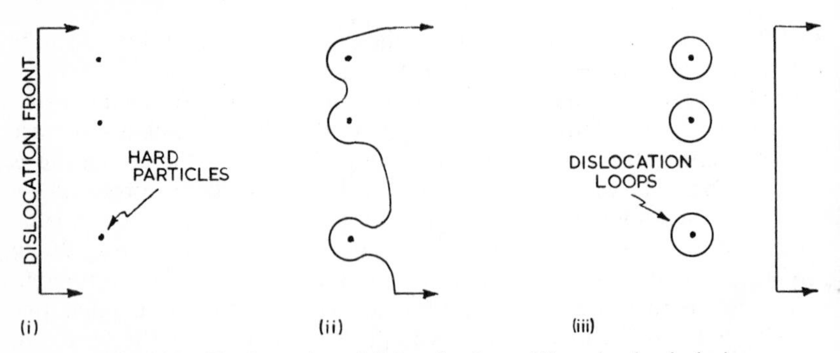

Fig. 14.2—The formation of dislocation loops (dispersion hardening).

Fibre-reinforced composites

14.2 Strong solids generally fail by the propagation of cracks from the surface. Once a crack has developed it travels quickly due to stress concentrations produced and the material fails. If the strong material is in the form of a bundle of fibres then the cracks which matter can travel only a very short distance since they are limited by the diameter of a single fibre. Thus a solid made up of a large number of such fibres embedded in a suitable matrix will have a greater fracture strength than would a single piece of strong material of similar total diameter.

The fibres must be held in position by some suitable means. Long hemp fibres can be twisted together to form strands which are then coiled as rope. Here frictional forces are sufficient to hold the fibres together since the fibres are relatively long. Glass fibres on the other hand are so susceptible to surface damage that they cannot be twisted together and must instead be embedded in a suitable matrix such as a polyester resin which produces a rigid material and, at the same time, protects the surface of the glass. GRP (glass-reinforced plastic) is a very strong and light material which is now widely used.

14.2.1 Extremely strong composites can be produced by embedding very hard materials such as silicon carbide in a metallic matrix. The function of the matrix is threefold:

(1) It must protect the surface of the fibre from abrasion damage which would precipitate fracture;

(2) It must adhere to the fibre surface to such an extent that the applied force is transmitted to the fibre;

(3) It must separate individual fibres sufficiently so that transverse cracks cannot be propagated from one fibre to the next.

It is shown that optimum results are obtained if the fibres exceed a certain critical length, L_c, and if the length-to-diameter ratio, R, is such that:

$$R > \frac{5L_c}{d}$$

where d is the diameter of the fibre.

Table 14.1—Properties of fibre materials.

Material	*Tensile strength* (kN/mm^2)	*Relative density*	*Young's modulus* (kN/mm^2)	*Specific strength* (kN/mm^2)	*Specific modulus* (kN/mm^2)
'S' glass	4·5	2·50	88	1·8	35
Steel (drawn wire)	4·2	7·74	200	0·54	26
Tungsten	4·1	19·4	410	0·21	21
'E' glass	3·5	2·55	74	1·4	29
Carbon (high-strength)	3·0	1·74	230	1·8	130
Boron	2·8	2·36	390	1·2	160
Molybdenum	2·2	10·2	360	0·22	36
Carbon (high-modulus)	2·1	2·00	420	1·1	210
Beryllium *	1·1	1·85	310	0·57	170
Aluminium *	0·5	2·70	70	0·18	26

* For bulk materials.

As indicated above the object of fibre reinforcement is to produce a composite of high tensile strength and high modulus of elasticity. Such properties are often quoted as 'specific values', i.e. as a ratio of tensile strength (or modulus of elasticity) to relative density. This presents the property in terms of what used to be called the 'strength-to-weight ratio'. Some of the properties of important fibre materials are shown in Table 14·1, in which the materials are listed in order of normal tensile strength. It will be noted, however, that this is by no means the order of either specific strength or specific modulus.

14.2.2 Extremely high strengths are attainable in materials produced in the form of 'whiskers'. These are hair-like single crystals, between $0{\cdot}5 \times 10^{-3}$ and $2{\cdot}0 \times 10^{-3}$ mm in diameter and as much as 20 mm long, grown under controlled conditions and generally having a single dislocation running along the central axis. This relative freedom from dislocations means that their yield strength is nearer to the theoretical for the

Table 14.2—Properties of some 'whiskers'.

Material	*Tensile strength* (kN/mm²)	*Relative density*	*Young's modulus* (kN/mm²)	*Specific strength* (kN/mm²)	*Specific modulus* (kN/mm²)
Alumina	21	3·96	430	5·3	110
Silicon carbide	21	3·21	490	6·5	150
Graphite	20	1·66	710	12·0	430
Boron carbide	14	2·52	490	5·6	190
Silicon nitride	14	3·18	380	4·4	120

material (Table 14.2). Thus a graphite whisker may have a strength of as much as 21 kN/mm² whilst the strength of carbon fibre is of the order of only 3 kN/mm². Unfortunately whiskers are normally produced only under laboratory conditions and their great strength is difficult to utilise under ordinary production processes. Thus the high cost of forming a whisker composite generally rules out its use at present and ordinary fibre composites are used instead.

14.2.3 *Uses of fibre-reinforced composites.* Glass fibres are relatively cheap and have been used for some years for the reinforcement of resins. Unfortunately the surface of glass is sensitive to moisture and to the presence of defects, both of which cause a considerable fall in tensile strength. An even greater disadvantage, however, is the relatively low elastic modulus of glass fibre. GRP has nevertheless been in use for a considerable time for the manufacture of such articles as fishing rods, loudspeaker cones, vaulting poles, archery bows, boat hulls and masts. It is also used as an automobile engineering material and in the building trades. No doubt as CFRP (carbon fibre reinforced plastic) is developed and becomes cheaper it will replace GRP in cases where extra strength and stiffness are required.

Boron fibres and carbon fibres are being used in the establishment of a completely new generation of high-strength composites. Moreover whiskers of alumina in a matrix of nickel provide a composite which is strong at high temperatures and it is possible that such materials will be developed for use as turbine rotor blades. A particular difficulty in the use of many composites at high temperatures is that chemical reactions occur between the component materials of the composite leading to a change in structure and generally to failure.

Because of the high production costs of advanced fibre materials and composites generally, such materials only find application where conditions are extremely arduous. Components for the aircraft and aerospace industries require the maximum in terms of specific strength and specific modulus. Boron/resin and carbon/resin components are being made available for use in aircraft structures with a weight saving in the order of 25%.

Particle-hardened composites

14.3 These are commonly materials in which the ductility and toughness of a metallic matrix is combined with the hardness and strength of ceramic particles. Usually known as *cermets* these materials are widely used as cutting-tool tips, drilling bits, friction materials for brakes and in the compounding of fuel elements for nuclear power production. The principles of particle hardening were widely investigated in the production of materials for use in gas turbines. Such materials require a high creep-resistance at high temperatures. Unfortunately cermets proved to be generally inferior to 'Nimonic' alloys (11.5.2) in respect of relative brittleness.

Cemented carbides are possibly the best known materials of this type and have been used for many years. They consist of hard particles of tungsten or titanium carbide in a matrix of tough, ductile cobalt. The product is manufactured by compressing a mixtue of carbide powder and cobalt powder to the extent where cold-welding occurs at points between the cobalt particles. This process is followed by 'sintering'—that is, heating the component at some temperature above that of recrystallisation for the cobalt so that a continuous tough matrix of cobalt is formed (6.5.1).

14.3.1 Cermets are also made by causing a liquid metal to infiltrate around particles of a solid ceramic. Ideally a strong positive bond should be formed at the metal/ceramic interface. This may be achieved either by a chemical reaction between metal and ceramic, resulting in the formation of a film of intermetallic compound bridging the two phases, or by a simple solid solution of the metal and ceramic forming at the interface. Since intermetallic compounds are generally weak and brittle solid solution formation would seem to be the more effective.

In order that any form of bonding may occur it is necessary for the

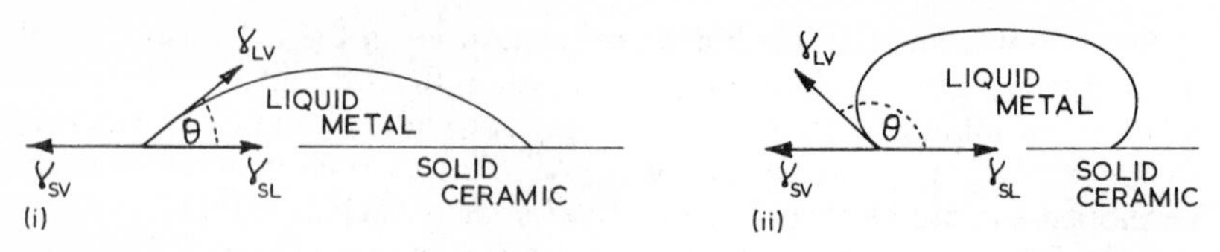

Fig. 14.3—Wetting of a ceramic surface. (i) Good wettability. (ii) Poor wettability.

molten metal to wet the surface of the ceramic. This wetting ability is related to the surface tension of the liquid and to the surface energy of the ceramic, and is a function of the angle of contact, θ, between liquid and solid (Fig. 14.3). The relation between the surface energies is given by the equation:

$$\gamma_{SV} - \gamma_{SL} = \gamma_{LV} \cos \theta$$

Wettability is at a maximum in the ideal case, viz. when $\theta \rightarrow 0$, but this state of affairs is never approached in practice. The quality and nature of the solid surface is generally the limiting factor and films of oxide or other matter cause a progressive increase of θ and a consequent reduction of adhesion energy between the surfaces.

Typical cermets are based on hard carbides, oxides, borides or nitrides bonded with a suitable tough strong metal. The main groups of cermets are listed in Table 14.3.

Cermets may contain up to 80% of the ceramic phase with the remainder bonding metal.

Dispersion-hardened materials

14.4 Some composite materials containing particles which are roughly spherical in form are developed for tensile strength rather than for hardness. Of primary importance in dispersion strengthening is alumina, Al_2O_3, in the form of very small particles dispersed throughout a suitable matrix. Possibly the easiest material in which to achieve this state of affairs is aluminium itself. If aluminium is produced in a powdered form and then ground, in the presence of oxygen under pressure, much of the surface film of oxide so formed disintegrates and forms very fine powder intermingled with the aluminium particles. The mixture is then sintered by powder-metallurgy techniques to give a more or less homogeneous aluminium matrix containing about 6% of alumina particles This product is known as 'sintered aluminium powder'—or SAP—and shows a definite advantage over ordinary aluminium in terms of tensile strength particularly at high temperatures. The alumina has a dispersion-hardening effect (14.1.5).

The high strength attained in some aluminium alloys as a result of precipitation hardening (11.3.2) is unfortunately lost at high temperatures because of the formation of large particles of non-coherent precipitates. In this respect SAP shows an advantage in retaining strength

Table 14.3—Principal cermet groups.

Group	*Ceramic*	*Bonding metal or alloy*	*Uses*
Carbides	Tungsten carbide WC	Cobalt	Cutting tool bits fabricated by powder-metallurgy
	Titanium carbide TiC	Molybdenum, cobalt or tungsten	
	Molybdenum carbide Mo_2C	Cobalt	
	Silicon carbide SiC	Cobalt or chromium	
Oxides	Aluminium oxide Al_2O_3	Cobalt, iron or chromium	Notable for their high-temperature properties—sparking-plug bodies, rocket motor and jet engine parts. Also 'throw-away' tool bits.
	Magnesium oxide MgO	Magnesium, aluminium, cobalt, iron or nickel	
	Chromium oxide Cr_2O_3	Chromium	
	Uranium oxide UO_2	Stainless steel	Nuclear fuel elements
Borides	Titanium boride TiB_2	Cobalt or nickel	Mainly as cutting tool tips
	Chromium boride Cr_3B_2	Nickel	
	Molybdenum boride Mo_2B	Nickel or nickel–chromium	

at relatively high temperatures (Fig. 14.4). Alumina particles are utilised for the dispersion strengthening of other materials, in particular silver and nickel.

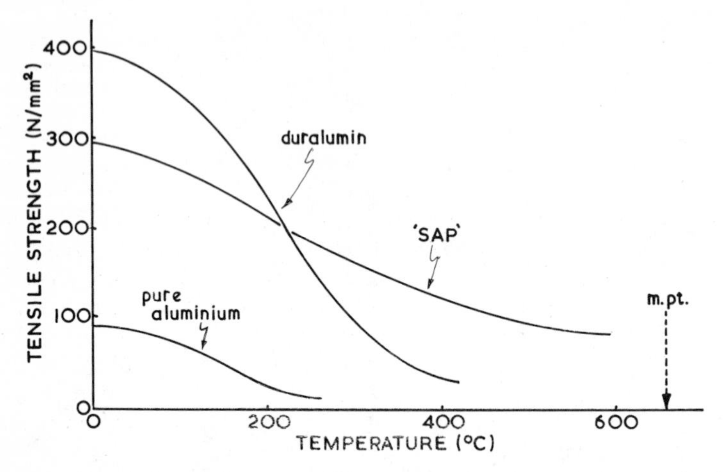

Fig. 14.4—The relationship between tensile strength and temperature for pure aluminium, duralumin and sintered aluminium powder (SAP).

Composite-bearing materials

14.5 A bearing material requires good wear-resistance and a low coefficient of friction combined with compressive strength, toughness and rigidity. As far as metallic bearings are concerned these properties do not occur together in a single-phase alloy. A solid solution, though tough and strong, has a poor wear-resistance and often a high coefficient of friction whilst an intermetallic compound, though hard and rigid and having a low coefficient of friction, is brittle and weak. Thus a common-bearing alloy—such as a Babbitt metal (11.7.1)—has a duplex structure consisting of hard particles of an intermediate phase embedded in a tough solid solution. The latter wears, leaving the particles of compound standing in relief and providing the low-friction bearing surface, whilst the wearing of the solid solution matrix provides channels which assist the flow of liquid lubricant traditionally used in orthodox engineering practice. Friction between surfaces is due largely to adhesion and this is at a maximum with metals which form a solid solution in each other. Often wear-resistant surfaces exhibit high coefficients of friction and only a few solid materials give both low friction and resistance to wear. Chief of these are graphite and some other lamellar solids and some polymers.

14.5.1 A large number of modern mechanisms particularly in aerospace engineering, require *solid* lubrication in some form. This may be due to the fact that liquid lubricants are unstable at high temperatures or that their transmission to the working surfaces may be erratic or, alternatively, expensive to promote. Some materials such as graphite are intrinsically low-friction substances but the most successful method of using the solid lubricant is to incorporate it in some form of composite so that the lubricant is in effect 'built in'. Metallic or polymeric matrices containing either lamellar solids (graphite or molybdenum disulphide) or polymers (such as PTFE) are the most widely suggested.

Sintered bronze bearings containing small amounts of graphite have been used for many years as have various mixtures of graphite and metals employed as brush materials in electrical machinery. The function of graphite as a lubricant depends upon its lamellar structure (3.8.2). Since only weak van der Waals forces operate between the layered structure of carbon atoms the latter slide over each other with ease and provide the lubrication effect. Other solids of lamellar structure include the sulphides, selenides and tellurides of molybdenum, tungsten, niobium and tantalum. Of these the best-known is molybdenum disulphide which has been used as a high-pressure lubricant for some time. A number of MoS_2/metal composites have been used but possibly the most successful so far are those containing MoS_2 bonded with up to 16% iron and 4% platinum. MoS_2/nickel composites give possibly the best results in terms of low friction ($\mu = 0{\cdot}03 - 0{\cdot}2$) though nickel is rather a poor binder. Composites containing calcium fluoride, CaF_2, as the solid lubricant in a nickel/chromium matrix are also promising, having a coefficient of friction of about 0·2 and a working temperature up to 800° C.

The coefficient of friction of PTFE (Table 12.2) is lower than that of any other solid material ($\mu = 0{\cdot}03 - 0{\cdot}1$) but its mechanical properties as a single-bearing material are poor. For this reason it is best used in a composite. Bronzes impregnated with PTFE are in use though the maximum working temperature of such bearings is limited by PTFE to about 300° C. Nylon is also widely used as a bearing material but its main disadvantage is its dimensional instability resulting from absorption of water.

Cement, mortar and concrete

14.6 The Ancient Greeks fixed together their stone building blocks using iron clips. Other early civilisations relied on closely fitting stones to produce a 'dry' wall of adequate stability—the Gallerus Oratory in the west of Ireland is such an example which was roofed by this method and is still perfectly sound as a structure after many centuries. Ultimately various inorganic cementing materials were developed in order to hold the building blocks firmly together. These are generally materials which can be mixed with water to produce a mouldable slurry and which ultimately 'set' due to the onset of a chemical reaction.

Limestone, $CaCO_3$, is a cheap plentiful material which, when 'calcined' at a high temperature forms *quicklime*:

$$\underset{\text{Limestone}}{CaCO_3} \rightarrow \underset{\text{Quicklime}}{CaO} + CO_3$$

When this quicklime is mixed with water—or 'slaked'—calcium hydroxide or 'slaked lime' is produced:

$$CaO + H_2O \rightarrow \underset{\text{Slaked lime}}{Ca(OH)_2} + \text{heat}$$

Slaking is accompanied by a considerable evolution of heat and an increase in volume. The slaked lime subsequently hardens slowly as it reacts with carbon dioxide from the atmosphere:

$$Ca(OH)_2 + CO_2 \xrightarrow{H_2O} CaCO_3 + H_2O$$

This reaction is accelerated when an excess of water is present in the cement. Since both carbon dioxide and calcium hydroxide are soluble in water a quicker reaction results than if a direct gas/solid reaction were involved. As $CaCO_3$ is much less water-soluble than $Ca(OH)_2$ it precipitates in crystalline form and the interlocking crystals of $CaCO_3$ convert the soft mass to a hard solid.

14.6.1 *Lime mortar*, obtained by mixing slaked lime with sand and water, was the principal building adhesive before the advent of Portland cement and is in fact still widely used for many purposes.

14.6.2 *Portland cement* is the principal building cement now in use. It is essentially a calcined mixture of lime-bearing and clay-base materials. The lime-bearing material may be ordinary limestone or even dredged marine shells, whilst the clay-bearing material comprises suitable clays or shales. These materials are pulverised and then mixed in the correct proportions prior to being calcined in rotary kilns at about 1500° C. The resultant clinker is then ground with small amounts of gypsum and water and a greenish-grey powder—the well-known Portland cement—is the result.

When mixed with water the cement 'sets' and hardens. The chemical reactions responsible for this change are quite complex but hardening seems to be caused by the combination of various silicates and aluminates with water of crystallisation to form a rigid crystal structure containing such compounds as $3CaO \,.\, 2SiO_2 \,.\, XH_2O$ and $3CaO \,.\, Al_2O_3 \,.\, YH_2O$.

14.6.3 *Concrete* is produced from a mouldable mixture of stone 'aggregates', sand, Portland cement and water. When hard it has many of the attributes of natural stone and is extensively used in civil engineering and building. It is very easy to mould in the wet state and is one of the cheapest constructional materials available in Britain because of the availability of the necessary raw materials.

A variety of materials may be used as aggregate. Stone and gravel are the most widely used and also the most satisfactory. Other substances which may be available cheaply, e.g. broken brick, blast-furnace slag, can be used where a lower-quality product will suffice. Sand is also included in the aggregate and to obtain a dense product a correct stone/sand ratio is essential. By mixing various 'grades' (sizes) of aggregate in suitable proportions (Fig. 14.5) a relatively small quantity of cement is all that is necessary to provide an adherent film between the aggregate particles.

Fig. 14.5—A satisfactory concrete structure.

The proportion of cement to aggregate used is governed by the strength required in the product and can vary from 1:3 for a 'rich' mixture to 1:10 for a 'lean' one. The ratio commonly specified is 1:6 and this produces very satisfactory concrete if the aggregate material is sound, well-graded, properly mixed with the cement paste and, finally, adequately consolidated.

Other things being equal the greater the size of the large aggregate the smaller the cement ratio required to produce concrete of a given strength. This is to be expected since a smaller area of interface will need to be coated for a given volume of aggregate. Concrete made with Portland cement usually hardens in about a week but rapid curing methods can be used to shorten this time.

14.6.4 *Plain concrete* is suitable for many purposes such as retaining walls, dams and structures which rely for their stability on great mass. Modern motorways, and foundations where large excavations have to be filled, are also examples where concrete is cast *in situ*.

At the other end of the scale small pre-cast parts can be manufactured by casting concrete into moulds, e.g. 're-constituted stone' blocks and various items of garden furniture.

14.6.5 *Reinforced concrete.* Most ceramic materials are relatively strong in compression but relatively weak in tension and concrete is no exception to this rule. For structural members such as beams the use of an adequate section of plain concrete would be uneconomic and often unsuitably cumbersome (Fig. 14.6 (ii)). By using steel reinforcement rods (Fig. 14.6 (iii)) in that part of the beam which will be in tension, full advantage can be taken of the high compressive strength of the concrete

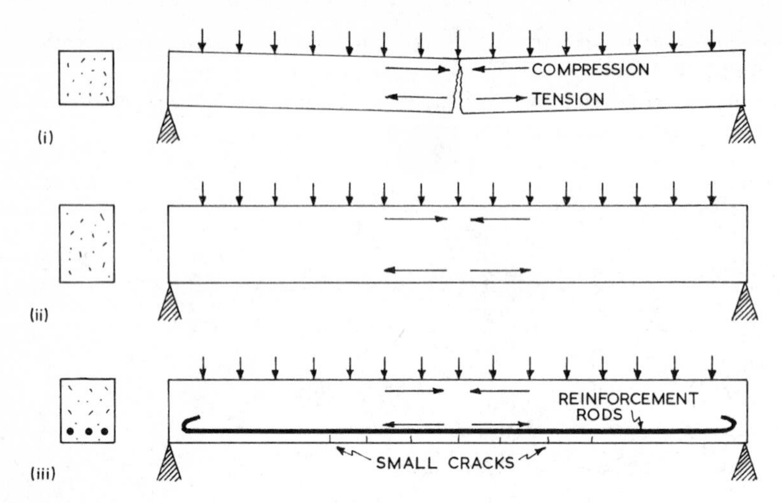

Fig. 14.6—The principle of reinforced concrete. In (i) a plain beam can fail when micro-cracks, acting as stress raisers, are formed in the edge which is in tension. A plain beam which is strong enough (ii) will be uneconomically bulky. In (iii) steel reinforcement-rods support that section of the beam which is in tension. Small cracks may, however, form in the concrete (14.6.6).

and the cross-section reduced accordingly. The ends of the rods are so shaped that they are firmly gripped by the rigid concrete.

14.6.6 *Pre-stressed concrete.* A source of weakness in reinforced concrete is indicated in Fig. 14.6 (iii) which shows the formation of small cracks in the concrete on that side of the beam which is in tension. Since steel has a much greater elasticity than concrete the steel rods will still be stretching within their elastic limit when the concrete has already begun to crack. The presence of small cracks which absorb moisture would lead to the gradual exfoliation of the concrete due to the action of frost. One method by which the situation can be met is to use the elastic properties of steel to apply an initial *compression* to the concrete, that is, to 'pre-stress' it.

A steel cable is stretched within its elastic limit and whilst it is in tension a concrete beam is cast around it (Fig. 14.7). When the concrete has completely hardened the externally applied tension in the steel cable is released (Fig. 14.7 (ii)). As the cable attempts to contract it exerts compressive forces on the concrete. Assuming that the cable is in the correct position relative to the cross-section of the beam these compressive forces in the concrete will exactly balance the tensile forces caused by the load applied to the beam during service (Fig. 14.7 (iii)). Since there is no resultant tension in the concrete cracks will not develop in it.

This method obviously depends upon the formation of a good concrete/steel bond without which the steel members would tend to slip through the concrete when the external tensioning force was released. Bonding between steel and concrete is not of a chemical nature but is more likely

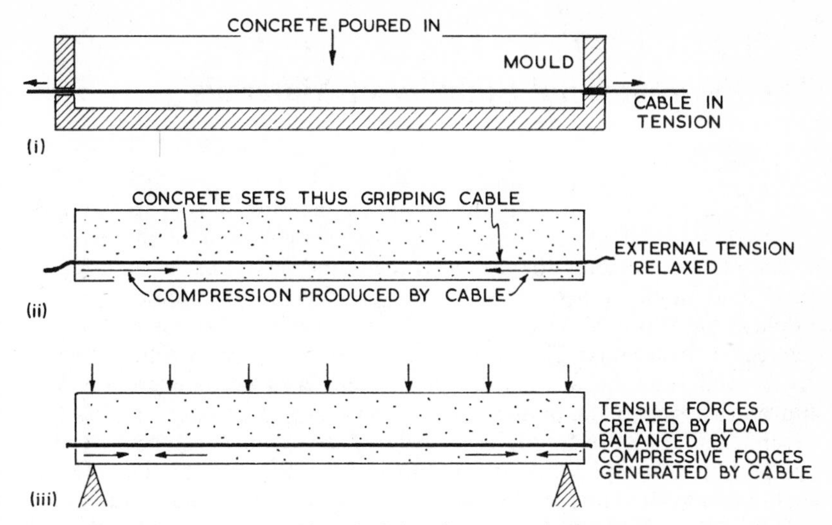

Fig. 14.7 The principles of prestressed concrete.

to be due to surface irregularities in the reinforcement steel. This bonding is improved by a thin layer of rust on the steel and assisted by the shrinkage of the concrete which occurs during hardening.

Tarmacadam and allied substances

14.7 Modern methods of road making were originated in the early years of the nineteenth century by Scots engineer John McAdam. The method he originated, that of coating suitable aggregate with tar, is basically that used today. Two types of material are now used as the bituminous matrix: tars and asphalts. Tars used are the residues derived from the destructive distillation of coal. Some asphalts occur naturally, e.g. 'Trinidad Lake', but most are residues obtained during petroleum refining processes.

The bituminous material is mixed with a suitable aggregate such as blast-furnace slag for coarse foundation work and fine gravel for the finishing layer. Mixtures are designed to meet specific requirements such as stability, durability, skid resistance and resistance to penetration by water. Those mixtures using harder grades of asphalt are normally processed in a 'hot-mix' plant at about 135° C, whilst those containing the more fluid asphalts may be mixed at ambient temperatures.

The resultant mixture is tough and crack-resistant because of the bituminous matrix whilst it is hard-wearing because of the presence of protruding pieces of aggregate. In many ways this resembles the principles upon which a bearing metal is designed (14.5) and is a good example with which to finish a chapter discussing the diverse properties which can be brought together by the use of composite materials.

Chapter Fifteen
The Environmental Stability of Materials

15.1 It is common knowledge that many engineering alloys corrode when exposed to moist atmospheres. The rusting of steel is a phenomenon which plagues not only the owner of the domestic 'tin lizzie' but costs us in Britain alone something in excess of £500 million annually in the application of protective measures. Other materials also possess varying degrees of instability. Thus rubber 'perishes' particularly when exposed to air and sunlight, and some plastics materials become progressively more brittle under the influence of ultra-violet light. Concrete and other ceramic materials are gradually eroded by frost and the chemical effects of our polluted atmosphere. A large group of materials, both metallic and non-metallic, may be damaged to some degree by various forms of radiation.

It might seem therefore that the only way to protect materials with certainty is to enclose them in lead-lined vacuum chambers. The engineer, however, must follow a rather more practical course and choose materials which will be reasonably stable in the particular environment in which they must work.

For materials working at high temperatures sufficient strength and rigidity are often difficult to attain and other physical properties like thermal coefficient of expansion and thermal conductivity may be important. Thus furnace refractories are liable to spall—or flake away—if the rate of heating or cooling is too great. Moreover chemical attack by the atmosphere is nearly always a problem with materials used at high working temperatures.

Dry corrosion or the oxidation of metals

15.2 Whilst many metals tend to oxidise to some extent at all temperatures, most engineering metals do not scale appreciably except at high temperatures. When iron is heated strongly in an atmosphere containing oxygen it becomes coated with a film of black scale. A chemical reaction has taken place between atoms of iron and molecules of atmospheric oxygen:

$$2Fe + O_2 \rightarrow 2FeO$$

Although the reaction can be expressed by the above simple equation in fact atoms of iron have been oxidised whilst atoms of oxygen have been reduced. These processes are associated with a transfer of electrons from one atom to the other.

Oxidation $Fe \rightarrow Fe^{++} + 2$ electrons (e^-)
Reduction $O + 2$ electrons $(e^-) \rightarrow O^{--}$

It should be noted that, in the chemical sense, the terms oxidation and reduction have a wider meaning than the combination or separation of a substance with oxygen. Thus a substance is oxidised if its atoms lose electrons whilst it is reduced if its atoms (or groups of atoms) gain electrons. For this reason we say that iron is oxidised if it combines with sulphur, chlorine or any other substance which will accept electrons from the atoms of iron.

15.2.1 In some cases the formation of the oxide film protects the metal from further oxidation yet in many instances oxidation and scaling continue. For this to occur either molecules of oxygen must pass through a very porous film of oxide, or *ions* of either oxygen or the metal must migrate *within* a continuous film.

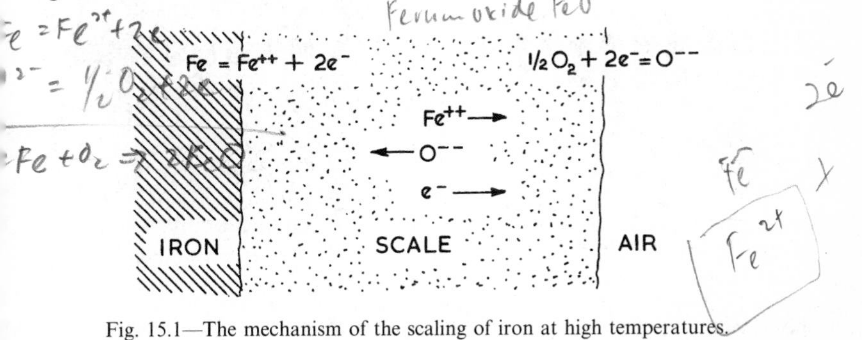

Fig. 15.1—The mechanism of the scaling of iron at high temperatures.

Fig. 15.1 indicates the state of affairs which exists within the film of iron oxide scale. The positively charged Fe^{++} ions are attracted outwards towards the cathodic regions, i.e. those rich in negative charge, whilst the negatively charged O^{--} ions are attracted inwards towards the region which is richer in positive charge, i.e. that near the metal surface. Since metallic ions are generally smaller than oxygen ions, the diffusion of the metal ions outwards is quicker than the diffusion of oxygen ions inwards. The rate of oxidation of the metal will depend partly on the mobility of the ions through the oxide film but also upon the rate of flow of electrons outwards. These mobilities are in turn affected by the nature and structure of the film, particularly in terms of 'vacancies', that is, positions from which ions are missing. The number of vacancies is affected by the presence of solute atoms such as chromium and ionic mobilities are reduced as a result.

As mentioned above oxidation rates also depend upon the porosity of the oxide film. Some metals form ions which are much smaller than the original atoms. Consequently as ions are formed there is a volume reduction of the scale which, as a result, becomes porous and permits easier access of oxygen to the metal surface. Contraction of the oxide film may also cause it to flake off thus exposing a fresh metal surface. In some cases expansion of the film may occur causing it to buckle and become detached.

15.2.2 An oxide film which adheres tightly to the surface of the metal generally offers good protection. Good adhesion is the result of coherence between the film and the metal beneath. In Fig. 15.2 (i) there is good 'matching' between the ions in the metal surface and the metallic ions in the oxide film such that the structure is virtually continuous, whilst in Fig. 15.2 (ii) matching is absent so that the two surfaces will be non-coherent and there will be little adhesion. The high degree of coherence between aluminium and its oxide leads to the effective protection of this metal especially when the film is artificially thickened by anodising.

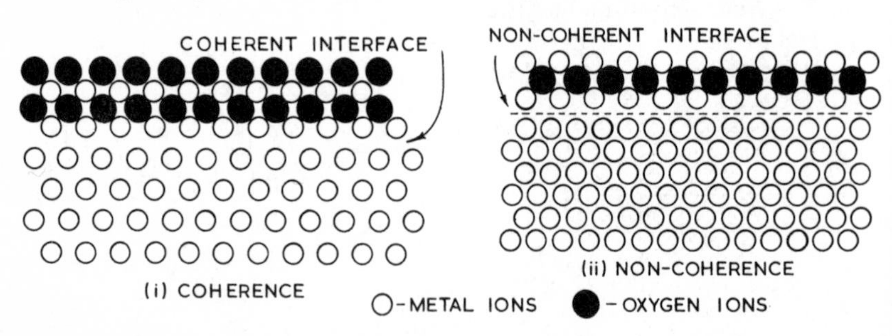

Fig. 15.2—Adhesion between a metal surface and the covering oxide film. In (i) the row of metal ions at the interface 'matches up' with those in the oxide film, but in (ii) ions are 'out of step' at the interface.

Some alloys are prone to attack by atmospheres containing sulphurous gases. Since nickel readily forms a sulphide at high temperatures it is particularly liable to be oxidised by gases containing sulphur. Heat-resisting steels used in such conditions must be of the high-chromium type, preferably containing little or no nickel.

Wet corrosion or electrochemical oxidation

15.3 Iron does not rust in a completely dry atmosphere nor will it rust in completely pure, oxygen-free water, but in a moist atmosphere the well-known reddish-brown deposit of ferric hydroxide soon begins to develop. The overall chemical reaction representing rusting can be expressed by a simple chemical equation:

$$4Fe + 6H_2O + 3O_2 \rightarrow 4Fe(OH)_3$$

This general result, however, is achieved in a number of stages and the fundamental principle involved is that atoms of iron in contact with oxygen and water are oxidised, that is they lose electrons and enter solution as ferrous ions (Fe^{++}):

$$Fe \rightarrow Fe^{++} + 2 \text{ electrons } (2e^-)$$

These ferrous ions are ultimately oxidised further to ferric ions (Fe^{+++}) by the removal of another electron:

$$Fe^{++} \rightarrow Fe^{+++} + e^{-}$$

As iron goes into solution in the form of ions the corresponding electrons are released. These electrons immediately combine with other ions so that overall equilibrium is maintained.

The ease with which a metal can be oxidised in this way depends upon the ease with which valency electrons can be removed from its atoms. Thus metals like calcium, aluminium and zinc hold their valency electrons comparatively loosely and can therefore be oxidised more easily than iron; but the noble metals gold, silver and platinum retain their valency electrons more strongly and are therefore much more difficult to oxidise than iron.

Copper holds on to its valency electrons more strongly than does iron and, under suitable conditions, copper ions will 'steal' valency electrons from atoms of iron. Thus when a pen-knife blade is immersed in copper sulphate solution the blade becomes coated with metallic copper. Copper ions at the surface of the blade have removed electrons from the atoms of iron there so that the resultant ferrous ions have gone into solution thus replacing the copper ions which have been deposited as copper atoms

$$Fe \rightarrow Fe^{++} + 2e^{-}$$
$$Cu^{++} + 2e^{-} \rightarrow Cu$$

Copper sulphate solution is sometimes used to coat the surface of steel with a thin layer of copper as an aid to 'marking out', in the tool room.

15.3.1 Thus different metals have different *oxidation potentials* since the amount of energy required to remove their valency electrons varies from metal to metal. The ionisation process builds up an electrical potential called an *electrode potential* and this depends upon the nature of the solution as well as upon the nature of the metal. Electrode potentials are measured with reference to hydrogen (Fig. 15.3) and are useful in assessing the tendencies of the metal to corrode. To do this a standard hydrogen electrode is used as indicated and some form of potentiometer employed to measure the potential difference (in volts) produced between the hydrogen electrode and the metal electrode. Since hydrogen is a gas the electrode used consists of a platinum tube (which is relatively inert) into which hydrogen is passed. Hydrogen dissolves interstitially in platinum and in this way forms an electrode. By using similar methods an *electrochemical series* can be prepared for the metals as indicated in Table 15.1.

15.3.3 *Galvanic action involving dissimilar metals.* In Fig. 15.3 the electrode which will supply electrons to the external circuit (when the latter is

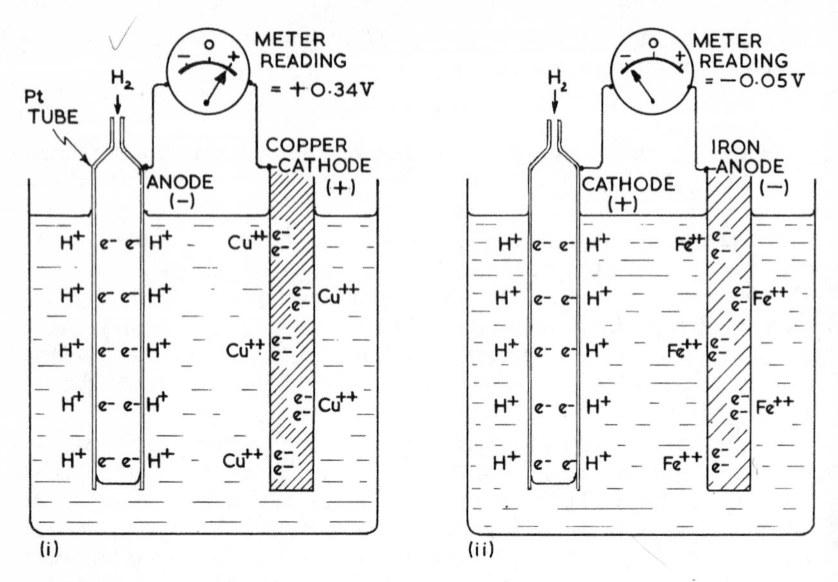

Fig. 15.3—The measurement of electrode potential. Since copper (i) produces a higher electrode potential than does hydrogen it becomes the cathode, whilst hydrogen is the anode. However, hydrogen (ii) produces a higher electrode potential than does iron so in this case hydrogen is the cathode whilst iron is the anode.

Table 15.1—Electrode potentials for some metals (standard molar solutions of ions at 25° C).

	Metal ion	*Electrode potential* (volts) E_h
(Noble)	Gold (Au^{+++})	+1·50 (Cathodic)
	Platinum (Pt^{++++})	+0·86
	Silver (Ag^{+})	+0·80
	Copper (Cu^{+})	+0·52
	(Cu^{++})	+0·34
	Hydrogen (H^{+})	0·00 (Reference)
	Iron (Fe^{+++})	−0·05
	Lead (Pb^{++})	−0·13
	Tin (Sn^{++})	−0·14
	Nickel (Ni^{++})	−0·25
	Cadmium (Cd^{++})	−0·40
	Iron (Fe^{++})	−0·44
	Chromium (Cr^{+++})	−0·74
	Zinc (Zn^{++})	−0·76
	Aluminium (Al^{+++})	−1·66
	Magnesium (Mg^{++})	−2·37
(Base)	Calcium (Ca^{++})	−2·87 (Anodic)

'closed') in each case is called the *anode* and the electrode which will receive electrons via the external circuit is called the *cathode*. Thus in Fig. 15.3 (i) the hydrogen electrode is the anode whilst in Fig. 15.3 (ii) the iron plate is the anode.

At this juncture it is necessary to mention an important point connected with electrochemical nomenclature and one which may otherwise puzzle the reader. The Greek derivation of the term 'anode' implies 'something which is being built up' and in fact the anode was so named in the early days of the study of electricity because it was assumed that positively charged 'particles of electricity' were flowing through the external circuit from the cathode to the anode. It is now realised that what we call an electric current is in fact a flow of *negatively charged particles* —or electrons—in the opposite direction, i.e. from the anode to the cathode via the external circuit. Thus, instead of *receiving* positively charged particles from the external circuit the anode *supplies* negatively charged electrons to the external circuit and is consequently left with positively charged particles (ions) which it immediately releases into solution. We therefore visualise the anode as being negatively charged because it is the source of electrons to the external circuit. In the illustrations which follow, electrons will be seen to be flowing in the opposite direction to that which is ascribed to 'current' by electrical engineers.

If an external electrical contact is made between the electrodes of the system shown in Fig. 15.3 (ii) as indicated in Fig. 15.4, the lower electrode potential of the iron will cause it to lose electrons which will flow away from the anode through the external circuit to the cathode, the

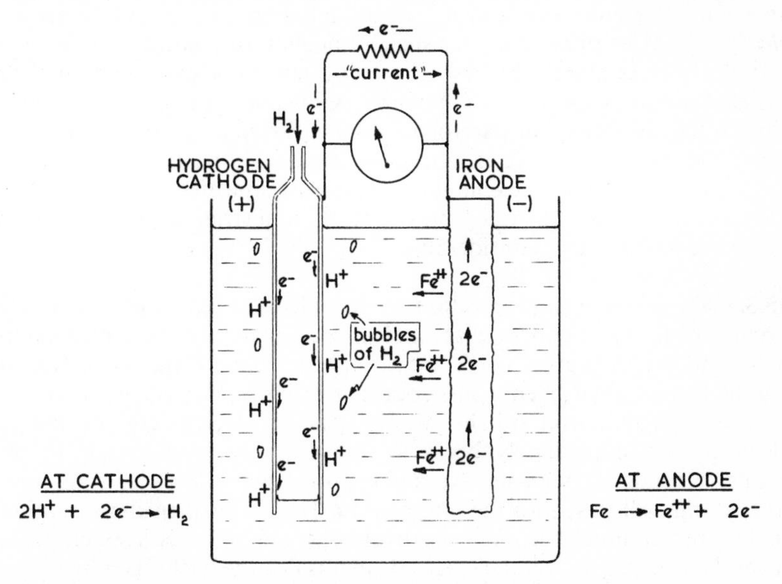

Fig. 15.4—The effect of closing the external circuit in the hydrogen-iron cell.

resultant Fe^{++} ions entering solution. The electrons which arrive at the cathode combine with H^+ ions in the surrounding electrolyte and form molecular hydrogen which subsequently escapes as small bubbles:

$$H^+ + e^- \rightarrow H$$
$$H + H \rightarrow H_2$$

This reaction causes further electrons to flow from the iron anode so that more iron atoms ionise liberating electrons and entering the electrolyte as Fe^{++} ions. The 'driving force' behind this reaction is the difference in electrode potential between iron and hydrogen. These reactions continue spontaneously, iron dissolving from the anode and hydrogen precipitating at the cathode.

The hydrogen ions (H^+) present in the electrolyte are formed by the ionisation of water molecules:

$$H_2O \rightleftharpoons H^+ + OH^-$$

As H^+ ions are 'neutralised' on receiving electrons to form hydrogen atoms (and subsequently molecules) more water ionises spontaneously. The concentration of hydrogen ions in *pure* water is extremely low—of the order of 10^{-7} gram-atoms/litre—and so the formation of hydrogen atoms and the simultaneous solution of iron would be very slow. In practice some soluble impurities in water lead to a much higher degree of ionisation of the water and so galvanic corrosion takes place at a greater rate. A solution in which free ions are present, and which will therefore allow an electric current to be transmitted in an external circuit, is termed an *electrolyte*. Most organic liquids such as alcohol, turpentine, oils, etc. are non-electrolytes since the presence of covalent bonds throughout their molecules makes ionisation impossible. Galvanic corrosion is not possible in the absence of an electrolyte. As mentioned above pure water is a very weak electrolyte since an extremely small proportion of its molecules are ionised, but when it contains dissolved impurities such as carbon dioxide or sulphur dioxide (prevalent in industrial atmospheres) its degree of ionisation is considerable

15.3.4 We are now in a position to consider the reactions which will occur when, say, a copper plate and an iron plate are connected externally and are in mutual contact with an electrolyte (Fig. 15.5). Iron is anodic towards hydrogen whilst copper is cathodic towards hydrogen. It follows then that iron will be strongly anodic towards copper and so electrons will immediately begin to flow from the iron anode to the copper cathode through the external circuit. Having lost electrons some iron atoms will pass into the electrolyte as Fe^{++} ions. In the meantime the copper cathode becomes negatively charged as it receives electrons from the iron anode, via the external circuit. Positively charged hydrogen ions, H^+, present in the electrolyte will be attracted to the cathode where

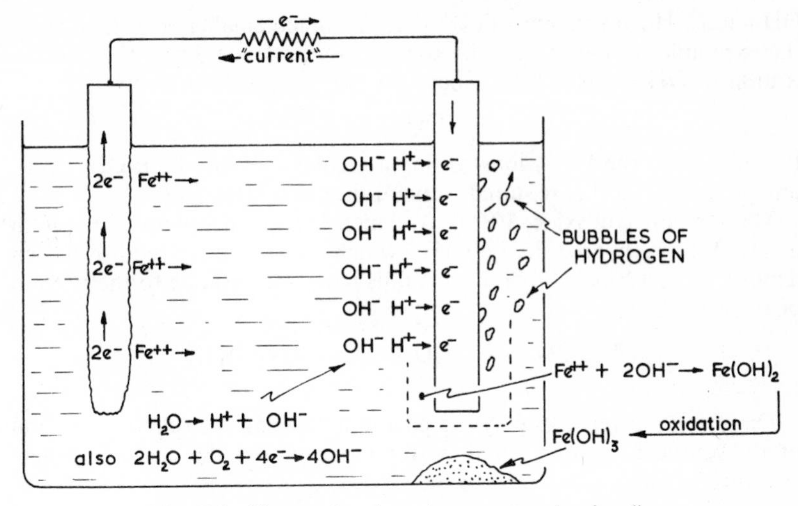

Fig. 15.5—The reactions in an iron-copper galvanic cell.

they will combine with electrons to form hydrogen atoms and, subsequently, molecules:

$$2H^+ + 2e^- \rightarrow 2H \rightarrow H_2$$

thus restoring the balance.

As more and more H^+ ions are neutralised by electrons from the anode in this way to form hydrogen atoms, so more and more water molecules will ionise to restore the balance between molecules and ions. The concentration of hydroxyl ions (OH^-), also formed by the ionisation of water molecules, will build up in the region of the cathode (Fig. 15.6).

Although Fe^{++} ions are released into the electrolyte at the anode they do not remain there. There is a mutual attraction between them and the

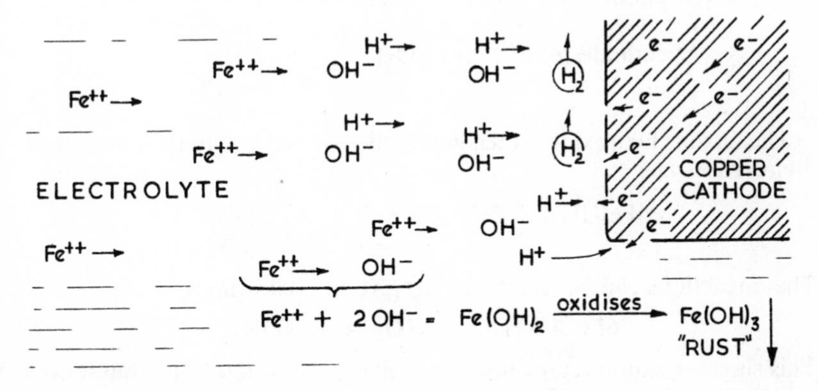

Fig. 15.6—Reactions at the cathode.

OH^- ions. However, since Fe^{++} ions are much smaller than OH^- ions they are able to migrate rapidly towards the cathode where the concentration of OH^- ions is high. There the two types of ion react:

$$Fe^{++} + 2(OH^-) \rightarrow Fe(OH)_2$$

In addition to the Fe^{++} ions having a greater mobility than OH^- ions, only one Fe^{++} ion is required to neutralise two OH^- ions.

The ferr*ous* hydroxide, $Fe(OH)_2$, which is formed rapidly oxidises to insoluble ferr*ic* hydroxide, $Fe(OH)_3$, which collects as the reddish-brown deposit we call 'rust'. Much of this falls from the cathode to the bottom of the cell.

$$4Fe(OH)_2 + 2H_2O + O_2 \rightarrow \underset{\text{'rust'}}{4Fe(OH)_3\downarrow}$$

Dissolved oxygen also plays its part in increasing the rate of corrosion of the anode by reacting with water molecules at the cathode (where electrons are available):

$$O_2 + 2H_2O + 4e^- \rightarrow 4OH^-$$

The rate of formation of hydroxyl ions is thus increased.

15.3.5 *Summary of reactions*

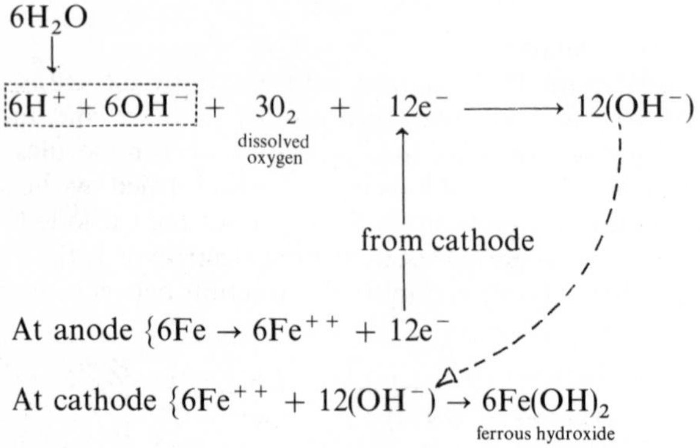

$6H_2O$
↓
$[6H^+ + 6OH^-] + \underset{\text{dissolved oxygen}}{3O_2} + 12e^- \longrightarrow 12(OH^-)$

↑ from cathode

At anode $\{6Fe \rightarrow 6Fe^{++} + 12e^-$

At cathode $\{6Fe^{++} + 12(OH^-) \rightarrow \underset{\text{ferrous hydroxide}}{6Fe(OH)_2}$

The ferrous hydroxide is then gradually oxidised due to the presence of dissolved oxygen:

$$6Fe(OH)_2 + 3H_2O + 1\tfrac{1}{2}O_2 \rightarrow \underset{\text{ferric hydroxide ('rust')}}{6Fe(OH)_3}$$

These reactions can be 'combined' to give the following:

$$6Fe + 9H_2O + 4\tfrac{1}{2}O_2 \rightarrow 6Fe(OH)_3$$

This single equation represents chemically the overall reaction but it does not explain the steps by which the result is achieved. In many of the examples which follow the electro-chemistry has been simplified by assuming that *Fe^{+++} ions are formed at the anode* to give a resultant deposit of $Fe(OH)_3$ near to the cathode.

15.3.6 *Common examples of galvanic corrosion involving dissimilar materials.* Readers who are also practising engineers will by now have realised why it is generally considered bad practice to use dissimilar metals in contact with each other and in mutual contact with an electrolyte albeit the latter is only condensed atmospheric moisture. Nevertheless, in domestic plumbing one often does not need to look far to discover examples of electrochemical malpractice. For example, most cold-water cisterns are of galvanised iron and whilst such material is quite adequate when properly used, the life of the zinc coating can only be shortened if a copper inlet pipe enters the water adjacent to the side of the tank. Zinc is strongly anodic to copper and since the water is efficiently aerated at the inlet, the zinc may be expected to corrode, ultimately exposing the iron beneath. Iron too is anodic to copper and so a leaking tank will be only a matter of time. Such leaks usually develop when one is away on the annual vacation, offering a dismal scene on one's return.

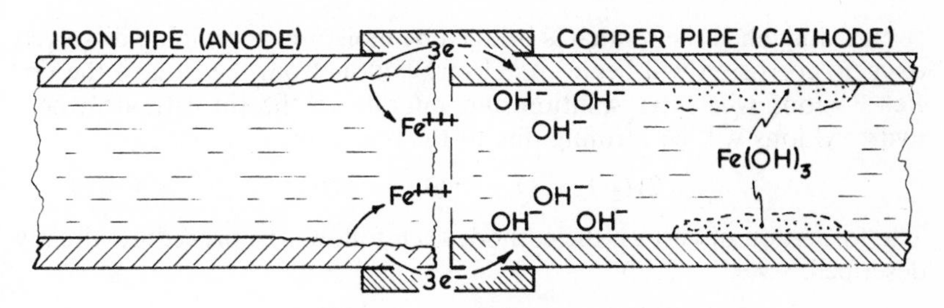

Fig. 15.7—An example of bad plumbing.

Mixing of lengths of iron and copper for carrying water leads to the accelerated corrosion of the iron pipe near to its joint with the copper (Fig. 15.7). Examples of this type of bad practice are often provided by those who should know better. The motor-car industry has a habit of sticking bits of stainless-steel 'trim' on to the mild-steel body-work. This is quite safe so long as the mild steel is completely coated with paint but invariably the paintwork around the fixing hole is damaged during the fixing process and the design is such that water collects in the region of the joint thus constituting a galvanic cell. Since the mild-steel body is anodic to the stainless-steel 'trim' the body-work ultimately rusts away, no doubt to the ultimate satisfaction of the manufacturer who finds such caprices good for trade.

Iron and steel must generally be protected from both atmospheric and aqueous corrosion by coating with some protective film. Painting, electroplating, tinning and galvanising are processes which can be used to this end. However, when metallic coatings are used the effect of possible film damage must be considered, particularly as to whether the film will be anodic or cathodic to the surface of the iron beneath.

Thus a damaged or discontinuous tin coating can actually accelerate the rusting of the mild steel it is meant to protect. Tin is cathodic to iron and so the galvanic corrosion of iron will take place as indicated in Fig. 15.8. The tin and iron are already in electrical contact in the 'external

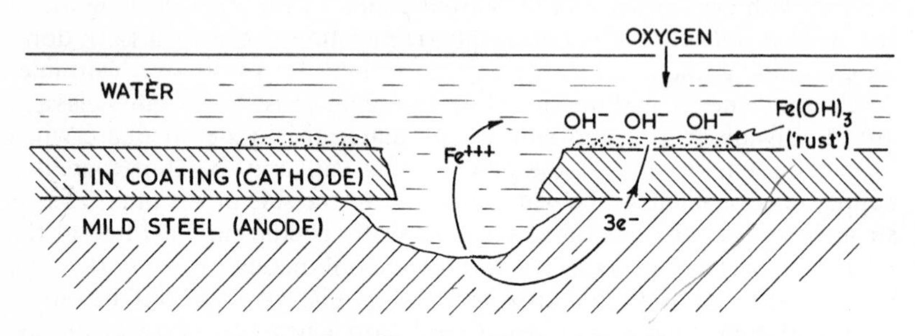

Fig. 15.8—The disastrous effect of a defective tin coating on mild steel.

circuit' and when the scratch is bridged by moisture (containing dissolved oxygen) corrosion will begin. As electrons flow from the iron to the tin, Fe^{+++} ions will enter solution and migrate to the tin cathode where hydroxyl ions will be forming due to the reaction:

$$2H_2O + O_2 + 4e^- \rightarrow 4OH^-$$

The formation of rust on the cathode then follows the procedure already described, viz.

$$Fe^{+++} + 3OH^- \rightarrow Fe(OH)_3 \downarrow$$

Tin is thus effective as a protective coating only if the coating is perfect. Since it is an expensive metal tin is now only used as a coating in the canning industry and for coating food-processing equipment where its high corrosion-resistance to animal and vegetable substances is utilised, and the fact that any compounds which may be formed are non-toxic.

15.3.7 For engineering purposes either zinc or aluminium are more effective since both are anodic to iron. Zinc is the more widely used of the two since it is easier to apply to mild steel either by galvanising (hot-dipping of the steel into molten zinc); spraying (as in the Schoop process used to coat the Forth Road Bridge); sherardising (a vapour condensation/-cementation process); or electroplating.

Consider a damaged zinc coating as indicated in Fig. 15.9. Since zinc is anodic towards iron it will send electrons through the 'external circuit' to the surface of the iron and as a result Zn^{++} ions enter solution. Hydroxyl ions (OH^-) will form at the surface of the iron as a result of

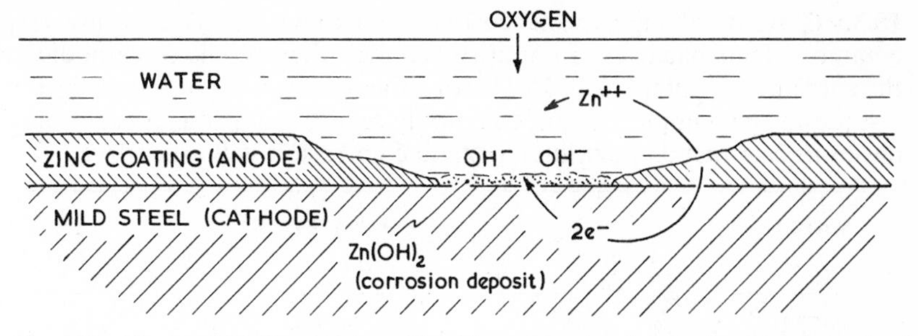

Fig. 15.9—The 'sacrificial' protection offered by a defective zinc coating. Iron is rendered cathodic so that the zinc goes into solution preferentially.

the standard reaction involving oxygen, water and electrons. Zn^{++} ions and OH^- ions will then react:

$$Zn^{++} + 2OH^- \rightarrow Zn(OH)_2 \downarrow$$

Thus, $Zn(OH)_2$ precipitates as a white corrosion product on the surface of the iron cathode.

In this case it is the coating metal which corrodes in preference to the mild steel. This form of corrosion is often termed 'sacrificial', since iron will not rust significantly as long as zinc remains in the vicinity of the initial scratch. It would be wrong to assume, however, that there is any lasting merit in producing a poor-quality porous coating. As with all methods of surface protection the aim should be to produce a perfect coating which will give no opportunity for galvanic action. With a zinc coating, however, the position is less serious if the coating is in fact defective since some measure of temporary sacrificial protection is afforded the mild steel.

15.3.8 On the question of the galvanic effect of surface coatings on iron the influence of oxide scale cannot be neglected. Ordinary 'black' sheet will rust more quickly than will clean sheet because oxide scale is cathodic to the metal beneath. Corrosion of the metal will then follow the usual pattern (Fig. 15.10).

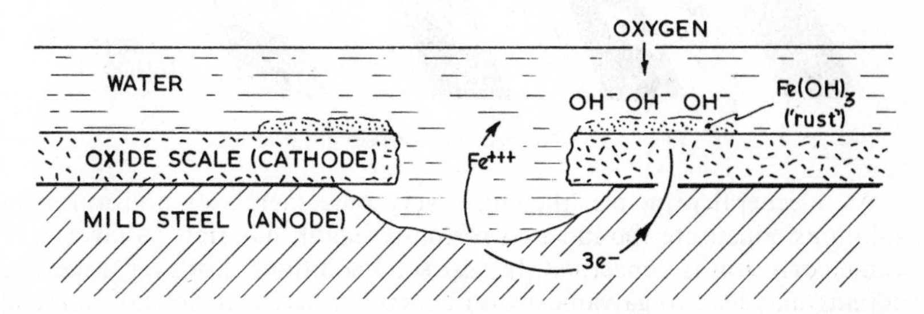

Fig. 15.10—The cathodic effect of oxide scale in accelerating the rusting of mild steel sheet.

15.3.9 Cases involving galvanic corrosion on a microscopic scale are very common. Thus particles of impurity may be either anodic or cathodic to the surrounding metal (Fig. 15.11). In either case pitting of the metal will result and since impurities are commonly segregated at the crystal boundaries in cast metals then intercrystalline corrosion will take place.

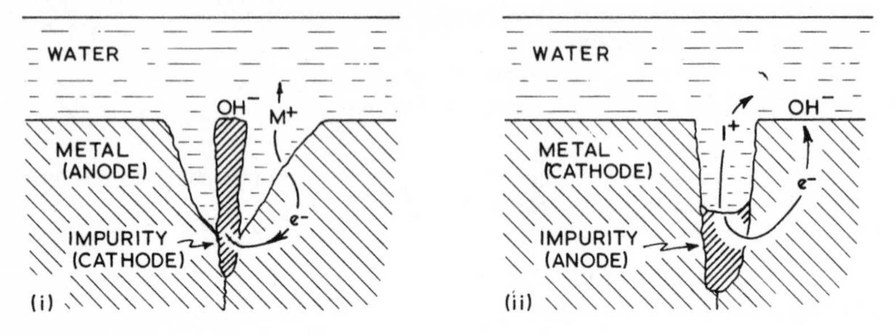

Fig. 15.11—The effect of grain-boundary impurities on galvanic corrosion. In (i) the impurity is cathodic to the metal, whilst in (ii) the impurity is anodic to the metal.

Alloys of duplex structure can be expected to corrode more quickly than single-phase materials of similar composition. Carbon steel in the pearlitic condition rusts far more quickly than it does in the fully hardened state. A steel in the pearlitic condition contains a structure in which the ferrite lamellae are anodic to the cementite lamellae. Consequently the ferrite tends to form Fe^{+++} ions and corrosion proceeds in the usual manner (Fig. 15.12). A steel of similar composition in the quenched state will be wholly martensitic and so galvanic corrosion will be at a minimum because of the uniform nature of the structure. Tempering will increase the tendency towards corrosion as cementite is progressively precipitated.

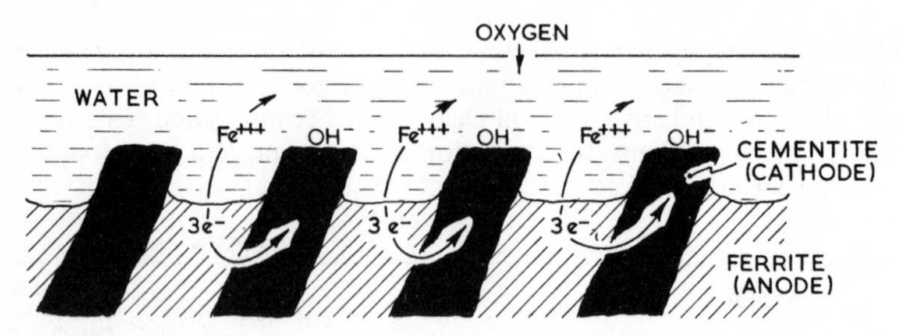

Fig. 15.12—The galvanic corrosion of pearlite.

As a general principle, therefore, very pure metals or uniform solid solutions constitute the most corrosion-resistant materials as far as galvanic corrosion is concerned. In *cast* solid solutions, however, excessive coreing may lead to galvanic action between different regions of a crystal due to differences in chemical composition.

Pourbaix diagrams

15.4 The extent to which a metal corrodes depends not only upon its electrode potential relative to other substances in the environment but also upon the chemical nature of the electrolyte and in particular its pH.

The connection between electrode potential, E_h, and the hydrogen ion concentration, pH, was investigated by M. Pourbaix at the Belgian Centre for the Study of Corrosion. He formulated diagrams which connect these values with the electrochemical equilibria of metals in contact with water. Both electrode potential and pH will influence the transfer of

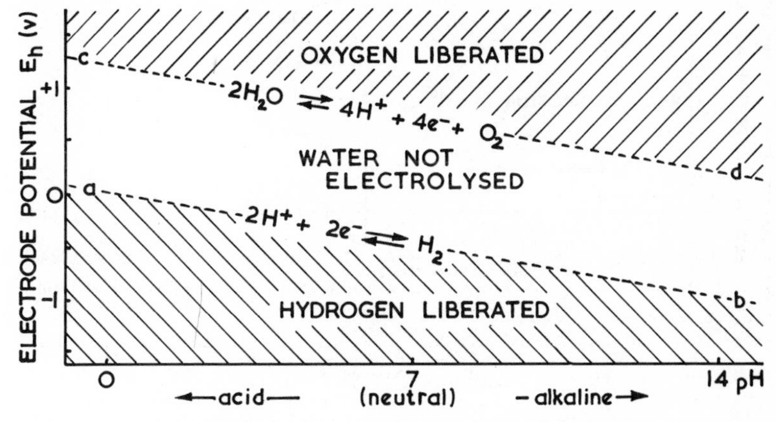

Fig. 15.13—The E_h/pH diagram for water.

electrons in the system so that hydrogen evolution and oxygen absorption are dependent on these factors. These reactions are the basis of electrolytic corrosion. Pourbaix diagrams resemble phase-equilibrium diagrams in so far that they do not give any information about the *rate* at which a reaction takes place.

15.4.1 Fig. 15.13 illustrates the effects of E_h and pH on the extent to which water will liberate either hydrogen or oxygen by what we usually term 'electrolysis'. Hydrogen evolution is only possible under conditions of E_h and pH which are represented by a point below *ab* and oxygen can be evolved only if E_h and pH are represented by a point above *cd*. Therefore there is a *domain* between *ab* and *cd* on the E_h/pH diagram where the electrolysis of water is thermodynamically impossible. Thus the electrolysis of water will not take place, regardless of the pH of the electrolyte, unless the electrodes used have a minimum potential difference of more than 1·2 volts, that is the distance apart in terms of E_h of *ab* and *cd*.

15.4.2 In Fig. 15.13 the equilibria of hydrogen and oxygen reduction reactions are represented on the E_h/pH diagram. However, equilibria for a metal in conjunction with water can also be represented. Since the corrosion of iron is most important from a technical standpoint this is represented in terms of E_h and pH in Fig. 15.14. The complete Pourbaix diagram for iron is complex since a number of equilibria are involved.

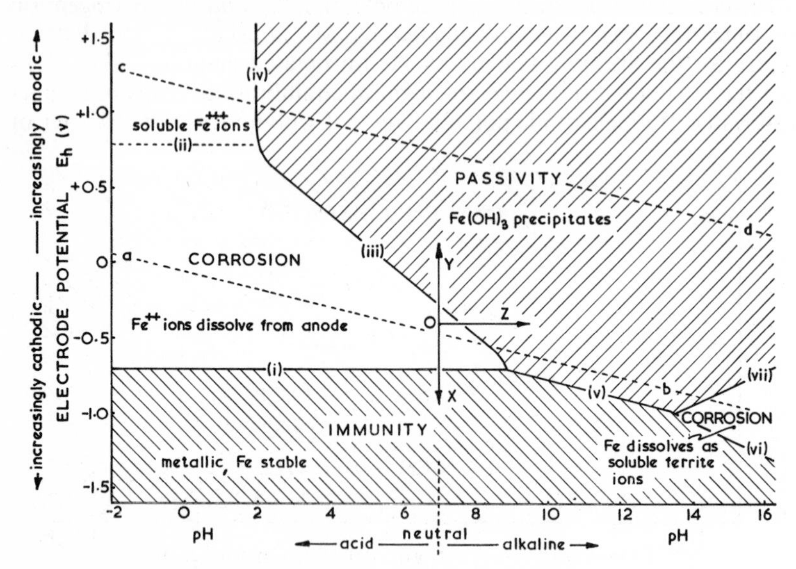

Fig. 15.14—The simplified Pourbaix diagram for the Fe/H_2O system.

Fortunately a simplified diagram can be drawn from a study of the following equilibria, a number of which we have already encountered in this chapter:

(*i*) $Fe \rightleftharpoons Fe^{++} + 2e^-$ Iron atoms form soluble Fe^{++} ions.

(*ii*) $Fe^{++} \rightleftharpoons Fe^{+++} + e^-$ Fe^{++} ions are oxidised to Fe^{+++} ions.

(*iii*) $Fe^{++} + 3OH^- \rightleftharpoons Fe(OH)_3 + e^-$ Insoluble ferric hydroxide precipitates.

(*iv*) $Fe^{+++} + 3H_2O \rightleftharpoons Fe(OH)_3 + 3H^+$ Insoluble ferric hydroxide precipitates.

(*v*) $Fe + 3H_2O \rightleftharpoons Fe(OH)_3 + 3H^+ + 3e^-$ Insoluble ferric hydroxide precipitates.

(*vi*) $Fe + 2H_2O \rightleftharpoons FeO_2H^- + 3H^+ + 2e^-$ Ferrite ions go into solution.

(*vii*) $FeO_2H^- + H_2O \rightleftharpoons Fe(OH)_3 + e^-$ Soluble ferrites precipitate as ferric hydroxide.

If the electrode potential E_h of mild steel is measured in pure water (pH = 7) it will be represented on the diagram (Fig. 15.14) by point O. This indicates that corrosion will take place under such conditions of E_h and pH.

The diagram also suggests three ways in which the extent of corrosion can be reduced. First, the potential can be changed in the negative direction so that it enters the domain of immunity (*X*). This is the normal cathodic protection such as is offered by zinc, either as a coating or as a separate anode (15.7.4). Secondly, the potential can be changed in a positive direction into the domain of passivity (*Y*) by applying a suitable external e.m.f. Thirdly, the electrolyte can be made more alkaline (z) so that the metal becomes passive. *Inhibitors* have this effect—amongst others—but need to be carefully controlled since with some metals the passivity domain extends over a very limited pH range and continuous surveillance of the electrolyte composition would be necessary.

Electrochemical corrosion due to internal mechanical stresses

15.5 Cold-worked materials of *uniform composition* throughout may nevertheless suffer severe electrochemical corrosion due to different energy levels prevailing in different parts of the material. If part of a component is heavily cold-worked it will contain 'locked-up' stresses which are internally balanced and which will produce some elastic strain.

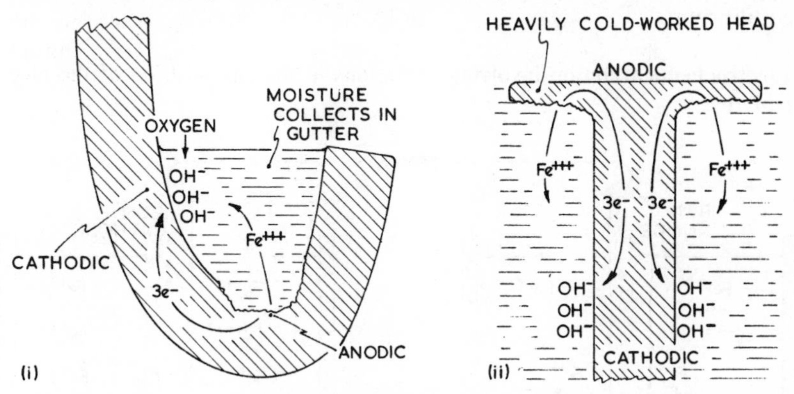

Fig. 15.15—Examples of electro-chemical corrosion due to local cold-work. (i) The rim of a motor car 'wing'. (ii) The cold-forged head of an iron nail.

The internal energy associated with this state of affairs causes the heavily worked material to corrode more quickly than the less heavily worked material, that is, the heavily worked material is *anodic* to the remainder of the component.

15.5.1 Consider the cold-formed rim of a motor-car 'wing' (Fig. 15.15 (i)). The forming operation will have cold-worked the rim more heavily than the remainder of the wing. If the paintwork is defective in this relatively inaccessible position corrosion can be expected to occur fairly rapidly in the heavily worked and, hence, anodic region. Similarly the cold-forged head of an ordinary nail will be anodic to the remainder of the nail and will rust more quickly as a result (Fig. 15.15 (ii)).

It is obvious that heat treatment can be useful in limiting galvanic corrosion of this type. Simple stress-relief annealing processes will be sufficient to remove energy arising from locked-up stress and will thus reduce local corrosion. Such heat treatment is not always possible since cold-working is often a means of stiffening and strengthening a component as in the case of both of those mentioned above.

Intercrystalline corrosion associated with the presence of impurities at grain boundaries has already been mentioned (15.3.9). However, grain boundaries are also regions of higher energy level, even in very pure metals, possibly due to the irregular arrangement of ions there (Fig. 3.22). Consequently corrosion tends to occur more quickly at grain boundaries. Hard-drawn α-brasses are prone of a phenomenon known as 'season-cracking' in which corrosion follows the crystal boundaries until the component, unable any longer to sustain internal stresses, cracks extensively along these boundaries. The tendency towards this behaviour can be eliminated by a low-temperature anneal which, although it does not cause recrystallisation, results in the removal of locked-up stresses as atoms move small distances into positions nearer to equilibrium.

Electrochemical action involving variations in the composition of the electrolyte

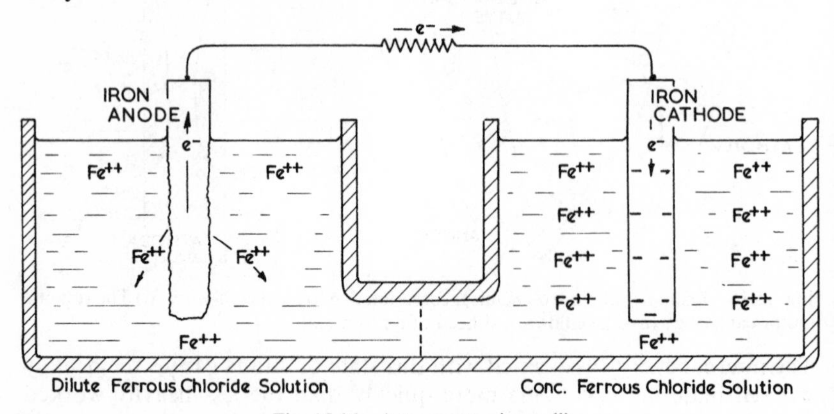

Fig. 15.16—A 'concentration cell'.

15.6 Fig. 15.16 illustrates a concentration cell in which both of the electrodes are of exactly the same material but in which the concentration of the electrolyte in contact with each electrode is different.

The general reaction in the cell is represented by:

$$Fe \rightleftharpoons Fe^{++} + 2e^{-}$$

and the direction in which this reaction proceeds is governed, as in many chemical reactions, by the concentration of the reactants. Thus on the left-hand side of the cell a low concentration of Fe^{++} ions will cause the reaction to proceed to the right:

$$Fe \rightarrow Fe^{++} + 2e^-$$

so that iron atoms will tend to go into solution as Fe^{++} ions and release electrons into the external circuit. On the right-hand side of the cell the already high concentration of Fe^{++} ions tend to make the reaction move in the other direction:

$$Fe \leftarrow Fe^{++} + 2e^-$$

Thus the left-hand electrode becomes the anode and the right-hand electrode the cathode of the galvanic system as indicated by the direction of electron movement in the external circuit.

15.6.1 Variations in the amount of oxygen dissolved in water will lead to corrosion of this type taking place in many different situations. In the

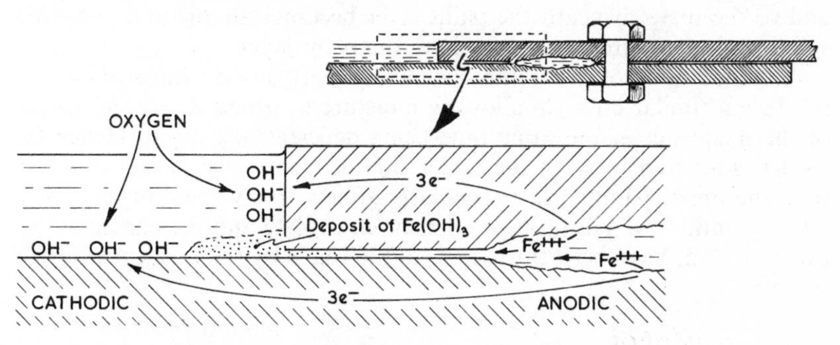

Fig. 15.17—The 'crevice corrosion' of steel plate.

case illustrated in Fig. 15.17 the two steel sheets are of the same composition and in the same condition but the composition of the electrolyte varies since dissolved oxygen is unlikely to penetrate into the further recesses of the crevice between the plates although water will have been drawn there by capillary action. Consequently the more exposed parts of the sheet become cathodic since electrons will travel to these areas and react with oxygen and water present to form hydroxyl ions:

$$O_2 + 2H_2O + 4e^- \rightarrow 4OH^-$$

Thus, *the region of metal exposed to electrolyte containing less dissolved oxygen will become anodic and go into solution.* This situation is helped by the fact that Fe^{+++} ions have greater mobility than OH^- ions and so rust forms freely outside the crevice. This phenomenon is often termed 'crevice corrosion' and is typified by the fact that corrosion will be more severe in the regions which one might have expected to be the better protected. The general axiom is that *the oxidation/concentration cell accelerates corrosion but it accelerates it where the concentration of oxygen is lower.*

15.6.2 This state of affairs explains why corrosion takes place with increasing rapidity when once the paintwork of a motor-car has been

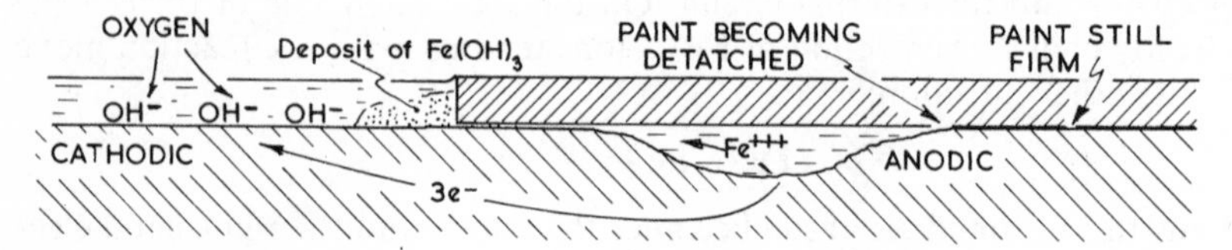

Fig. 15.18—The sub-paint rusting of a motor-car body—a further manifestation of crevice corrosion.

scratched—or was initially porous as is so often the case (Fig. 15.18). Water begins to penetrate beneath the paint layer but oxygen does not and so the metal beneath the paint layer becomes anodic and goes into solution thus progressively detaching the paint layer.

Accumulations of otherwise chemically inert dirt on the surface will produce a similar effect in allowing moisture to penetrate to the surface of the metal but at the same time being deficient in oxygen. Hence the reader is advised to wash the under side of his car even more carefully than the upper paintwork, particularly after a frosty spell during which large quantities of salt (a most efficient electrolyte and the car manufacturer's friend) have been thrown on to the roads by the Local Authority.

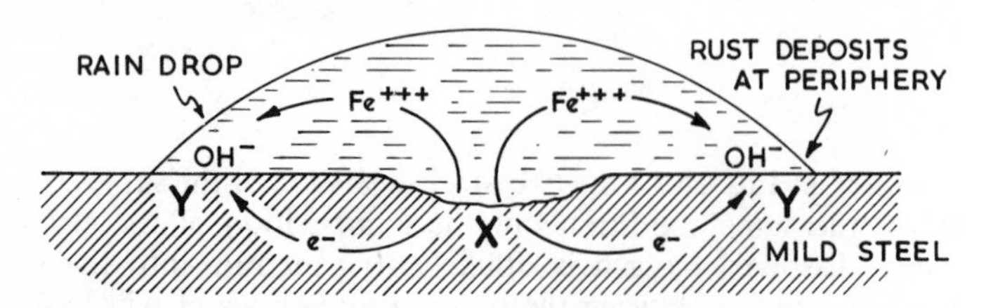

Fig. 15.19—The rusting of iron beneath a rain drop, due mainly to variations in the concentration of dissolved oxygen between centre and rim of the drop.

The fact that the salt does *not* penetrate crevices accelerates rather than diminishes the rate of corrosion since Cl^- ions will congregate along with OH^- ions outside the crevice (or dirt patch) and greatly increase the concentration gradient within the electrolyte. Cynics may claim that we get the rotten cars we deserve but the situation is certainly not helped by road salt.

Finally, we will consider the corrosive effect of a simple rain drop on the surface of mild steel (Fig. 15.19). Here the concentration of dissolved oxygen in water will obviously be higher at Y than at X where the depth of water is greater. A 'concentration cell' will therefore be set up so that at *X* iron atoms will go into solution as Fe^{+++} ions, releasing electrons which pass through the external circuit to react with oxygen and water at *Y*, producing OH^- ions:

$$O_2 + 2H_2O + 4e^- \rightarrow 4OH^-$$

The Fe^{+++} ions released will migrate towards the OH^- ions forming a deposit of $Fe(OH)_3$—'rust'—mainly at the periphery of the rain drop. But for the variation in concentration of oxygen in the rain drop the rusting of mild steel by its presence would be a much slower process since then electrolytic corrosion would depend entirely on the small variation in composition within the mild steel.

The limitation of corrosion

15.7 Corrosion is a phenomenon which is very expensive to eliminate completely. At best we can limit its action by one of the following methods:

15.7.1 *Choosing a material in which galvanic attack is unlikely.* Pure metals and completely uniform solid solutions offer the greatest freedom from galvanic corrosion assuming that steps are taken to prevent 'crevice corrosion' by ensuring that all such fissures are adequately sealed.

Either ferritic or austenitic stainless steels (10.5) offer good resistance to corrosion largely because of the protective film of chromium oxide which adheres to the surface. At the same time, uniformity of the solid solution structure almost eliminates the chance of galvanic corrosion even in the presence of very strong electrolytes. To attain this uniform solid solution structure it is generally necessary to quench stainless steel from 1050° C. This has no effect on the hardness of these very low-carbon steels but has the effect of retaining in solution carbon which would otherwise precipitate as carbides. Slow cooling will allow chromium carbide to precipitate at grain boundaries (Fig. 15.20) thus depleting the region immediately adjacent to the grain boundaries with respect to chromium. These low-chromium regions become anodic to the chromium-rich matrix so that, in the presence of a suitable electrolyte, inter-granular corrosion would occur.

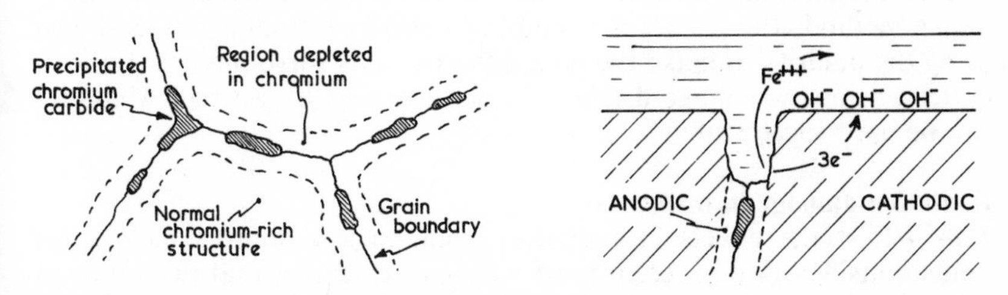

Fig. 15.20—Carbide precipitation and the resultant corrosion in stainless steel.

15.7.2 *The use of protective coatings.* The protection of a surface from galvanic corrosion by sealing off that surface from contact with a potential electrolyte has already been mentioned. Painting, enamelling, various methods of tinning and galvanising—or even coating with a non-reactive grease—all have this fundamental aim in the protection of steel but are seldom completely effective under commercial conditions. The anodising of aluminium in which the natural oxide film is artificially thickened is also a process of this type.

15.7.3 *The avoidance of potential 'crevices'.* Fissures are best sealed or filled in some way to prevent the entrance of thin films of electrolyte which, in the presence of external oxygen, may lead to this form of attack.

15.7.4 *The use of galvanic protection.* In some cases the close proximity of two dissimilar metals in mutual contact with an electrolyte is unavoidable. For example, ships' propellors are commonly of 'manganese bronze' (a high-tensile brass—Table 11.1) which is strongly cathodic towards the steel hull. In the presence of sea water which is a strong

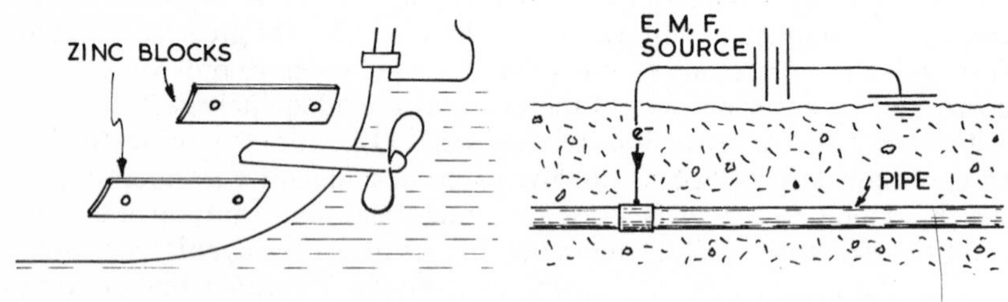

Fig. 15.21—The galvanic protection of a ship's hull.

Fig. 15.22—Protection of a buried steel pipe by using an impressed e.m.f.

electrolyte, the hull could be expected to rust quickly due to the close proximity of the manganese-bronze propellor. This tendency is largely eliminated by fixing some zinc slabs to the hull near to the propellor (Fig. 15.21). Since zinc is anodic both to iron and manganese bronze it will corrode sacrificially and can be replaced as required.

Buried iron pipes can be protected in a similar way by burying slabs of zinc or magnesium adjacent to the pipe at suitable intervals. An alternative method (Fig. 15.22) is to employ a small external e.m.f. as shown using the polarity to make the pipe cathodic to its surroundings. Either a battery or a low-voltage d.c. generator (operated by a small windmill) can be used successfully.

Radiation damage in materials

15.8 Whilst the effects of radiation from nuclear sources have come under consideration in recent years we must remember that the effects of other forms of radiation are also significant. Radiation can be classified into two main groups: (1) electromagnetic radiations which include radio 'waves'; light; X-rays and γ-rays. These are now commonly considered in terms of small 'energy packets' or *photons*. (2) 'radiations' which in fact consist of moving particles. They include fast-moving electrons (originally called β-rays or cathode rays); protons (once known as 'positive rays'); helium nuclei (the original α-rays); and neutrons.

Of these forms of radiation the effects of γ-rays and fast-moving neutrons are of possibly the greatest significance, particularly to the nuclear-power engineer, but in everyday life the effect of ultra-violet light on modern plastics materials cannot be completely neglected.

15.8.1 *Radiation damage in metals*. For too long the metallurgist visualised metallic crystals as being relatively perfect in form and under these circumstances many of the mechanical and physical properties of metals remained unexplained. Thus it was not until the 1940s that plasticity and malleability were explained in terms of the presence of dislocations, whilst diffusion in solid solutions was ultimately explained by postulating the existence of vacant sites. In a study of these aspects of solid-state physics lattice defects are of greater significance than the lattice patterns themselves.

Consider the irradiation of a metallic crystal by neutrons arising from nuclear fission (19.5). These may collide with the nuclei of atoms with a consequent transfer of kinetic energy. If the recoil energy is great enough the ion will be displaced (Fig. 15.23) from its lattice site and projected

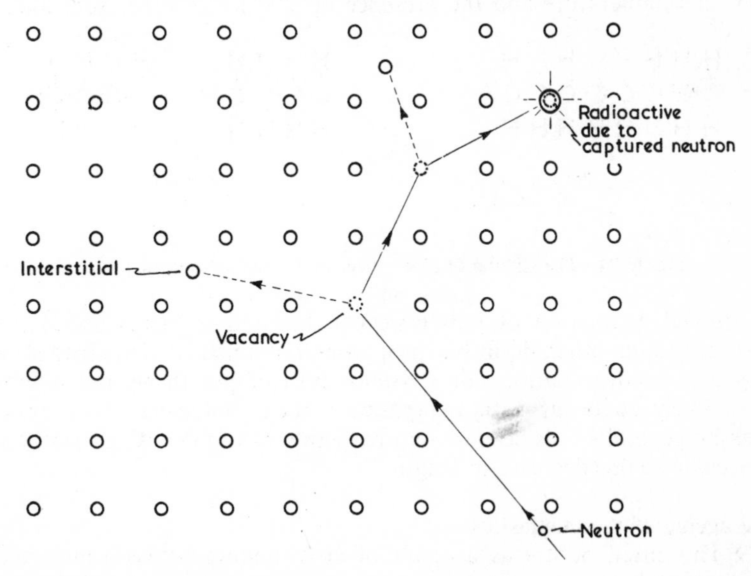

Fig. 15.23—The displacement of atoms—or, more correctly, 'ions'—by neutron bombardment.

through the crystal, possibly displacing other ions from their sites. The end point of this succession of ion collisions will be a number of vacant sites and an equal number of displaced ions, jammed into interstices in the lattice. Such defects are termed 'vacancies' and 'interstitials' respectively.

At sufficiently high temperatures these defects can migrate—as in an annealing process—and those of opposite type, i.e. vacancies and interstitials, will neutralise each other. Those of the same type may congregate and form something in the nature of a coherent precipitate. Under these circumstances the mechanical properties of the metal will be altered. Yield stress usually increases with an accompanying reduction in ductility, the results being comparable with those of work hardening.

However, one must also consider the ultimate destiny of the moving neutron which started these movements within the crystal. As its *KE* is diminished by collisions it may ultimately enter a nucleus and remain there forming an isotope atom of the element being bombarded. This isotope may prove to be radioactive and so emit dangerous γ-rays, for which reason neutron bombardment is rather limited as a method of strengthening metals.

15.8.2 *Radiation damage in polymers.* Irradiation of polymer materials with either neutrons or γ-rays may lead to cross-linking, degradation, chain branching or polymerisation. Which of these processes occurs depends upon the nature of the irradiated polymer, the amount of radiation, the temperature and the presence or absence of other substances.

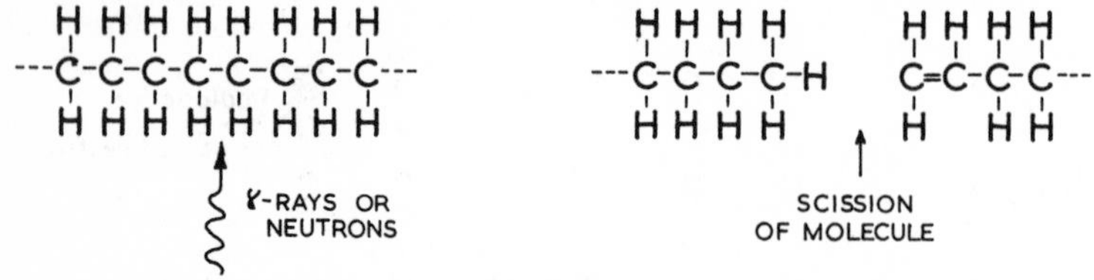

Fig. 15.24—The scission of a polythene molecule due to irradiation.

Controlled irradiation of polythene can be used to effect useful side branching and other desirable properties but uncontrolled irradiation may lead to degradation—or *scission*—even of polythene and is even more likely to occur with many other linear polymers. As a result strength generally falls because of a reduction in van der Waals forces as molecular chains decrease in length.

The ageing of thermoplastics

15.9 This often occurs as a result of cross-linking between molecules which contain unsaturated bonds. For example polythene will become progressively more brittle due to the formation of covalent bonds by oxygen atoms which link together adjacent molecules in this way (Fig. 15.25). Since the covalent cross-link prevents slip between the chain molecules polythene becomes progressively harder and more brittle.

This type of cross-linking with oxygen is likely to occur more readily in

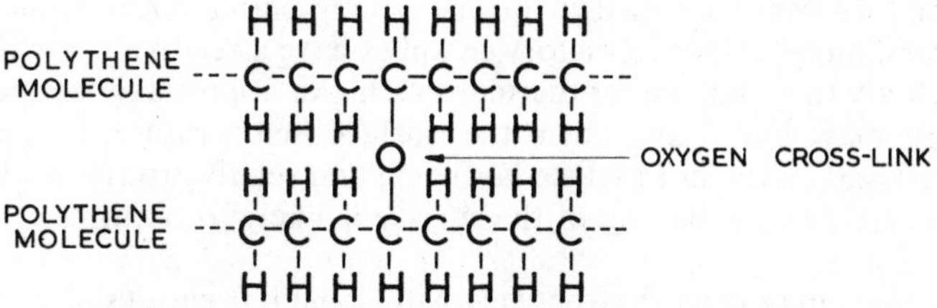

POLYTHENE MOLECULE
OXYGEN CROSS-LINK
POLYTHENE MOLECULE

Fig. 15.25—Cross-linking between polythene molecules.

the presence of ultra-violet light (strong sunlight) which catalyses the reaction. It can therefore be limited by making the material opaque by adding some carbon black but a more appropriate method is to include an anti-oxidant, usually 0·1 to 0·2% of phenol or an amine.

Poor quality rubber 'perishes' in a similar way due to the formation of oxygen cross-links, particularly at those unsaturated bonds not already used during the vulcanisation process. Ultimately it becomes brittle as degradation generally follows the fall in elasticity produced by cross-

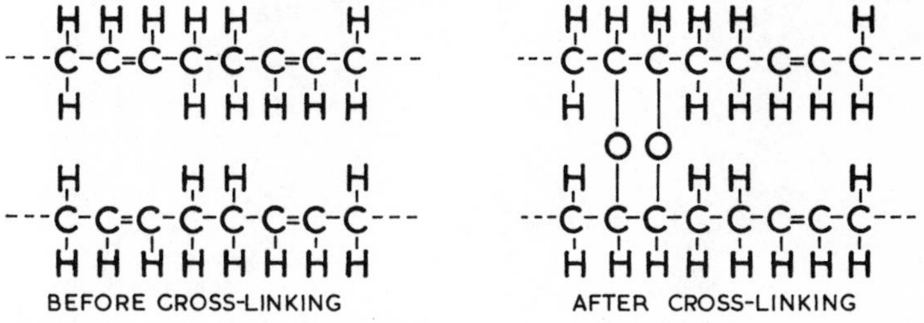

BEFORE CROSS-LINKING
AFTER CROSS-LINKING

Fig. 15.26—The perishing of polybutadiene rubber.

linking. Again intensity of light and also the presence of ozone will accelerate perishing. The reader may have noticed that continental car owners often cover their tyres to protect them from strong sunlight. Again oxidation can be supressed by adding carbon black as well as small amounts of aromatic amines or phenol derivatives as antioxidants.

Deterioration of ceramic materials

15.10 Most ceramic materials are chemically stable when in contact with ordinary atmospheres, though some natural stones used for building suffer attack by industrial atmospheres containing sulphur dioxide. The latter combines with atmospheric moisture to form sulphurous acid, which, condensing on masonry, will ultimately oxidise to sulphuric acid and attack many types of ceramic building material.

The weathering action of frost is also destructive of many building ceramics. Surface pores which absorb moisture will become disrupted when the moisture freezes and, as a result, expands.

Refractory ceramics used in furnace lining suffer both physical and chemical damage. Since refractory ceramics have a relatively low thermal conductivity they may suffer thermal shock. An appreciable temperature gradient may build up when the surface temperature is suddenly increased and as the hot surface layers expand relative to the cold layers beneath, *spalling*, or flaking of the surface is likely to occur.

15.10.1 Slags may react destructively with some refractories at high temperatures. Sometimes a slag will have a simple solvent action on the refractory but usually such reactions are of an acid/basic nature. Thus silica refractories—which are acid—will be attacked by basic slags containing lime and other metallic oxides.

Since the product of the reaction often has a low melting point fluxing of the furnace lining takes place. The general principle in choosing a refractory material which is to work in contact with a slag is that it should have a similar chemical nature to that of the slag. Thus a basic slag will demand a basic lining of dolomite, magnesite or zircona, whilst an acid slag will require a furnace lining of silica brick or some silica derivative.

Chapter Sixteen
The Failure of Materials in Stress

16.1 Engineering components must in most cases be designed in such a manner that any stresses involved during service are insufficient to give rise to plastic deformation. Whilst in tensile or compressive tests the specimen is subjected only to uni-axial stress, that is, stress in one direction, in practice a multi-axial stress system must be allowed for. Moreover, it is not realistic to use the simple yield stress as the criterion in engineering design and practical engineers have long been in the habit of using a suitable *factor of safety*, *f*, in order to arrive at a maximum allowable stress. Generally this maximum allowable stress is given by Yield Stress/*f*, though in some cases the value, Ultimate Stress/*f*, is used.

Be that as it may, many components which have been designed such that the stresses arising from applied loads are insufficient to lead to plastic deformation, nevertheless fail in service. There are many causes for such failure. These are now generally recognised by designers and are considered, as far as is possible, in subsequent design. Amongst the more important potential causes of failure which can be examined quantitatively are brittleness, creep, fatigue and the chemical influence of the environment in which the component is operating. The first three of these will be considered briefly in this chapter whilst the subject of corrosion was discussed in Chapter Fifteen.

As will be apparent from a study of Chapter Four, the results associated with such different tests as the simple tensile test and the impact test are not comparable. Thus strength must not be confused with toughness since the two values are not interrelated. The Izod impact test assesses different properties from those examined in the tensile test which is in any case a relatively 'long-time' method of loading when compared with the impact test. Even longer periods of static loading, particularly at high temperatures, may involve creep (or gradual extension) which leads to failure; whilst the presence of stresses which are continually varying in magnitude or direction (though well within the yield stress) may cause failure by fatigue.

Fracture

16.2 The application of stress to any material will lead to the production of elastic and/or plastic strain and if the stress is increased progressively fracture will ultimately occur. This fracture can be classified as either 'brittle' or 'ductile' and the mode of fracture produced is governed by the stress at which it occurs in relation to the elastic/plastic properties of the material (Fig. 16.1).

In the case of brittle materials failure occurs *before* any appreciable

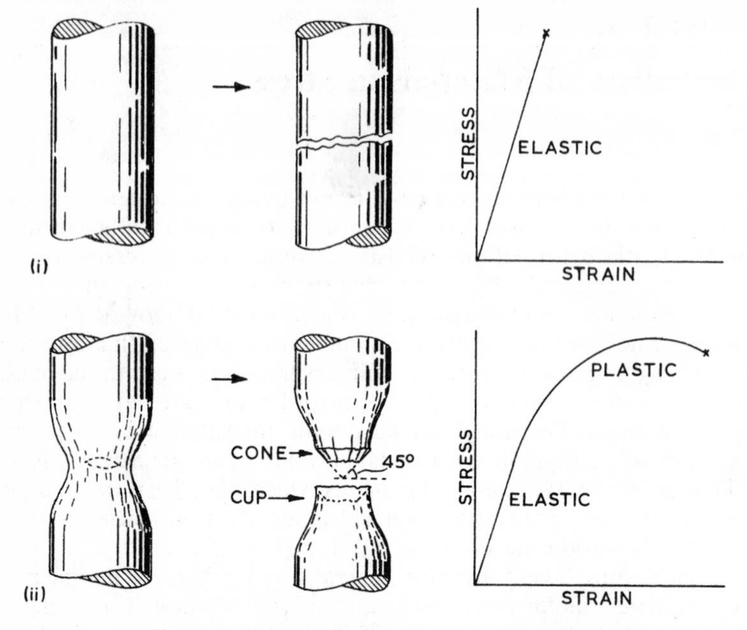

Fig. 16.1—Types of fracture: (i) brittle; (ii) ductile.

plastic deformation can take place by shear. This is typified in materials like cast iron, glass and concrete. Fracture of this type—also termed 'cleavage fracture'—follows paths between adjacent crystallographic planes which are generally those with the weakest atomic bonding. In addition to the materials mentioned, cleavage fracture is prevalent in all metals with BCC and CPH structures rather than in those which are FCC. Brittle fractures generally appear bright and 'granular' because of the reflection of light from the sections of individual crystals.

Ductile fracture takes place at some stress *above* the shear strength of the material so that some plastic flow precedes fracture. As the neck begins to form (Fig. 16.1 (ii)) the stress within the necked region obviously increases and reaches a level where small cavities begin to propagate in the central region of the section. Such cavities will nucleate more easily if a second phase or particles of some impurity are present for which reason very pure metals are more ductile than impure ones. The cavities which have formed join up to form a crack which is roughly normal to the axis of stress of the test piece. Final fracture then occurs by simple shear at an angle of 45° to the axis of stress, shear stress being at a maximum in this direction. The resultant fracture is of the well-known cup-and-cone variety and occurs particularly in FCC metals.

The type of fracture produced is dependent mainly on the nature of the material and in particular on its lattice structure. It is also affected by other factors such as the condition of the material in respect of the

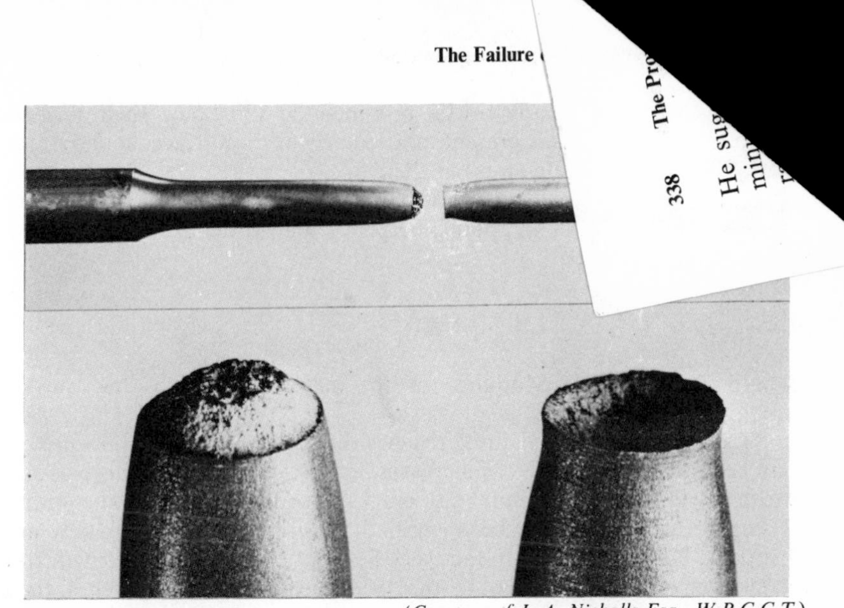

(*Courtesy of J. A. Nicholls Esq., W.B.C.C.T.*)

Plate 16.1—A typical 'cup-and-cone' fracture in a ductile carbon steel, obtained during a normal tensile test. (The specimen was a 0·1% carbon steel, normalised at 900° C.)

amount of cold-work it may have received; the quality of its surface finish; its environment and temperature; and the rate of application of stress. The latter factor is illustrated by the behaviour of a material in a simple tensile test as compared with its behaviour during the Izod test where impact loading allows insufficient time for the adequate movement of dislocations to take place, so that brittle failure tends to result under some circumstances.

16.2.1 *Griffith's Crack Theory*. The fact that the real strength of a metal is but a very small fraction of the theoretical value (derived from a knowledge of inter-atomic forces) has already been noted (5.2.3). A similar state of affairs exists for non-metallic materials and one of the earliest successful attempts to explain this was made by A. A. Griffiths in 1920.

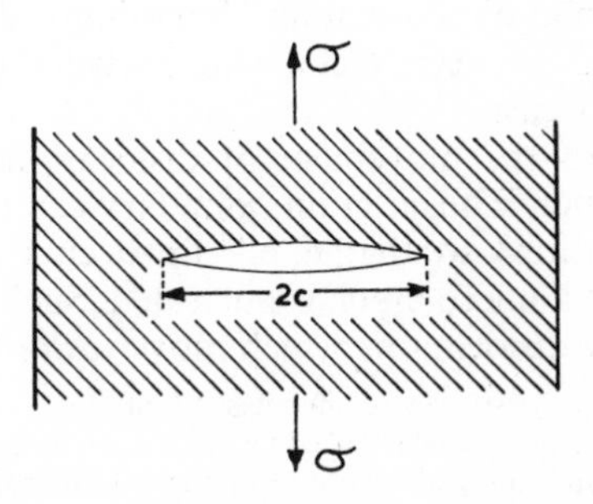

Fig. 16.2.

gested that in any brittle non-metallic substance such as glass ute cracks or fissures present, particularly at the surface, act as stress-aisers by concentrating stress at their tips.

In the case of a small crack of elliptical section (Fig. 16.2) it can be shown that for a stress, σ, applied perpendicular to the major axis (length $2c$) of the crack, then:

$$\sigma = \sqrt{\frac{2\gamma E}{\pi c}}$$

where E is Young's Modulus for the material and γ is the 'surface energy'* per unit area.

As soon as the applied stress reaches the value, σ, small cracks present can begin to propagate. As c increases, σ must decrease and this can only be achieved by the rapid spread of cracks to produce catastrophic failure.

Griffith's Theory has been verified experimentally on glasses and brittle ceramic materials. In fact very fine glass fibres freshly drawn from a melt have strengths near to the theoretical value. However if these fibres come into contact with any object, or even with the atmosphere for short periods, these strengths are considerably reduced. This suggests that the strength of the fibre is very dependent on surface perfection and that anything which may initiate even the most minute surface irregularities will weaken it.

The Theory cannot be applied directly to ductile materials such as metals. Provided, however, that allowance is made for the effects due to plastic deformation some degree of application is possible.

16.2.2 *Factors leading to crack formation.* It is reasonable to suppose that crack initiation in metals occurs most frequently in the vicinity of an included particle. When slip in a metal takes place past a rigid inclusion dislocations will pile up near to the inclusion/metal interface. Assuming that the inclusion itself does not shear, a minute fissure will begin to develop at this interface (Fig. 16.3). Obviously fissures can develop most easily when there is little adhesion between the inclusion and the metal containing it. Thus fissures can form more easily at a copper/cuprous oxide interface than at an aluminium/aluminium oxide interface since the degree of adhesion between metal and particles of its oxide is much greater in the latter case.

Other barriers to the movement of dislocations can initiate micro-cracks in this manner. Thus a grain boundary can act as a barrier to the passage of dislocations so that a pile-up occurs and they become so closely packed that a micro-crack is nucleated. Such a crack grows by the addition of further dislocations which move towards it.

* That a free surface possesses energy manifests itself in surface tension. Thus a globule of mercury tends to become spherical since in a sphere maximum volume is contained in a minimum surface area and so surface energy is minimised. In producing a crack as outlined, new free surfaces must be generated and energy must be supplied to achieve this.

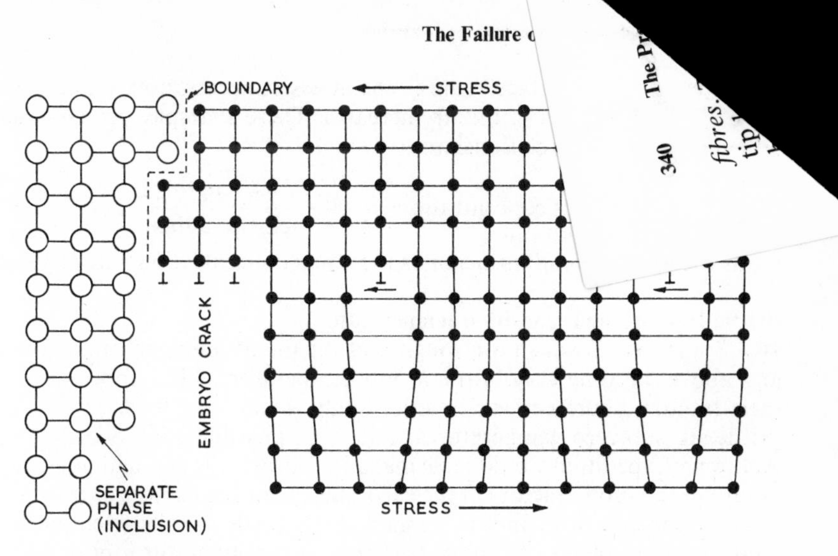

Fig. 16.3—A 'pile-up' of dislocations at an inclusion, leading to the formation of a fissure which will be propagated as the stress increases.

Crack formation also takes place by the movement of dislocations along close-packed or other slip planes within a crystal (Fig. 16.4 (i)). These dislocations pile up at the intersection of the slip planes and so initiate a micro-crack (Fig. 16.4 (ii)).

In some cases the presence of a discontinuity can arrest the propagation of a crack by reducing the stress concentration which is always present at the tip of a crack. Thus the propagation of a crack in a plate-glass window can be stopped by carefully drilling a hole in front of the

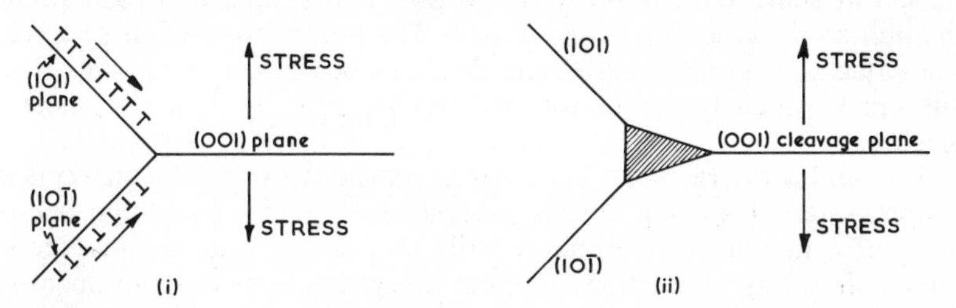

Fig. 16.4—The initiation of a micro crack by the running together of dislocations (after Cottrell).

advancing crack. Similarly brittle fracture in welded-steel ships (16.3) was arrested by including dummy rivet holes to arrest the spread of cracks. Some fibrous materials—notably the now-obsolete wrought iron—are tough along the direction of the fibres because any crack formed at right angles to the stress will then have to change direction on meeting a fibre barrier. If the interface between fibre and metal is weak this can actually *increase* the overall strength of the composite *in the direction of the*

The weak interface reduces the stress concentration at the crack because the radius of the tip increases. These examples are governed by the approximate relationship:

$$\text{Stress concentration} = \sqrt{\frac{\text{Crack length}}{\text{Crack-tip radius}}}$$

i.e. as the crack-tip radius increases the stress concentration will decrease.

Brittle fracture and transition temperature

16.3 Some metals which are tough and ductile at ambient temperatures and above, become very brittle at low temperatures. The appearance of their fractured surfaces varies accordingly. Mild steel tends to become brittle at sub-zero temperatures and it is possible that Scott's 1912 Antarctic Expedition would have met a less disastrous end had not brittle failure in the snow tractors in the early stages put the burden of transport solely on horses, dogs and, later, men alone. Little serious attention was paid to the problems of brittle fracture at low temperatures prior to the catastrophic failure of some of the all-welded 'Liberty' ships during the Second World War.

Under normal circumstances the stress required to cause cleavage is higher than that necessary to produce slip, but if for some reason slip is suppressed, brittle fracture will occur when internal stresses increase to the value necessary to produce failure. Tri-axial stresses acting in the material may be residual from some previous treatments and the presence of points of stress concentration may aggravate the situation. The 'Liberty' ships, mentioned above, were fabricated by welding together steel plates to form a continuous hull. Cracks often, though not always, started at sharp corners or arc-weld spots and propagated right round the hull so that the ship broke in half. The propagation of these cracks was often very rapid. Such failure does not occur in riveted ships since any crack which begins to run will terminate in the first rivet hole it reaches.

Though brittle fracture often starts at some obvious fault which acts as a stress-raiser this is not always the case and it seems likely that one or more of a number of possible contributory causes lead to this type of failure. It seems likely, however, that sluggishness in the movement of dislocations at low temperatures is common to all cases of brittle failure. Plastic flow depends upon the movement of dislocations and this requires some finite time. If stress increases very rapidly it is possible that it cannot be relieved adequately by slip and a momentary increase in stress to a value above the shear strength can cause failure.

As temperature decreases the movement of dislocations becomes more difficult so that internal stresses may exceed the shear strength at some instant. Brittle fracture is therefore more common at low temperatures, a fact supported by the failure of the 'Liberty' ships during service in the cold North Atlantic.

16.3.1 Metals with FCC structures are generally more ductile than those with either BCC or CPH structures because of the greater number of available slip directions associated with the FCC structure (5.2.1). At low temperatures this becomes even more apparent so that BCC metals like tungsten, molybdenum, chromium and iron all exhibit low-temperature brittleness, as do CPH metals like zinc and beryllium. BCC ferrite, the principal constituent of mild steel, is particularly susceptible to brittle fracture which follows a transcrystalline path along the (100) planes. As indicated in Fig. 16.5, this occurs at a fairly low temperature and the temperature at which ductile fracture is replaced by brittle fracture is termed the *transition temperature.* This transition temperature can be more precisely defined as the temperature at which 50% of the failed surface exhibits a ductile fracture with the remaining 50% showing brittle failure.

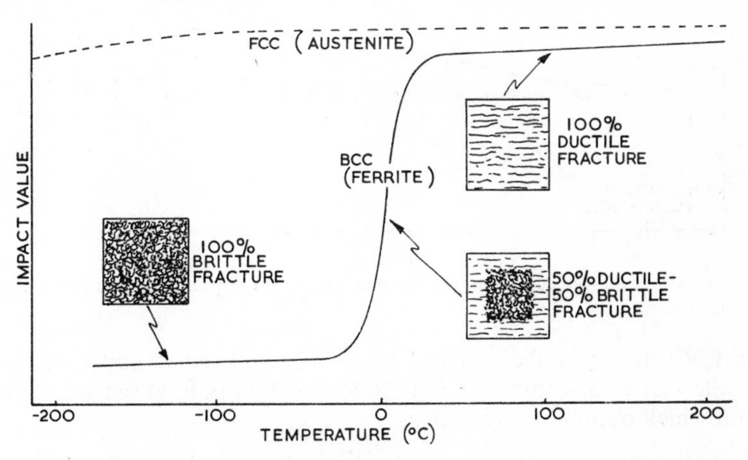

Fig. 16.5—The relationship between brittle fracture and temperature for ferritic and austenitic steels.

Some elements, notably phosphorus, carbon and nitrogen raise the transition temperature in mild steel for which reason plates used in the building of ships' hulls must be low in phosphorus. Other elements, but in particular manganese and nickel, depress the transition temperature of mild steel. Thus the least expensive way of reducing its transition temperature is to increase the manganese/carbon ratio. A suitable steel, in which the transition temperature has been reduced to a safe low limit, contains 0·14% C and 1·3% Mn. When lower temperatures are involved a low-nickel steel must be used.

Creep

16.4 Creep can be defined as the continuing deformation, with the passage of time, in materials subjected to constant stress. This deformation is plastic and occurs even though the acting stress is *below the yield stress* of the material. At temperatures below 0·4*T* (where *T* is the melting point of the material on the Kelvin scale) the rate of creep is very small but at temperatures higher than this it becomes increasingly important. For this reason creep is commonly regarded as being a high temperature phenomenon associated with steam plant and gas-turbine technology. Nevertheless with some of the soft, low-melting point metals and alloys creep will occur significantly at ambient temperatures. It has been reported that some ancient lead roofs are measurably thicker at the eaves than at the apex, the lead having crept over the centuries even under its own weight.

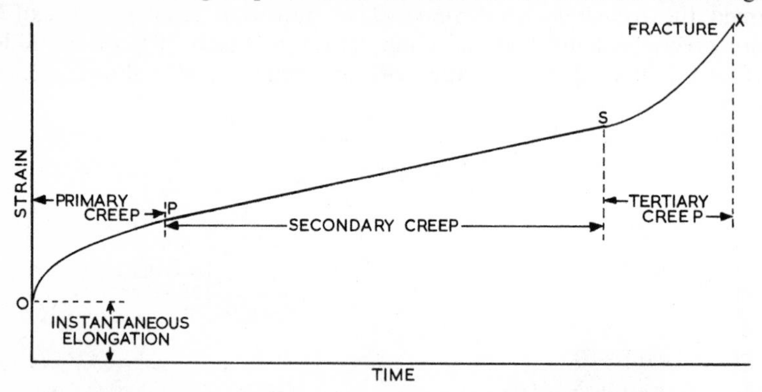

Fig. 16.6—Typical creep curve showing the three stages of creep during a long-time, high-temperature creep test.

16.4.1 When a metallic material is suitably stressed it undergoes immediate elastic deformation (Fig. 16.6) and this is followed by plastic strain which occurs in three stages:

(*i*) Primary, or transient, creep, OP, beginning at a fairly rapid rate which then decreases with time as strain-hardening sets in.

(*ii*) Secondary, or steady-state, creep, PS, in which the rate of strain is fairly uniform and at its lowest value.

(*iii*) Tertiary creep, SX, in which the rate of strain increases rapidly so that fracture occurs at X. This stage coincides with necking of the member.

Creep in polymer materials below the glass transition temperature follows roughly the same pattern as it does in metals (12.5.6).

The relationship which exists between stress, temperature and the resultant rate of creep is shown in Fig. 16.7. At a low stress and/or a low temperature some primary creep may occur but this falls to a negligible value in the secondary stage due presumably to strain hardening of the material. With increased stresses and/or temperatures (curves *B* and *C*) the rate of secondary creep also increases leading into tertiary creep and inevitable failure.

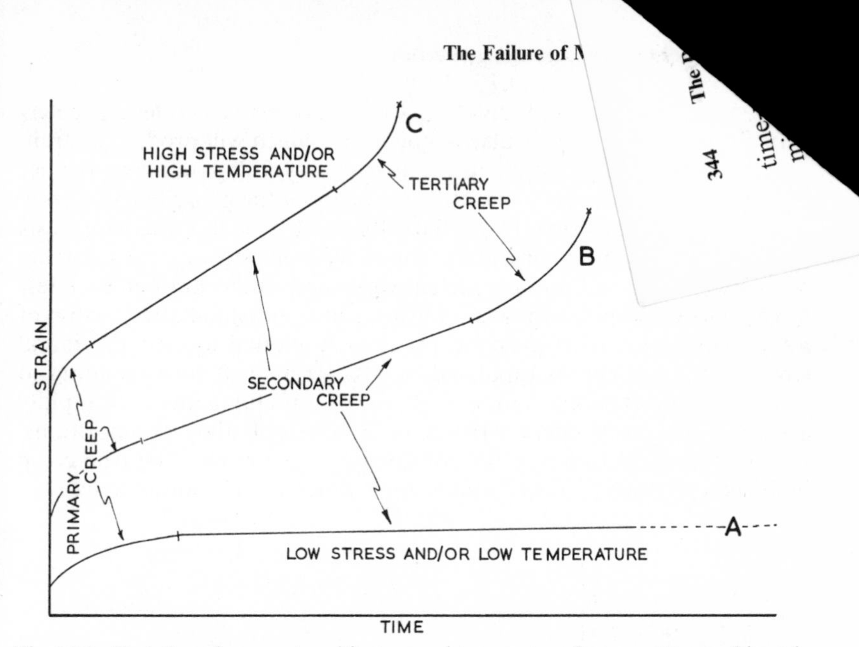

Fig. 16.7—Variation of creep rates with stress and temperature. In curve *A* creep ultimately becomes negligible, due presumably to work hardening. In curve *C* the secondary creep rate is higher than in curve *B* because of the use of a higher stress and/or a high temperature.

16.4.2 As will have been gathered from the above it is now generally accepted that formal short-time tensile tests do not give reliable information for the design of structures which must carry static loads over long periods of time at elevated temperatures. Strength data determined from long-time creep tests are therefore desirable. In creep testing a test piece, in form similar to a tensile-test piece, is enclosed in a thermostatically controlled electric tube furnace which can be maintained accurately over long periods at the required temperature. A suitable tensile force is applied and the extension of the test piece determined at suitable time intervals by means of some form of sensitive extensometer. Generally a set of identical test pieces will be tested at the same temperature but at different stresses and from a study of the results it will be possible to arrive at a maximum stress which can be applied and which will result in no *measurable* creep taking place. This stress is known as the *limiting creep stress* or *creep limit.* Obviously this is not a very satisfactory measurement since it is dependent largely upon the sensitivity of the extensometer equipment. Moreover, formal creep tests can occupy long periods of time running into weeks or months depending upon the combination of stress and temperature involved.

For these reasons it is often more satisfactory to use a testing method which involves measurement of the stress which gives rise to some arbitrary measurable creep rate, that is, strain/time during the secondary stage of uniform deformation. This creep rate will be equivalent to the slope of the curve (Fig. 16.6) during the secondary stage of creep. The Hatfield yield value is derived from one such short-term creep test. It determines the stress, at a particular temperature, which will produce a strain of 0·5% of the gauge length in the first twenty-four hours and further strain of not more than 10^{-6} metres per metre of the gauge length during the next forty-eight hours. In the Barr–Bardgett creep test a series of tests is taken at the same temperature but at different stresses, an assembly being used which will produce virtually the same initial strain in each test piece. Due to creep the stress diminishes and is measured after a lapse of forty-eight hours. This decrease in stress is plotted against the initial stress (Fig. 16.8) and extrapolated to indicate a stress corresponding to zero creep in forty-eight hours. This affords a useful method for rapidly assessing the creep characteristics of a series of alloy compositions. Although tests of this type do not attempt to determine the true creep limit they produce practical values upon which design can be based.

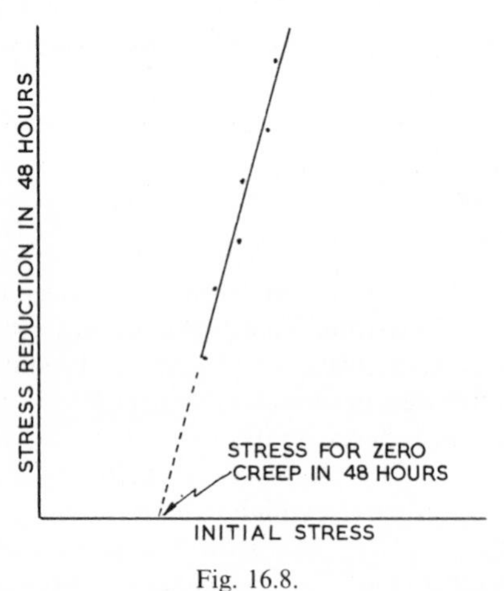

Fig. 16.8.

16.4.3 *The mechanism of creep*. Study of creep phenomena suggests that creep involves two separate types of plastic deformation:

(*i*) that due to the normal movement of dislocations and which occurs within crystalline materials;
(*ii*) that which is of a viscous character and is associated with the non-crystalline grain-boundary regions.

In the primary stage of creep dislocation movements are initially rapid and various barriers such as solute atoms are easily overcome. New barriers are introduced as dislocation pile-ups occur at grain boundaries and elsewhere but thermal activation enables dislocations to surmount such barriers, though at a decreasing rate, so that the creep rate is reduced. Thus, as deformation continues, so does work-hardening, but since the temperature is high, recovery processes are also actively assisted by dislocation climb (5.8) and cross-slip which is the motion of a dislocation from one slip plane into another intersecting plane. Both of these dislocation motions will oppose work hardening. Hence creep is essentially a process in which work-hardening is balanced by thermal softening. At low temperatures (curve *A* in Fig. 16.7) recovery does not take place and so unrelieved work-hardening leads to a reduction of the creep rate almost to zero.

During the secondary creep stage the fact that work-hardening and recovery apparently balance each other gives rise to a constant rate of deformation by creep. It is generally believed that the climb of edge dislocations is the process which controls the rate of creep in this stage since it requires the higher activation energy.

In addition to deformation by dislocation movement, deformation by viscous flow in the non-crystalline grain boundary region also occurs during this secondary stage. These movements in the boundary material lead to the formation of vacant sites which, in turn, will assist the process of dislocation climb. The relationship between grain boundaries and creep is verified by the fact that at high temperatures fine-grained materials creep more than coarse-grained materials of the same composition, presumably because fine-grained materials contain more grain-boundary region per unit volume. At low temperatures, however, when grain-boundary 'viscosity' is very high fine-grained materials are more creep resistant.

The tertiary stage of creep coincides with the initiation of micro-cracks at the grain boundaries. Some of these may form as the result of migration of vacancies to points of high stress concentration. The formation of these cavities leads to necking and the consequent rapid failure of the member.

16.4.4 *Creep resistance.* Increased knowledge of the mechanism of creep has enabled the materials technologist to develop creep-resistant materials with more confidence than was possible a few decades ago. Since creep depends upon the movement of dislocations it is obvious that any feature which reduces the movement of these dislocations, and also limits the formation of new ones, will effectively oppose creep. Generally those metals with close-packed crystal structures (FCC and CPH) are most suitable and their creep resistance may be increased by one or more of the following methods:

(*i*) the addition of an alloying element which will form a solid solution with the parent metal. This will only be really effective if the solute atoms have a *low mobility*. If on the other hand they diffuse freely with thermal activation they will also allow dislocations to travel so that recovery—and hence further creep—can proceed.

(*ii*) the addition of an alloying element which will give rise to dispersion hardening. Coherent and small non-coherent precipitates are generally produced by precipitation treatment and it is essential that at the service temperature such particles should remain finely dispersed and not coagulate. Finely dispersed precipitates form effective barriers to the movement of dislocations, as in the Nimonic series of alloys (11.5.2).

(*iii*) treatment of the alloy to ensure large grain when this is possible, since this reduces the total amount of grain boundary per unit volume of material and so reduces the formation of vacancies which in turn would otherwise assist dislocation movement.

Fatigue

16.5 Practical engineers have long been aware of the fact that 'live' loads and alternating stresses of quite small magnitudes can cause failure in a member which could carry a considerable 'dead' load. Under the action of changing stresses a material may become *fatigued*. Thus, whilst creep is a phenomenon associated with extension under a steady force acting over a long period, usually at high temperatures, fatigue refers to the failure of a material under the action of repeated or fluctuating stresses.

Early investigation into fatigue was carried out in 1861 by Sir William Fairbairn. He found that a wrought-iron girder which could safely sustain a static mass of 12 tonnes, nevertheless broke if a mass of only 3 tonnes was raised and lowered on to it some 3×10^6 times. From further investigation he concluded that there was some mass below 3 tonnes which could be raised and lowered an infinite number of times on to the girder without causing failure. Some years later the German engineer Wöhler did further work in this direction and developed the useful fatigue-testing machine which bears his name.

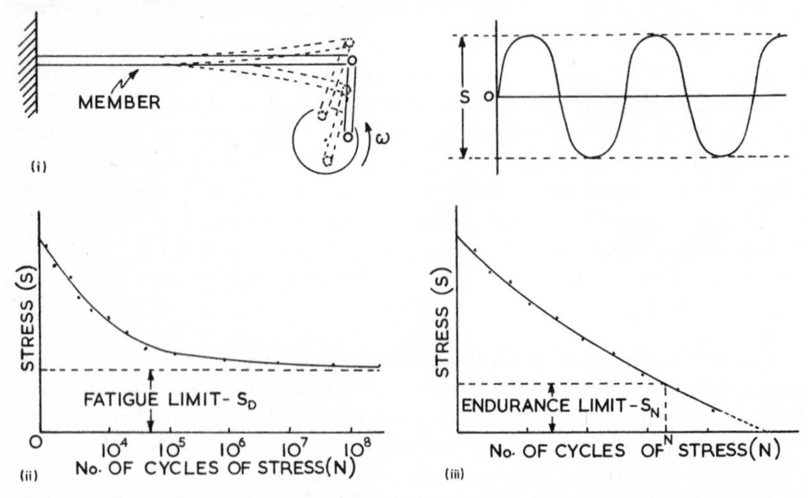

Fig. 16.9—(i) Represents a member suffering, in this case, alternating stress of range S and mean value zero. (ii) Shows a typical S–N curve for steel. (iii) Shows an S–N curve typical of many non-ferrous alloys and also some steels operating under corrosive conditions.

16.5.1 The yield strength of a material is a measure of the *static* stress it can resist without permanent deformation and is applicable only to structures which operate under static loading. Materials subjected to *fluctuating* or repeated forces fail at *lower* stresses than do similar materials under the action of 'dead' or steady loads. Rotating shafts and aircraft wings are obvious examples of members coming under the action of such alternating or fluctuating stresses. Less obvious examples are structures subjected to alternating stresses as a result of vibrations set up within them. Some aircraft disasters have been a result of vibrational stresses transmitted unexpectedly to parts of wings or fuselage so that these alternating stresses, *though well below the normal yield point,* have led to fracture after a relatively short period of service.

16.5.2 A typical relationship between the number of stress cycles (N) and the stress range (S) for a steel is shown in the S–N curve (Fig. 16.9 (ii)). Here it is seen that if the stress on the material is reduced then it will endure a greater number of stress cycles and a point is reached where the curve becomes virtually horizontal indicating that for the corresponding stress the member will endure an infinite number of cycles. The stress endured under these conditions is called the *fatigue limit,* S_D. For steels the fatigue limit is generally about one-half of the value of the tensile strength as measured in a static test.

Many non-ferrous metals and alloys, and also some steels operating under corrosive conditions, give S–N curves of the type shown in Fig. 16.9 (iii). Here there is no fatigue limit as such and any member will fail if it is subjected to the appropriate number of stress reversals even at extremely small stresses. In the case of these materials which show no fatigue limit an *endurance limit,* S_N, is quoted instead. This is defined as the maximum stress which can be sustained for N cycles of stress. Components in materials of this type must of course be designed with some specific life in mind and 'junked', as Americans so descriptively put it, before the number of cycles N (for stress S_N) has been attained.

16.5.3 Non-metallic materials are also liable to failure by fatigue. As in the case of metals the number of cycles of stress required to produce

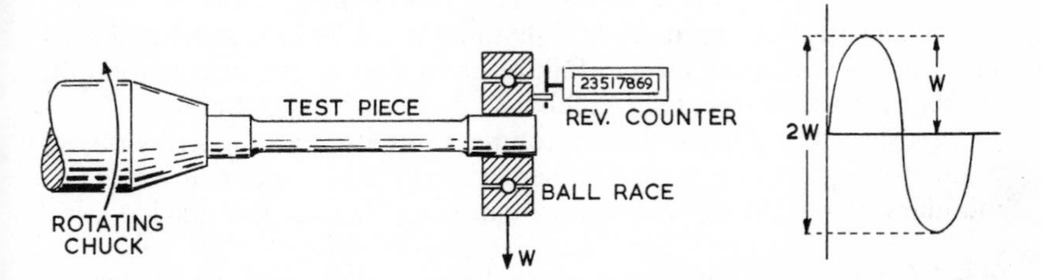

Fig. 16.10—The principle of the Wöhler fatigue-testing machine. In this case the stress range $S = 2W$.

fracture increases as the maximum stress in the loading cycle decreases but there is generally no fatigue limit and some form of endurance limit must be applied. Difficulties arise in the actual practical methods available for testing materials like polymers. Since they have low thermal conductivities, energy is not dissipated quickly enough during testing and an increase in temperature of the polymer test piece is inevitable. The number of cycles N required to produce failure will then decrease.

16.5.4 *Some causes of fatigue failure.* The fatigue characteristics of most materials are now known or can be accurately measured. It would therefore be reasonable to suppose that fatigue failure due to a lack of suitable allowances in design should not occur. This is indeed true, yet fatigue failure *does* occur from time to time even in the most sophisticated pieces of engineering equipment and such failure must be ascribed to the action of unforeseen factors either in design; to the prevailing environment; or, of course, to the quality of the material used. For example, the author was once called upon to investigate the failure by fracture of a stout copper tube which had carried, cantilever fashion, a small pressure gauge of but a few grams in mass. After only a few days' service the tube had broken off without warning. The fracture was of the fatigue type and when the investigator visited the scene of failure the reason for this was immediately apparent. The gauge had been attached by means of the copper tube to an air compressor the vibrations from which could be experienced at a distance of several hundred metres. Since copper is represented by an S–N curve of the type shown in Fig. 16.9 (iii) it is obvious that extremely small stresses will cause failure when the number of stress cycles reaches N. The disasters to the early Comet airliners followed a similar pattern in that fluctuations in cabin pressures in this case caused fatigue failure. Many causes of fatigue failure are due to unforeseen and sometimes undetected vibrations giving rise to reversals of stress at values *above* S_D (or S_N). Such vibrations are often sympathetic.

Another common cause of fatigue failure is the presence of some fault which will act as a stress-raiser thus increasing the local stress to a value above S_D. A keyway in a shaft; a sharp radius; or the results of poor workmanship in the shape of tool marks on a surface, can cause such stress concentrations to arise. Minute quench cracks in heat-treated steel shafts are a source of fatigue failure as of course are microstructural defects such as micro-cracks, inclusions and other discontinuities.

Corrosive conditions in the environment can also provide stress-raisers in the form of etch-pits and other surface intrusions such as grain-boundary attack. Such stress-raisers can nucleate a fatigue crack.

16.5.5 Fatigue failure will of course take place if the maximum stress is above the fatigue limit. Nevertheless this stress is still well below the normal static yield stress for the material yet it has been found that some *plastic* deformation by slip does take place during continued cyclic stressing. Such slip bands (5.2) as do appear on the surface are either *intrusive* or *extrusive* (Fig. 16.11). Although such an intrusion is generally very small, being of the order of 1 μm, it can of course act as a stress-raiser and initiate a fatigue crack.

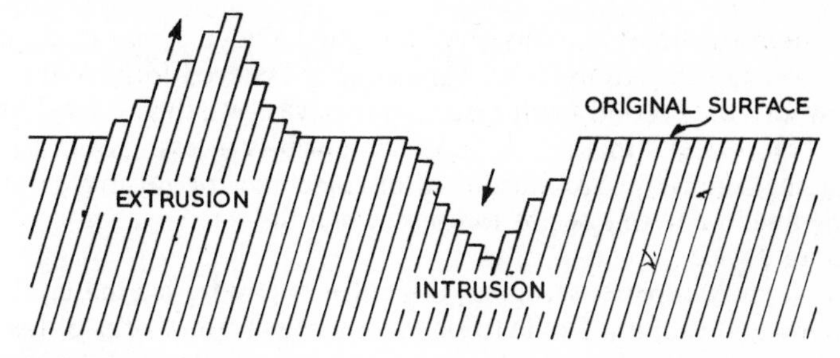

Fig. 16.11—Local slip giving rise to extrusions and intrusions which can initiate fatigue cracks.

A fatigue fracture is generally considered to develop in three stages—nucleation, crack growth and ultimate failure (Fig. 16.12). The resultant fractured surface has a characteristic appearance and a fatigue failure is therefore easy to identify. As the crack propagates *slowly* from the source the fractured surfaces rub together due to the pulsating nature of the

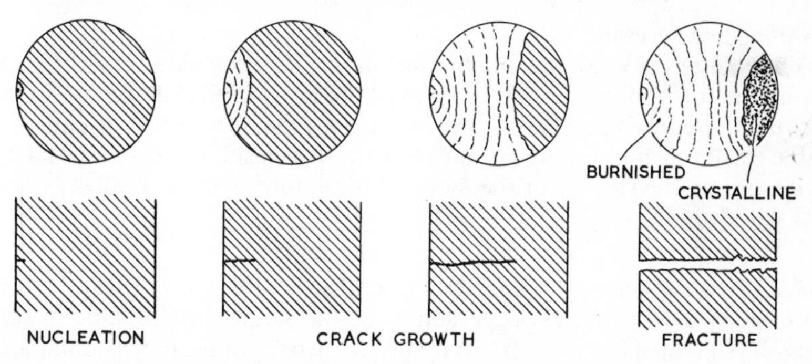

Fig. 16.12—The stages of fatigue failure. A fatigue fracture is generally easy to identify as the crack-growth region is burnished by the mating surfaces rubbing together as stresses alternate. The ultimate fracture is crystalline.

stress and so the surfaces become burnished. Often conchoidal markings are present showing the direction of spread of the fatigue crack. Ultimately the member is no longer able to carry its load and final fracture occurs, this freshly fractured surface being typically crystalline in appearance.

Chapter Seventeen
Methods of Joining Materials

17.1 At about the time that Diogenes, the original Cynic, was living in his tub, a fellow Greek, Archytas of Tarentum, is believed to have invented the screw. By the second century B.C. this device was being used extensively by the Greeks in their wine presses, whilst a less exotic use was made of the screw by Archimedes in his machine for raising irrigation water. Strangely, no one seems to know when the screw was first used as a fastening device.

Primitive man experienced great difficulty in joining his materials. He used sinews and thongs of hide to tie together the frameworks of his huts and also to tie on the heads of his stone axes. By 3000 B.C., however, the coppersmiths of Egypt were producing nails whilst, a little later, the woodworkers of that region were using quite sophisticated six-layer plywoods cemented by animal glues. Almost certainly by 3000 B.C. the smith had mastered the art of pressure welding using hammer and anvil but it was less than a century ago that the development of welding technology, as we now understand it, really began.

Nailing and screwing

17.2 Nailing is used mainly in the joining of soft woods in building construction and is effective because of the fairly high elasticity of wood. A nail driven wedge-fashion into wood gives rise to a region of considerable elastic strain around the nail. In turn this elastic strain causes lateral pressure on the surface of the nail and resultant forces of friction oppose its withdrawal.

17.2.1 Screwing is more satisfactory because greater strain energy can be produced by the spiral-wedge action of the screw thread. This in turn involves greater forces of friction between wood and screw as compared with those between wood and nail. Moreover the force required to separate the joined members will be greater since the forces opposing friction between screw and wood is only that component of the separating force which is resolved along the 'inclined plane' of the screw. For this reason metals can also be joined successfully by screwing.

Riveted joints

17.3 Despite the increased use of welding as a means of joining materials riveting is still important, particularly in shipbuilding and structural engineering. Most readers who have studied elementary applied mechanics will know that the design of a riveted joint involves equating

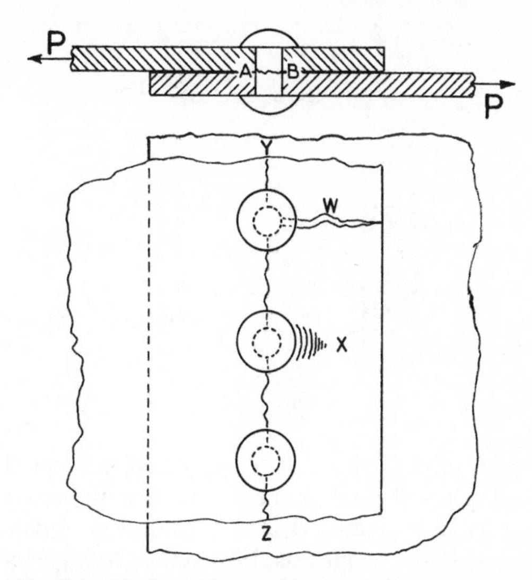

Fig. 17.1—Modes of failure of a simple riveted lap joint.

the strength of the rivets themselves with the strength of the pierced plates.

Consider a simple riveted lap joint (Fig. 17.1). This may fail in any one of the following ways:

(*i*) the plate may split beneath the rivet (at *W*). To prevent this type of failure it is customary to make the distance from the centre of the hole to the edge of the plate not less than 1·5 times the diameter of the rivet;

(*ii*) the plate may 'crush' at *X*;

(*iii*) if the rivet holes are too close together one of the plates may tear through the line *YZ*;

(*iv*) if the rivets are too small they may shear along *AB*.

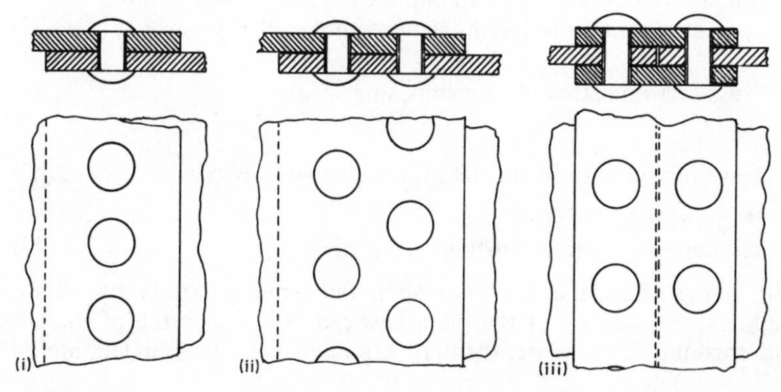

Fig. 17.2—Some types of riveted joint. (i) single riveted lap joint; (ii) double riveted lap joint; (iii) single riveted butt joint. Both lap and butt joints may be single or double riveted.

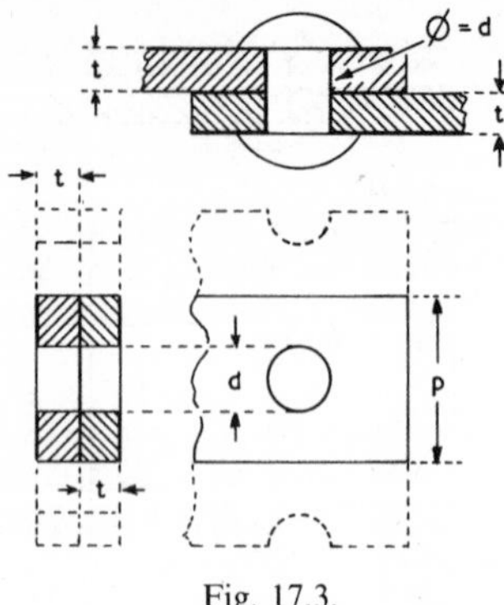

Fig. 17.3.

17.3.1 To design a joint in the most economical manner it should be on the point of failing in all four of these ways simultaneously. Lap joints may be single or double riveted (Fig. 17.2) but it is unusual to use more than two rows of rivets. The *pitch* is the distance between centres of rivets measured along a row. The strength of the joint may be calculated by considering a strip equal in width to the pitch (Fig. 17.3).

Let p = pitch of rivets;

d = diameter of rivets;

t = thickness of plates;

σ_t = tensile strength of plate material;

τ = shear strength of plate material;

γ_c = crushing strength of plate material.

Hence for the single-rivetted lap joint (Fig. 17.3):

Smallest cross-sectional area of plate in tension $= (p - d) \,.\, t$

$\therefore$ Resistance of joint to tearing between holes $= \sigma_t(p - d)t$ (1)

Cross-sectional area of rivet opposing shear $= \dfrac{\pi \,.\, d^2}{4}$

$\therefore$ Resistance of joint to shearing through a rivet $= \tau \,.\, \dfrac{\pi d^2}{4}$ (2)

'Projected area' of rivet $= t \,.\, d$

$\therefore$ Resistance of joint to crushing $= \sigma_c \,.\, td$ (3)

It is desirable that the joint shall fail simultaneously by all three methods, viz. tearing of the plate between holes, shearing of the rivets and crushing of the plate, then (1), (2) and (3) can be equated, and:

$$\sigma_t(p - d)t = \tau \,.\, \frac{\pi d^2}{4} = \sigma_c \,.\, td$$

$$\text{using} \quad \tau \, . \frac{\pi d^2}{4} = \sigma_c \, . \, td$$

$$\text{then} \quad d = \frac{4t}{\pi} \, . \frac{\sigma_c}{\tau}$$

from which the *diameter* of the rivet can be calculated.

The *pitch* can then be derived from:

$$\sigma_t(p - d)t = \tau \, . \frac{\pi d^2}{4}$$

$$p - d = \frac{\tau}{\sigma_t} \, . \frac{\pi d^2}{4t}$$

$$p = \frac{\tau}{\sigma_t} \, . \frac{\pi d^2}{4t} + d$$

17.3.2 *Efficiency of riveted joints.* The efficiency of a riveted joint is taken as the ratio of its actual strength to that of the unpierced plate. It is assumed that the joint has been designed such that failure is equally likely to occur by tearing, shearing or crushing.

$$\text{Resistance of joint to tearing} = \sigma_t(p - d) \, . \, t$$

$$\text{Resistance of unpierced plate to tearing} = \sigma_t \, . \, p \, . \, t$$

$$\text{Efficiency of joint} = \frac{\sigma_t(p - d) \, . \, t}{\sigma_t p \, . \, t}$$

$$= \frac{p - d}{p}$$

Example

Two steel boiler plates 10 mm thick are to be joined by a single riveted lap joint using rivets 20 mm in diameter. Determine the pitch of the rivets if the tensile stress in the plate is not to exceed 90 N/mm^2 and the shear stress in the rivets is not to exceed 60 N/mm^2.

$$p = \frac{\tau}{\sigma_t} \, . \frac{\pi d^2}{4t} + d$$

$$= \frac{60}{90} \, . \, \pi \, . \frac{400}{40} + 20 \text{ mm}$$

$$= 21 + 20 \text{ mm}$$

$$= \underline{\underline{41 \text{ mm}}}$$

Adhesion and adhesives

17.4 When the surfaces of two solids are brought sufficiently close together the interaction of surface forces will give rise to some degree of adhesion where the two surfaces are within roughly atomic range. This adhesion often manifests itself as friction if the two surfaces are made to slide relative to each other and the heat energy produced is a measure of the work done against these inter-surface forces of attraction. Rubber tyres do not skid if they can make close contact with a dry road surface.

17.4.1 Two *freshly drawn* glass fibres will adhere to each other if they are brought into contact. This can be demonstrated if the fibres are held between the finger and thumb of each hand (Fig. 17.4) and the upper fibre (*X*) then gently drawn across the lower (*Y*). The motion will be found to be 'snatchy', indicated by the deflection and jerking of the lower fibre. If the two fibres are now drawn through the fingers and the experiment repeated the snatching motion is reduced almost to zero.

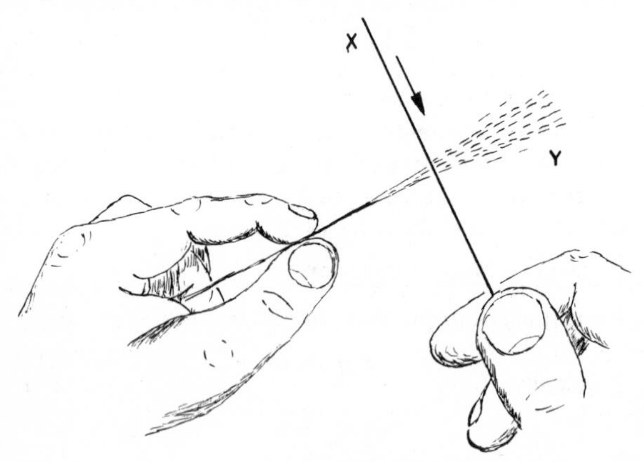

Fig. 17.4—Adhesion between *clean* glass fibres. If *X* is drawn lightly across *Y*, the fibres adhere causing vibration of *Y* due to the pulling action of *X*.

Presumably grease from the fingers produces a lubricating film spacing the glass surfaces apart so that adhesion is no longer possible. Similarly if the freshly drawn fibres are allowed to remain in contact with the atmosphere for some time they will no longer adhere, apparently due to contamination of the surfaces with dust, moisture and fumes.

17.4.2 In practical terms an adhesive may be regarded as any substance capable of holding materials together by surface attachment. Typical adhesives include glue, mucilage, paste and a wide range of polymeric cements. Generally the mechanism by which surface attachment occurs is:

(*i*) *specific adhesion* which depends principally upon the action of van der Waals forces (2.4) between molecules in the adhesive and the surface layers of the materials (*adherands*) being joined;

(*ii*) *mechanical adhesion* in which the adherands are to some extent keyed together by molecular filaments of the adhesive which penetrate into the substrates* of the adherands.

Since materials like glass and metal can be successfully bonded by adhesives it is reasonable to suppose that specific adhesion is the more important of these two possibilities. Generally speaking the larger the molecules involved the greater the adhesion since van der Waals forces will then be greater. Consequently most adhesives are organic compounds composed of very large and complex molecules having a high density of polar groups. The more common polar groups met with in adhesives include hydroxyl, carbonyl, nitro, nitrile, amide, sulphonic and some others. It is reasonable to suggest that the stickiness of sugar is due to the presence of hydroxyl groups in sugar molecules.

The strength of the adhesive bond is derived from the sum of all the van der Waals attractions between the molecules of the adhesive and neighbouring molecules in the substrate of the adherand. The molecules themselves will be strong due to the covalent bonds operating within them. However, not all large polymer molecules provide good adhesion. For example the fluoro-carbon grouping is associated with such poor adhesive properties that (PTFE Table 12.2) is used as a non-stick coating and also as a bearing material. Needless to say *cohesion* within the adhesive itself is just as important as *adhesion* between the adhesive and the adherand, otherwise failure in the adhesive layer will be likely. For this reason mobile liquids such as alcohol and nitrobenzene, whilst having suitable polar groups, are useless as adhesives since there is little cohesion between their relatively small molecules. In adhesives molecules are often so large as to have molecular masses of several hundred thousands or even millions. The sum total of polar bonds each molecule then provides is very great.

17.4.3 Another important feature is that the adhesive must mould itself perfectly to the surfaces being joined in order that its molecules will remain as close as possible to the molecules in these surfaces. For this reason ordinary glue must be applied hot whilst it is sufficiently mobile. Many polymeric adhesives are of the cold-setting type in which two mobile liquids are mixed together and then proceed to harden by polymerisation after the joint has been made.

Soldering and brazing

17.5 These processes differ from fusion welding in that the solder which is used melts at a *lower* temperature than the metal parts being joined so that no direct fusion of the latter occurs. It is essential, however, that a solder shall 'wet' the metallic surfaces it is meant to join, that is form a liquid solution with them. On solidification this liquid solution should be

* This term refers to the surfaces and those regions near to the surfaces of the adherands.

Table 17.1—Adhesives.

Group	*Adhesive substance (or source)*	*Materials joined*
Animal glues	Animal hides or bones, Casein (from milk), Blood albumen — protein materials	Wood, paper, leather and fabrics
Vegetable glues	Starch, Dextrines — carbohydrates	Paper and fabrics
	Soya beans—protein	'Sizing' of paper
Elastomer materials	Natural rubber and its derivatives—latex	Rubber, sealing strips (automobiles), gaskets, weatherproofing, veneers, linoleum tiles; footwear industries
	Synthetic rubbers—styrene/butadiene; butylene; nitrile; neoprene and carbonylic elastomers	Nylon; footwear; metal/metal; adhesive tapes; tyre construction; cardboard industries; aircraft industries; bonding of abrasives in grinding wheels
Thermoplastic resins	Vinyls, e.g. polyvinyl acetate	Wood furniture assemblies; paper industry
	Cellulose derivatives, e.g. cellulose nitrate	Glass; balsa wood; paper; china
	Acrylics, e.g. polymethyl methacrylate	Optical elements
Thermosetting resins	Phenol formaldehyde, resorcinol formaldehyde	Weather-proof plywoods; cellulose acetate; nylon
	Epoxies	Optics; glass; polystyrene; nylon; metals;—very versatile adhesives for dis-similar adherands

replaced by a solid solution if maximum toughness of the joint is to be attained, together with crystalline continuity across the interface. The formation of an intermetallic compound on the other hand will lead to brittleness. For example, the use of a tin-base solder to join copper may produce a brittle joint since at some point between the pure copper parts and the centre of the solder joining them, the metals copper and tin will be present in the correct proportions to form very brittle compounds such as $Cu_{31}Sn_8$ and Cu_6Sn_5. If such a joint is torn apart it will show a bluish fracture due to the presence of $Cu_{31}Sn_8$ in the surface through which failure has occurred. For this reason copper is soldered with a lead–silver solder in high-class work. Copper does not form intermetallic compounds with either lead or silver.

During soldering and brazing processes oxide films tend to act as a barrier to successful 'wetting' of the work pieces by the solder. These oxide films are usually dissolved by some suitable flux in order to expose a clean metallic surface to the alloying action of the solder. Nevertheless some metals, such as aluminium, which have tenacious oxide films are particularly difficult to solder.

17.5.1 In ordinary 'soft' soldering tin–lead base alloys are used. Ideally tin-man's solder should be of eutectic composition (62% Sn–38% Pb) so that it solidifies sharply at 183° C thus reducing to a minimum the time required to produce a joint and also the temperature to which the work piece is exposed. Since tin is extremely expensive relative to lead, commercial solders often contain 50% or more lead. Plumber's solder should, of course, solidify over a range of temperature to enable the plumber to 'wipe' a joint in a fractured pipe. For this reason plumber's solder contains about 67% Pb–33% Sn and solidifies between about 265° and 183° C.

Soft soldering may be accomplished with the aid of some form of soldering 'iron'. In large-scale production, however, a mixture of solder and flux may be preplaced in the joints and the assemblages heated *en masse* usually by high-frequency electrical currents.

17.5.2 In brazing the soldering iron is replaced by a gas torch. Alternatively brazing solder/flux mixture can be pre-placed and the work pieces heated by high-frequency induction; salt baths; furnace or dip methods. Brazing solders contain between 50 and 66% Cu (balance–Zn). This produces a low-melting point alloy but one which would be extremely brittle were it not for the fact that, during brazing, some zinc is lost so that the resultant joint consists of a tough, ductile α brass instead of a hard, brittle γ structure (Fig. 11.1 (i)). Silver solders contain varying quantities of silver in order to reduce the melting range and so produce a brazing alloy with a melting point as low as 690° C.

Welding

17.6 The joining of pieces of metal by first heating them to a temperature above that of recrystallisation and then forging them together is a *pressure*-welding process which the blacksmith has operated for many centuries. A number of mechanical pressure-welding processes are now used, but in the meantime *fusion* welding has gained the ascendency in the ever-

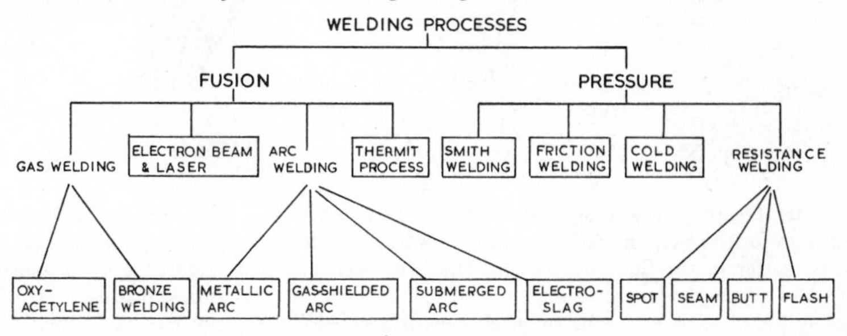

Fig. 17.5—A classification of welding processes.

widening range of processes which continue to be introduced and developed.

Fusion-welding processes

17.6.1 *Gas welding* using the oxy-acetylene torch is now used mainly for jobbing and repair work. By using suitable mixtures of oxygen and acetylene temperatures in the region of 3000° C can be attained so that ordinary metals and alloys are quickly melted. This is necessary in order to overcome the tendency of metallic sheets and plates to conduct heat away from the weld zone so quickly that fusion is impossible.

Metal for the weld joint is, of course, supplied from a welding rod which is normally of a composition similar to that of the work pieces. In addition to metals, some thermoplastics such as rigid PVC are gas-welded.

17.6.1.1 *Bronze welding* resembles brazing in that little or no fusion of the work pieces occurs since the bronze filler metal (generally 4·5% Sn; 0·5% P; balance—Cu) melts at a much lower temperature than the work pieces which are usually mild steel, cast iron or nickel alloys. At the same time bronze welding differs from brazing in the application of the filler metal which is in rod form whereas in brazing powdered solder is generally preplaced.

17.6.2 *Electric arc welding processes* now constitute by far the most important fusion-welding processes. The earliest electric-welding process employed a carbon electrode from which an arc was struck to the work itself. Weld metal was supplied from a filler rod, the end of which was held in the arc. This method is now virtually obsolete.

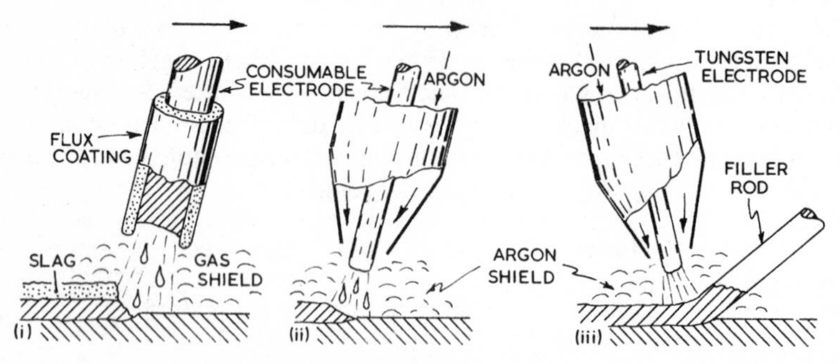

Fig. 17.6—Electric-arc welding processes. (i) The metallic arc—here molten flux provides protection for the weld, helped to some extent by a gas shield provided by burning carbonaceous matter in the flux coating. (ii) The MIG (metallic inert gas) process in which a blanket of argon gas protects the weld. (iii) The TIG (tungsten inert gas) process in which a separate filler rod is used, the tungsten electrode providing the arc but being itself non-consumable.

17.6.2.1 The *metallic arc process* is the most widely used of the arc-welding operations. It makes use of a consumable electrode, that is, the electrode is a rod of suitable composition so that when an arc is struck between it and the work piece, the rod melts and provides the necessary weld metal (Fig. 17.6 (i)). The electrode is coated with a flux which not only provides a protective slag cover for the weld but also assists in stabilising the arc. Although it is used principally for steel, metallic-arc welding is also applicable to many non-ferrous metals and alloys.

17.6.2.2 *Submerged-arc welding* is essentially an automatic process used for straight-line joining of metals. It is a variation of the metallic-arc process in which powdered flux is fed into the prepared joint just in advance of the electrode. As the flux melts it envelops the melting end of the electrode and so covers the arc. The weld area is thus protected by a coating of slag so that a smooth clean weld is produced.

17.6.2.3 *Gas-shielded arc processes* are used for welding those metals and alloys in which oxidation is troublesome. In these processes the electrode protrudes from the end of a tube from which an inert gas also issues. In the UK the noble gas argon is generally used whilst in the USA helium, a by-product of the oil-wells, is also available.

The *metallic-inert-gas* (or MIG) process employs a consumable electrode (Fig. 17.6 (ii)) as in ordinary metallic-arc welding, whilst in the *tungsten-inert-gas* (or TIG) process a permanent tungsten electrode is used to strike the arc and an external filler rod used to supply the weld metal (Fig. 17.6 (iii)). These processes were developed during the Second World War in order to weld alloys containing magnesium, a metal which reacts with most gases, including nitrogen. For this reason relatively expensive inert gases must be used to provide the protective atmosphere. These processes are now used more widely for the welding of other metals and alloys.

The CO_2 *process* was developed more recently and has become very popular for welding steel. It uses an atmosphere of carbon dioxide but since this can oxidise steel at high temperatures, filler rods containing deoxidants such as manganese and silicon are employed.

Although nitrogen tends to be regarded as an unreactive gas it dissolves in steel and gives rise to brittleness. It also reacts with several non-ferrous metals, particularly magnesium, and is therefore of no interest in gas-shielded welding processes.

17.6.2.4 *Electro-slag welding* was introduced in the USSR in 1953 for the joining of heavy steel plates in a vertical *position*. The molten metal is delivered progressively to the vertical gap between the plates rather as in an ingot-casting process. The plates themselves form two sides of the 'mould' whilst two travelling water-cooled copper shoes dam the weld pool until solidification is complete.

17.6.2.5 *Plasma-arc welding* is a relatively new process in which a suitable gas such as argon is passed through a constricted electric arc so that it dissociates into positive and negative ions, this ion mixture being termed *plasma*. Outside the arc the ions re-combine and the heat-energy released produces an extremely hot 'electric flame'. Temperatures up to 15 000° C can be produced in this manner. Plasma is used as a high-temperature heat source in cutting, drilling and spraying very refractory materials like tungsten, molybdenum and uranium as well as in welding.

17.6.3 *Thermit welding* is used mainly to repair large iron and steel castings. A simple mould is constructed around the parts to be joined, and above this is a crucible containing the Thermit mixture. This thermit mixture consists of powdered ferrous-ferric oxide, Fe_3O_4, and aluminium dust in molecular proportions. The chemical affinity of aluminium for oxygen is greater than the affinity of iron for oxygen hence, on ignition, aluminium reduces the iron oxide:

$$3\ Fe_3O_4 + 8\ Al \rightarrow 4\ Al_2O_3 + 9\ Fe \quad \Delta_h = -15{\cdot}4\ \text{MJ/kg}$$

The heat of reaction is so intense that molten iron is produced and this is tapped from the crucible into the prepared joint.

17.6.4 *Electron beam welding.* When a stream of fast-moving electrons strikes a suitable target X-rays are generated (20.8.2). At the same time a considerable amount of heat is produced as a result of the conversion of the kinetic energy of the fast-moving electrons. If the electron beam is focused to impinge at a point very high temperatures can be produced. To be really effective the operation must be carried out *in vacuo* otherwise many of the electrons will lose their kinetic energy as a result of collisions with relatively massive molecules of oxygen and nitrogen present in the air space between electron 'gun' and target.

Electron-beam welding is used principally for the refractory metals tungsten, molybdenum and tantalum and also for reactive metals such as beryllium, uranium and zirconium which cannot oxidise if welded *in vacuo*.

17.6.5 *Laser welding* has been used so far mainly for micro spot-welding and, to some extent, for drilling and cutting. A *laser* (*L*ight *A*mplification by *S*timulated *E*mission of *R*adiation) device consists essentially of a generator of light pulses along with a suitable active substance so that a beam of light of very high intensity is emitted.

Pressure-welding processes

17.7 The smith has been practising his art for at least 6000 years. In order to weld together two pieces of iron he first heats them to a high temperature and then brings them into contact under pressure by hammering them together on an anvil. Deformation above the recrystallisation temperature (5.8.2) induces recrystallisation which occurs across metal/metal interfaces so that a strong continuous joint is produced.

Most modern pressure-welding processes are similar in principle to smith welding except that the welding temperature is generally localised to the region where joining is to be accomplished.

17.7.1 *Electric resistance-welding processes.* These processes are used to produce either 'spot', 'seam' or 'butt' welds.

17.7.1.1 *Spot welding.* In this process the work pieces are overlapped and gripped between heavy metal electrodes (Fig. 17.7 (i)). A current sufficient in magnitude to produce local heating of the work pieces to a

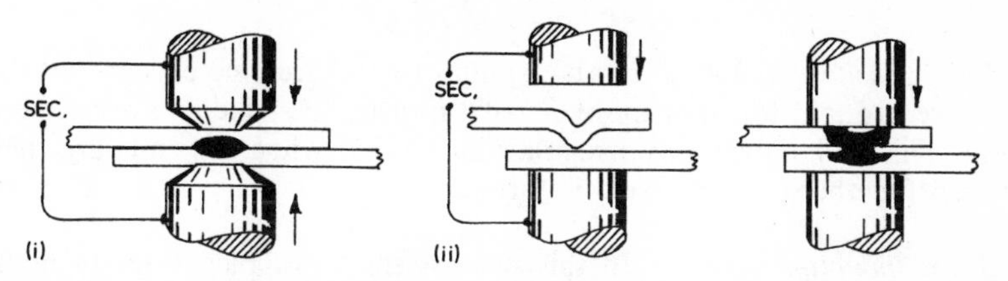

Fig. 17.7—(i) Spot welding; (ii) projection welding. (SEC = secondary of welding transformer.)

plastic state is then passed through the electrodes. Since the metal between the electrodes is under considerable pressure a spot weld is produced. This method is particularly suitable for 'tacking' together plates and sheets, often preparatory to producing a more permanent weld.

17.7.1.2 *Projection welding* is a modified form of spot welding in which the current flow and hence, the resultant heating effect, are localised by embossing one of the parts to be joined. This method (Fig. 17.7 (ii)) is very suitable for joining heavy sections where spot welding would be difficult because of the big heat losses by conduction and consequently heavy currents required.

17.7.1.3 *Seam welding* (Fig. 17.8 (i)) resembles spot welding in principle but produces a continuous weld by using wheel electrodes. Plastics can be joined by both spot and seam welding, the electrodes being heated by high-frequency induction since these polymer materials are of course non-conductors of electricity.

17.7.1.4 *Butt welding* (Fig. 17.8 (ii)) is used for joining lengths of wire, rod or tubes. The ends are pressed together and heated at the joint by a current which is passed through the work. Since there will inevitably be a higher electrical resistance at the point of contact of the two work pieces, more heat will be generated there.

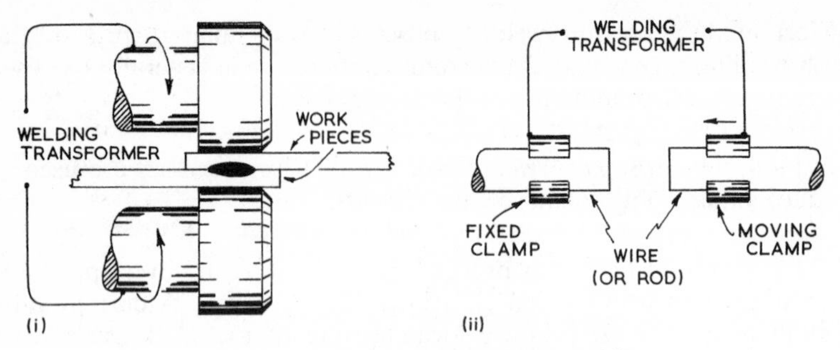

Fig. 17.8—(i) Seam welding. (ii) Butt welding.

17.7.1.5 *Flash welding* is somewhat similar to butt welding except that the ends are brought together momentarily in order to strike an arc between them. This melts away any irregularities so that when the ends are then pressed together a sound weld is formed.

17.7.1.6 *Induction welding.* In this process the work pieces are brought into contact under pressure and a high-frequency induction coil placed around the prospective joint. HF currents induced in the work heats it to welding temperature and the applied pressure causes welding to take place.

17.7.2 *Other pressure-welding processes.* Several other pressure-welding processes have been introduced during the last two decades or so.

17.7.2.1 *Friction welding* (Fig. 17.9 (i)) can be used to join some materials in rod form. The rods are gripped in chucks, one rotating and the other stationary. As the ends are brought together heat is generated by friction between the slipping surfaces, ultimately sufficient to cause welding. Rotation is stopped automatically as welding commences. A modification of this process, used to join plastics, is referred to as spin welding.

17.7.2.2 *Cold welding* in the form of lap, seam or butt welding involves making the two surfaces slide relative to each other under great pressure. Oxide films are disrupted and a cold weld forms between the

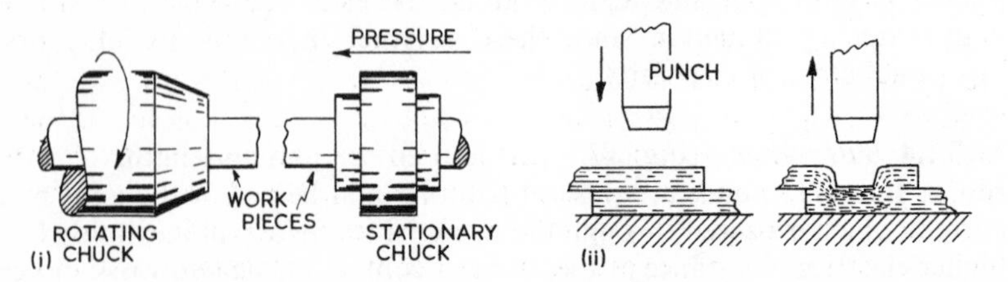

Fig. 17.9—(i) Friction welding. (ii) Cold spot-welding.

surfaces. Spot welds are made by punching the two work pieces together (Fig. 17.9 (ii)).

17.7.2.3 *Explosive welding* can be used to 'clad' one sheet of metal with another. The 'flyer' plate, topped with a sheet of high explosive, is suspended at a slight angle to the target plate. On detonation of the explosive the flyer plate strikes the target at speeds of up to 300 m/s and with inter-face pressures of up to 7000 N/mm^2. As a result the material in the region of the impact behaves for a short time as a fluid of quite low viscosity. The flyer plate virtually 'jets' along the surface of the target so that cold welding takes place (Fig. 17.10 (i)).

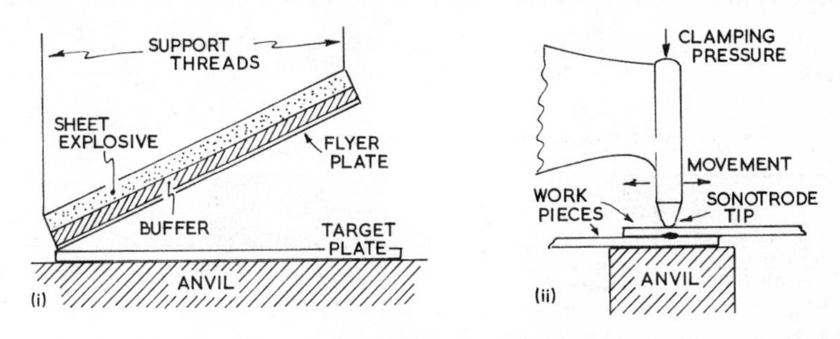

Fig. 17.10—(i) Explosive welding. (ii) Ultrasonic welding.

17.7.2.4 *Ultrasonic welding* employs ultrasonic vibrations of the order of 20 000 Hz to produce slip between two surfaces which are being pressed together. The vibrations lead to disruption of the surface films of oxide so that the surfaces can cold-weld under the pressure used. The sonotrode tip (Fig. 17.10 (ii)) clamps the work pieces to an anvil and vibrates laterally.

The structure and properties of welded joints

17.8 The common welding processes involve the use of high temperatures so that the region containing the weld will be relatively coarse-grained in contrast to the remainder of the work piece which will generally be of a wrought, and consequently, fine-grained structure. If a fusion method of welding has been used the weld metal itself will show a coarse as-cast type of structure. Here the crystals will be large and other features such as coring and segregation of impurities at crystal boundaries may also be present. These features give rise to brittleness which results from weakness at the crystal boundaries. The effect of such faults can be minimised if it is practicable to hammer the weld whilst it is still hot. Not only will this induce fine grain on recrystallisation but it will redistribute more evenly those impurities segregated as a result of coring.

17.8.1 Poor welding technique may result in gas porosity and/or the presence of oxide particles trapped in the weld metal. Gas porosity may arise either from gas dissolved by the molten weld metal or from chemical reactions between such gases and constituents already present. Oxide particles are generally due to reactions between atmospheric oxygen and constituents of the weld metal or those dissolved from the surfaces of the work pieces.

Specific difficulties are associated with the welding of some metals and alloys. Thus, as the carbon content of plain carbon steels increases so do the difficulties in welding them. Carbon tends to oxidise during the welding process and porosity is likely to arise due to the gases released. Thin sections in carbon steel may cool so rapidly as to cause brittleness whilst

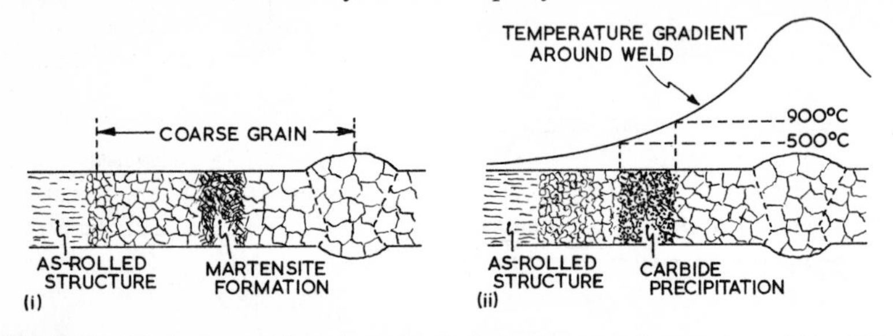

Fig. 17.11—(i) Grain growth and martensite formation in and around a weld in 13% Cr stainless steel. (ii) Carbide precipitation in a non-stabilised 18–8 stainless steel. This will lead to 'weld decay'.

much thicker sections in oil- or air-hardening steels may become brittle near to the weld as a result of martensite formation. The 13% Cr air-hardening cutlery steel is particularly difficult to weld satisfactorily since brittleness arises not only from excessive grain growth around the weld but also the formation of martensite in that region (Fig. 17.11 (i)).

17.8.2 The 'weld decay' of some 18–8 austenitic steels has been mentioned elsewhere (10.5.1). If they have not been 'proofed' against such an eventuality these steels are normally quenched from 1050° C in order to retain any carbon in solid solution in the austenite. Heating in the range 500°–900° C will cause the precipitation of carbide particles and this will lead to corrosion (15.7.1). During a welding process some region either side of the joint will be maintained in this temperature range for long enough for some such precipitation to occur and corrosion will subsequently take place in that region. Many items of plant used in the chemical industries are fabricated by the welding of stainless-steel sheets because highly corrosive conditions are involved. It is then even more important to reduce the tendency towards weld decay. Assuming that the welded structure cannot be solution-treated from above 1050° C—and obviously this will rarely be possible—one of the following means of limiting the precipitation of chromium carbide must be used:

(*i*) the use of stainless steel with a carbon content *below* 0·03%;
(*ii*) the stabilisation of carbides under all conditions by adding niobium or titanium. No dissolved carbon is then available to precipitate as chromium carbide since any carbon present has already been 'tied up' by the strong carbide-forming niobium (or titanium).

17.8.3 Cast irons may be successfully welded if filler rods rich in deoxidants such as manganese and silicon are used. Even so the weld metal may absorb considerable amounts of carbon from the work pieces and since cooling will inevitably be rapid a very brittle white iron structure may be formed (10.6.2). For this reason cast iron is often joined by bronze welding, though subsequent electrolytic corrosion between the bronze weld and the surrounding cast iron must be expected if moisture is present in the environment. Cast iron, being anodic to the bronze weld, would quickly rust.

17.8.4 Aluminium alloys oxidise readily so that during ordinary metallic arc welding they must be protected adequately by a flux. For this reason the bulk of aluminium welding is now carried out with the gas-shielded arc, a process developed to deal with the even more difficult magnesium alloys.

Chapter Eighteen

Electrical and Magnetic Properties of Materials

18.1 Inevitably a book of this type is concerned mainly with such mechanical properties of materials as concern the engineer but it is also appropriate to make brief mention of some of the more important electrical and magnetic properties, as well as to describe nuclear properties as they affect radioactivity and also chemical properties as they influence the rate of corrosion of a material.

Electrical conductivity

18.2 It was suggested early in this book that what we refer to as a 'current of electricity' in a metal is in fact a movement of electrons. Thus metals, in which the valency electrons may become de-localised and relatively free to move among the metallic ions which constitute the crystal structure of the metal, are consequently good conductors of electricity. Since electrons will repel each other they tend to move from regions of high concentration to regions of low concentration. Thus Ohm's Law governing the electric current follows the same principle as Fick's Law of Diffusion (7.4.2) governing the movement of solute ions.

Ionic crystals, however, are non-conductors since valency electrons are held rigidly by ions constituting the crystal. Similarly materials such as polymers, in which all chemical bonding is of the covalent type, are also non-conductors since shared valency electrons are unable to move out of their particular molecular orbitals.

18.2.1 Whilst an electric current in metals is due to a movement of electrons, in other materials positive or negative ions may be involved. Thus ions are responsible for conduction through electrolytes.

The nature of an electric current

18.3 The 'electron cloud' theory is a convenient way of explaining some of the properties of metals but in some respects we must regard it as an oversimplification of the true state of affairs. Such shortcomings can be partially explained when quantum restrictions and Pauli's exclusion principle (1.3.3) are applied to the behaviour of electrons.

18.3.1 In a single atom the electrons occupy a number of discrete energy levels and these energy levels become closer together the further they are from the nucleus. In so far as properties like electrical conductivity are concerned it is the outer-shell or valency electrons which interest us. Thus the sodium atom has 11 electrons arranged in 4 shells (i.e. the second shell is divided into two sub-shells, $2s$ and $2p$). Shells $1s$, $2s$ and $2p$ are

Table 18.1—Electrical resistivity of some materials.

Material	*Resistivity* (10^{-8} Ω.m) at 20° C	*Temperature coefficient of resistivity* (× 10^3) (0–100° C)
Aluminium	2·69	4·2
Chromium	12·9	2·1
Copper	1·67	4·3
Gold	2·3	3·9
Germanium	46 × 10^6	—
Iron	9·71	6·5
Lead	20·6	3·4
Magnesium	3·9	4·2
Mercury	95·8	0·9
Nickel	6·84	6·8
Platinum	10·6	3·9
Silicon	23 × 10^{10}	—
Silver	1·6	4·1
Tin	12·8	4·2
Zinc	5·92	4·2
80 Ni–20 Cr	109	—
18–8 stainless steel	70	—
55 Cu– 45 Ni	49	—

filled but the valency sub-shell 3*s* could hold another electron according to Pauli's exclusion principle. The electrons in each sub-shell occupy specific levels (Fig. 18.1 (i)). If sodium atoms are now brought into close association, that is a crystal of sodium is considered, they begin to influence one another and the electrons are compelled to seek slightly different energy levels since—again according to Pauli's exclusion principle—not more than two electrons can occupy the same quantum state. Within even a small crystal, containing nevertheless many millions of atoms, the energy *levels* will be replaced by densely filled *bands* (Fig. 18.1 (ii)). The sodium atom has only one valency electron so that only half of the energy states in the 3*s* (or valency) band are filled. For this reason the

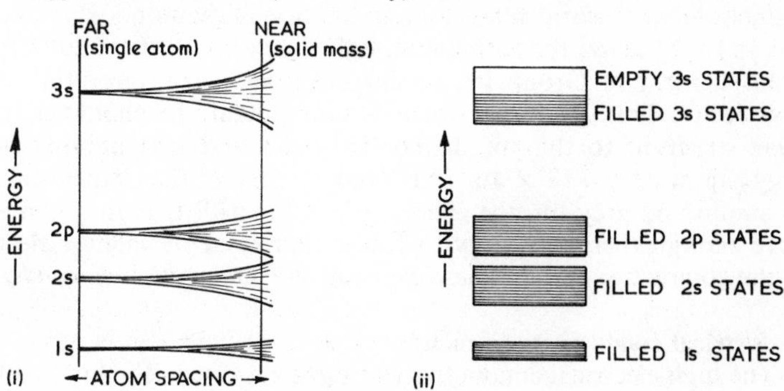

Fig. 18.1—Energy states of electrons in the sodium atom.

energy required to raise a valency electron to an empty state is negligible. This enables it to move freely within the crystal and so to conduct electricity.

In the case of, say, magnesium the 3*s* band is filled but since it actually overlaps with the 3*p* band (Fig. 18.2) the valency electrons may have their

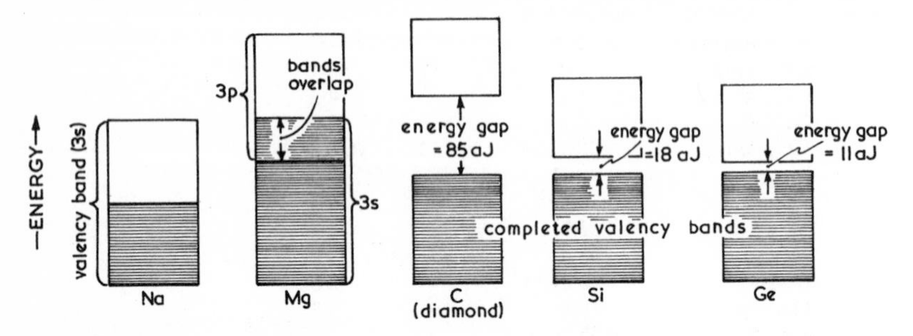

Fig. 18.2—Energy bands in metals and metalloids which control electrical conductivity.

energies increased within the 3*p* band so that magnesium conducts electricity very easily. However, in some materials the adjacent energy bands do not overlap in this way and energy is then required to move electrons across this 'energy gap'. Thus in the diamond allotrope of carbon the valency band is filled by the four outer-shell electrons and a large energy gap exists between this band and the next possible energy band.

18.3.2 In the case of diamond the magnitude of the energy gap is a measure of the amount of energy needed by an electron to break away from the covalent bond and become a free electron. In diamond this energy gap is great (about $8{\cdot}5 \times 10^{-19}$ J at 20° C) so that it has a very high electrical resistivity. The other elements of Group IV have similar valency electron structures and crystallise in patterns like that of diamond, that is with four covalent bonds per atom. In silicon and germanium, however, the energy gap between the 'valency band' and the 'conduction band' is much less than in the case of diamond. For silicon it is $1{\cdot}8 \times 10^{-19}$ J and for germanium only $1{\cdot}1 \times 10^{-19}$ J (at 20° C). Tin, also an element of Group IV, occurs as an allotrope, 'grey tin', stable below 18° C. This allotrope which is non-metallic in character, has a similar structure to that of diamond, silicon and germanium and an energy gap of only $0{\cdot}13 \times 10^{-19}$ J. Thus, in each of the elements silicon, germanium and grey tin, the energy gap is so small that they can easily receive enough thermal energy to cause the release of valency electrons into the conduction band. These elements are known as *semiconductors*.

The electrical conductivity of metals

18.4 The high electrical conductivity of metals is due to the high mobility of the outer-shell or valency electrons which, in turn, is dependent on the

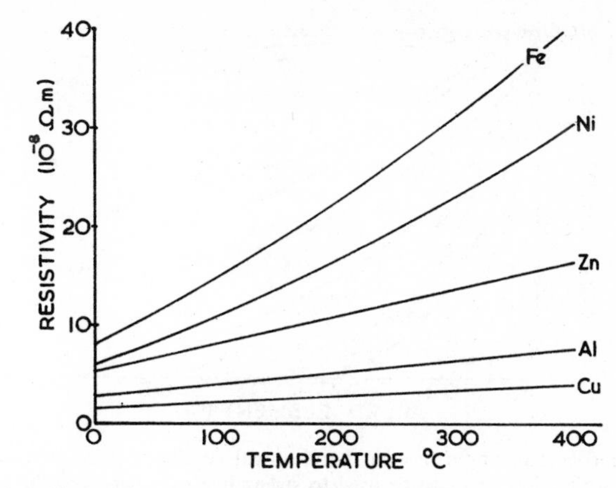

Fig. 18.3—The effects of temperature on resistivity. This is almost linear except near Absolute Zero where the resistivity of some metals falls very rapidly (18.4.2).

ease with which these electrons can pass from the 'valency' band to the 'conduction' band.

Since an electric current is a flow of electrons amongst the positive ions in a metallic crystal any factor which impedes this movement will reduce conductivity. Thus an increase in temperature introduces greater thermal agitation as ions vibrate about their mean positions. This obviously reduces the mean free path which electrons can follow and consequently their mobility, resulting in a decrease in conductivity (Fig. 18.3).

Similarly the presence of solute atoms, whether substitutional or interstitial, will increase the possibility of collisions between moving electrons and the ions constituting the crystal structure. Thus alloys are always less conductive than the pure metals from which they are made (Fig. 18.4).

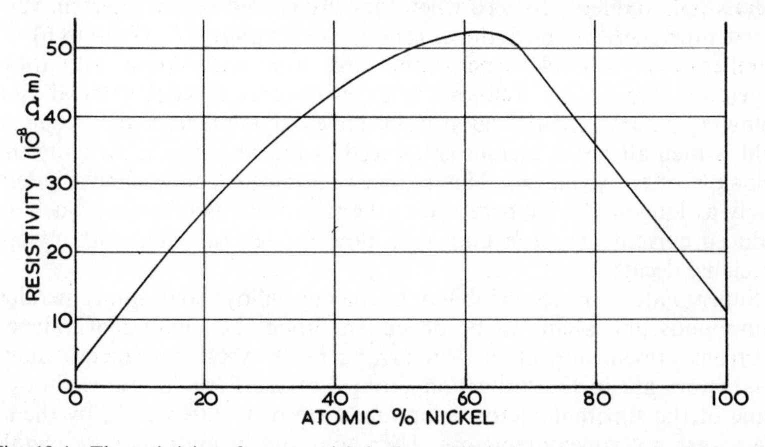

Fig. 18.4—The resistivity of copper–nickel alloys. Alloys have lower conductivities than pure metals.

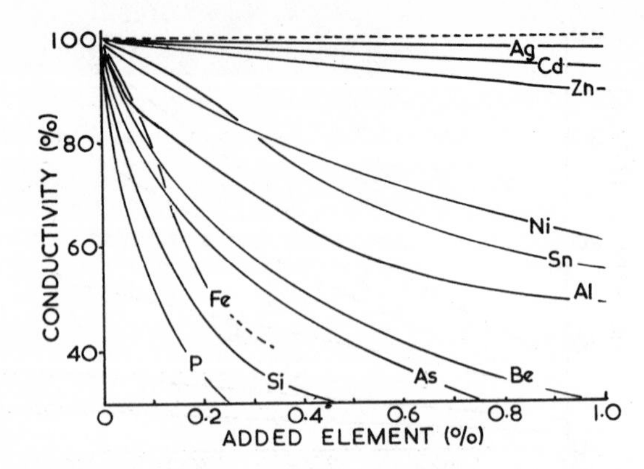

Fig. 18.5—The effects of impurities on the electrical conductivity of copper. Silver and cadmium have little effect and can be used to strengthen over-head conductors (e.g. telephone lines), whilst arsenic and phosphorus must not be present in copper destined for electrical purposes.

The conductivity of copper may be reduced by as much as 40% by the presence of only 0·05% phosphorus (Fig. 18.5).

18.4.1 The presence of lattice faults such as dislocations will also reduce conductivity by increasing the risk of collisions of the mobile electrons with atom cores in the lattice. For example heavily cold-worked copper is some 10% less conductive than annealed copper.

18.4.2 *Superconductivity*. The electrical resistivity of a metal decreases as its temperature falls for reasons suggested above. However, it was discovered as long ago as 1912 that the electrical resistivity of some metals falls suddenly to zero when they are cooled below a certain very low temperature, termed the *transition temperature*, T_c (Fig. 18.6). A number of remarkable experiments have been carried out with these superconductors. One well-known experiment is to cool a metal ring below T_c in a magnetic field at right angles to its plane. If the magnetic field is then altered a current is induced in the ring (this is virtually the principle of the dynamo). This induced current will flow almost indefinitely as long as the temperature of the ring remains below T_c. Such an induced current has been known to flow for several years without appreciable decay.

Superconductivity is exhibited by metals, alloys and some metallic compounds and seems to be dependent upon the number of valency electrons present and upon their arrangement. Most pure-metal superconductors are in the 'transition groups' of the Periodic Table though some of the superconductor compounds are of metals which, by themselves, are not superconductors. Thus gold and bismuth as pure metals are not superconductors whilst Au_2Bi is superconductive.

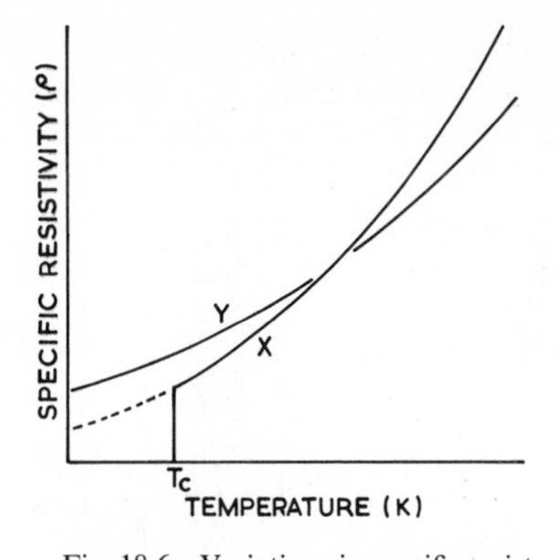

Fig. 18.6—Variations in specific resistivity with temperature for a superconductor (X) and a metal which does not exhibit this phenomenon (Y).

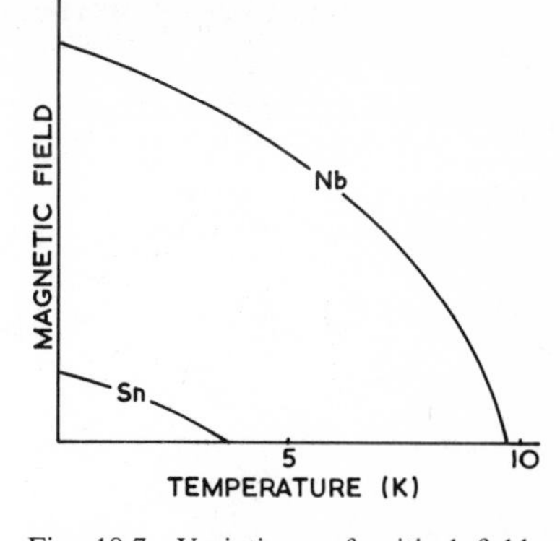

Fig. 18.7—Variations of critical field, H_c, with absolute temperature for a 'hard' superconductor (Nb) and a 'soft' superconductor (Sn).

Theories which attempt to explain superconductivity are generally complex and beyond the scope of this book, but very briefly, some modern theories suggest that with superonductive metals and alloys the 'conduction electrons' become 'paired' when the temperature falls below T_c. This means that an electron with 'spin' in one direction becomes paired with an electron with 'spin' in the opposite direction. As a result of this pairing the electrons produce no resultant field and, since they suffer *no resistance*, can move freely through the lattice.

Under the influence of a suitably strong magnetic field the superconductivity of a material below T_c is destroyed and the material reverts to being a normal conductor. Since magnetism is dependent principally upon the spin of electrons (18.7.1) it is reasonable to suppose that superconductivity is also connected with the movement and spin of electrons, possibly as outlined above.

Superconductivity below T_c is also destroyed when a high current density is employed. Consequently superconductivity disappears:

(*i*) if the temperature of the material is raised above T_c;
(*ii*) if a sufficiently strong magnetic field is applied;
(*iii*) when a high current density prevails in the conductor.

Critical field strength, H_c, transition temperature, T_c, and current density, J_c, are interdependent. The general relationship between field strength, H_c, and temperature, T_c, as this affects superconductors is indicated in Fig. 18.7. Some of the transition metals as typified by niobium, are termed *hard* superconductors whilst those with low values like tin are called *soft* superconductors. Properties of some superconductive metals and alloys are given in Table 18.2. Some of these compounds and alloys are able to sustain large direct currents in strong magnetic fields at the boiling point of liquid helium (4·2 K), and for this reason are

Table 18.2—Properties of some superconductors.

Material		Critical temperature (T_c) (K)	Critical field strength (Tesla)	
Pure metals	Technetium	11·2	—	at 0 K
	Niobium	9·46	0·194	
	Lead	7·18	0·080	
	Vanadium	5·30	0·131	
	Tantalum	4·48	0·083	
	Tin (white)	3·72	0·031	
	Aluminium	1·19	0·010	
Alloys and compounds	Nb_3Sn	18·1	>25·0	at 4·2 K
	V_3Ga	16·8	>30·0	
	Mo–Re alloy	10·0	1·6	
	Pb–Bi eutectic	7·3–8·8	1·5	

important in some experimental work. They can be fabricated, though with difficulty because of their general brittleness, into wire and tape which can be wound as solenoids capable of generating and maintaining very strong magnetic fields.

Insulators

18.5 Non-conducting materials—or insulators—are those in which bonding is either covalent or ionic. In such materials the valency electrons are firmly retained in fixed orbitals and are unable to move freely as they are in metals. Valency bonds in insulators are completely filled and there is a wide energy gap between the valency bond and the next possible energy band. Under such circumstances a large input of energy would be required in order that an electron could cross the gap. A very small number of electrons may acquire such energy if the material is subjected to a high potential difference and these insulators possess very low but nevertheless measurable conductivities. At high temperatures individual electrons will possess greater energies and so the chances of them 'escaping' is increased. Conductivity will therefore increase. Insulating materials may break down under the pressure of an electrical potential difference which is high enough to raise the energies of large numbers of electrons to the point where they can cross the energy gap and become 'free'. Such a potential difference is generally termed the 'breakdown voltage'.

Semiconductors

18.6 During the author's schooldays a three-valve radio receiver was housed in a large cabinet which occupied considerable space in the family living-room. Its bulk was due in no small measure to the size of the thermionic valves and the volume of surrounding air space required to cool these large glass 'bottles'. A radio receiver of equivalent performance can now be housed in a matchbox. This was made possible by the development of 'transistors' in the late 1940s. The complex electronics industry which has developed since then is dependent upon these semiconductors.

18.6.1 Like carbon, both silicon and germanium are Group IV elements and resemble the carbon allotrope, diamond, in that they crystallise in structures in which each atom is covalently bonded to four similar atoms. However, both silicon and germanium differ from diamond in that the

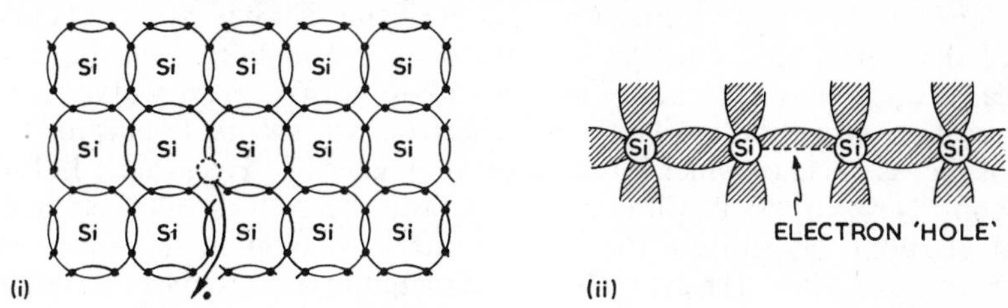

Fig. 18.8—(i) This diagram represents silicon atoms covalently bonded in a *flat* plane. In fact they will exist in a three-dimensional tetragonal system like diamond (Fig. 3.24); (ii) a simplified representation.

energy gap between valency and 'conduction' bands is relatively small compared with that in the diamond atom so that it is easier to free an electron into the conduction band. The transfer of this electron leaves a vacancy such that one atom in the lattice (Fig. 18.8) possesses only three valency bonds, instead of four. This deficiency is referred to as an 'electron hole'. Another valency electron is able to 'jump' into this 'hole' and so cause the movement of the hole to proceed in the opposite direction to that in which the electrons are moving. Consequently, since the electron hole is migrating in the opposite direction to the flow of electrons its movement can be regarded as equivalent to the motion of a positive

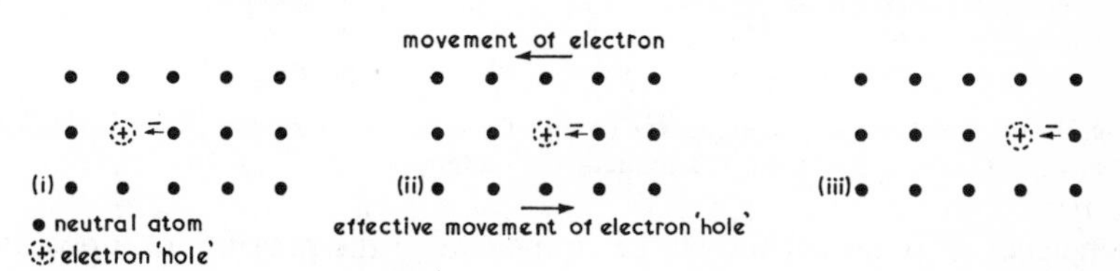

Fig. 18.9—Representing the relative movement of electrons and 'electron holes' (positive ions).

charge (Fig. 18.9). Materials in which conduction occurs in this manner are termed *intrinsic* semiconductors, that is, conduction is due to the combined flow of freed valency electrons and of positive holes in the opposite direction. Valency electrons can be freed through the energy gap and into the conduction band by supplying thermal energy. For this reason raising the temperature of an intrinsic semiconductor increases its conductivity. In normal conductors the opposite effect is obtained by raising the temperature (18.4).

18.6.2 Some semiconductors rely for their properties on the presence of a few 'impurity' atoms in the lattice. These are called *extrinsic* semiconductors. The trace elements concerned are generally those of Groups III or V, that is, those whose atoms contain either one less or one more valency electron than the Group IV atom concerned. Suppose an atom of the Group V element phosphorus is present in silicon. Since its atomic radius is of the right order it will fit into the crystal structure of silicon, but because silicon has a valency of four and phosphorus a valency of five, an extra electron will be available. This electron will still be bound to the phosphorus nucleus (which has an atomic number of 15 as against 14 for silicon). Nevertheless it will require very little energy to break this bond and at room temperature the thermal energy present in the lattice is sufficient to do this. The extra electron is therefore freed to move through the silicon lattice. An extrinsic semiconductor of this type which utilises the movement of freed electrons is called an *n*-type and the element (in this case phosphorus) which provides the electron is referred to as a *donor*.

18.6.3 If, on the other hand, an impurity atom of a Group III element such as boron or aluminium is present in the silicon lattice then this will provide only three valency electrons for the four bonds involved. This

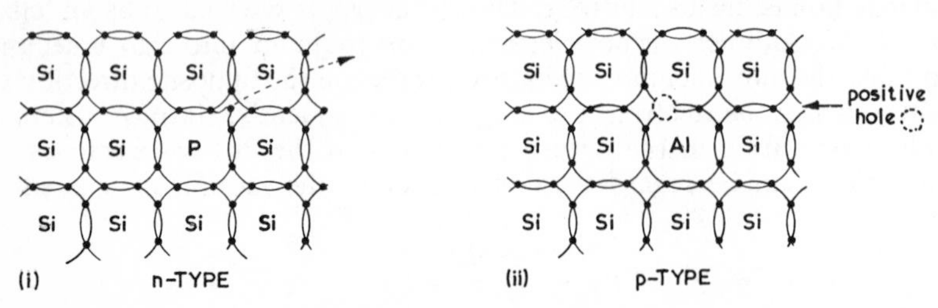

Fig. 18.10—Extrinsic semi-conductors. (Again the structure is shown two-dimensionally instead of three-dimensionally as a tetragonal structure.)

deficiency of one electron will be equivalent to the presence of a positive hole. This form of semiconductor is referred to as the *p*-type and the 'impurity' element involved as the *acceptor* since it provides a hole which will capture electrons. By controlling the small amounts of impurity atoms present in either silicon or germanium crystals, semiconductors with specific properties in either *n*- or *p*-types can be produced.

18.6.4 *The p–n junction: diodes and transistors.* A semiconductor in which a section of *p*-type material shares a common interface with a piece of *n*-type material is known as a *p–n* junction. It is upon the properties of this

p–n interface that the diodes and transistors of modern radio and other electronic equipment depend. Such devices are not made by joining separate p- and n-crystals but by taking a pure silicon (or germanium) crystal and then 'doping' appropriate parts of it with donor and acceptor atoms, generally by diffusion. Supposing, for the sake of argument, that a p-type region and an n-type region are brought together. Although each

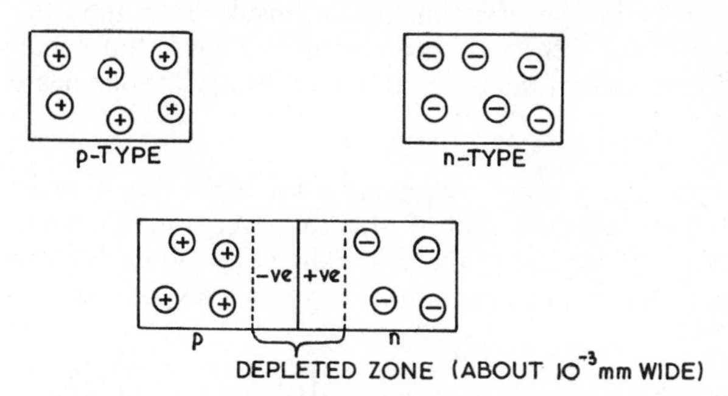

Fig. 18.11—The formation of a p–n junction.

section is initially electrically neutral,* some migration of electrons and positive holes will occur near to the interface such that they tend to neutralise each other. Thus charge carriers are lost in this depleted zone. Inside the depleted zone acceptor atoms (p-type) have captured electrons and so become negatively charged ions, whilst donor atoms (n-type) have lost free electrons to become positively charged ions. Thus the depleted layer has become charged, holding a positive charge on one side of the interface and a negative charge on the other side. Such a charged zone will discourage charged particles from passing through it and so constitute an effective barrier to the passage of electricity.

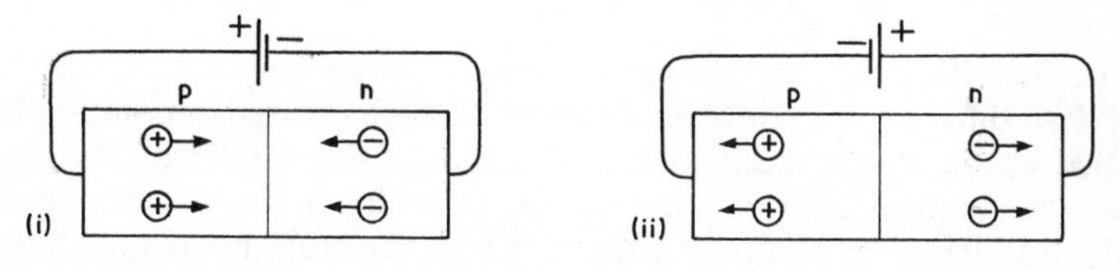

Fig. 18.12.

Consider a potential difference applied to a crystal containing a p–n junction as in Fig. 18.12 (i). Here the positive holes in the p-type are repelled to the right whilst the free electrons in the n-type are driven to the left. That is the electrons and the positive holes are driven towards each other and a current will flow in the circuit. However, in the arrangement

* There are equal numbers of protons and electrons in the semiconductor as a whole.

shown in Fig. 18.12 (ii) the reverse situation obtains; the free electrons and positive holes are attracted *away* from each other by the polarity of the applied potential difference leaving a region around the interface which is insulating since it contains neither free electrons nor positive holes.

This device has the properties of a rectifier, that is, it is a good conductor of electricity in one direction but an insulator in the other. Thus it may be used in the same way as a thermionic diode but occupies only a fraction of the space of the latter and requires no cathode heater to make it operate.

18.6.5 The current–voltage relationship for a *p–n* junction is shown in Fig. 18.13. This indicates that in practice a very small current will flow when a voltage is applied in the $n \rightarrow p$ direction (so-called 'reverse bias', i.e. V is negative in the diagram). However, this reverse current is so

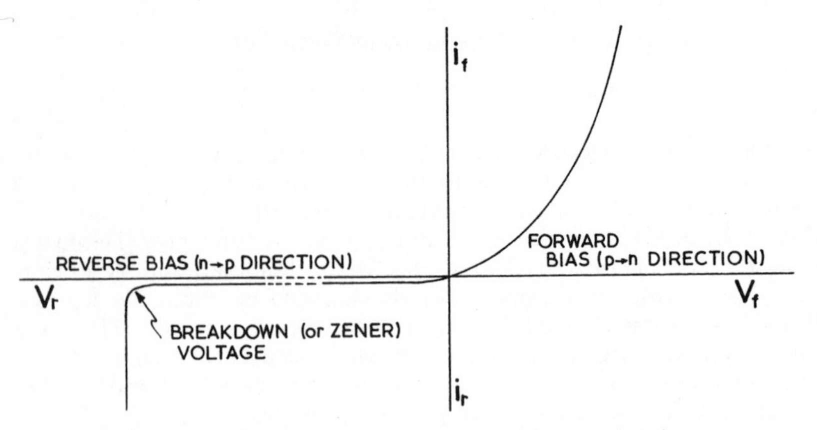

Fig. 18.13—Current as a function of voltage for a *p–n* junction.

minute that the junction behaves virtually as a one-way conductor, that is as a rectifier. If reverse bias continues to be increased a point will be reached where, under the influence of the increasing electrical field, electrons will acquire sufficient energy to move from the valency to the conduction band. The current then increases very rapidly at this 'breakdown', or *Zener*, voltage.

Magnetism

18.7 Iron is the metal generally associated with magnetism but three other 'transition' metals—cobalt, nickel and gadolinium—are also strongly magnetic, or *ferromagnetic*, as it is usually termed. This distinction is necessary because many other elements and materials possess weak magnetic properties. Most metals are *paramagnetic*, that is they are weakly attracted by a strong magnetic field, whilst some metals and all of the non-metals are *diamagnetic*, that is they are repelled by a strong magnetic field.

Only the ferromagnetic metals have characteristics which make them technologically useful as magnets, either alloyed with each other or with other suitable materials.

18.7.1 *Ferromagnetism.* What we refer to as a 'magnetic field' associated with a ferromagnetic material is produced by the 'spin' of electrons. In fact we can produce an electromagnetic field by passing a stream of electrons through an electrical conductor. In a stable atom the electrons may be regarded as moving particles rotating in orbitals around the nucleus. At the same time each electron behaves as though it were spinning on its own axis and it is this spin which generates the magnetic field, that is, it is in effect a minute magnet.

We have seen that in a stable atom, whether it exists as a single unit or is part of a crystal, not more than two electrons are permitted to occupy the same energy level. These two electrons have spins which are in opposite directions. Consequently in the majority of elements as many electrons will have spins in one direction as in the other and individual fields produced by electrons will cancel out. Hence there will be little, if any, resultant magnetic field associated with the atom. Such elements tend to

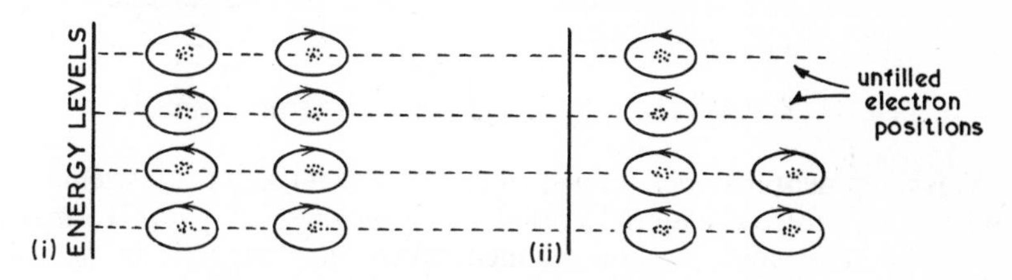

Fig. 18.14—Electron 'spin' as the basis for magnetism.

be either paramagnetic or diamagnetic (Fig. 18.14 (i)). Some of the transition elements, however, have unfilled *sub*-valency shells so that *unpaired* electrons will be present. As a result, in any atom, more electrons will spin in one direction than in the other (Fig. 18.14 (ii)) and such an atom will have a resultant magnetic moment. In the elements α-iron, cobalt, nickel and gadolinium the magnetic moments of the atoms are great enough and the ions sufficiently close together that when the atoms become magnetically aligned—during 'magnetisation'—a powerful magnetic field is produced.

18.7.2 Magnetic properties are affected critically by atomic spacing. If atoms are too far apart then forces between them are too weak to resist the effects of thermal agitation which throws spins out of alignment so that individual fields cancel. If atoms are too close together then interatomic forces are too great to allow alignment of spins. Conditions appear to be favourable when the atomic radius is between 1·5 and 2·0 times the radius of the shell containing the un-paired electrons. The elements α-iron, cobalt, nickel and gadolinium fulfil these requirements but other transition elements such as titanium, chromium and manganese are only just outside the range. Thus if manganese contains interstitial nitrogen atoms it becomes ferromagnetic due to a modification in the atomic spacing. Similarly interatomic spacing in such ceramics as Ba-$Fe_{12}O_{19}$ makes them available as magnetic materials. Alloys containing various proportions of the metals iron, chromium, tungsten, cobalt, nickel, aluminium and titanium (Table 18.3) are particularly suitable as permanent magnets.

Contained within the crystals which comprise a ferromagnetic material are small regions known as magnetic *domains.* These are regions, possibly not more than 0·05 mm in diameter, in which atoms are aligned so that their orientations are in the same direction. In the unmagnetised state these domains are randomly orientated so that there is no resultant field, but if these domains are aligned by applying an external magnetic field the material becomes magnetised (Fig. 18.15).

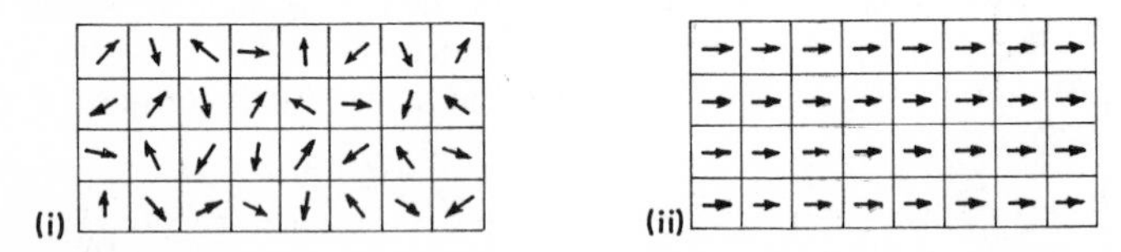

Fig. 18.15—Alignment of domains: (i) unmagnetised; (ii) magnetised.

This alignment may be retained when the magnetising field is removed and in this case the material is said to be *magnetically hard.* In many cases the magnetism is not retained when the magnetising field is removed and the material is said to be *magnetically soft.* In fact magnetic hardness usually coincides with physical hardness, that is, a hardened steel is often suitable as a permanent magnet, whilst a soft steel is also magnetically soft and therefore unsuitable as a permanent magnet. Possibly the residual strains present in the hardened material oppose *randomisation* of the domains once the material has been magnetised. If a magnetised material is heated it loses its magnetism because the thermal energy supplied provides the necessary activity which permits randomisation of domain alignment.

18.7.3 A permanent magnet is designed to provide a source of magnetism which does not involve the use of energy from an external source. It must also be able to retain its magnetism for an unlimited period even under the action of limited demagnetising influences. The actual magnetic strength, or *flux*, in a permanent magnet is dependent upon its design and the material from which it is made. The essential characteristics of a permanent-magnet material after it has been magnetised are:

(*i*) *coercive force*;
(*ii*) residual magnetism or *remanence*;
(*iii*) *energy product value*.

The coercive force is a measure of the material's resistance to demagnetisation by electrical means, whilst remanence relates to the intensity of magnetism remaining in the magnet after the magnetising field has been switched off. The energy product value is virtually the quantity of magnetic energy stored in the magnet following magnetisation. It is derived from the demagnetisation curve *RU* (Fig. 18.16).

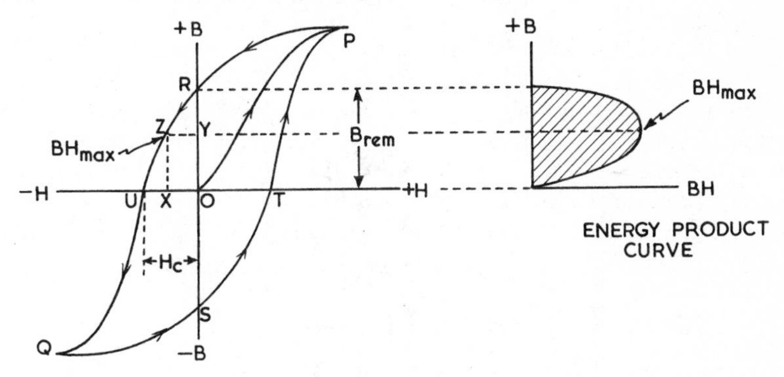

Fig. 18.16—A typical hysteresis curve for a ferromagnetic material.

18.7.4 Fig. 18.16 shows what is known as a *hysteresis loop* for a magnetically hard material. Here the induced magnetic flux density (measured in *teslas* and denoted by *B*) is plotted against the magnetising field (expressed in *ampere-turns per metre*, *H*). Such information is obtained when a ferromagnetic material is placed in a solenoid through which a current can be passed. Increased values of current will produce an increased magnetic field strength *H* which, in turn, will induce an increasing flux density *B* in the magnet.

Beginning with a small value of *H* and increasing it gradually, corresponding values of *B* and *H* are plotted to give the curve *OP*. This is the magnetisation curve which ultimately 'levels out' at *P* where the material reaches magnetic saturation. If the magnetising field is now reduced to zero the corresponding fall in the induced magnetic flux density is indicated by *PR*. Thus *OR* represents the value of the magnetic flux density remaining in the material when all external magnetising influence is removed. This is the remanent magnetism or remanence, B_{rem}.

To demagnetise the material and at the same time measure the force required to do this, a current is passed through the solenoid in the reverse direction and gradually increased. This produces the curve *RU* which is the demagnetisation curve. Intercept *OU* represents the force required to demagnetise the material completely and is called the coercive force. To

Table 18.3—Some permanent-magnet alloys.

	Composition (%) (balance–Fe)									Magnetic properties		
Type of alloy	C	Cr	W	Co	Al	Ni	Cu	Nb	Ti	B_{rem} (T)	H_c (A/m)	BH_{max}
Hardened carbon steel	1·0	—	—	—	—	—	—	—	—	0·9	4 400	1 560
35% Co steel	0·9	6·0	5·0	35·0	—	—	—	—	—	0·9	20 000	7 800
'Alnico'	—	—	—	12·0	9·5	17·0	5·0	—	—	0·73	44 500	13 500
'Alcomax III'*	—	—	—	24·5	8·0	13·5	3·0	0·6	—	1·26	51 700	38 000
'Hycomax III'*	—	—	—	34·0	7·0	15·0	4·0	—	5·0	0·88	115 400	35 200
'Columax'	—	—	—	24·5	8·0	13·5	3·0	0·6	—	1·35	58 800	52 800
'Feroba III'*	Fired ceramic—13.0 Ba and 27·0 oxygen									0·34	200 000	19 000

* Anisopropic materials, properties measured along the preferred axis.

complete the hysteresis loop the operations outlined are carried on to 'negative' saturation and then repeated in the reverse sense. The hysteresis loop represents the lagging of the magnetic flux induced behind the magnetising force producing it. The area within the loop is proportional to the quantity of energy lost when the material is magnetised or demagnetised and, for permanent magnet materials, the area of the loop must be as large as possible.

18.7.5 The ultimate criterion of a permanent magnet is the maximum product of B and H obtained from a point Z on the demagnetisation curve, i.e. the position of Z is such that the product of XZ and YZ is a maximum. This value, BH_{max}, corresponds to the maximum energy the magnet can provide in a circuit external to itself.

Hardened carbon steels were originally used as permanent magnets but these were later replaced by alloy steels containing chromium, tungsten and cobalt. Modern magnetic alloys are of the iron–aluminium–nickel–cobalt type (Table 18.3). Often the choice of a permanent magnet material relates to whether high remanence or a high coercive force is required. Thus 'Columax' has a high remanence whilst 'Hycomax III' and the ceramic 'Feroba III' both have a high coercive force. Some of the materials are *anisotropic*, that is they have superior magnetic properties along a preferred axis. This effect is obtained by heating the alloy to a high temperature and allowing it to cool in a magnetic field. As thermal agitation is reduced groups of atoms become orientated along the field and this orientation is 'frozen in'.

18.7.6 Whilst magnetically hard materials are required for permanent-magnet applications such as loudspeakers, magnetic chucks, generators, small electric motors and the like, such components as transformer cores and dynamo pole-pieces need to be magnetically soft but of high

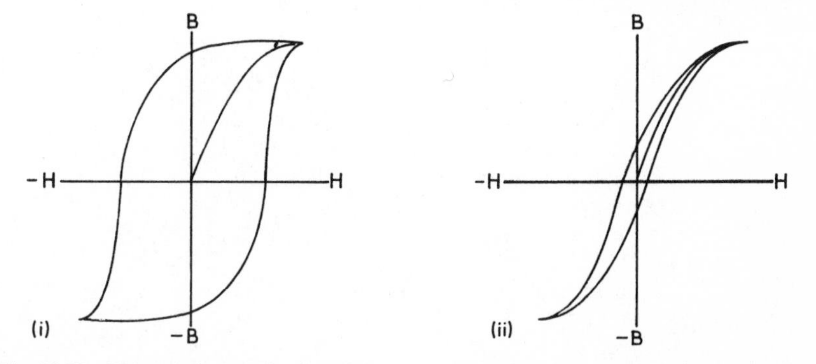

Fig. 18.17—Magnetic hysteresis. (i) This is a curve typical of a magnetically hard material. (ii) Shows a curve for a magnetically soft material. Here both B_{rem} and H_c are very small so that little energy will be lost in overcoming these forces when the material is used for transformer cores.

magnetic permeability. Permeability (μ) is given by the ratio B/H and its value alters with the magnetising field, reaching a maximum and then falling. Since a transformer core is under the influence of alternating current, the core is first magnetised, then demagnetised and then remagnetised with reverse polarity. Such a cycle produces the closed hysteresis loop PRUQST (Fig. 18.16) and the area of this loop represents the amount of energy wasted in overcoming remanent magnetic forces each time the current is reversed (for example 50 times per second). This energy is lost as heat and it is therefore necessary to use materials which give very narrow hysteresis loops and consequetly small 'hysteresis loss' (Fig. 18.17 (ii)).

18.7.7 Soft iron and iron-silicon alloys (4% Si) have very low hysteresis and the latter is widely used for stampings for transformer cores. Some iron–nickel alloys are also important as low-hysteresis, high-permeability alloys. Thus 'Mumetal' (74 Ni; 5 Cu; 1 Mn; balance Fe) is used in communications engineering particularly as a shielding material.

Chapter Nineteen
Introductory Nuclear Science

19.1 By the processes of ordinary chemical reactions atoms regroup themselves to form molecules or crystals as outlined in Chapters Two and Three of this book. During these reactions only certain of the electrons present in each atom undergo some rearrangement or redistribution during the formation of new bonds.* The nuclei of the atoms, however, remain inactive and do not change in character or composition. Thus chemical changes can be regarded in simple terms as processes involving only the surfaces fringes of atoms.

Some atoms do undergo changes which cannot be explained in terms of alterations in the arrangements of outer shells of electrons. Such changes were first noticed towards the end of the nineteenth century.

Radioactivity

19.2 The natural phenomenon which we call radioactivity was discovered in 1896 by Henri Becquerel who found that uranium compounds could fog photographic plates even when the latter were wrapped in black paper. Although uranium had been known as a heavy metallic element for many years before this its very weak radioactivity had remained undetected. Systematic examination of both uranium and thorium minerals led to the discovery of other radioactive elements, most of which were present in minute quantities but exhibited intense activity. Best known amongst these elements is radium separated in extremely small quantities from the uranium mineral *pitchblende* by Mme Curie in 1910. Altogether some forty nuclei—if we include isotopes—with radioactive properties occur naturally.

The two particular properties of radioactive elements are their emission of some form of characteristic radiation and their simultaneous transformation into other elements which may or may not be radioactive. Thus when an atom of radium, with an atomic mass number of 226, undergoes radioactive disintegration it transforms to an atom of the gas, radon, with an atomic mass number of 222. At the same time both α and γ radiations are emitted (Fig. 19.1).

Radiations

19.3 All radioative elements emit radiations which will 'fog' a photographic film. These radiations also have the power of ionising air and hence of making it a conductor of electricity. That these radiations were

* Covalent, electrovalent or metallic bonds depending upon the nature of the element.

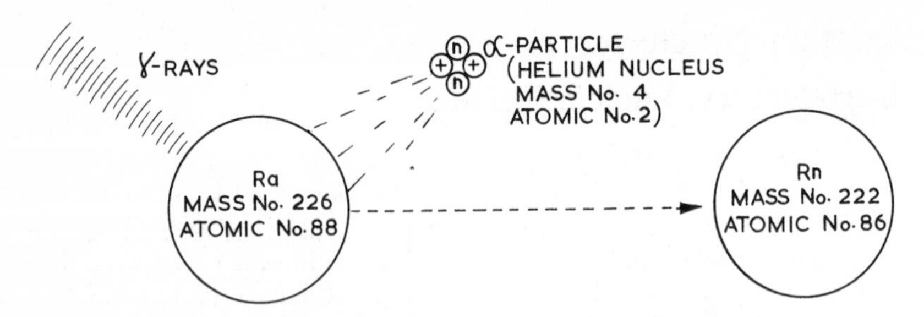

Fig. 19.1—The transformation of an atom of radium, $^{226}_{88}Ra$.

complex was realised quite early and ultimately Rutherford and others established that they were of three kinds:

19.3.1 *Alpha* (α) *'rays'*. These were found to consist of a stream of particles of matter travelling at, on average, about one-twentieth of the velocity of light. The particles themselves are helium nuclei, that is helium atoms lacking their two electrons. Hence an α-particle consists of a close association of two protons and two neutrons and, since it is travelling at a high velocity, possesses considerable kinetic energy. Consequently an α-particle is capable of producing considerable local effects on collision. Nevertheless it rarely travels more than 100 mm in air (at NTP) due to collisions with molecules of nitrogen or oxygen. Similarly α-particles are effectively trapped by a sheet of thin paper (Fig. 19.2). Since an α-particle is in fact a helium nucleus it carries a resultant positive charge.

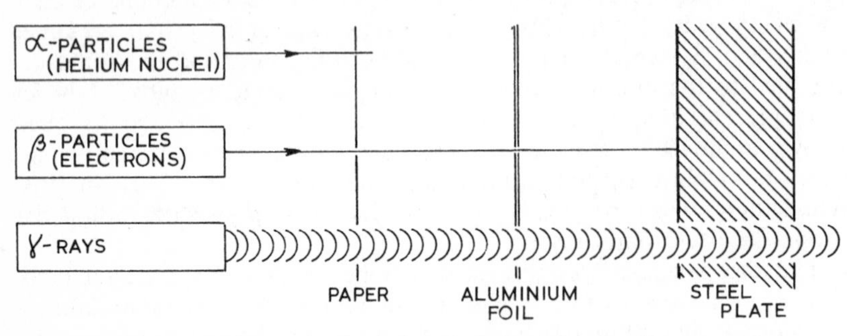

Fig. 19.2—The relative penetrating powers of α, β and γ rays.

19.3.2 *Beta* (β) *'rays'*. These consist of particles similar to ordinary electrons, as far as mass and magnitude of charge are concerned, projected at enormous velocities approaching closely that of light. However, the electrical charge carried may be either negative or positive. When the charge is negative the β-particle is of course identical with the ordinary electron, but when a β-particle carries an equal but *positive* charge it is called a *positron*. It may be regarded as a 'positive electron'. Since a positron and an electron tend to annihilate each other on contact the positron has little effect on ordinary chemical properties for which reason it has received no previous mention in this book.

19.3.3 *Gamma* (γ) *rays*, like both light and X-rays, are electromagnetic vibrations. They are of extremely short wavelength and have great powers of penetrating matter (Fig. 19.2). This makes them of value in the radiography of metals (20.9). At the same time biological tissue is generally damaged by penetration by γ-rays so that for the sake of safety of operators all γ-ray sources must be adequately screened.

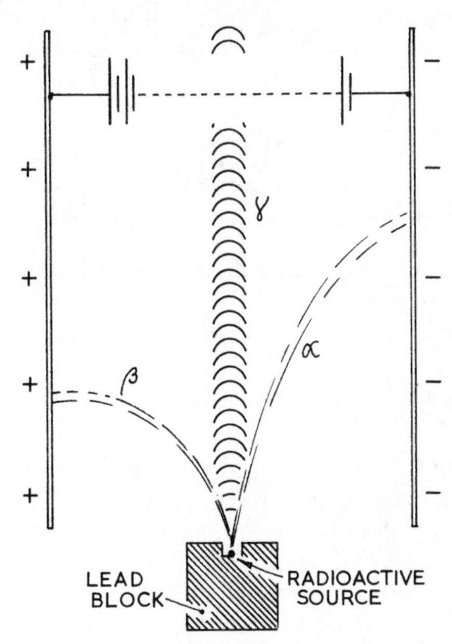

Fig. 19.3—The effects of an electrical field on α, β and γ radiations.

The nature of these forms of radioactive emanation was first investigated qualitatively by passing them between two charged metal plates (Fig. 19.3). Pieces of photographic film attached to each plate were used as radiation detectors. The emissions known as α-rays were attracted towards the negatively charged plate whilst those referred to as β-rays were attracted towards the positive plate. Some radiation passed on without being deflected by either of the charged plates. These were the γ-rays.

Since the α-rays were attracted towards the negative plate it was realised that they carried a positive charge, whilst the β-rays being attracted to the positive plate were negatively charged. Further investigation led to the realisation that these α and β 'rays' were in fact streams of fast-moving particles as outlined above—debris produced by the decay of uranium and radium atoms. The β-particles, being much less massive than the α-particles, were deflected more easily by the electrostatic field (Fig. 19.3). Thus γ-rays are the only form of *electromagnetic radiation* involved here.

19.3.4 *Detection of radiation.* The principal methods used for detecting the presence of α- and β-particles depend upon the ionisation which occurs when these particles pass through air. Rapidly moving α and β particles strip electrons from atoms of nitrogen and oxygen with which they collide. In each case the resultant particle is a positively charged ion and air which becomes enriched with these ions will conduct electricity in exactly the same way as will an electrolytic solution of, say, sodium chloride.

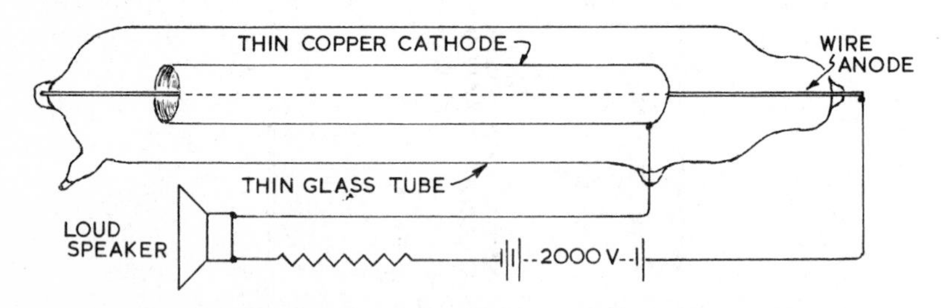

Fig. 19.4—The principle of the Geiger–Müller Counter.

The Geiger–Müller Counter virtually measures the degree of such gaseous ionisation, which, in turn, is proportional to the intensity of radiation. It consists of a cylinder of *thin* glass filled with air or argon at only a few cm pressure, and containing a hollow metal cylinder along the axis of which is a wire. Cylinder and wire are electrically insulated from each other but are connected in series with an e.m.f. of about 2000 V, a high resistance and a small loudspeaker (or some form of electrical counting device). The e.m.f. is regulated so that a 'threshold' is reached where a discharge just fails to pass between wire and cylinder. If an α- or β-particle enters the tube momentary ionisation of the gas occurs. Ions so formed are accelerated by the strong electrostatic field towards the charged plates to such an extent that, as they collide with other atoms, further ionisation is produced. The original ionisation is therefore greatly amplified and the discharge of all these ions at the plates is marked by an impulse of current through the loudspeaker. This discharge reduces the potential difference across the plates so that the accelerating force is diminished and the pulse dies immediately. Thus each α- and β-particle entering the tube produces a single report in the loud speaker.

γ-rays do not ionise gases directly but a small fraction of the γ-rays passing through a gas is absorbed by the atoms of the gas with the release of electrons. These secondary electrons cause ionisation so that the presence of γ-rays can be detected indirectly.

Nuclear stability

19.4 The relationship between the size and make-up of the atomic nucleus and its stability was dealt with earlier in this book (1.4.2). Briefly the maximum size of nucleus which is possible is dependent upon a state of balance existing between the special binding forces operating between *all nucleons* and the electrostatic forces of repulsion between *protons.* However, the *total* binding force of all nucleons in a massive nucleus is *weaker* than it is in a light nucleus. Thus in a massive nucleus the total binding force is less able to hold the nucleus together *against* the total repulsion force between the positively charged protons. This is one reason why 'extra' neutrons are present in the massive nuclei of uranium and radium—to provide additional binding force. Neutrons, having no charge, add to the binding force but do not contribute to electrostatic repulsion.

Nevertheless the 'extra' neutrons make the nucleus even more massive and the average binding force per nucleon proportionally smaller still, so that, despite the extra 'cementing force' provided by these neutrons, the nucleus becomes unstable with a tendency to break up.

Fig. 19.5 shows the relationship between the mean binding energy per nucleon and the size of the nucleus. It indicates that the mean binding force per nucleon is at a maximum with an atomic mass of about 60 but then decreases as A increases so that atoms with atomic masses of more than 200 are likely to be radioactive.

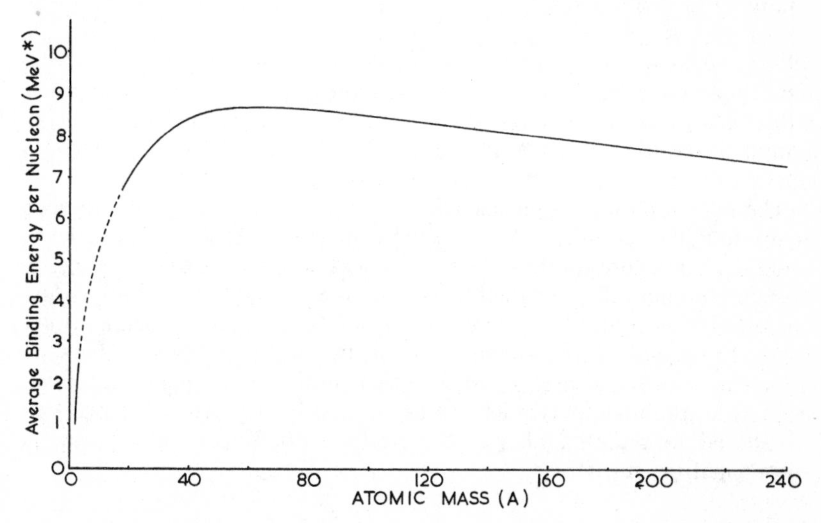

Fig. 19.5—Average Binding Energy per nucleon for the most stable nuclides of each Atomic Mass Number.

The nucleus reduces its mass by ejecting α-particles. Since an unsuitable neutron/proton ratio will also cause nuclear instability 'daughter elements' may also be radioactive as neutrons change to protons by means of β-decay thus reducing the amount of neutrons present. By these means the nucleus is reduced to more stable proportions. Typically radium gradually 'decays' in a number of stages to produce stable lead.

As indicated earlier the presence of a nuclear binding force gives rise to potential energy within the nucleus and when such a nucleus breaks down even partially some of the energy is released. One manifestation of this is the powerful γ-rays already mentioned, whilst fast-moving α- and β-particles also possess considerable kinetic energy.

19.4.1 *Radioactive series.* There are three natural 'decay series' in which massive nuclei diminish in size and ultimately become stable. The radium series starts with the uranium isotope ${}^{238}_{92}U$; the actinium series starts with the uranium isotope ${}^{235}_{92}U$; and the thorium series which starts with the isotope ${}^{232}_{90}Th$. In each case the isotope decays by particle emission through a number of intermediate radioactive stages until a stable isotope of lead remains. Here we shall deal only with the radium series (Fig. 19.6).

The decay of uranium ${}^{238}_{92}U$ to the stable lead isotope ${}^{206}_{82}Pb$ takes place in stages as indicated. Here both α- and β-decays are involved, a diagonal arrow indicating α-decay and a horizontal arrow indicating β-decay. As already mentioned it is not only the over-massive nuclei which tend to break down but also nuclei which contain the wrong ratio of neutrons to protons and the acceptable ratio can often be restored by neutrons changing to protons.

19.4.2 *Alpha* (α) *decay* refers to the emission of α-particles, which as described above, are helium nuclei containing two protons and two neutrons. They tend to be emitted by the more massive nuclei in their attempts to reduce themselves in size, though they are also ejected from the nuclei of some of the unstable isotopes of the lighter elements.

The rejection of an α-particle from a nucleus reduces its mass number, A, by four, i.e. $2p + 2n$. At the same time the character of the nucleus and the chemical properties of the atom as a whole are changed since the atomic number, Z, is reduced by two, i.e. $2p$. Thus radium, which chemically resembles the calcium group of fairly reactive metals, transforms to the gas radon which is chemically inert since it is in the same group as helium and argon and the other 'noble' gases. Hence, chemical activity and radioactivity must not be confused. The former is a property associated with the stability of the electron 'shells', the latter with the stability of the nucleus.

19.4.3 *Beta* (β) *decay* of course refers to the emission of β-particles and is associated with the change:

$$\text{neutron} \underset{\beta^+}{\overset{\beta^-}{\rightleftharpoons}} \text{proton}$$

Thus if a nucleus contains too many neutrons the balance is restored as a neutron changes to a proton by emitting a negatively charged β-particle (identical with an ordinary electron). Since the number of protons in the nucleus has increased by one the *atomic number*, Z, will have increased by one though the *atomic mass number*, A, will remain the same.

If the nucleus contains too many protons then the balance may be restored by rejection of a positively charged β-particle, or *positron*. In this way a proton changes to become a neutron. Obviously there is a strong

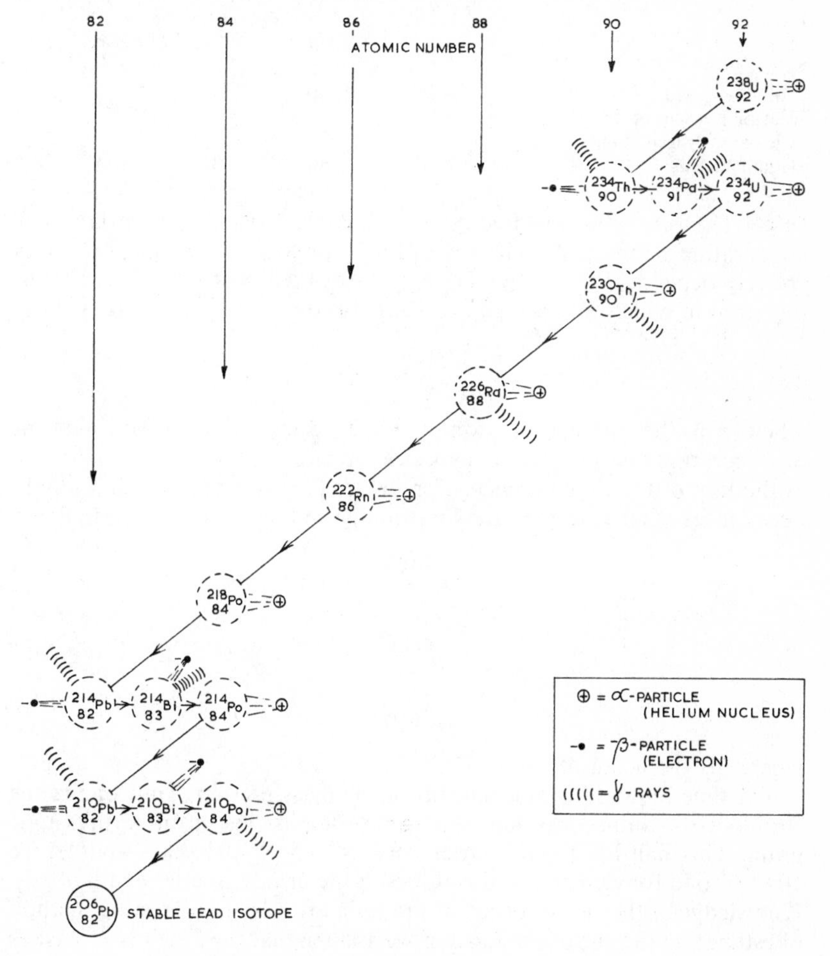

Fig. 19.6—The natural decay series of uranium, $^{238}_{92}U$.

force of attraction between any ejected positrons and the electrons surrounding the nucleus. Consequently collisions between positrons and electrons are inevitable. When this occurs the two particles annihilate each other and the very short wave length electromagnetic rays we call γ-rays are emitted.

Table 19.1—Sub-atomic particles associated with nuclear reactions.

Particle	*Mass number*	*Charge* (electron = −1	*Symbol*	*Symbol, with mass number* (A) *and atomic number* (Z) i.e., $^{A}_{Z}S$
Electron	0	−1	β^-	e^- (or $^{0}_{-1}e$)
Positron	0	+1	β^+	e^+ (or $^{0}_{1}e$)
Neutron	1	0	n	$^{1}_{0}n$
Proton	1	+1	p	$^{1}_{1}H$
Deutron (nucleus of the heavy-hydrogen atom)	2	+1	d	$^{2}_{1}H$ (or $^{2}_{1}D$)
Helium nucleus	4	+2	α	$^{4}_{2}He$

19.4.4 The *rate* of radioactive decay is independent of such variables as temperature or pressure, whereas ordinary physical or chemical changes are very dependent upon such factors. Rate of decay is a characteristic of the atom in which the change occurs. It is governed by the expression:

$$-\frac{dn}{dt} = \lambda n$$

where n is the number of atoms of the 'species' present and λ is the disintegration constant for that species. Integration of the expression and evaluation of the time necessary for *half* of the atoms present initially to decay gives what is termed the 'half-life period' (t) of the species. Thus:

$$t = \frac{\log 2}{\lambda}$$

$$= \frac{0{\cdot}6931}{\lambda}$$

$$= 0{\cdot}6931\ l$$

where l is the mean life.

The time required is independent of the mass present. Thus one tonne shrinks to $\frac{1}{2}$ tonne in exactly the same time as one gram shrinks to $\frac{1}{2}$ gram. This half-life period varies between $4{\cdot}5 \times 10^9$ years and $1{\cdot}5 \times 10^{-4}$ second for elements and isotopes in the uranium series (Table 19.2). Knowledge of the rate of decay of uranium affords an interesting method of estimating the age of the Earth if we assume that the Earth is at least as old as the minerals in its crust. Since the half life of uranium is $4{\cdot}51 \times 10^9$

years one gram of uranium present $4{\cdot}51 \times 10^9$ years ago would by now have decayed to produce 0·5 g uranium, 0·4326 g lead and 0·0674 g helium. Thus by comparing the relative amounts of uranium and lead now present in uranium-bearing minerals the age of the rocks can be estimated. The result so derived is about $2{\cdot}6 \times 10^9$ years which agrees reasonably well with the value obtained by other means.

Table 19.2—Details of the radioactivity of the uranium series.

Note that the element is characterised by its atomic number (Z) whilst its mass number may vary, i.e. it exists as different isotopes (1.4.1). Thus whilst the lead isotope $^{206}_{82}Pb$ is stable, the isotopes $^{214}_{82}Pb$ and $^{210}_{82}Pb$ are unstable and therefore radioactive.

Element	*Symbol*	*Atomic number* (Z)	*Isotope mass number* (A)	*Radiation*	*Half-life*
Uranium	U	92	238	α	$4{\cdot}51 \times 10^9$ years
Thorium	Th	90	234	β, γ	24·5 days
Protoactinium	Pa	91	234	β, γ	1°14 minutes
Uranium	U	92	234	α	$2{\cdot}33 \times 10^5$ years
Thorium	Th	90	230	α, γ	$8{\cdot}3 \times 10^4$ years
Radium	Ra	88	226	α, γ	1590 years
Radon	Rn	86	222	α	3·825 days
Polonium	Po	84	218	α (β)	3·05 minutes
Lead	Pb	82	214	β, γ	26·8 minutes
Bismuth	Bi	83	214	β, γ (α)	19·7 minutes
Polonium	Po	84	214	α	$1{\cdot}5 \times 10^{-4}$ seconds
Lead	Pb	82	210	β, γ	22 years
Bismuth	Bi	83	210	β	5 days
Polonium	Po	84	210	α, γ	140 days
Lead	Pb	82	206	—	Stable

The changes in atomic number (Z) indicated in the transformations in the uranium-lead series set out in Fig. 19.6 confirm the statement made earlier in this chapter that radioactivity is produced by changes taking place in the positively charged nucleus of the radioactive atom. A radioactive change is therefore a nuclear change.

19.4.5 A special form of equation is used to represent nuclear changes. Thus the change from radium to radon can be written:

$$^{226}_{88}\mathrm{Ra} \rightarrow {}^{222}_{86}\mathrm{Rn} + \underset{(\alpha\text{-particle})}{^{4}_{2}\mathrm{He}} + \text{energy}$$

This indicates that the atomic mass of the radium atom has fallen by four units which is, of course, the mass of the α-particle (or helium nucleus). Since the α-particle contains two protons the atomic number (Z) falls from 88 (Ra) to 86 (Rn).

Similarly the decay of the radioactive isotope of bismuth can be expressed:

$$^{210}_{83}Bi \rightarrow ^{210}_{84}Po + \underset{(\beta\text{-particle})}{^{0}_{-1}e} + \text{energy}$$

In this case a neutron changes to a proton with the emission of a negative β-particle (electron) and so the atomic number (Z) *increases* by one. Table 19.2 and Fig. 19.6 indicate that certain isotopes of lead and bismuth are radioactive. This is because their nuclei, formed during the process of radioactive decay, contain too many neutrons and so further decay occurs until a stable isotope of lead, $^{206}_{82}Pb$, remains.

Initially nuclear radiations could only be obtained from natural sources such as radium (and radon arising from its decay) in sufficient intensities for medical purposes such as the treatment of cancer. Radium was extremely expensive to extract from its ores. Fortunately, as we shall see later, radioactivity can be induced in many elements by bombarding their nuclei with neutrons in an 'atomic pile'. To a large extent therefore radium has lost its importance as an intense source of radioactivity.

'Artificial' nuclear reactions

19.5 The dream of the alchemists of the Middle Ages was to transmute base metals into gold but it was not until 1919 that Rutherford saw nuclear transmutation as a practical possibility. He was able to demonstrate that some nuclear change did in fact occur when nitrogen gas was exposed to bombardment by α-particles from a radium source. In 1925 Blackett showed that about 0·000 3% of the α-particles passing through the nitrogen found a target in a nitrogen nucleus. When such a moving α-particle collides with a nitrogen nucleus the following reaction occurs:

$$^{14}_{7}N + \underset{(\alpha\text{-particle})}{^{4}_{2}He} \rightarrow ^{17}_{8}O + ^{1}_{1}H$$

Thus an oxygen isotope of mass number 17 was formed along with a proton (hydrogen nucleus, $^{1}_{1}H$).

These early experiments by Lord Rutherford and other workers laid the foundations of nuclear science as we know it today, that is, a science in which particles are projected at other atoms or molecules. Although these particles are small enough to penetrate objects of nuclear dimensions they must, in many cases, also possess sufficient kinetic energy in terms of their velocities, to bring about changes in the nuclei with which they collide.

In the early 1930s other particles such as the proton and the deutron—the nucleus of the 'heavy hydrogen' atom, $^{2}_{1}H$ (1.4.1)—were added to the list of projectiles which could bring about nuclear reactions.

All of these projectiles so far mentioned (α-particles, protons and deutrons) are positively charged. They are, therefore, naturally repelled by all atomic nuclei since the latter are also positively charged. In the case of very large nuclei the total charge is sufficient to repel these positively charged particles completely but lighter nuclei of small total charge suffer collision.

19.5.1 In 1932 Chadwick discovered the existence of the neutron,* an uncharged particle with a mass number roughly the same as that of the proton. These neutrons are produced during many nuclear reactions but one of the most important of these reactions is between α-particles and the nuclei of the light metal beryllium:

$$^{9}_{4}\text{Be} + \underset{(\alpha\text{-particle})}{^{4}_{2}\text{He}} \rightarrow {^{12}_{6}\text{C}} + {^{1}_{0}n}$$

Originally it had been suspected that the carbon isotope $^{13}_{6}C$ along with γ-rays had been the product of this reaction but Chadwick explained the penetrating power of the 'rays' in terms of the new particle, the neutron, which had a very high velocity and this, coupled with its lack of charge, enabled it to penetrate other nuclei due to the absence of any Coulomb repulsion. Hence high-velocity neutrons are very effective nuclear projectiles.

Also a source of neutrons is the reaction between deutrons (the nuclei of heavy-hydrogen atoms) and deuterium (heavy-hydrogen) atoms themselves. These deuterium atoms are most conveniently contained in 'heavy ice' or frozen deuterium oxide, D_2O.

$$\underset{\text{atom}}{^{2}_{1}\text{H}} + \underset{\text{nucleus}}{^{2}_{1}\text{H}} \rightarrow {^{3}_{2}\text{He}} + {^{1}_{0}\text{n}}$$

The product is a helium isotope $^{3}_{2}He$ and a neutron. Note that in the above equation the symbol $^{2}_{1}H$ represents both the deuterium (heavy-hydrogen) atom and the deutron.

The various nuclear projectiles are used for three different purposes:

(*i*) the production of artificially radioactive elements to replace scarce elements like radium;

(*ii*) the release of energy by nuclear chain reactions;

(*iii*) the experimental transmutation of elements for research and other purposes.

19.5.2 *Artificial radioactivity.* The ordinary chemical properties of an element depend upon its atomic number (Z) which is the number of protons present in the nucleus and which, in a stable free atom, is balanced by an equal number of electrons arranged in their appropriate shells. Any pure element, however, may consist of two or more isotopes

* Previous to the discovery of the neutron, nuclear mass which carried no charge had been explained in terms of 'nuclear electrons' which neutralised an equal number of protons present there. Even in the author's undergradute days, some years after the time of Chadwick's discovery, many scientists disregarded the neutron and clung to the idea of 'nuclear' and 'extra-nuclear' (or 'planetary') electrons. It was a tidy world of two simple particles—the electron and the proton—which, alas, soon had to end as a multiplicity of new particles were discovered.

in which the mass number (A) will vary from one atom to another because the number of neutrons in the nucleus also varies. The instability of very large nuclei such as those of uranium, thorium and actinium, leading to their natural radioactivity has already been dealt with (19.4.1). The naturally occurring isotopes of the lighter elements are on the other hand generally stable,* but radioactivity can be induced in many of these by projecting particles at their nuclei so that new, unstable isotopes are produced in which the neutron/proton ratio is too high. The neutron is frequently used for this purpose but its velocity on impact often determines the nature of the nuclear reaction which follows.

Thus if a *slow-moving* neutron is fired into the nucleus of the common aluminium isotope $^{27}_{13}Al$, a radioactive istope of mass 28 is produced:

$$^{27}_{13}Al + ^{1}_{0}n \rightarrow ^{28}_{13}Al + \gamma\text{-rays}$$

The excess energy in this case is liberated as γ-rays and the new istope $^{28}_{13}Al$ is radioactive. It emits negative β-particles (electrons) leaving a stable isotope of silicon:

$$^{28}_{13}Al \rightarrow ^{28}_{14}Si + \underset{(\beta\text{-particle or electron})}{^{0}_{-1}e}$$

When a *fast-moving* neutron penetrates an atom of the same aluminium isotope, an α-particle is emitted and a radioactive isotope of sodium remains:

$$^{27}_{13}Al + ^{1}_{0}n \rightarrow ^{24}_{11}Na + \underset{(\alpha\text{-particle})}{^{4}_{2}He}$$

Again β-decay follows so that a lower neutron/proton ratio is restored:

$$^{24}_{11}Na \rightarrow ^{24}_{12}Mg + \underset{(\beta\text{-particle or electron})}{^{0}_{-1}e}$$

A stable isotope of magnesium remains.

The fast-moving particles referred to above are often produced by accelerating them in some form of machine such as a cyclotron or a synchrotron. The initial use of such techniques was adopted by Cockroft and Walton in 1933. They used a comparatively simple voltage generator in order to accelerate protons sufficiently so that they would enter the nuclei of lithium atoms:

$$^{7}_{3}Li + \underset{(\text{proton})}{^{1}_{1}H} \rightarrow ^{4}_{2}He + ^{4}_{2}He$$

The kinetic energies of the two α-particles produced in this way are considerable. Naturally, particle accelerators which depend upon the application of electromagnetic fields can only be used to accelerate charged particles, e.g. protons, deutrons and α-particles. Neutrons would be unaffected by the application of an accelerating field but, on the other hand, have the advantage that they are not deflected by nuclear charges either.

* An exception is carbon $^{14}_{6}C$ which has a long half-life period of 5568 years. Since $^{14}_{6}C$ is absorbed biologically along with the ordinary $^{12}_{6}C$ it can be used to 'date' objects which are organic in basis, e.g. ancient manuscripts and mummies.

Radioactive isotopes are commonly produced by irradiating the normal element with neutrons in an atomic pile (19.6.3). A nucleus struck by a neutron absorbs it and then possesses an excess of energy which is subsequently released as one or more forms of radiation. The ultimate uses of an artificially produced radioactive isotope may depend upon a number of factors such as its half-life, its chemical properties and of course the type of radiation emitted. Thus 'cobalt 60', i.e. $^{60}_{27}Co$ has a convenient half-life of about 5·3 years and is a powerful γ-ray source. It is used instead of radium (or radon) in the treatment of tumours into which 'seeds' of the isotope can be planted. It is also used as a γ-ray emitter in the radiography of metals (20.9.1).

19.5.3 Radioactive isotopes of elements which can be absorbed biologically are often used in internal treatments. Thus radioactive 'iodine-131', i.e. $^{131}_{53}I$, which will concentrate in the thyroid can be used to treat overactivity in that organ, whilst 'phosphorus-32', i.e. $^{32}_{15}P$, administered orally in the form of a safe phosphate, can be used to treat a form of blood cancer in which a surplus of red cells is produced. Some β-emitters can be placed in contact with the skin to cure skin cancers.

The release of nuclear energy in nuclear chain reactions

19.6 The 1933 experiment of Cockroft and Walton in which they 'split' the lithium atom was mentioned above. Not only was this the first truely artificial transmutation of an element but also the first man-made thermonuclear reaction. The reaction can be written more fully:

$$\,^{7}_{3}Li + \underset{\text{(proton)}}{^{1}_{1}H} \rightarrow \,^{8}_{4}Be \rightarrow \underset{(\alpha\text{-particles})}{^{4}_{2}He + \,^{4}_{2}He}$$

The fast-moving proton enters the lithium nucleus to form the beryllium isotope, $^{8}_{4}Be$. This is unstable and immediately 'fissions' or breaks up to produce two rapidly moving α-particles the kinetic energy of which is equivalent to the mass annihilated during the reaction and can be calculated as follows:

$$\begin{aligned} \text{Mass of nucleus } ^{7}_{3}Li &= 7{\cdot}015\,99 \text{ a.m.u.} \\ \text{Mass of proton } ^{1}_{1}H &= \underline{1{\cdot}007\,83} \text{ a.m.u.} \\ \text{Total} &= \underline{8{\cdot}023\,82} \text{ a.m.u.} \end{aligned}$$

$$\begin{aligned} &\text{Mass of two helium nuclei} \\ &\qquad (\alpha\text{-particles}) = 2 \times 4{\cdot}002\,60 \text{ a.m.u.} \\ &\qquad\qquad\qquad\quad = 8{\cdot}005\,20 \text{ a.m.u.} \\ &\text{Loss in mass due to reaction} = 8{\cdot}023\,82 - 8{\cdot}005\,20 \text{ a.m.u.} \\ &\qquad\qquad\qquad\quad = 0{\cdot}018\,62 \text{ a.m.u.} \end{aligned}$$

This loss in mass, as converted by the Einstein equation, $E = mc^2$ (1.4.4), is the equivalent of about $2{\cdot}72 \times 10^{-12}$ J of energy, the bulk of which is in the kinetic energy of the two α-particles.

In 1938 two German scientists, Hahn and Strassman, had shown that when uranium was bombarded with slow-moving neutrons certain of its atoms underwent fission into two separate fragments. Amongst the fission products identified were istopes of the elements barium, krypton, lanthanum, cerium and molybdenum—all nuclei more stable than that of uranium. The fission process resulted in some loss of mass which was of course released as energy. Of all the *naturally occurring* elements only the uranum isotope, $^{235}_{92}U$, is fissionable by *slow* neutrons. The outcome of this particular branch of nuclear chemistry received its first macabre public demonstration over Hiroshima on 6th August, 1945.

By the capture of a neutron the nucleus of $^{235}_{92}U$ becomes unstable so that it breaks down to form two smaller nuclei and some neutrons. These smaller nuclei are often unstable so that they undergo β-decay. One such reaction is:

$$^{235}_{92}U + ^{1}_{0}n \rightarrow ^{236}_{92}U \rightarrow ^{141}_{56}Ba + ^{92}_{36}Kr + 3^{1}_{0}n + 3{\cdot}2 \times 10^{-11}\ J$$

In this instance the unstable $^{235}_{92}U$ atom is shown as fissioning to form barium, $^{141}_{56}Ba$ and krypton $^{92}_{36}Kr$ but other elements may be formed, the nuclei of which are always unequal in mass with a ratio of approximately 5:7. Thus fission of $^{235}_{92}U$ can produce forty or more different elements in at least 160 isotopes.

Because the neutron/proton ratio for the massive nuclei which undergo fission is greater than the ratio for the nuclei of the products, then excess neutrons are ejected during fission. The average number per $^{235}_{92}U$ nucleus is between two and three. Even so the resultant nuclei still have a high neutron/proton ratio so that these fission products generally undergo β-decay, in which electrons are ejected as neutrons change to protons to adjust the neutron/proton ratio.

19.6.1 The main feature of the fission of $^{235}_{92}U$ is that a much greater mass is converted into energy than occurs during the fission of lighter nuclei like $^{7}_{4}Li$ (19·6). The amount of mass so converted can be calculated to be

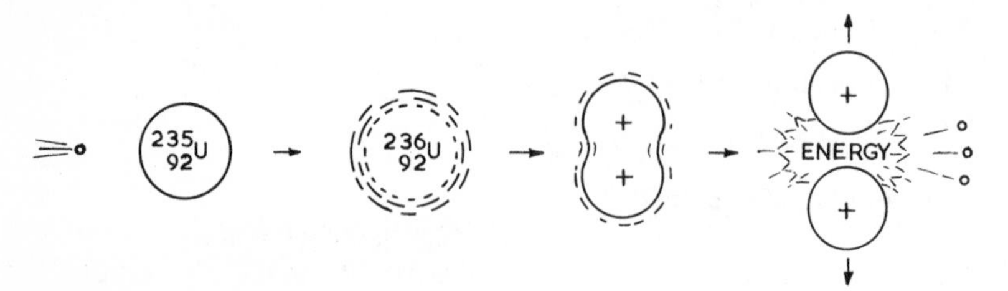

Fig. 19.7—The 'water drop' analogy of the fission of $^{235}_{92}U$. A heavy nucleus has been compared with a spherical drop of incompressible liquid which is in a state of oscillation. Deviations from spherical symmetry can be large enough to cause deformation to a 'dumb-bell' shape. The coulomb force of repulsion between the two halves can overcome the nuclear binding forces so that disruption follows (in a liquid drop the forces of attraction manifest themselves as surface tension).

about 0·215 a.m.u. per $^{235}_{92}U$ nucleus. Since one atomic mass unit is equivalent to $1{\cdot}5 \times 10^{-10}$ J (by $E = mc^2$) then $0{\cdot}215 \times 1{\cdot}5 \times 10^{-10}$ J $= 0{\cdot}323 \times 10^{-10}$ J is released by each $^{235}_{92}U$ nucleus which fissions. Roughly 80% of this energy manifests itself as kinetic energy of the fission fragments and the bulk of the remainder is released ultimately as radioactivity. Or, in more practical terms, one kilogram of $^{235}_{92}U$ undergoing fission may release as much energy as the burning of 2500 tonnes of coal.

It has been noted that the fission of an atom of $^{235}_{92}U$ releases either two or three neutrons depending upon what fission elements are produced. These neutrons then enter neighbouring $^{235}_{92}U$ nuclei and so a

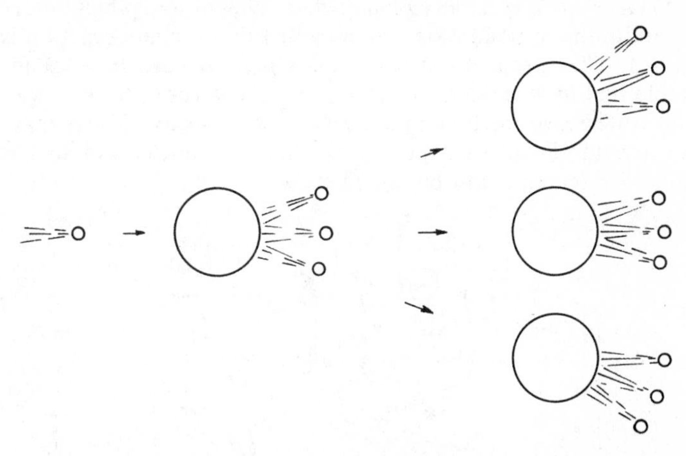

Fig. 19.8—The chain-reaction fission of $^{235}_{92}U$ under neutron bombardment.

rapid chain-reaction is set up. By the use of substances which arrest the reaction by absorbing some of the neutrons, the chain reaction can be controlled to avoid the danger of its getting out of hand. In this way the nuclear chain reaction can be controlled so that energy is produced at a rate which enables it to be used as an industrial source of power.

19.6.2 *The atomic bomb.* In a small piece of fissionable material the proportion of neutrons which escape is so large that a chain reaction is *not* set up. When two small pieces of fissionable material are brought close together quickly to form a larger body with a total mass greater than the *critical mass* the relative number of neutrons which escape decreases and a chain reaction, leading to the complete fission of most of the nuclei present, is set up.

Hence an atomic bomb (of the 'uranium type') consists essentially of several kilograms of pure $^{235}_{92}U$ in the form of two sub-critical masses, *widely separated,* and a suitable explosive device for compressing the two halves together so that the resultant mass 'goes critical'. A source of fast-moving neutrons is also required to trigger the reaction since slow-moving neutrons do not produce the catastrophic disintegration required.

Naturally occurring uranium consists principally of the isotope $^{238}_{92}U$ and only about 0·7% of $^{235}_{92}U$ is present. Obviously both isotopes undergo similar chemical reactions since their electron structures are similar. For this reason the separation of the two isotopes by chemical means is impossible and a laborious and therefore expensive process must be used which depends upon the different rates of diffusion of their gaseous fluorides through porous barriers.

19.6.3 *Nuclear chain reactors* (*atomic piles*). The atomic pile is a device in which conditions of fission are so controlled that a chain reaction is just maintained. The speed of the neutrons is first reduced by encasing the fissionable fuel in a 'moderator' which may be carbon, 'heavy' water or ordinary water. Slowing down the neutrons in this way ensures that most of those available will be captured by the $^{235}_{92}U$ nuclei and so steadier control of the reaction will be possible.

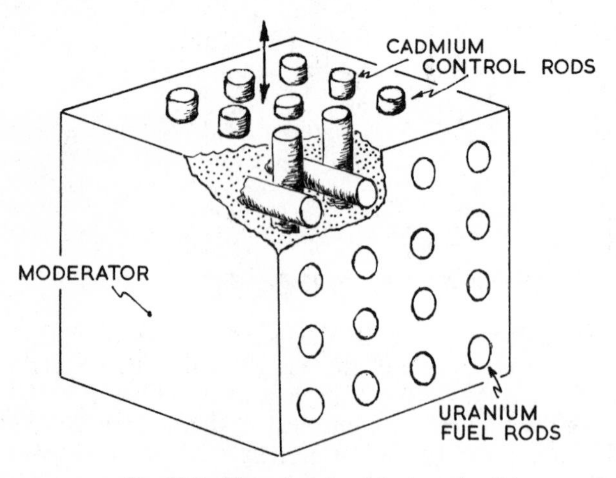

Fig. 19.9—The principle of the 'atomic pile'.

The *rate* of the chain reaction is controlled by introducing rods of cadmium (or other suitable material) into the pile (Fig. 19.9) so that the desired number of neutrons will be 'blotted up' by these rods. Whilst the rods are in the pile the chain reaction is interrupted. If the rods are withdrawn a point is reached at which the chain reaction proceeds at a suitable rate. The rods are held near to this position by an automatic device. If such a pile got out of control and became too hot it would break down into masses of subcritical size and would not therefore be liable to explode.

The energy liberated during fission within the pile is in the form of heat.* This is drawn off to operate turbo generators. A number of different designs are in use whereby heat is transferred from the pile to the generator. Piles can be either gas cooled (generally by carbon dioxide) or liquid cooled (using water or molten sodium). The principle of the 'magnox' reactor, such as is used at Calder Hall, is shown in Fig. 19.10. This is

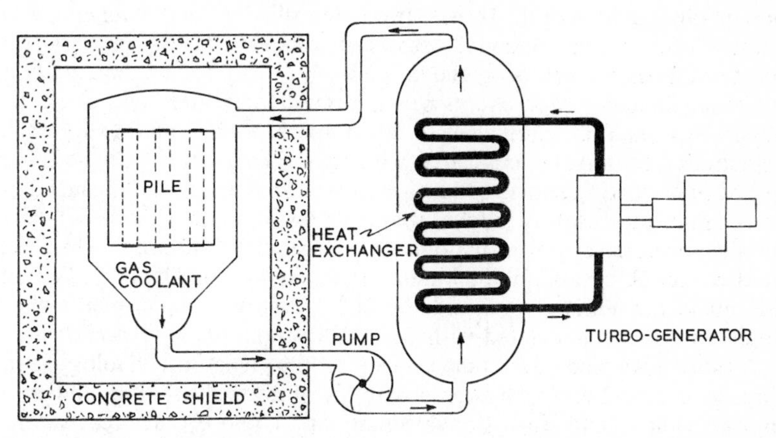

Fig. 19.10—The principles of the Magnox type of gas-cooled, graphite-moderated reactor.

a gas-cooled graphite-moderated reactor using natural uranium as fuel. Carbon dioxide, heated within the reactor pile, is then circulated through a heat exchanger.

19.6.4 *The production of new elements.* In the above brief mention of the 'magnox' reactor it was stated that *natural* uranium is used as the fuel. Here, of course, it is the small proportion of $^{235}_{92}U$ which is the fissionable isotope releasing energy. At the same time neutrons ejected during fission are also absorbed by the abundant $^{238}_{92}U$ atoms:

$$^{238}_{92}U + ^{1}_{0}n \rightarrow ^{239}_{92}U$$

This uranium isotope $^{239}_{92}U$ has a short half-life of only 23 minutes and undergoes β-decay:

$$^{239}_{92}U \xrightarrow{\text{23 minutes}} ^{239}_{93}Np + ^{0}_{-1}e$$

It will be noted that the product has an atomic number of 93 which is greater than that of any naturally occurring element. It is thus a new element manufactured by Man. Originally produced in 1940 by McMillan and Abelson who bombarded $^{238}_{92}U$ with deutrons, the new element was called 'neptunium'. Also radioactive and with a half-life of

* Derived from the KE of the fission fragments.

2·3 days, neptunium transforms by β-decay to yet another new element plutonium which has an atomic number of 94:

$$^{239}_{93}Np \xrightarrow{2\cdot 3\ \text{days}} {}^{239}_{94}Pu + {}^{0}_{-1}e$$

Plutonium, $^{239}_{94}Pu$, is also fissionable like $^{235}_{92}U$ so that the uranium pile serves as a source of further fissile material as well as of heat. However, plutonium is generally regarded as a sort of Frankenstein's monster of the metallurgical world. It has been described as a 'fiendish metal' because of its intense toxicity. A total of some 0·000 006 g in mass constitutes the maximum dose the human body can accumulate without suffering damage. Moreover since it accumulates and *remains* in the bones, this quantity, equivalent to a tiny speck of dust, could very easily be acquired by anyone associated with its working unless very stringent precautions were adopted. Plutonium is also a troublesome metal to work and 'goes critical' above about 10 kg. It made possibly its first public appearance as the fissionable charge in 'Thin Man' which exploded over Nagasaki on 9th August, 1945. As the late Dr. R. E. Wilson, US nuclear scientist, once said; 'If there ever was an element which deserved a name associated with Hell it is plutonium.'

A more advanced development in the field of reactor technology is the 'breeder reactor' whereby large quantities of natural uranium and thorium are converted into fissionable material. These breeder reactors are constructed with a blanket of 'fertile' material ($^{238}_{92}U$ or $^{232}_{90}Th$) surrounding a core of concentrated fissionable material ($^{235}_{92}U$ or $^{239}_{94}Pu$). Extra neutrons arising from fission of the core material are captured by the blanket material to 'breed' more fissionable atoms. Thus the breeder reactor produces as much fissionable fuel as it consumes and at the same time generates heat. Naturally conservation is maintained in the sense that the fertile atoms are being used up to produce fissionable ones.

In view of the current concern about a possible energy crisis in the foreseeable future these breeder reactors are arousing much interest. At the same time *safe* disposal of the strongly radioactive wastes constitutes a problem, particularly since some of the substances have half-lives running into thousands of years. Storage in stainless-steel tanks encased in concrete seems to be the current, rather uneasy, solution, whilst fusion into glass blocks has been suggested. One cannot help wondering, however, whether we are not creating a problem for future generations.

Other new elements have also been produced by similar artificial means. They have atomic numbers ranging from 95 to 106, and have been named, in order, americum (Am); curium (Cm); berkelium (Bk); californium (Cf); einsteinium (Es); fermium (Fm); mendelevium (Md); nobelium (No) and lawrencium (Lw). These elements are known collectively as the *transuranic elements*. So far names have not been allocated to elements 104–106.

Nuclear fusion

19.7 It has been shown that the nuclear binding energy of heavy atoms may be increased by fission into two fragments of lower mass numbers and that this nuclear fission is accompanied by the conversion of some mass into energy.

When very light nuclei are converted into more massive ones the reaction is also accompanied by the conversion of mass into energy, these reactions being known as *nuclear fusion*.

19.7.1 *The hydrogen bomb.* This device depends upon the nuclear reaction between the nuclei of the two heavy isotopes of hydrogen (1.4.1), the deutron, ${}^{2}_{1}H$, and the triton, ${}^{3}_{1}H$. These undergo fusion at very high temperatures:

$${}^{2}_{1}H + {}^{3}_{1}H \rightarrow {}^{4}_{2}He + {}^{1}_{0}n + \text{energy}$$

so that a helium nucleus and a neutron are produced. The reaction is accompanied by the conversion of a portion of the mass into energy. Since a very high temperature is required for this reaction a small fission bomb (uranium or plutonium) is exploded inside the hydrogen bomb to provide the temperature of several million degrees necessary for nuclear fusion to occur in hydrogen.

19.7.2 *Solar energy.* So far the large-scale *controlled* fusion of these heavy isotopes of hydrogen has not been perfected. If and when this occurs the vast supplies of hydrogen in the oceans will provide enormous reserves of energy for the future. One particular problem, of course, is being able to contain physically the products of nuclear fusion at the enormously high temperatures generated.

It is generally accepted that the energy radiated by the sun is derived from the fusion of four hydrogen nuclei, ${}^{1}_{1}H$:

$$4{}^{1}_{1}H \rightarrow {}^{4}_{2}He + 2{}^{0}_{1}e + \gamma\text{-rays} + \text{energy}$$

That is, four hydrogen nuclei (protons) combine to produce a helium nucleus along with two positrons and γ-radiation. During this process there is a loss of mass of approximately 0·7% and this is released as energy.

The sun radiates energy at the rate of approximately 4×10^{13} kilowatts. Nevertheless if we assume that this energy is produced in the manner suggested above the process should continue, at the present rate of energy release, for a further 10^{11} years. However, it is unlikely that Man will still be here on Earth to receive it, having doubtless annihilated himself long before then in some other ingenious manner.

Chapter Twenty
Non-destructive Testing

20.1 No material produced under industrial conditions is ever perfect. The incidence of imperfections in any batch of material will depend not only upon the degree of care and attention paid to production by the operator and the extent to which plant has received maintenance, but also upon the type of process. Thus a casting may suffer from one or more of a large number of defects such as incorrect composition; the presence of blow holes, shrinkage cavities, non-metallic inclusions, segregation, surface flaws and cracks; and so on. Many of these defects will not submit to simple visual examination and an apparently acceptable casting may contain a number of internal defects. Rolled, forged or extruded metal, however, is less likely to be defective, assuming that no surface defects are apparent. Had the internal structure been unsatisfactory or its composition incorrect it is likely that the ingot would have cracked during the initial stages of the working process, or that gas-filled cavities would have given rise to blisters near the surface.

20.1.1 Considerable lengths of a wrought section may be produced from a single cast ingot. Assuming that the piped portion of the ingot has been adequately cropped the remainder can be expected to have produced material of reasonably uniform properties and quality. Hence examination by analysis and by mechanical or physical tests of samples taken from the batch will generally be adequate to control quality. The same principles, however, do not apply to components made *individually* by casting, welding and other forms of fabrication involving mainly manual skill. A large number of variable influences often prevails during the manufacture of such a component. Moreover, since it is in finished form ordinary mechanical tests of a destructive nature cannot be used, other than the making of test pieces from the runners or risers of a casting. Hence some form of appropriate non-destructive test must be adopted so that the quality of the finished component can be assessed. The proportion of components thus tested can vary. Whilst for cheap consumer articles the representative proportion which is tested may be very small, in the aircraft industries it will often be 100%.

The objectives of non-destructive testing
20.2 The general aims of non-destructive testing (which we shall refer to subsequently as NDT) may be classified as follows:

20.2.1 *Improved productivity*. Defective material may be isolated at an early stage in production so that further manpower is not frustrated in futile processing. Shop capacity is also saved in this way and there is less useless wear and tear on plant. These are obvious advantages of testing at an early stage basic work pieces such as castings and welds. Equally important, the knowledge that NDT is being applied often leads to the production of less scrap due to indifferent workmanship. In the case of welding, instances have been reported where a 40% scrap rate fell to less than 5% when it was announced that radiographic examination was to be installed.

20.2.2 *Increased serviceability.* Here NDT is used to detect faults which are likely to reduce the service life of the finished component. For example, some defects whether occurring at the surface or below the surface, can initiate fatigue failure (16.5.4). Quench cracks in heat-treated steel components are in this category. The problem of safety must also be considered in this connection. Failure can in some cases lead to disaster particularly in aircraft structures.

Types of defect

20.3 Defects in metallurgical materials may be inherent in the original casting or they may arise due to faulty technique during subsequent processing. They may also develop when the resultant component is in service. It is convenient to classify defects into two groups:

(*i*) those occurring at the surface. These include *cracks* resulting from unsatisfactory conditions during cooling of the casting; cracks produced during heat-treatment processes; surface *scale* and defects due to the rolling in of *laps* and *seams.*

(*ii*) those occurring below the surface. The more common defects here include *porosity*; *laminations*; *inclusions* of various types; *segregation*; *blowholes*; *internal cracks* and other less obvious faults such as *coarse grain.* Welds may suffer from lack of penetration and unsatisfactory bonding as well as many of the defects already mentioned such as porosity, inclusions and cracking.

Visual inspection

20.4 This is undoubtedly the most widely used, yet least costly and very effective method of NDT. Whatever other method of inspection is applied it will invariably be supplemented by visual examination. For example thorough visual inspection of a weld by an experienced operator can reveal the following faults:

(*i*) lack of penetration by the weld metal (Fig. 20.1 (ii));
(*ii*) badly oxidised or rough surfaces;
(*iii*) the presence of large cracks and their position relative to the weld;
(*iv*) surface porosity;

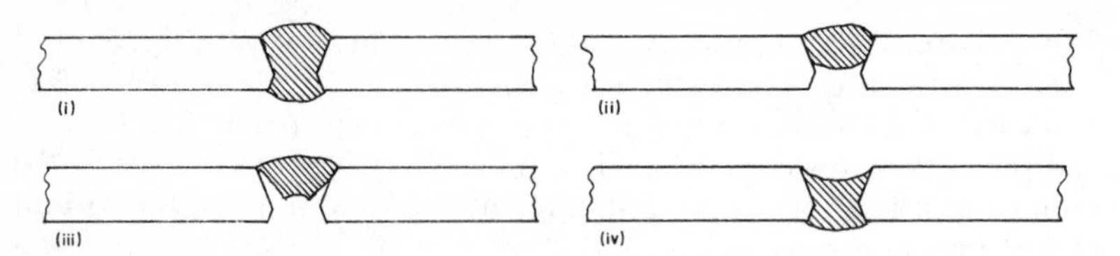

Fig. 20.1—Welding defects which can be assessed by visual inspection. (i) Satisfactory weld. (ii) Lack of penetration by weld metal. (iii) Inadequate fusion. (iv) Insufficient weld metal.

(*v*) inadequate fusion of the weld metal as shown by 'undercutting' and concavity of the weld (Fig. 20.1 (iii));
(*vi*) the use of insufficient weld metal (Fig. 20.1 (iv)).

The principal requirement—other than employing an intelligent operator who knows what he is looking for—is to illuminate the component adequately so that it can be examined by the eye or in a few cases by some light-sensitive device such as a photo-electric cell.

20.4.1 Visual examination can be assisted when necessary by using a small hand magnifier consisting of a simple convex lens (Fig. 20.2). The magnification obtained by such a lens is given by:

$$\text{Magnification} = \frac{250}{f}$$

where f is the focal length of the lens in mm. The value 250 represents the minimum distance (in mm) at which an object can be seen clearly by the

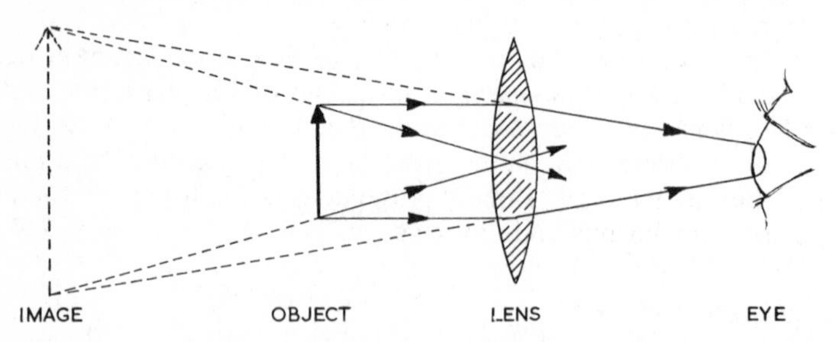

Fig. 20.2—The principle of a hand magnifier.

unaided eye. Thus a lens with a focal length of 100 mm will give a magnification of 2·5 (usually written 2·5$^{\times}$). The focal length of a lens and its working distance from the object are about equal.

Liquid penetrants

20.5 Inspection based on the use of liquid penetrants can be employed to detect cracks and other fissure flaws occurring at the surface of a material. Though most faults of this type could be detected by careful visual inspection the use of the penetrant method provides a more positive aid to detection and makes it less dependent on the human element as well as less time consuming.

Penetrant methods can be used to locate cracks in castings, weld cracks, grinding cracks, pinhole porosity *at the surface* of welds and castings, fatigue cracks, seams and laps in rolled stock, forging laps and many other cavity defects providing they break out at the surface.

20.5.1 The earliest penetrant tests employed oil and whiting. The surface to be examined was coated with a penetrating oil (often paraffin) which was usually heated so that its viscosity was reduced. Moreover the hot oil caused any surface cracks to expand so that oil would be drawn into them. After allowing sufficient time for the oil to penetrate any surface defects the excess was wiped off and the surface thoroughly dried. A thin coating of whiting, either as a dry powder or as a suspension in alcohol (industrial 'methylated spirit') was then applied to the surface and allowed to dry. As the surface cooled any cracks contracted again so that the oil was forced out to stain the coating of whiting.

20.5.2 More sophisticated modifications of this simple test were developed subsequently. In most of these tests a brightly coloured dye is carried in some suitable penetrant, the colour being chosen to produce a high degree of contrast with the coating of 'developer'. In most cases this developer is based on whiting, talc or some such powder suspended in

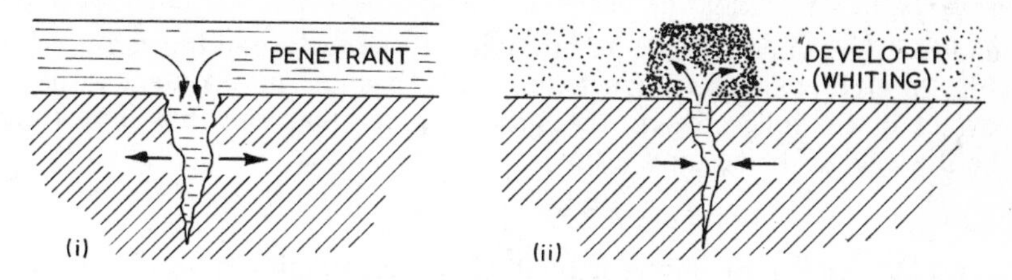

Fig. 20.3—Crack detection by dye penetrant.

water. Light-alloy castings are examined for surface cracks by first immersing them in a boiling solution containing a brilliant red dye. As the cracks expand the penetrant dye is drawn into them. The castings are then removed from the penetrant tank, rinsed to remove excess penetrant from the surface and then dipped into a hot suspension of whiting in water. On withdrawal of the casting its surface soon dries and as any cracks contract dye is squeezed out into the layer of whiting 'developer' (Fig. 20.3).

20.5.3 Fluorescent penetrants are also used. Most of these substances are oil soluble so that a penetrating oil is used as the vehicle. Dip, brush or spray methods are employed to coat the penetrant on to the test surface.

After excess penetrant has been cleaned from the surface the latter is viewed under ultra-violet light (approximately $3{\cdot}65 \times 10^{-7}$ m wavelength). This is outside the visible spectrum and so, assuming that the test surface is inspected in a darkened cubicle, it remains completely dark. The action of ultra-violet light on the fluorescent substance in the penetrant, however, causes it to fluoresce so that any cracks containing the penetrant appear as bright yellow-green lines on a dark background.

Magnetic methods

20.6 Discontinuities such as cracks, inclusions and cavities in a magnetic material produce distortion in a magnetic field which may be induced in that material. The direction of the magnetic lines of force is altered because the discontinuities have much lower magnetic permeabilities than the main body of the component. Consequently if a magnetic component is carrying a high magnetic flux density any discontinuity will

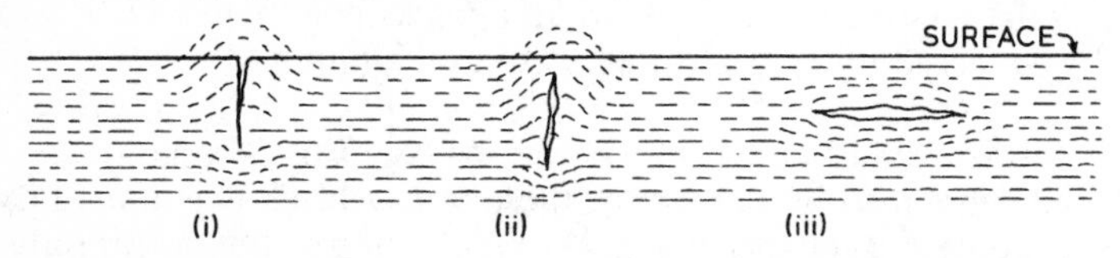

Fig. 20.4—The effect of cracks and sub-cutaneous cavities on the direction of a magnetic field. Flaws at right angles to the lines of force ((i) and (ii)) will be revealed, whilst those parallel to the lines of force (iii) will not.

tend to cause 'lines of force' to break out at the surface of the component (Fig. 20.4). If some fine magnetic powder is brought into contact with the surface it will collect where the lines of force break out thus indicating the presence of some fault at or near to the surface.

20.6.1 A number of different methods are available for producing the necessary high flux density in the component. The method chosen will depend to some extent on the direction in which the lines of force are required to pass. To obtain maximum sensitivity the lines of force should be perpendicular to the line of the flaw as already indicated in Fig. 20.4. Two ways in which a desired field direction can be produced are indicated in Fig. 20.5.

The magnetic particles can be applied to the surface either by wet or dry techniques. In the 'dry' method finely divided ferromagnetic particles having a high permeability are used and these can be simply shaken on to the surface from a pepper pot. They will align themselves along any lines of force emerging from the surface of the specimen in the region of the flaw. The surplus can be gently blown away. In the 'wet' process a suspension of finely divided magnetic iron oxide in a light petroleum 'fraction' is generally used. This can be applied by spraying or by immer-

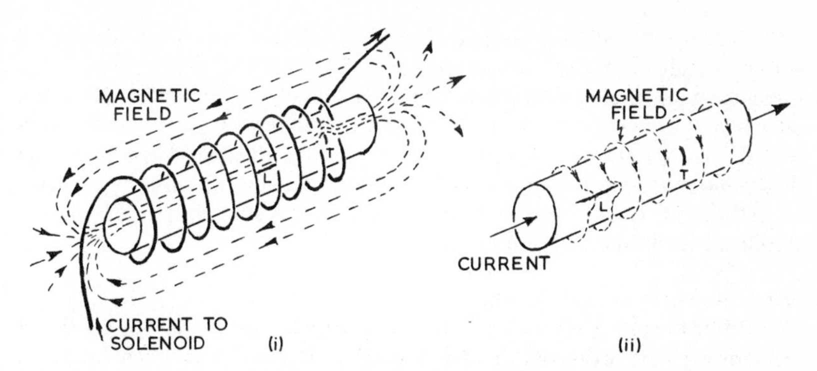

Fig. 20.5—In (i) the component is magnetised by means of a solenoid (or by placing it across the poles of an electromagnet). Here the field will be parallel to the axis so that the transverse crack, *T*, will be detected whilst the longitudinal crack, *L*, will not. In (ii) the component is magnetised by passing a current through it so that the induced magnetic field will be perpendicular to the axis of the component. In this case the longitudinal crack, *L*, will be detected whilst the transverse crack, *T*, will not.

sion. Magnetic particles with a fluorescent coating can be used in which case the surface will be observed under suitable ultra-violet light (20.5.3). This method is useful when dealing with rough surfaces.

These methods of fault detection are of course applicable only to ferromagnetic materials but they afford very positive means of locating cracks and crack-like faults so long as they are not more than about 10 mm beneath the surface.

Thermal and electrical tests

20.7 A number of non-destructive tests based on thermal properties have been devised from time to time. Many of these rely on the high thermal conductivity of metal and upon the fact that any discontinuity *beneath* the surface will act as a barrier to heat transfer. Thus if the surface of the component is coated with some heat-sensitive indicator and the component then heated at a suitable uniform rate, any build-up of heat will be revealed by an appropriate change in the indicator. Areas on the surface where heat has been prevented from escaping inwards, by the presence of some subcutaneous defect, will be indicated by melting (or colour change) of the indicator used. In some of these tests the surface is first 'frosted' by spraying it with a coating of some suitable indicator such as acenaphthene.* The component is then passed at a uniform speed through an induction coil. Eddy currents induced in the *outer skin* of the component generate heat which is prevented from diffusing into the core if any discontinuities of low thermal conductivity exist between the skin and the core. This type of test has been used to examine the degree of continuity between uranium nuclear fuel rods and their protective cladding cans.

* Acenaphthene, $C_{12}H_{10}$, m.pt. 95° C, is derived from the anthracene fraction of coal tar.

20.7.1 Various electrical methods have been developed to test for continuity of structure in conducting materials. Most of these depend upon the measurement of electrical resistance or, alternatively, potential difference across a standard section of material. For example railway rails can be tested by attaching electrodes at each end of a measured length of rail. The potential drop is then measured. The presence of discontinuities such as fissures will cause an increase in electrical resistance, or a decrease in potential drop across the section.

X-ray methods of radiography

20.8 Metals and many other solids are opaque to 'visible' light, that is electromagnetic radiation of such wavelength as will affect our organs of

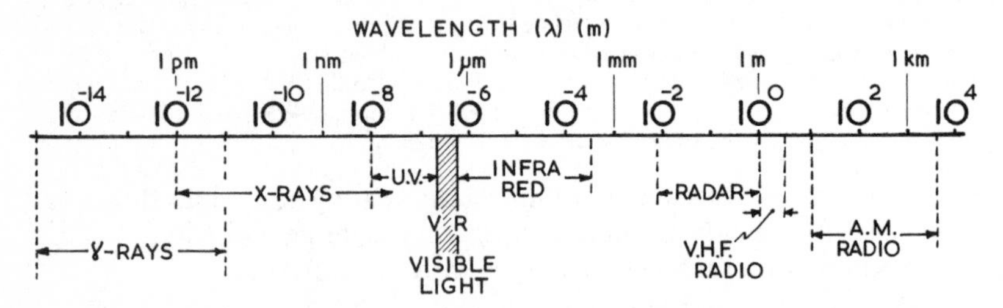

Fig. 20.6—The electromagnetic spectrum.

sight, but are transparent to γ-rays and also to those X-rays of short wavelength. Hence radiography is one of the oldest non-destructive testing methods having been introduced quite early in the 1920s.

20.8.1 Like light, X-rays travel in straight lines and can therefore be used to produce a reasonably sharp image on a photographic film, the emulsion of which is affected in much the same way by X-rays as it is by visible light. X-rays are absorbed logarithmically:

$$I = I_o e^{-\mu d}$$

where I_o and I are the intensities of the incident and emergent rays respectively, d the thickness and μ the linear coefficient of absorption of radiation for the medium through which the rays are passing. μ is dependent upon the wavelength of the radiation and is lower for that of shorter wavelength. Since X-rays are thus absorbed differentially the density of the photographic image will vary with the thickness of the material through which the rays have passed (Fig. 20.7). Those rays which pass through a region containing a cavity will be less effectively absorbed than those which travel through the entire thickness of material. Consequently the cavity will show as a dark patch on the resultant photographic negative in a similar way to which greater intensity of visible light

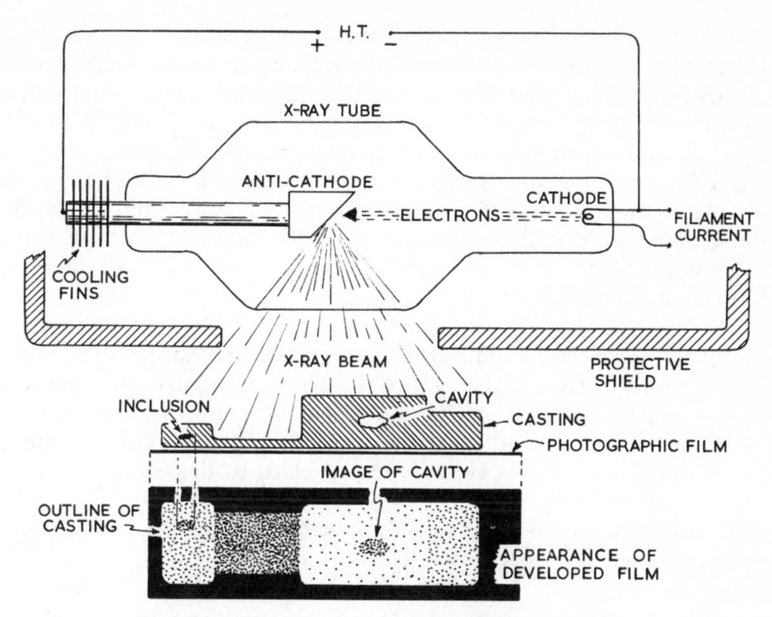

Fig. 20.7—The principles of *X*-radiography of a casting.

produces a darker area on the negative in ordinary monochrome photography.

Although a photographic emulsion is the most sensitive means of detecting both X-rays and γ-rays other means such as Geiger counters and substances which fluoresce under the action of these radiations can also be used. Fluoroscopy is obviously cheaper and quicker than photography but is less sensitive and its use is limited to the less dense metals and alloys.

20.8.2 X-rays are produced when high-velocity electrons strike a metal target. Hence the basic requirements for producing a beam of X-rays are:

(*i*) a suitable source of electrons;
(*ii*) some means of accelerating these electrons to a high velocity;
(*iii*) a suitable metal target.

In an X-ray tube the electrons are generated from a heated metal filament by passing a current of electricity through it—electrons from the current become so thermally activated that they 'escape' from the filament. The filament also acts as the cathode of the high-tension circuit (Fig. 20.7) used to accelerate the electrons. The target—or anti-cathode—is of tungsten which will resist melting at the high temperatures involved and is sealed into the other end of the tube. The tube is evacuated since the presence of relatively massive gas molecules would obstruct the passage of electrons. A high potential difference between cathode and target will cause the negatively charged electrons to be accelerated away from the cathode and towards the positively charged target. By the time the electrons reach the target they will have acquired considerable velocity and, hence, kinetic energy and this is absorbed in collision, X-rays being generated. In addition to X-rays, much heat is also liberated and this fact is utilised in electron-beam welding (17.6.4). Here, however, heat is an undesirable by-product and must be conducted away from the target even though the latter is of high melting point tungsten. Internal water cooling of the target or air cooling by a system of radiator fins external to the tube can be used.

20.8.3 X-rays used in metallurgical radiography are 'harder' than those used in medicine. That is, they are of shorter wavelengths which are able better to penetrate metals. At the same time these properties make them more dangerous to human body tissues and the plant producing radiations of this type must be very carefully shielded so that stray radiations do not reach the operators.

The wavelength, λ, of X-rays produced is governed by the accelerating voltage, V, applied between cathode and target:

$$\lambda \propto \frac{1}{V}$$

so that as the potential difference across the tube is increased the wavelength of the X-rays produced becomes shorter.

The absorption of X-rays is a function of the density of the material being penetrated and for this reason inclusions as well as cavities can be detected. At the same time the increased absorption of X-rays by the denser metals means that only thin sections of such metals can be effectively examined. Consequently X-rays are used mainly in the radiography of light-alloy castings and sections. The maximum sensitivity of this method of examination is of the order of 2%, that is a defect about 0·2 mm in diameter may be detected in a sectional thickness of 10 mm.

γ-ray methods of radiography

20.9 γ-rays (19.3.3) are electromagnetic radiations of very short wavelength (Fig. 20.6) and are able to penetrate considerable thicknesses of many metals. In practice γ-ray methods are used to examine steel from 25 to 250 mm or more in thickness. A further advantage of the use of γ-rays is the very small size of source as compared with that of an X-ray unit. Moreover the cost of the γ-ray source is comparatively low.

20.9.1 Natural γ-ray sources, chiefly radium and radon, have been used for industrial radiography since the late 1920s but their use never became common due to the scarcity and high cost of radium and, in the case of radon, a decay product of radium (19.4.5), its very short half-life. The development of the nuclear reactor, however, has provided new and more effective γ-ray sources. These are generally produced by irradiation of some suitable element in a nuclear reactor. For example when cobalt $^{59}_{27}Co$ is bombarded with slow-moving neutrons in a nuclear reactor the radioactive isotope $^{60}_{27}Co$ is formed:

$$^{59}_{27}Co + ^{1}_{0}n \rightarrow ^{60}_{27}Co$$

$^{60}_{27}Co$ then undergoes decay:

$$^{60}_{27}Co \rightarrow ^{60}_{28}Ni + ^{0}_{-1}e \ (\beta\text{-particle}) + \gamma\text{-rays}$$

The half-life of $^{60}_{27}Co$ is a convenient 5·3 years.

20.9.2 Radioactive isotopes are also formed as a result of fission but in most cases a mixture of different isotopes is produced and this is generally unsuitable for use as a γ-ray source because, for example, of the rapid decay of some of the components relative to others.

In choosing a suitable radioactive isotope for this type of work the following criteria are important: (i) its half-life, (ii) the γ-ray energy output, (iii) the characteristics of the material to be radiographed. The half-life of some radioisotopes is but a few days, hours—or even seconds. These will generally be less suitable for this type of work since corrections would need to be applied continually to compensate for their decrease in intensity of radiation. Consequently isotopes with a half-life measured in years or months will be more suitable assuming that its γ-ray emission energy is adequate.

Table 20.1—Half-value thicknesses (mm) of some materials.

Material	$^{60}_{27}Co$	$^{192}_{77}Ir$
Lead	12·2	4·8
Steel	22·1	11·2
Aluminium	55·8	30·4
Concrete	68·6	48·2
Water	134·6	81·3

In addition to cobalt $^{60}_{27}Co$ already mentioned, other artificially produced radioisotopes which are useful in this work include $^{192}_{77}Ir$, $^{137}_{55}Cs$ and $^{170}_{69}Tm$ (Table 20.2).

The absorption of γ-rays by matter follows the same general pattern as for X-rays (20.8.1), i.e.:

$$I = I_o e^{-\mu d}$$

The 'half-value thickness' of a material is sometimes quoted. This is the thickness of the absorber which will reduce the intensity of the beam to one-half of its incident value (Table 20.1).

20.9.3 The extreme danger of exposure of the human body to γ-rays and to 'hard' X-rays is well known but whereas the latter are not emitted once the tube has been switched off, a γ-ray source continues to emit as long as the source remains active ($^{60}_{27}Co$ has a half-life of 5·3 years). Therefore γ-ray sources must be very carefully stored and shrouded at all times so that operators are protected. A typical storage container is shown in Fig. 20.8 (i).

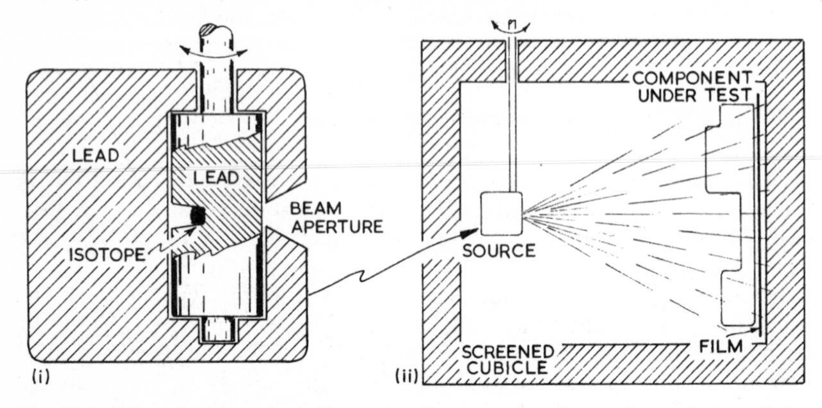

Fig. 20.8—The principles of γ-radiography: (i) a container for storing and shrouding a radioactive source; (ii) outline of a system for γ-radiography.

The general procedure in γ-radiography is similar to that using X-rays in that the component to be radiographed is interposed between the γ-ray source and a photographic film (Fig. 20.8 (ii)). The whole process will be conducted in a radiation-proofed cubicle the source being uncapped by remote control in order that the operator shall be fully protected. A disadvantage in the use of γ-rays is that the wavelength of the radiation (which determines the penetrating power) cannot be varied whereas with X-rays wavelength can be altered to suit the material and also its thickness so that a radiograph of suitable photographic contrast can be obtained.

20.9.4 Table 20.2 shows the main characteristics of the more important γ-ray sources and indicates thicknesses which can be radiographed. These ranges of thickness are only approximate and greater thicknesses can be dealt with provided that exposure times are suitably increased. The lower limit of thickness in each case is that associated with adequate photographic quality. Lesser thicknesses would produce negatives of lower contrast which may not reveal clearly the defects sought.

The relatively small size of the γ-ray sources and their associated handling equipment make γ-rays useful for radiography on sites inaccessible to bulky X-ray equipment. The initial cost of these γ-ray sources is also much lower than that of X-ray equipment but the results obtained may be inferior to those of X-rays because of the lack of flexibility of wavelength which is obtainable with X-rays.

Table 20.2

Source	*Half-life*	*γ-ray output* (rhm/curie) *	*Minimum and maximum thickness for which source is suitable* (mm)
Radium, Ra	1590 years	0·83	50–150 (steel)
Radon, Rn	3·8 days	0·83	50–250 (steel)
Cobalt, ^{60}Co	5·3 years	1·35	50–200 (steel)
Iridium, ^{192}Ir	74 days	0·5	6·5–90 (steel)
Caesium, ^{137}Cs	33 years	0·35	25–100 (steel)
Thulium, ^{170}Tm	127 days	0·0045	1·5–12·5 (aluminium)

* This is equivalent to a dose rate of one röntgen/hour measured one metre from a source of one curie.

Ultrasonic testing

20.10 Sonic methods for measuring the depth of the sea and so protecting ships from running aground have been practised for many years. Here a sound is emitted from the bottom of the ship and the echo received and suitably amplified. The interval which elapses between transmitting the sonic impulse and receiving its echo is a measure of the distance the impulse had to travel, that is to the floor of the ocean and back. Sonic methods have also been used to test metals for the presence of flaws. The railway wheel-tapper was for some strange reason always 'good for a laugh' in the old-time music-hall but he was none the less doing a very useful job of NDT. However, ordinary sound waves, audible to the human ear, are of relatively long wavelength and cover a frequency of about 30–16 000 Hz (according to the more optimistic hi-fi experts who probably waste a lot of money producing sounds they cannot hear). Such long wavelengths tend to by-pass small defects in a metallic component and will therefore not reveal them. Consequently ultrasonic frequencies of more than 20 000 Hz are used in this form of testing—generally 0·5 to 15 MHz for metals inspection.

20.10.1 Though here we are concerned principally with the detection of faults these methods are also suitable for measuring wall thicknesses of vessels where other methods are impossible. Thickness of sheet and plate and the thickness of cladding can also be measured in this way.

When an ultrasonic vibration is transmitted from one medium to another some reflection occurs at the interface. Any defect such as an

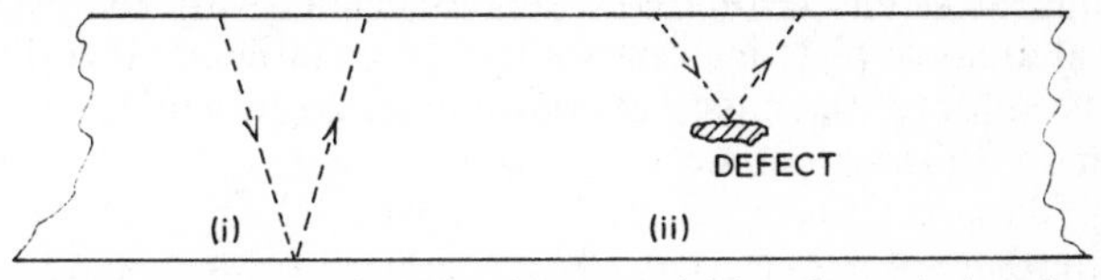

Fig. 20.9—The detection of a fault in plate material by ultrasonic impulses. In (i) the impulse is reflected from the lower surface of the plate whilst in (ii) it is reflected from the defect. Measurement of the time interval between transmission of impulse and reception of echo determines the depth of the defect.

inclusion or an internal cavity will therefore provide a reflecting surface for ultrasonic impulses (Fig. 20.9). The ultrasonic vibrations are produced in a piezo-electric crystal by stimulating it with an electrical impulse. Since ultrasonic vibrations are dissipated very quickly in air there must be close contact between the face of the crystal probe and the surface under test and this is most easily achieved by spreading a film of oil between the two. The crystal used is generally either of quartz

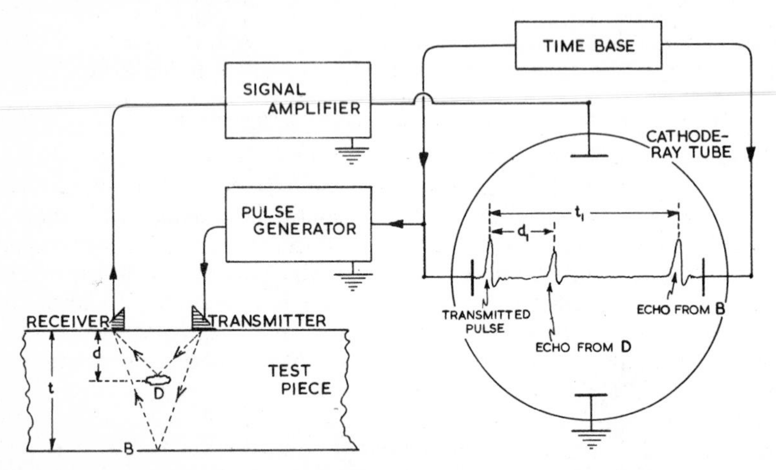

Fig. 20.10—The principles of an ultrasonic testing apparatus. The values d and t in the test piece are indicated by the values d_1 and t_1 on the CRT.

or barium titanate. In some systems (Fig. 20.10) separate crystals are used as transmitter and receiver and the time interval between transmission of pulse and reception of echo is recorded on a cathode-ray tube (CRT).

20.10.2 In many modern testing devices a single crystal probe is used so that it acts both as transmitter and receiver. Since the crystal is very highly 'damped' a pulse of extremely short duration is emitted. Hence the crystal is dormant and able to receive the ultrasonic echo and so convert it into an electrical signal which is subsequently amplified and displayed on the CRT.

Although ultrasonic testing is closely associated with the exploration of large flat areas of plate materials it is very versatile in other directions. It can be used for testing most materials and in great depths, for example, 10 m or more thickness of steel.

Dynamic testing

20.11 It was once the rather gruesome practice in the East to sacrifice a village virgin into the molten metal destined for casting large temple bells. If the bell turned out to be unsatisfactory it was assumed that the 'virgin' had not been all that she should have been and her father was severely ostracised for not exercising adequate surveillance over her. Fortunately the progress of metallurgical science has outlawed such inhumanity.

The test for quality of such a bell was, and is, simple. When struck it should ring, that is vibrate on the required frequencies which will be a combination of its fundamental frequency and certain harmonics and sub-harmonics. Any fault present will cause a damping of vibrations whilst a cracked bell may also vibrate on a different frequency to that intended because its dimensions have in effect been changed.

20.11.1 The principle of stimulating vibrations in a body and measuring the frequency of these vibrations is now used in research as an accurate means of determining the elastic properties of a material. Longitudinal, transverse or torsional vibrations can be induced in rods or bars and each specimen has its own natural frequency of vibration which is dependent on size, shape, mass and elastic properties. Since size, shape and mass are easy to determine this provides an accurate means of deriving information about elastic properties, since for specimens of simple shape it is possible to derive expressions which relate the various parameters and the frequencies for simple modes of vibration.

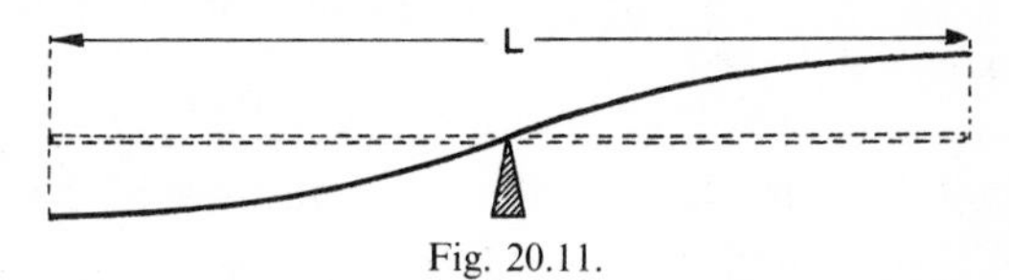

Fig. 20.11.

Consider a bar supported as in Fig. 20.11 and vibrating longitudinally. In a standing wave, nodes and antinodes must always separate each other and the distance between adjacent nodes (or antinodes) is always one-half of a wavelength. Hence (in Fig. 20.11):

$$\lambda = 2L \tag{1}$$

where λ is the wavelength of the vibration.

Since $$f\lambda = V$$
(where V = the velocity of propagation of longitudinal waves and f their frequency)

then $$f = \frac{V}{2L} \tag{2}$$

also $$V = \left(\frac{E}{\rho}\right)^{\frac{1}{2}} \tag{3}$$

where E is Young's modulus of elasticity for the bar material and ρ is the relative density of the bar material.

Substituting (3) in (2):

$$f = \frac{1}{2L}\left(\frac{E}{\rho}\right)^{\frac{1}{2}}$$

and for the general case:

$$f = \frac{n}{2L}\left(\frac{E}{\rho}\right)^{\frac{1}{2}}$$

where n is an integer the value of which will depend on the mode of vibration, i.e. the number of nodes and antinodes relative to the length, L (Fig. 20.12).

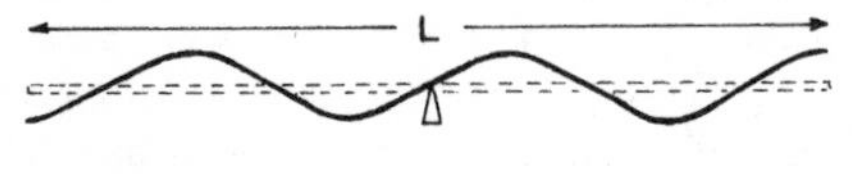

Fig. 20.12.

Using methods such as this it is possible to determine elastic constants of a material within an accuracy related to 0·001% error.

Some applications of NDT

20.12 NDT is associated particularly with the following manufactured shapes and materials:

20.12.1 *Castings.* The incidence of defects in castings is probably greater than in most other manufactured components. This is not surprising when one considers the number of variable factors which prevail during the casting processes. These include melting conditions in so far as they affect gas solution or surface oxidation; casting temperature; rate of pouring; mould temperature and conditions; the design of the mould itself and many other factors.

The principal internal defects likely to arise include:

(*i*) gas porosity arising from gas dissolved during melting and precipitated during solidification. Bubbles of entrapped air may also be present if the mould was inadequately vented to allow air to escape as the mould was being filled.

(*ii*) shrinkage cavities formed when molten metal is unable to feed into regions where solidification, and consequently shrinkage, is taking place.

(*iii*) internal inclusions due to entrapped slag and dross.

These internal defects are revealed most effectively by X-radiography particularly in the case of light alloys. For steel castings γ-radiography may be used.

Some of the surface defects in castings include:

(*i*) cracks formed by uneven cooling of the casting during or after solidification. Sometimes cracks are initiated by sand cores which are too strong and do not collapse under the pressure of the solidifying metal.

(*ii*) cold 'shuts', generally formed when two streams of molten metal fail to unite at an interface because of the presence of surface oxide.

(*iii*) surface inclusions of moulding sand, dross or slag.

Extensive hot cracks which consist of several parallel fissures branching to follow grain boundaries may also occur internally during cooling. These, and surface fissures, may show on X-ray film as irregular dark lines. Surface cracks, however, are more easily and cheaply revealed by one of the hot liquid penetrant methods.

20.12.2 *Welds*. The investigation of welded joints has been well developed and most of the orthodox NDT methods are useful.

Radiography, using both X-rays and γ-rays, will detect cavities, inclusions and those surface irregularities which result in a change in thickness of the metal. These defects will also be recorded by ultrasonic testing methods as also will incomplete penetration, lack of fusion and piping.

Cracks and other surface and subsurface imperfections in *magnetic* materials are very reliably detected by using a magnetic method of examination. In non-magnetic materials surface imperfections can be revealed using a fluorescent penetrant method or one of the other penetrant methods, assuming that the defects actually break out to the surface.

20.12.3 *Plate and similar material.* Such materials may suffer from the presence of inclusions, laminations and other discontinuities elongated in the direction of rolling. Such defects are most conveniently detected using ultrasonic methods which are very sensitive even when applied to a considerable thickness of material.

20.12.4 *Concrete*. Concrete structures *in situ* may be tested by some form of pulse method. The pulse may be provided by a simple hammer blow or by some form of electro-acoustic transducer. As with ultrasonic testing methods already described, the time taken for the pulse to travel from one boundary surface to the other—or to be reflected back to the source—is measured. Obstructions or imperfections interfere with the passage of the wave as in the case of ultrasonic testing.

20.12.5 *Ceramics, plastics materials and other non-metallic materials.* Surface cracks are likely to occur in materials such as these as well as in vitreous enamel coatings and glazes of various types. Such faults can be detected by some form of penetrant method. Special methods which make use of the weak electrostatic field surrounding minute cracks in vitreous enamel coatings have been developed. When fine calcium carbonate powder is blown on to such a surface it collects in the region of the field associated with the fissure.

20.12.6 NDT methods often have to be devised for special cases and many such methods are very ingenious. For example, in detecting leaks in very long water or oil pipe lines a radioactive method is often used. A quantity of radioactive isotope (in solution as one of its compounds) is injected into the liquid flow at the input end. After sufficient time has passed for the isotope to have travelled about 2·5 km a battery-operated radiation detector/recorder is passed into the line so that it floats along in the liquid stream. Points along the line where the radioactive isotope has leaked out, and is therefore still present in the surrounding sand, are located and recorded by the detector. Small γ-ray sources are sited along the outside of the pipeline at fixed intervals to act as distance calibration markers. The detector/recorder is recovered from the outlet end of the line so that the record can be examined.

Appendix I

The Preparation of Specimens for Metallographic Examination

A1.1 *Macro*scopic metallography which does not generally involve the use of any optical instrument other than a 'hand magnifier' (20.4.1), originated at the beginning of the nineteenth century with Aloys Beck von Widmanstätten's classical observations of the surface structures of meteorites. *Micro*scopical metallography was introduced in about 1840 but did not begin to flourish until the 1860s with Professor Henry Sorby's detailed investigations into the structures of steels.

Modern microscopical metallography involves not only the use of high-quality optical microscopes of special design but also those instruments in which beams of electrons replace rays of light in order to obtain much higher magnifications of the order of millions rather than hundreds. Since metallography is a very important field of metallurgical investigation, a very brief introduction to the preparation of metallographic specimens will be attempted here.

Selecting the specimen

A1.2 To be of any value a specimen must be representative of the material being studied. Clearly the structure and composition of a large body of an alloy such as a cast ingot, may vary considerably from one region to another. In such cases it is necessary to select *several* specimens—from different parts of the ingot—in order adequately to represent the structure. Even with small components, e.g. a case-hardened gear or a flame-hardened shaft, the structure will obviously vary from one point to another over a cross-section.

A1.2.1 Some materials exhibit *directionality* in structure. Thus a rolled mild steel will show 'stringers' of manganese sulphide elongated in the

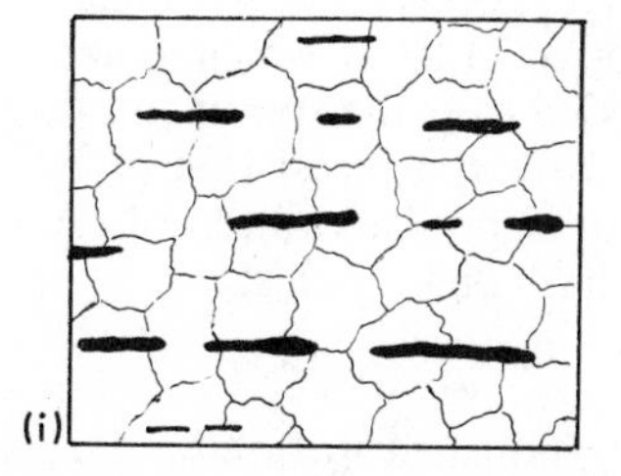

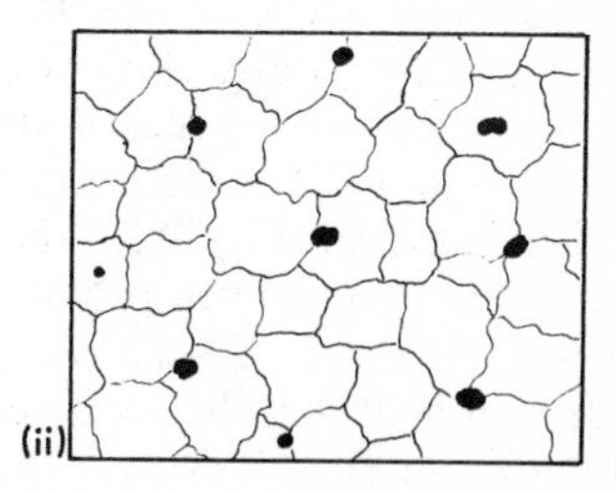

Fig. A1.1—(i) 'Stringers' of manganese sulphide in mild steel as they appear in a longitudinal section. (ii) A transverse section will give the misleading impression of the presence of manganese sulphide in the form of globules.

direction of rolling if a longitudinal section is examined whilst these manganese sulphide inclusions will appear misleadingly to be rounded globules if a transverse section is examined (Fig. A1.1). Thus at least two specimens are required properly to represent the structure of a wrought material.

Mounting the specimen

A1.3 A specimen approximately 20 mm diameter (or 20 mm square) is a convenient size for the beginner to handle. If the specimen is smaller than this, then it is advisable to mount it in some inert plastics material which will neither have a galvanic effect leading to uneven etching nor dissolve in any de-greasing agent used. The specimen, whatever its size, will also need to be mounted if a section through its surface is to be examined, otherwise the surface edge will be lost due to bevelling during the polishing process. Similarly, surface cracks are best examined by cutting a specimen so that it contains the crack, mounting it in bakelite (or other plastics compound) and then grinding it down until the crack becomes exposed.

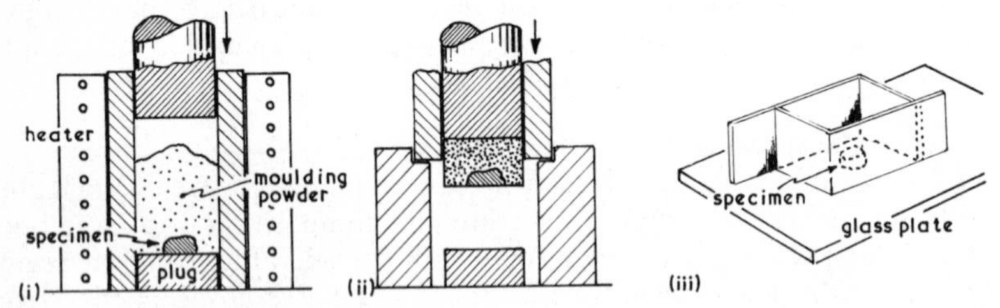

Fig. A1.2—Mounting a specimen. (i) Using a small press to mount the specimen either in 'bakelite' or in a thermoplastics material. (ii) Using retaining pieces for mounting the specimen by moulding it in a cold-setting resin.

Specimens are mounted in bakelite (or a thermoplastics material) with the aid of a small press and a mould of the type shown in Fig. A1.2 (i) and (ii). Most of the plastics substances used as moulding materials flow at about 150° C which is too low a temperature to cause any structural change in most metallic alloys. Alternatively the specimen may be mounted in some cold-setting compound (12.3.2) which generally consists of a white powder which is sprinkled over the specimen followed by a colourless liquid. The whole is retained in a metal ring or by two L-shaped angle pieces (Fig. A1.2 (iii)), until polymerisation of the powder and liquid are complete to provide a hard rigid mount.

Grinding and polishing the specimen

A1.4 In addition to achieving a smooth polished surface on the specimen it is essential that this surface be *absolutely flat* since when using high magnifications with a microscope the depth of field which is in focus at any time is extremely small. For this reason the surface must first be filed absolutely flat and this is best achieved by rubbing the specimen to and fro on the file instead of attempting to file the surface flat in the orthodox manner. When the saw marks have been filed out, the specimen—and the operator's hands—should be washed to remove coarse filings and other abrasive particles.

A1.4.1 Intermediate and fine grinding is then carried out using emery papers of progressively finer grades. These should be of the finest quality with regard to uniformity of particle size. With modern papers not more than five grades are necessary (150; 220; 320; 400 and 600—from coarse to fine) and since these are manufactured with a water-proof base, wet grinding is now generally used. Most modern laboratories are equipped with rotary grinding machines so that the feed water carries away swarf, loose grit and other particles likely to produce deep-seated scratches on the specimen surface. Alternatively stationary grinding decks may be used on which strips of paper are clamped side by side on a sloping sheet of plate glass over which a current of water flows. The specimen is then rubbed to and fro on each sheet in turn.

Whichever method is used the specimen is first ground on the 150-grade paper, the specimen being held such that the grinding 'furrows' formed are at right angles to the 'furrows' produced originally by the file. In this way it will be possible to see when the file marks have in fact been ground out. When the file marks have disappeared the specimen is again washed and then transferred to the next finer grade of paper (220), again being turned through 90° and ground until the previous grinding marks (from the '150' paper) have disappeared. This process is repeated with grades 320, 400 and 600 papers, finally washing the specimen free of any grit, when an apparently scratch-free surface has been obtained.

A1.4.2 So far the operation has been purely one of grinding and at this stage if the surface of the specimen is examined at, say $100^{\times}$, it will be seen to be covered by a series of parallel grooves cut by the '600' emery paper. The final operation—polishing—is somewhat different in character since it removes the ridged surface by means of a 'burnishing' operation. This 'drags' ions of the metal across the surface to produce an amorphous or 'flowed' layer which effectively hides the crystal structure beneath. The amorphous layer must therefore be dissolved by a suitable etching reagent (Fig. A1.3).

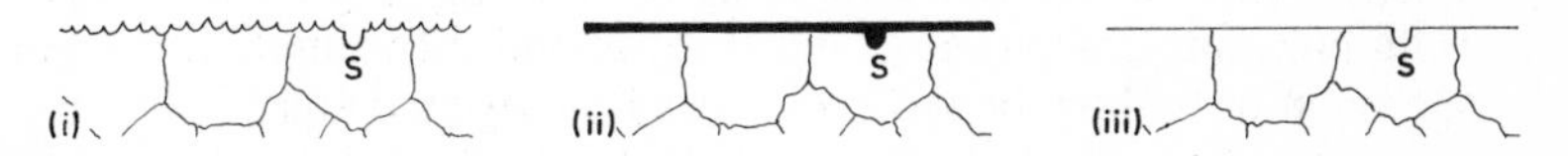

Fig. A1.3—(i) Grooves produced in the metal by the final grinding operation. A deep scratch, S, has been produced by a particle of coarse grit on the '600' paper. (ii) Polishing produces a 'flowed layer' which may hide the scratch, *S*. (iii) Etching dissolves the flowed layer revealing the crystal structure—and also the deep scratch—beneath.

A1.4.3 Irons and steels are polished on a rotating cloth pad impregnated with some suitable polishing medium. Various cloths are now sold specifically for this purpose being much harder wearing than the original 'Selvyt' cloth which was produced rather for polishing the family 'silver'. Proprietary polishing compounds containing diamond 'dust', carried in a 'cream' base, are now generally used for this task, a few spots of an oil-based fluid being used as a lubricant. Preferably two polishing wheels should be used, the first charged with 6-μ grade diamond paste and the second with 1-μ grade. Although these diamond compounds are expensive only a very small quantity is required to 'charge' the pad and since an oily lubricant is used the pad does not become dry during intermittent usage. The older, water-lubricated polishing media like 'gamma-alumina'; magnesium oxide (which 'carbonated' and formed 'scratchy' particles); and jeweller's rouge (very messy) all dried on the pad and thus necessitated frequent changing of expensive polishing cloths. Today a cloth charged with diamond paste can be used for several weeks or until it is worn out—always assuming that some clumsy oaf does not tear it or contaminate it with foreign particles of abrasive grit.

A1.4.4 Non-ferrous specimens are often polished by hand on a small square of 'Selvyt' cloth held on plate glass, a few spots of 'Silvo' being used as the polishing medium. A circular sweep of the hand should be employed rather than the back and forth motion used in hand grinding.

In any polishing process *very light* pressure should be used as this will give a lower incidence of scratches. Moreover heavy pressure on a rotating pad may 'cut' the cloth and invoke the justifiable wrath of the laboratory superintendent. Polishing should be accomplished in only two or three minutes. If a longer period is necessary this indicates that deep-seated scratches are present and it is wiser to return to, say, 400-grade grinding paper. A specimen polished for too long may suffer a rippled surface. Moreover any scratch which may have been successfully filled by prolonged polishing will usually make an unwelcome reappearance during subsequent etching (Fig. A1.3).

Etching

A1.5 In order to reveal the crystal structure the 'flowed' layer produced by polishing must be dissolved by a suitable etchant. First, however, it is essential to remove every trace of dirt and grease from the specimen otherwise uneven etching and staining will be the result. Unmounted specimens can be de-greased very easily by first washing off any loosely adhering polishing compound under the tap, and then simply immersing the specimen in boiling alcohol ('industrial methylated spirit') for about one minute.* By using hot alcohol in this way the specimen dries immediately on withdrawal from the bath and does not stain. Specimens

* The alcohol container should of course be immersed in an electrically heated water bath and NOT heated over a naked flame.

mounted in thermoplastics materials must generally be degreased by swabbing with cold alcohol as the hot liquid has a solvent action on some of these polymer materials. Rapid drying to avoid staining is then accomplished by holding the specimen in a current of warm air—an ordinary domestic hair drier is quite adequate for this.

A1.5.1 The specimen may be examined in this unetched condition for the presence of slag and other non-metallic inclusions which will be the more easily visible at this stage. The crystal structure however is still hidden by the 'flowed layer' which adheres to the metallic but not to the non-metallic regions.

A1.5.2 Having degreased the specimen it should not be touched by the fingers but handled with tongs until after it has been etched, otherwise traces of grease may cause staining. First the specimen is cooled in running water and then immediately immersed in the etching reagent, *agitating it vigorously* until etching is complete. This is usually denoted by the surface of the specimen becoming uniformly dull, often with a slight change in colour. Etching times vary from a few seconds with plain carbon steels and brasses to several minutes for a stainless steel. A fair amount of practice is generally necessary before a good etch can be obtained as a matter of course. Practising a *near-clinical standard of cleanliness* in the later stages of degreasing and etching will be found to be the best recipe for success.

Many different etching reagents are used in metallographic laboratories but some of these have only specialised uses. The beginner will require only those shown in Table A1.1 yet these few etchants will probably deal with at least 90% of the metals and alloys he will encounter.

As soon as the specimen has been adequately etched it must be washed in running water without delay, otherwise the surface will begin to stain due to the combined effects of etchant and atmospheric oxygen. Hence the dish of etchant should be situated at the side of the sink not more than a few inches from running water, thus facilitating quick transfer and thorough washing. Having washed the specimen it is then quickly dried in alcohol as previously described.

A1.5.3 Both polishing and etching can be carried out electrolytically. This involves setting up a small electrolytic cell in which the surface of the specimen is made anodic. By choosing a suitable electrolyte and current conditions the surface of the specimen can be selectively dissolved to the required finish. Nevertheless the traditional methods of polishing and etching outlined above will deal successfully with most metallic materials.

Finally the specimen must be mounted for microscopical examination. In order that the prepared surface should be 'normal' to the optical

Table A1.1— A 'short list' of etching reagents.

	Etchant	Composition	Characteristics and uses
Steels	Nital	2 cm³ nitric acid; 98 cm³ alcohol (industrial methylated spirit)	The best general etching reagent for irons and steels. Etches pearlite, martensite and tempered structures. Attacks ferrite grain boundaries. To resolve pearlite etching must be light. Also suitable for most cast irons. Etches tin-base and lead-base alloys
	Alkaline sodium picrate	2 g picric acid; 25 g sodium hydroxide; 100 cm³ water	The sodium hydroxide is dissolved in the water and the picric acid then added. The mixture is heated in a boiling water bath for 30 minutes and the clear liquid poured off. The specimen is etched for 5–15 min in the boiling solution. Its main use is to distinguish between primary ferrite and primary cementite. The latter is stained black
	Mixed acids and glycerol	10cm³ nitric acid; 20 cm³ hydrochloric acid; 20 cm³ glycerol; 10 cm³ hydrogen peroxide	Suitable for nickel–chromium alloys and most stainless steels. Also for high-chromium–high carbon-steels and high-speed steels. Warm the specimen in boiling water before immersion
Copper and its alloys	Ammonia-hydrogen peroxide	50 cm³ ammonium hydroxide (0·880); 50 cm³ water. Just before use add 20 cm³ 3% solution of hydrogen peroxide	The best general etchant for copper and most copper-base alloys. The hydrogen peroxide content can be varied to suit particular alloys. Must be freshly prepared as the peroxide decomposes rapidly. For swabbing or immersion
	Acid ferric chloride	10 g ferric chloride; 30 cm³ hydrochloric acid; 120 cm³ water	Produces a very contrasty etch on brasses and bronzes. Darkens the β-phase in brasses. Use at full strength for nickel-rich copper alloys. Dilute 1 part with 2 parts of water for copper-rich solid solutions in brasses and bronzes
Aluminium and its alloys	Dilute hydrofluoric acid*	0·5 cm³ hydrofluoric acid; 100 cm³ water	For aluminium and its alloys. A good general etchant. The specimen is best swabbed with cottonwool soaked in the etchant

*** On no account should hydrofluoric acid be allowed to come into contact with the skin or eyes.**

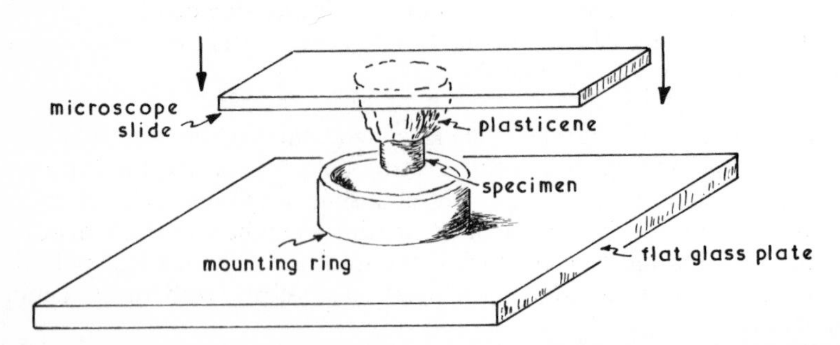

Fig. A1.4—The use of a simple device for mounting the specimen so that its surface will be normal to the optical axis of the microscope.

system of the microscope some form of mounting device is necessary. The specimen is attached to a microscope slide with a piece of modelling clay (plasticene) and then 'squared-up' using a mounting press or, if this is not available, a simple ring which parallel end faces (Fig. A1.4).

Appendix II
The Metallurgical Microscope and its Use

A2.1 Optically the metallurgical microscope is similar to any other microscope. It is in the method of illuminating the specimen where they differ. In microscopes used for biological examination a relatively simple illumination system will suffice since the transparent specimens generally used can be viewed by *transmitted* light. Hence a simple concave mirror is fixed beneath a hole in the microscope stage to 'collect' the light required to illuminate the specimen. Metallurgical specimens however are opaque and must therefore be examined by *reflected* light. Attempts to illuminate

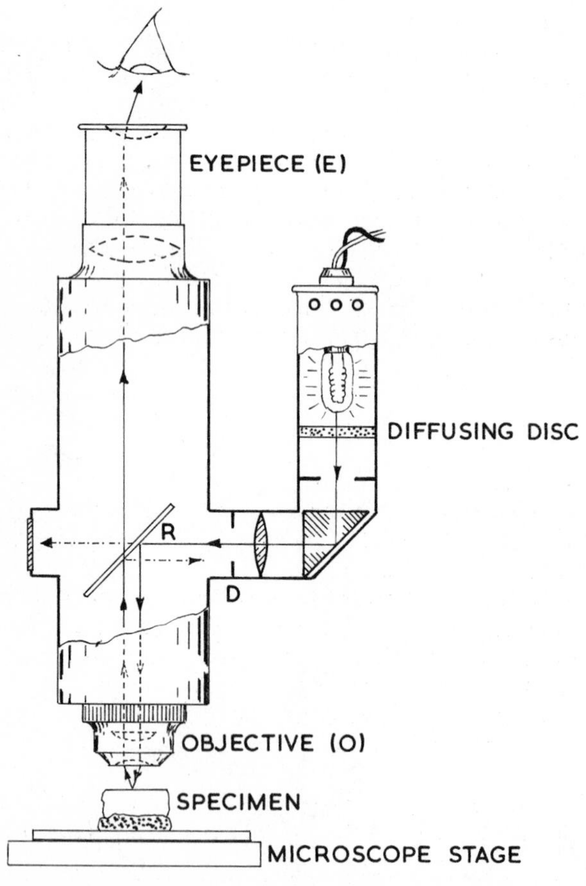

Fig. A2.1—The standard 'student's' metallurgical microscope.

the specimen from a source at the side of the microscope will generally fail (Fig. A2.2) since most of the light will be reflected such that it does not enter the front lens element of the objective (rays *A* and *B*) and only defects such as scratches will be usefully illuminated (ray *C*). Thus in the case illustrated scratches would appear as bright lines on a dark background. There are a few instances where this *dark-field illumination* is

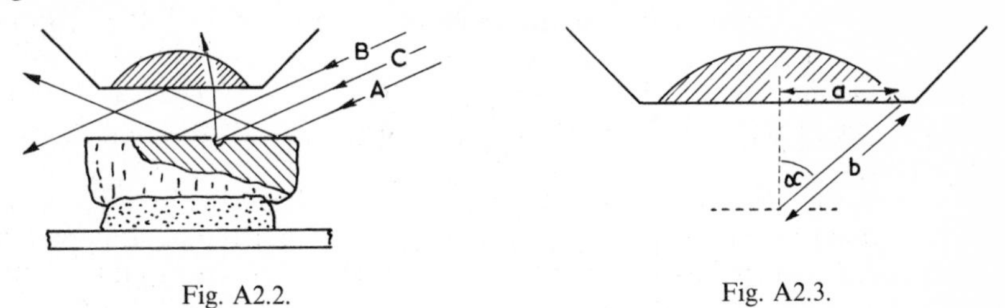

Fig. A2.2. Fig. A2.3.

useful but in the majority of cases normal illumination must be achieved by using a light source *inside* the microscope tube as indicated in Fig. A2.1. Unfortunately much light is lost by transmission as it reaches the plain glass reflector, *R*, and again by reflection when the ray returns from the objective on its way to the eyepiece. An illumination system of this type is only about 4% efficient. Nevertheless a small 3-watt bulb provides sufficient illumination for most purposes other than photomicrography.*

The diameter of the 'pencil' of light required is controlled by the iris diaphragm, *D*. This should be closed until the pencil of light is just sufficient to cover the rear component of the objective lens. (The diaphragm is closed until an image of it appears in the field of view. It is then opened until the image just disappears.) Excess light reflected within the microscope tube will give rise to 'glare' in the field of view with a consequent loss of contrast in the image.

A2.1.1 The optical system of any microscope consists basically of two lenses, the objective, *O*, and the eyepiece, *E*. The former is the more optically critical with regard to its influence on the quality of the image produced and it must 'resolve' fine detail. Objectives have to be 'corrected' to minimise the effects of spherical and chromatic aberrations, for which reason microscope objectives as well as camera and telescope lenses are of compound construction. *Achromatic* objectives are widely used because they combine reasonably adequate correction with low cost. *Apochromatic* objectives however are very expensive since they are highly corrected lenses of the finest quality.

A2.1.2 The *magnification* given by an objective depends upon its focal length—the shorter the focal length the higher the magnification. In

* Note that one speaks of a 'photomicrograph' and NOT a 'microphotograph'—which would imply a 'very small photograph'.

addition to magnification *resolving power* is very important. This is the ability of the objective to produce sharply defined separate images of two lines which are very close together. Resolving power depends upon the quality of the lens but also upon its *numerical aperture*. For 'dry' lenses, i.e. those which do not use oil immersion, this value is found from (Fig. A2.3):

$$\text{Numerical aperture (N.A.)} = \sin \alpha = \frac{a}{b}$$

Assuming that a narrow pencil of light is used the finest detail which can be observed as separate images is given by:

$$\text{Fineness of detail} = \frac{\lambda}{\text{N.A.}}$$

where λ = the wavelength of the illuminant light.

Working on this basis the maximum useful magnification of an optical microscope system is about $1000^{\times}$. Beyond this we move into the realm of what is usually called 'empty magnification', that is, magnification not accompanied by extra resolution of detail. The same effect is obtained by over-enlarging from a photographic negative taken with a cheap camera —the image may be magnified but it appears increasingly blurred because the extra detail is just not there in the negative (except of course in fictional spy-thrillers). Focal lengths of objectives commonly vary between 2 mm and 25 mm.

A2.1.3 The purpose of the eyepiece is to magnify the image produced by the objective and to focus this image so that it can be observed by the metallographer. Eyepieces are supplied in a number of 'powers', e.g. ×6; ×8; ×10 and ×15, and are of simple and inexpensive construction as compared with objectives. The overall approximate magnification given by a microscope system can be derived from:

$$\text{Magnification} = \frac{\text{Tube length (mm)} \times \text{power of eyepiece}}{\text{Focal length of objective (mm)}}$$

Example: A microscope has a tube length of 20 cm. What magnification will be obtained when using a 4 mm objective in conjunction with a ×8 eyepiece?

$$\text{Magnification} = \frac{200 \times 8}{4} = 400^{\times}$$

(Most 'bench' microscopes have a tube length of 200 mm.)

Using the microscope

A2.2 Having mounted the specimen so that its surface is normal to the optical axis it is placed on the stage of the microscope. The surface is then brought into focus by using first the coarse and then the fine adjustment. Small adjustments in the tube length may then be necessary to accommodate the eye of the operator. This is achieved by sliding the tube which carries the eyepiece up or down in its mount until the point of clearest vision is obtained. Finally the illumination system is adjusted by

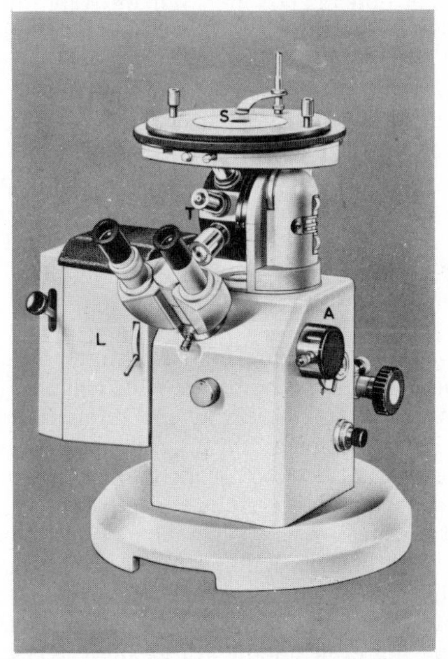

(*Courtesy of Messrs. C. Z. Scientific instruments Ltd., Borehamwood, Herts.*)
Plate A.2.1—A metallurgical microscope of the 'inverted' type designed for binocular viewing.

The specimen is held on the travelling table at S, the objectives being carried in the rotating turret (T). The system of illumination is housed in L. A viewing screen or a 35 mm camera for photography can be attached at the aperture revealed by removing the cover at A.

first closing iris *D* until its image appears in the field of view and then opening it until it *just* disappears. This ensures that a 'pencil' of light is being used which covers the rear element of the objective but no more, thus limiting reflection and glare within the tube. The iris diaphragm nearest to the bulb is then adjusted to give the desired intensity of illumination.

A2.2.1 The specimen should always be examined initially at a *low* magnification of not more than $100^{\times}$. In the case of cast structures a lower magnification of $50^{\times}$—or even $25^{\times}$—will be preferable as high magnifications will not give a true impression of the dendritic pattern. To resolve pearlite in wrought steels, however, a magnification of $500^{\times}$ is generally necessary but higher magnifications than this are not often used except to reveal very small features such as films and inclusions at grain boundaries.

The care of microscope optics

A2.3 Many types of optical glass are comparatively soft and will scratch very easily. For this reason dust and grit should be removed using a small 'squeeze blower' instead of being wiped off as this may cause abrasion of the soft surfaces when grit is present. Alternatively the surface may be brushed gently using a soft camel hair brush. On no account should optical surfaces be touched with the fingers. Accidental finger marks are removed by *very gently* wiping (*not rubbing*) the surface with a piece of well-washed linen barely moistened with xylol. A *good-quality* lens-cleaning tissue is also acceptable provided it has been stored in grit-free surroundings. Before the surface is wiped it should of course be freed from dust or grit by blowing or brushing.

A2.3.1 High-power objectives are of the oil-immersion type. That is a film of cedar-wood oil, interposed between the lens and the surface of the specimen, has the effect of increasing the refractive index of the gap between the lens and specimen. Since light rays are 'bent' towards the axis of the lens this has the effect of increasing the numerical aperture. The cedar wood oil should be carefully wiped from the surface of the lens as soon as possible using a well-washed linen cloth (or good-quality lens-cleaning tissue). Both of these materials are superior to chamois leather which is more likely to contain particles of grit and man-made fibre textiles which have a tendency to scratch the soft surfaces of optical glass. Such materials are not even suitable for cleaning NHS spectacles. If the cedar-wood oil has been allowed, by neglect, to harden on the lens surface it will need to be removed by wiping the surface with a linen cloth slightly moistened with xylol. There is always a danger, however, that xylol will penetrate between the lens components and dissolve the cement holding them together. Thus prevention is better than cure.

The electron microscope

A2.4 Although the bulk of microscopical metallography is carried out at magnifications between 50 and 500, higher magnifications are sometimes necessary or desirable and the optical instrument is incapable of providing useful magnification much above 1000—or 2000 at the very outside—for reasons already mentioned.

In order to obtain magnifications above 1000 and up to several hundreds of thousands, some form of microscope using electron streams rather than light rays is used. The high resolving power of the electron microscope arises from the fact that just as light can be considered as having both wave-like and particle-like properties so *moving* electrons seem to behave both as particles and waves. Thus, like light, electrons can produce interference and diffraction effects. However, the wavelength associated with a moving electron can be controlled since it depends upon its speed and also upon the p.d. causing it to accelerate. Since the resolving power of a microscope *increases* as the wavelength of the radiation reaching the system *decreases* (Fineness of detail = λ/N.A.) it follows that much finer detail will be resolved using a suitable electron beam ($\lambda = 3{\cdot}5 \times 10^{-12}$ m) than when using light ($\lambda = 6 \times 10^{-5}$ m on average). Thus whilst the optical microscope can resolve detail of about 10^{-6} m the electron microscope can produce resolution up to about 10^{-10} m.

Electron beams tend not to be reflected in the regular manner of light rays therefore specimens must be viewed by *transmitted* rather than by reflected electrons.* In the case of metallurgical specimens this involves making a thin plastic replica of the prepared surface. This replica is then carefully separated from the surface of the specimen and subsequently viewed by a transmitted electron beam. The thinner the replica the more electrons are able to penetrate it and so the contrast of the image is preserved.

The electron path in an electron microscope is very similar to the light path in an optical instrument (Fig. A2.4). In the case of the electron microscope the 'lenses' are of the electromagnetic type in which electromagnetic fields are used to focus the electron beam in order to form an image on a fluorescent screen (or a photographic film). The 'focal lengths' of these electromagnetic lenses are of course variable and depend upon the current passing through the coils. The body of the electron microscope must of course be evacuated otherwise molecules of nitrogen and oxygen would seriously obstruct the passage of the relatively small electrons.

Since the orthodox electron microscope, as outlined above, was developed several other instruments of allied design have made their appearance. For example the high-voltage electron microscope permits the use of thicker replica specimens and at the same time can produce greater magnifications.

Possibly the most important development, however, has been the scanning electron microscope. This is very different in principle from the orthodox electron microscope in that electrons reflected and generated from the *actual surface* of the specimen are used to form the image. The image is developed by scanning the surface in the manner employed in a synchronously scanned cathode-ray tube. It produces an image of very high resolution but possibly the most important feature is the great depth of field available, being at least three hundred times better than that of the optical microscope. This enables *un-prepared* surfaces, e.g. fractures, to be systematically explored at very high magnifications up to 50 000.

Also in a rather different category is the field-ion microscope in which

* Some specialised electron microscopes do make use of 'reflected electrons'. This involves maintaining the specimen surface at a 'negative potential' so that electrons are repelled.

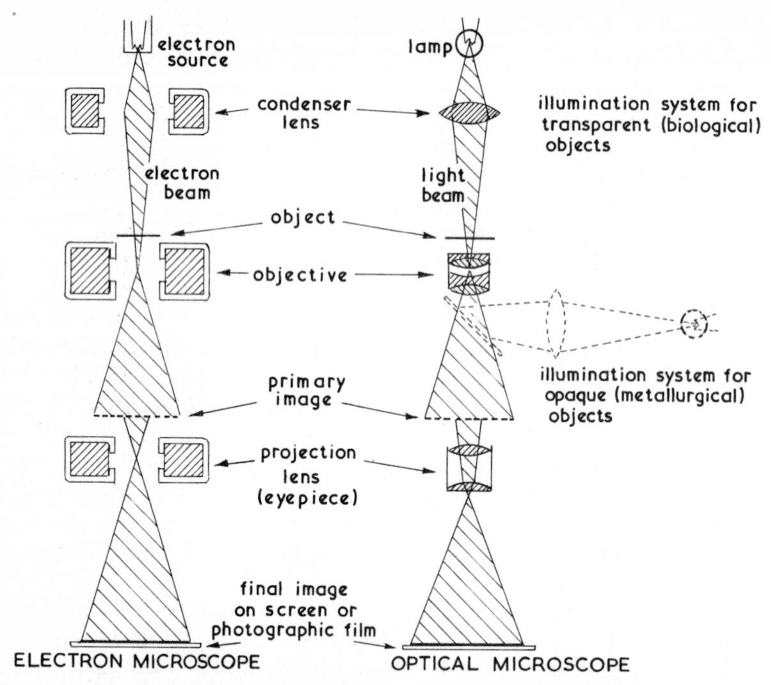

Fig. A2.4—The electron and optical microscopes compared. The optical system here is of the 'inverted type' (Plate A2.1). This is generally used for photomicrographic instruments since it gives greater mechanical stability—and hence freedom from vibration—during the relatively long photographic exposures which are necessary.

the image is formed by ionised gas atoms rather than by electrons. Here magnifications can be measured in millions, enabling details on the atomic scale such as dislocations, vacancies, grain-boundary defects, small precipitates and many other features to be observed. Single atoms are resolvable by this technique.

Appendix III
Some Properties of the Elements

Element	Symbol	Atomic number (Z)	Relative atomic mass	Melting point (°C)	Boiling point (°C)	Relative density	Crystal structures
Hydrogen	H	1	1·0080	−259·2	−252·5	0·0899 (g)	Hexagonal
Helium	He	2	4·0026	−272·2	−268·9	0·178 (g)	—
Lithium	Li	3	6·941	180·5	1329	0·534	$\text{CPH} \xrightleftharpoons{-196\,^\circ\text{C}} \text{BCC}$
Beryllium	Be	4	9·01218	1284	(2400)	1·85	$\text{CPH} \xrightleftharpoons{1200^\circ\text{C}} \text{BCC}$
Boron	B	5	10·811	(2300)	(2550)	2·3	$\text{Rhombohedral} \xrightleftharpoons{1200^\circ\text{C}} \text{simple rhombohedral}$
Carbon	C	6	12·01115	(5000) Gr.	(5000)	2·25 (graphite) 3·51 (diamond)	α-graphite (hexagonal); β-graphite (rhombohedral); diamond (cubic)
Nitrogen	N	7	14·0067	−210·0	−195·8	1·2506 (g)	Hexagonal; complex cubic
Oxygen	O	8	15·9994	−218·8	−183·0	1·429 (g)	$\propto$ (orthorhombic) $\rightleftharpoons \beta$ (rhombohedral) $\rightleftharpoons \gamma$ (cubic)
Fluorine	F	9	18·9984	−218·0	−188·0	1·69 (g)	—
Neon	Ne	10	20·179	−248·7	−245·9	0·9002 (g)	—
Sodium	Na	11	22·9898	97·8	883	0·97	$\text{CPH} \xrightleftharpoons{-222^\circ\text{C}} \text{BCC}$
Magnesium	Mg	12	24·305	650	1103	1·74	CPH
Aluminium	Al	13	26·9815	660	2480	2·699	FCC
Silicon	Si	14	28·086	1412	3240	2·4	Diamond structure (cubic)
Phosphorus	P	15	30·9738	44·1 (yellow)	280	1·82	Orthorhombic; complex cubic
Sulphur	S	16	32·06	112·8 (rhombic)	444·5	2·07	Orthorhombic; monoclinic; rhombohedral
Chlorine	Cl	17	35·453	−101·0	−34·1	3·214 (g)	—
Argon	Ar	18	39·948	−189·4	−185·8	1·78 (g)	—
Potassium	K	19	39·102	63·6	775	0·86	BCC
Calcium	Ca	20	40·88	843	1350	1·55	$\text{FCC} \xrightleftharpoons{448^\circ\text{C}} \text{BCC}$

Appendix III—*continued*

Element	*Symbol*	*Atomic number (Z)*	*Relative atomic mass*	*Melting point (°C)*	*Boiling point (°C)*	*Relative density*	*Crystal structures*
Scandium	Sc	21	44·9559	1538	(2870)	2·5	CPH $\overset{1334^\circ C}{\rightleftharpoons}$ BCC
Titanium	Ti	22	47·90	1667	3285	4·54	CPH $\overset{822\cdot5^\circ C}{\rightleftharpoons}$ BBC
Vanadium	V	23	50·9414	1920	3380	6·0	BCC
Chromium	Cr	24	51·996	1900	2690	7·19	BCC
Manganese	Mn	25	54·9380	1244	2020	7·43	Complex cubic $\overset{727^\circ C}{\rightleftharpoons}$ complex cubic $\overset{1095^\circ C}{\rightleftharpoons}$ FCC $\overset{1133^\circ C}{\rightleftharpoons}$ BCC
Iron	Fe	26	55·847	1536	(3070)	7·87	BCC $\overset{910^\circ C}{\rightleftharpoons}$ FCC $\overset{1400^\circ C}{\rightleftharpoons}$ BCC
Cobalt	Co	27	58·9332	1492	(2900)	8·90	CPH $\overset{400^\circ C}{\rightleftharpoons}$ FCC
Nickel	Ni	28	58·71	1453	(3000)	8·90	FCC
Copper	Cu	29	63·546	1083	2575	8·96	FCC
Zinc	Zn	30	65·37	419·5	907	7·133	CPH
Gallium	Ga	31	69·72	29·8	2420	5·91	Orthorhombic
Germanium	Ge	32	72·59	937	2700	5·36	Diamond structure (cubic)
Arsenic	As	33	74·9216	(817) (36 at.)	616	5·73	Rhombohedral
Selenium	Se	34	78·96	220·5	685	4·81	Hexagonal; monoclinic
Bromine	Br	35	79·904	−7·3	58	3·12	—
Krypton	Kr	36	83·80	−157	−152	3·708 (g)	—
Rubidium	Rb	37	85·4678	38·8	680	1·53	BCC
Strontium	Sr	38	87·62	770	1350	2·6	FCC $\overset{215^\circ C}{\rightleftharpoons}$ CPH $\overset{605^\circ C}{\rightleftharpoons}$ BCC
Yttrium	Y	39	88·9059	1500	3030	5·51	CPH $\overset{1490^\circ C}{\rightleftharpoons}$ BCC
Zirconium	Zr	40	91·22	1860	4400	6·5	CPH $\overset{865^\circ C}{\rightleftharpoons}$ BCC
Niobium (Columbium)	Nb (Cb)	41	92·9064	2468	4735	8·57	BCC
Molybdenum	Mo	42	95·94	2620	4650	10·2	BCC
Technetium	Tc	43	(99)	2720	5030	11·5	CPH
Ruthenium	Ru	44	101·07	2310	4120	12·2	CPH
Rhodium	Rh	45	102·9055	1960	(3760)	12·44	FCC
Palladium	Pd	46	106·4	1552	2940	12·0	FCC
Silver	Ag	47	107·868	960·8	2210	10·49	FCC
Cadmium	Cd	48	112·40	321	765	8·65	CPH
Indium	In	49	114·82	156·4	2075	7·31	Face-centred tetragonal
Tin	Sn	50	118·69	231·9	2623	7·298	Gray tin (cubic) $\overset{13^\circ C}{\rightleftharpoons}$ White tin (tetragonal)

Antimony	Sb	51	121·75	630·5	1610	6·62	Rhombohedral
Tellurium	Te	52	127·60	449·8	990	6·24	Hexagonal
Iodine	I	53	126·9045	113·6	183	4·93	—
Xenon	Xe	54	131·30	−112	−108	5·851 (g)	—
Caesium	Cs	55	132·9055	29·7	700	1·9	BCC
Barium	Ba	56	137·34	710	(1700)	3·5	BCC
Lanthanum	La	57	138·9055	920	(3420)	6·16	CPH $\xrightleftharpoons{330^\circ C}$ FCC $\xrightleftharpoons{864^\circ C}$ BCC
Cerium	Ce	58	140·12	804	3500	6·77	FCC $\xrightleftharpoons{-178^\circ C}$ CPH $\xrightleftharpoons{-10^\circ C}$ FCC $\xrightleftharpoons{725^\circ C}$ BCC
Praseodymium	Pr	59	140·9077	932	(3020)	6·77	CPH $\xrightleftharpoons{792^\circ C}$ BCC
Neodymium	Nd	60	144·24	1024	(3060)	7·01	CPH $\xrightleftharpoons{862^\circ C}$ BCC
Promethium	Pm	61	(147)	—	—	—	—
Samarium	Sm	62	150·4	1052	(1900)	7·54	Rhombohedral $\xrightleftharpoons{917^\circ C}$ BCC
Europium	Eu	63	151·96	826	1490	5·17	BCC
Gadolinium	Gd	64	157·25	1350	(3000)	7·87	CPH $\xrightleftharpoons{1264^\circ C}$ BCC
Terbium	Tb	65	158·9254	1356	(2500)	8·25	CPH $\xrightleftharpoons{\text{near m.pt.}}$ BCC
Dysprosium	Dy	66	162·50	1500	(2630)	8·56	CPH $\xrightleftharpoons{\text{near m.pt.}}$ BCC
Holmium	Ho	67	164·9303	1500	(2700)	8·80	CPH $\xrightleftharpoons{\text{near m.pt.}}$ BCC
Erbium	Er	68	167·26	1530	(2600)	9·06	CPH $\xrightleftharpoons{1370^\circ C}$ BCC
Thulium	Tm	69	168·9342	1525	(2400)	9·32	CPH $\xrightleftharpoons{\text{near m.pt.}}$ BCC
Ytterbium	Yb	70	173·04	824	(1800)	6·96	FCC $\xrightleftharpoons{798^\circ C}$ BCC
Lutecium	Lu	71	174·97	1700	(3500)	9·85	CPH $\xrightleftharpoons{\text{near m.pt.}}$ BCC
Hafnium	Hf	72	178·49	2220	4450	11·4	CPH $\xrightleftharpoons{1755^\circ C}$ BCC
Tantalum	Ta	73	180·9479	3010	5240	16·6	BCC
Tungsten	W	74	183·85	3380	(5500)	19·3	BCC
Rhenium	Re	75	186·2	3150	(5700)	20·0	CPH
Osmium	Os	76	190·2	3045	>5300	22·5	CPH
Iridium	Ir	77	192·22	2443	4400	22·5	FCC
Platinium	Pt	78	195·09	1769	3880	21·45	FCC
Gold	Au	79	196·9665	1063	(2950)	19·32	FCC
Mercury	Hg	80	200·59	−38·9	357	13·55	Rhombohedral; body-centred tetragonal
Thallium	Tl	81	204·37	303	1460	11·85	CPH $\xrightleftharpoons{234^\circ C}$ BCC
Lead	Pb	82	207·2	327·4	1740	11·34	FCC
Bismuth	Bi	83	208·9806	271	1680	9·80	Rhombohedral

Appendix III—*continued*

Element	*Symbol*	*Atomic number* (Z)	*Relative atomic mass*	*Melting point* (°C)	*Boiling point* (°C)	*Relative density*	*Crystal structures*
Polonium	Po	84	(209)	246	965	(9·2)	Simple cubic $\xrightleftharpoons{75^\circ\,C}$ rhombohedral
Astatine	At	85	(210)	(300)		—	—
Radon	Rn	86	(222)	−71	−61·8	—	—
Francium	Fr	87	(223)	(24)	(750)	—	—
Radium	Ra	88	226·0254*	960	(1140)	5·0	—
Actinium	Ac	89	(227)	(1050)	2927	10·07	FCC
Thorium	Th	90	232·0381*	1750	(4850)	11·5	FCC $\xrightleftharpoons{1400^\circ\,C}$ BCC
Protoactinium	Pa	91	231·0359*	(1425)	(4410)	15·4	Body-centred tetragonal
Uranium	U	92	238·029*	1130	3930	18·7	Orthorhombic $\xrightleftharpoons{660^\circ\,C}$ tetragonal $\xrightleftharpoons{775^\circ\,C}$ BCC
Neptunium	Np	93	237·0482*	637	3380	19·5	Orthorhombic $\xrightleftharpoons{278^\circ\,C}$ tetragonal $\xrightleftharpoons{540^\circ\,C}$ BCC
Plutonium	Pu	94	(244)	640	3454	(19·0)	Simple monoclinic $\xrightleftharpoons{122^\circ\,C}$ BC monoclinic $\xrightleftharpoons{206^\circ\,C}$ FC orthorhombic $\xrightleftharpoons{319^\circ\,C}$ FCC $\xrightleftharpoons{451^\circ\,C}$
Americium	Am	95	(243)	(995)	2600	11·7	Double CPH $\xrightleftharpoons{451^\circ\,C}$ BC tetragonal $\xrightleftharpoons{476^\circ\,C}$ BCC
Curium	Cm	96	(247)	—	—	—	—
Berkelium	Bk	97	(249)	—	—	—	—
Californium	Cf	98	(251)	—	—	—	—
Einsteinium	Es	99	(254)	—	—	—	—
Fermium	Fm	100	(257)	—	—	—	—
Mendelvium	Md	101	(256)	—	—	—	—
Nobelium	No	102	(254)	—	—	—	—
Lawrencium	Lw	103	(257)	—	—	—	—
		104					
		105					
		106					

(g) denotes density of the gas in g/dm^3.
* Relative atomic mass of the most commonly available long-lived isotope.
( ) values shown in parentheses are uncertain.

Index